Communications
in Computer and Information Science   **698**

Commenced Publication in 2007
Founding and Former Series Editors:
Alfredo Cuzzocrea, Dominik Ślęzak, and Xiaokang Yang

Editorial Board

Simone Diniz Junqueira Barbosa
  Pontifical Catholic University of Rio de Janeiro (PUC-Rio),
  Rio de Janeiro, Brazil
Phoebe Chen
  La Trobe University, Melbourne, Australia
Xiaoyong Du
  Renmin University of China, Beijing, China
Joaquim Filipe
  Polytechnic Institute of Setúbal, Setúbal, Portugal
Orhun Kara
  TÜBİTAK BİLGEM and Middle East Technical University, Ankara, Turkey
Igor Kotenko
  St. Petersburg Institute for Informatics and Automation of the Russian
  Academy of Sciences, St. Petersburg, Russia
Ting Liu
  Harbin Institute of Technology (HIT), Harbin, China
Krishna M. Sivalingam
  Indian Institute of Technology Madras, Chennai, India
Takashi Washio
  Osaka University, Osaka, Japan

More information about this series at http://www.springer.com/series/7899

Hanning Yuan · Jing Geng
Fuling Bian (Eds.)

Geo-Spatial Knowledge and Intelligence

4th International Conference
on Geo-Informatics in Resource Management
and Sustainable Ecosystem, GRMSE 2016
Hong Kong, China, November 18–20, 2016
Revised Selected Papers, Part I

 Springer

Editors
Hanning Yuan
Beijing Institute of Technology
Beijing
China

Fuling Bian
Wuhan University
Wuhan
China

Jing Geng
Beijing Institute of Technology
Beijing
China

ISSN 1865-0929   ISSN 1865-0937 (electronic)
Communications in Computer and Information Science
ISBN 978-981-10-3965-2   ISBN 978-981-10-3966-9 (eBook)
DOI 10.1007/978-981-10-3966-9

Library of Congress Control Number: 2017932437

Printed on acid-free paper

This Springer imprint is published by Springer Nature
The registered company is Springer Nature Singapore Pte Ltd.
The registered company address is: 152 Beach Road, #21-01/04 Gateway East, Singapore 189721, Singapore

Preface

The 4th Annual 2016 International Conference on Geo-Informatics in Resource Management and Sustainable Ecosystem (GRMSE 2016) was held in Hong Kong, China, during November 18–20, 2016. It aims to bring researchers, engineers, and students to the areas of geo-spatial information science, engineering, and systems in socioeconomic development, resource management, and sustainable ecosystem. GRMSE 2016 features unique mixed topics of spatial data mining, geographical information science, photogrammetry and remote sensing, data science, data engineering, cloud computing, deep learning, and recent applications in the context of building a smarter planet, healthier life, more enjoyable ecology and more sustainable resources.

We received a total of 311 submissions from various parts of the world. The international Program Committee worked very hard to have all papers peer-peer reviewed before the review deadline. The final program consisted of 118 papers. There were four key note speeches and five invited sessions. All the keynote speakers are internationally recognized leading experts in their research fields, who have demonstrated outstanding proficiency and have achieved distinction in their profession. The proceedings are published as a volume in Springer's *Communications in Computer and Information Science* (CCIS) series. Some excellent papers were selected and recommended to the special issue of *Journal of Environmental Science and Pollution*, a Science Citation Index Expanded journal. We would like to mention that, due to the limitation of the conference venue capacity, we were not able to include many fine papers in the program. Our apology goes to those authors.

We would like to express our sincere gratitude to all the members of international Program Committee and organizers for their enthusiasm, time, and expertise. Our deep thanks also go to the many volunteers and staff members for the long hours and hard work they have generously given to GRMSE 2016. We are very grateful to Professor Fuling Bian, Professor Hui Lin and Professor Yichun Xie for their support in making GRMSE 2016 possible. The generous support from Beijing Institute of Technology is greatly appreciated. Finally, we would like to thank all the authors, speakers, and participants of this conference for their contributions to GRMSE 2016.

January 2017 General Chair

Organization

The Advisory Committee

Hui Lin   Institute of Space and Earth Information Science
(ISEIS), The Chinese University
of Hong Kong, Hong Kong

Qingquan Li   Shenzhen University, Shenzhen, China

Honorary General Chair

Fuling Bian   Wuhan University, China

General Co-chairs

Shuliang Wang   Beijing Institute of Technology, China
Yong Xia   Northwestern Polytechnical University, China
Hongzhi Wang   Harbin Institute of Technology, China

International Program Committee Co-chairs

George Christakos   San Diego State University, USA
Yangge Tian   Wuhan University, China
Qingwen Xiong   Wuhan University, China
Quan Zou   Tianjin University, China

International Editorial Committee Co-chairs

Fuling Bian   Wuhan University, Wuhan, China
Hanning Yuan   Beijing Institute of Technology, Beijing, China
Jing Geng   Beijing Institute of Technology, China

International Program Committee

Tao Chen   Tsinghua University, China
Chau Yuen   Singapore University of Technology and Design
(SUTD), Singapore

Maytham Safar   Kuwait University, Kuwait
Alfrendo Satyanaga Nio   Nanyang Technological University, Singapore
Pengfei Zhang   Institute for Infocomm Research (I^2R), Singapore

Mohd Adib Bin Mohammad Razi	Universiti Tun Hussein Onn Malaysia, Malaysia
Hanning Yuan	Beijing Institute of Technology, China
Jing Geng	Beijing Institute of Technology, China
Huijun Yang	Northwest A&F University, China
Hongyi Li	Jiangxi University of Finance and Economics, China
Ismail Rakip Karas	Karabuk University, Turkey
Xianglin Zhan	Civil Aviation University of China, China
Ray-I Chang	National Taiwan University, China
Qunyong Wu	Fuzhou University, China
Qian He	Guilin University of Electronic Technology, China
Ken Chen	Chengdu University of Technology, China
Fuucheng Jiang	Tunghai University, Taiwan
Mohd Haziman Wan Ibriahim	Universiti Tun Hussein Onn Malaysia, Malaysia
Ho Pham Huy Anh	Ho Chi Minh City University of Technology (HUT), Vietnam
Le Sun	Victoria University, Melbourne, Australia
Xia Zhang	Wuhan University, China
Mojtaba Maghrebi	University of New South Wales, Australia
Maciej Zieba	Wroclaw University of Technology, Poland
Jianguo Sun	Harbin Engineering University, China
Ulas Akkucuk	Bogazici University, Turkey
Cheng-Yuan Tang	Huafan University, Taiwan
Mohammed A. Akour	Yarmouk University, Jordan
Chien-Hung Yeh	Feng Chia University, Taiwan
Yi-Kuei Lin	National Taiwan University of Science & Technology (Taiwan Tech), Taiwan
Zongyao Sha	Wuhan University, China
George Christakos	San Diego State University, USA
Ping Fang	Tongji University, China
Kuishuang Feng	University of Maryland, USA
Nanshan Zheng	China University of Mining and Technology, China
Changsheng Cai	Central South University, China
Zhenhong Li	University of Glasgow, UK
Yuqi Bai	Tsinghua University, China
Sabine Baumann	Technische Universität München, Germany
Qinghui Huang	Tongji University, China
David Forrest	University of Glasgow, UK
Arie Croitoru	George Mason University, USA
James Cheng	Manchester Metropolitan University, UK
Paul Torrens	University of Maryland, USA
Stephan Mäs	Technische Universität Dresden, Germany

Gina Cavan	Manchester Metropolitan University, UK
Jan Dempewolf	University of Maryland, USA
Bor-Wen Tsai	National Taiwan University, Taiwan
Yu Liu	Peking University, China
Xiaojun Yang	Florida State University, USA
Yan Liu	The University of Queensland, Australia
Jinling Wang	University of New South Wales, Australia
Xiaolei Li	Wuhan University, China
Pariwate Varnakovida	Prince of Songkla University, Thailand
Manfred F. Buchroithner	Technische Universität Dresden, Germany
Anthony Stefanidis	George Mason University, USA
Chaowei Yang	George Mason University, USA
Xiaoxiang Zhu	Technische Universität München, Germany
Matt Rice	George Mason University, USA
Jianjun Bai	Shaanxi Normal University, China
Yongmei Lu	Texas State University, USA
Alberta Albertella	Technische Universität München, Germany
F. Benjamin Zhan	Texas State University, USA
Huamin Wang	Wuhan University, China
Edwin Chow	Texas State University, USA
Lin Liu	University of Cincinnati, USA
Shuqiang Huang	JiNan University, China
Weihua Dong	Beijing Normal University, China
Mengxue Li	University of Maryland, USA
Wenwen Li	Arizona State University, USA
André Skupin	San Diego State University, USA
Alan Murray	Arizona State University, USA
Mike Worboys	The University of Maine, USA
Amirhossein Sajadi	Case Western Reserve University, USA
Chien-Hung Yeh	Feng Chia University, China
Helmi Zulhaidi Mohd Shafri	Universiti Putra Malaysia, Malaysia
Peng-Sheng Wei	National Sun Yat-Sen University, Taiwan
Maria Hallo	Notre Dame University, Belgium
Jingyu Yang	Shenyang Aerospace University, China
Zulkifli Mohd Rosli	Universiti Teknikal Malaysia Melaka, Malaysia
M. Thang Trung Nguyen	Ton Duc Thang University, Vietnam
Chan King-ming	Hong Kong, SAR China
Huan Yu	Chengdu University of Technology, China
Yong Xia	Northwestern Polytechnical University (NPU), China
Rosmayati Binti Mohemad	Universiti Malaysia Terengganu, Malaysia
Sedat Keleş	Çankırı Karatekin University, Turkey
Yanying Chen	Meteorological Science Institute of Chongqing, China
Xiukai Ruan	Wenzhou University, China

Guoqing Li	Institute of Soil and Water Conservation, CAS & MWR, China
Jinghu Pan	Northwest Normal University, China
Guodong Wang	South Dakota School of Mines and Technology, USA
Hongzhi Wang	Harbin Institute of Technology, China
Bin Liu	Dalian University of Technology, China
Xin Yan	Wuhan University of Technology, China
Ali Karrech	University of Western Australia, Australia
Syed Abdul Rehman Khan	Iqra University and Brasi School of Supply Chain Management, USA
Saouli Hamza	University Khider Mohamed, Algeria
Huey-Ming Lee	Chinese Culture University, Taiwan
Lily Lin	China University of Technology, Taiwan
Jolanta Mizera-Pietraszko	Opole University, Poland
Hanmin Jung	Korea Institute of Science and Technology Information (KISTI), South Korea
Chenfei Gao	AT&T Labs Research
Qiang Gao	Beihang University, Beijing, China
Ben-Shun Yi	Wuhan University, China
Yong Xia	Northwestern Polytechnical University, China
Yun-Xiao Zu	Beijing University of Posts and Telecommunications, China
Jen-Fa Huang	Electrical Engineering, National Cheng Kung University, Taiwan
Jian Wang	Wuhan National Laboratory for Optoelectronics, Huazhong University of Science and Technology, China
Tzong-Yi Lee	Yuan Ze University, Taiwan
Wei-Chiang Wu	Da-Yeh University, Taiwan
Wen-Tsai Sung	National Chin-Yi University of Technology, Taiwan
Faizal Mustapha	Universiti Putra Malaysia, Malaysia
Chin-Ling Chen	Chaoyang University of Technology, Taiwan
Nursabillilah Binti Mohd Ali	Universiti Teknikal Malaysia Melaka, Malaysia
Zhen-Dong Wang	Jiangxi University of Science and Technology, China
Sina Vafi	Charles Darwin University, Australia
Trong-Minh Hoang	Posts and Telecommunication Institute of Technology, Vietnam
Deng Chen	Wuhan Institute of Technology, China
Yuan-Long Cao	Jiangxi Normal University, China
Xi-Ming Fu	Tsinghua University, China
Tian-Hua Xu	University College London, UK
Malka N. Halgamuge	University of Melbourne, Australia

Jing-Yu Yang	Shenyang Aerospace University, China
Fang-Jun Huang	Sun Yat-sen University, China
Ying-Ji Zhong	Ohio State University, USA
Jian-Guo Sun	Harbin Engineering University, China
Yi-Fei Wei	Beijing University of Posts and Telecommunications, China
Chi-Wai Kan	Hong Kong Polytechnic University, SAR China
Shih-Chuan Yeh	De Lin Institute of Technology, Taiwan
Muh-Tian Shiue	National Central University, Jhongli, China
Sarmad Sohaib	University of Engineering and Technology, Taxila, Pakistan
Yasin Kabalci	Nigde University, Turkey
Tomasz Andrysiak	University of Science and Technology, Poland
Marcin Kowalczyk	Warsaw University of Technology, Poland
I-Shyan Hwang	Yuan Ze University, Chung-Li, China
Cheng-Yuan Tang	New Taipei City, Taiwan
Yu-Chen Hu	Providence University, Taiwan
Megat Farez Azril	Malaysian Institute of Information Technology, Universiti Kuala Lumpur, Malaysia
Chang-Yu Liu	South China Agricultural University, China
Prosanta Gope	Singapore University of Technology and Design, Singapore
Ming-Jian Li	University of Wisconsin Madison, USA
Choi Jaeho	Chonbuk National University, South Korea
Muhammed Enes Bayrakdar	Duzce University, Turkey
Rkia Aouinatou	Mohamed V Agdal Rabat, Rabat, Morocco
Najeeb Ullah Khan	CECOS University, KPK, Pakistan
Sadaqat Jan	University of Engineering & Technology, Peshawar, Pakistan
Yee-Jin Cheon	University of Science and Technology, Daejon, South Korea
K. Balakrishnan	Karpaga Vinayaga College of Engineering and Technology, Chennai, India
Imran Memon	Zhejiang University, China
Bongani Ngwenya	Solusi University, Zimbabwe
Alexey Nekrasov	Southern Federal University, Taganrog, Russia
Dmitry Popov	Moscow State University of Printing Arts, Russia
Qing-zheng Xu	Xi'an Communications Institute, China
Hsing-Chung Chen	Asia University, Taiwan
Muhammad Zeeshan	National University of Sciences & Technology, Pakistan
Chi-Wai Chow	National Chiao Tung University, Taiwan
Yair Wiseman	Holon Institute of Technology, Israel

Rong-Jong Wai	National Taiwan University of Science and Technology, Taiwan
Xiuyan Ma	Dalian University of Technology, China
Lamei Zhang	Harbin Institute of Technology, China
Jyh-Cheng Chen	National Yang-Ming University, Taiwan
Yupeng Hu	Hunan University, China
Ying-Chun Chuang	Kun Shan University, Taiwan
Ahmet H. Ertas	Karabuk University, Turkey
Jianxun Zhang	Chongqing University of Technology, China
Aleksandra Mileva	Goce Delchev University, Macedonia
Hui-Mi Hsu	National Ilan University, Taiwan
Hamidah Ibrahim	Universiti Putra Malaysia, Kuala Lumpur, Malaysia
Yingji Zhong	Ohio State University, USA
Yun Lin	Harbin Engineering University, China
Guoming Lai	Guangdong Polytechnic of Science and Technology, China
Yinghua Zhou	Chongqing University of Posts and Telecommunications, China
Guojun Mao	Central University of Finance and Economics, China
Kurban Ubul	Xinjiang University, China
Ruipeng Ning	East China Normal University, China
Duanduan Chen	Beijing Institute of Technology, China
Zhiting Lin	Anhui University, China
Weiyu Yu	South China University of Technology, China
Hongjun Li	Beijing Forestry University, China
Liping Yang	Huazhong Agricultural University, China
Farn Wang	National Taiwan University, Taiwan
Lain-Chyr Hwang	I-Shou University, Taiwan
Mahmood K. Ibrahem Al Ubaidy	Al-Nahrain University, Iraq
Juin-Ling Tseng	Minghsin University of Science and Technology, Taiwan
Biju T. Sayed Mohammed	Dhofar University, Oman
Tran Cao Quyen	University of Engineering and Technology, Pakistan
Bappaditya Mandal	Institute for Infocomm Research, Singapore
Simon K.S. Cheung	The Open University of Hong Kong, Hong Kong, SAR China
Megat Farez Azril	System and Networking Section Universiti Kuala Lumpur, Malaysia
Massila Kamalrudin	Universiti Teknikal Malaysia Melaka, Malaysia
Lee Beng Yong	Universiti Teknologi MARA Sarawak, Malaysia
Andy Shui-Yu Lai	Technological and Higher Education Institute of Hong Kong, SAR China
Carlos Humberto Salgado	Universidad nacional de San Luis, Argentina

Adam Glowacz	AGH University of Science and Technology, Poland
Nur Sukinah Aziz	TATI University College, Malaysia
Krzysztof Gdawiec	University of Silesia, Poland
Chien-Hung Yeh	Feng Chia University, Taichung, Taiwan
Bai Li	Zhejiang University, Zhejiang, China
Ming Ming Wong	Sarawak Campus, Malaysia
Kai Tao	Nanyang Technological University, Singapore
Jun Ye	Sichuan University of Science & Engineering, China
Quanyi Liu	Tsinghua University, China
Zhendong Wang	Jiangxi University of Science and Technology, Ganzhou, China
Zhu Tang	National University of Defense Technology, China
Najam ul Hasan	Dhofar University, Oman
Chengyu Liu	Shandong University, Jinan, China
Sanjeevikumar Padmanaban	University of Johannesburg, South Africa
Fengqi Tan	University of Chinese Academy of Sciences, China
Bing Wen	Xinjiang Institute of Ecology and Chinese Academy of Science, China
Qiang Ye	Nanjing Institute of Physical Education and Sports, China
Shuai Liu	Inner Mongolia University, China
Yuhua Wang	Wuhan University of Science and Technology, China
Fei Huang	Ocean University of China, China
Sen Bai	Chongqing Communication Institute, China
Fali Cao	Xi'an Jiaotong University, China
Binyi Liu	Tongji University, China
Bo Cheng	Earth Observation & Digital Earth Chinese Academy of Sciences, China
Chun Shi	Hainan Normal University, China
Weichun Pan	Zhejiang Gongshang University, China
Sathaporn Monprapussorn	Srinakharinwirot University, Thailand
Seethalakshmi Rajashankar	SASTRA University, India
Partha Pratim Ray	Sikkim University, India
Wenchen Hu	University of North Dakota, USA
K.M. Suceendran	Tata Consultancy Services, India
Siwei Chen	National University of Defense Technology, China
Wei Chen	China University of Mining and Technology, China
Chuanfei Xu	Concordia University, Canada
Ti Peng	Southwest Jiaotong University, China
Jianjiao Chen	Georgia Institute of Technology, USA
Jinzhu Gao	University of the Pacific, USA
Lifeng Wei	Beijing University of Civil Engineering and Architecture, China
Rui Sun	Beijing Normal University, China

Anhua He	China Earthquake Administration, China
Ning Zhang	Beijing Union University, China
Imran Memon	Zhejiang University, Pakistan
Qian Tang	Xidian University, China
Xiaofei Zhang	Nanjing University of Aeronautics and Astronautics, China
Lianru Gao	Chinese Academy of Sciences, China
Liang Yang	Guangdong University of Technology, China
Zhenjiang Dong	Nanjing University of Science and Technology, China
Shuo Liu	Institute of Remote Sensing and Digital Earth Chinese Academy of Sciences, China
Qingke Wen	Institute of Remote Sensing and Digital Earth Chinese Academy of Sciences, China
Fan Ning	Beijing University of Posts and Telecommunications, China
Bo Cheng	Beijing University of Posts and Telecommunications, China
Tianhong Li	Peking University, China
Xiaofeng Wang	Chang'an University, China
Shuqing Hao	China University of Mining and Technology, China
Xianchuan Yu	Beijing Normal University, China
Zhaoyang Li	Jilin University, China
Shengcheng Cui	Chinese Academy of Sciences, China
Baiqiu Zhang	Jilin University, China
Yongzhi Wang	Jilin University, China
Ying Li	Dalian Maritime University, China
Chaokui Li	Hunan University of Science and Technology, China
Behshad Jodeiri Shokri	Hamedan University of Technology, Iran
Anand Nayyar	KCL Institute of Management and Technology, India
Hongjun Cao	Ocean University of China, China
Hong Fan	Institute of Remote Sensing and Digital Earth Chinese Academy of Sciences, China
Hyunsung Kim	Kyungil University, South Korea
B. Shanmugapriya	Sri Ramakrishna College of Arts and Science for Women, India
Erfeng Ren	Qinghai University, China
Qianli Ma	University of California, USA
Elena Simona Lohan	Tampere University of Technology, Finland
Laura Mónica Vargas	National University of Córdoba, Argentina
Dionisio Machado Leite	Federal University of Mato Grosso do Sul, Brazil
Edwin Lughofer	Johannes Kepler University Linz, Germany
Alberto Cano	Virginia Commonwealth University, USA
Andrew Kusiak	The University of Iowa, USA
Wilfried Uhring	University of Strasbourg, France

Khor Shing Fhan	Universiti Malaysia Perlis (UniMAP), Malaysia
Jeonghwan Gwak	Gwangju Institute of Science and Technology, South Korea
Ashok Prajapati	IEEE Computer Society South-East Michigan, USA
Leszek Borzemski	Wroclaw University of Technology, Poland
Ramesh K. Agarwal	Washington University, USA
Oscar Esparza	Universitat Politècnica de Catalunya, Spain
Meng Xianyong	Zhuhai College of Jilin University, China
Shian-Chang Huang	National Changhua University of Education, Taiwan
Kuniaki Uehara	Kobe University, Japan
Anjali Awasthi	Concordia University, Canada
Guo-Shiang Lin	Da-Yeh University, Taiwan
Zhenguo Gao	Harbin Engineering University, China
Chunjiang Duanmu	Zhejiang Normal University, China
Iyad Al Khatib	Politecnico di Milano, Italy
Fengxiang Qiao	Texas Southern University, USA
Mehdi Ammi	University of Paris-Sud, France
Daniel Thalmann	Nanyang technological University, Singapore
Roberto Llorente	Universitat Politècnica de València, Spain
Lulu Wang	Hefei University of Technology, China
Cuicui Zhang	Tianjin University, China
Abdallah Makhoul	University of Bourgogne Franche-Comté, France
Alain Lambert	University of Paris-Sud, France
Tchangani Ayeley	University of Toulouse III, France
Bahareh Asadi	Islamic Azad university of Tabriz, Iran

International Steering Committee

Hui Lin	Institute of Space and Earth Information Science (ISEIS), The Chinese University of Hong Kong, SAR China
Qingquan Li	Shenzhen University, China
Zongyao Sha	Wuhan University, China
Xicheng Tan	Wuhan University, China
Pengfei Zhang	Institute for Infocomm Research (I^2R), Singapore
Wenzhong Shi	The Hong Kong Polytechnic University, Hong Kong, SAR China
Ismail Rakip Karas	Karabuk University, Turkey
Yonghui Zhang	Central South University, China
Lin-gun Liu	ATL, China
Chung-Neng Huang	National University of Tainan, Taiwan

International Editorial Committee

Abstracts of Keynote Speeches

Abstracts of Keynote Speeches

Name: Prof. Hui Lin
The Chinese University of Hong Kong, Hong Kong, China

Position held:
Chen Shupeng Professor of GeoInformation Science, Department of Geography and Resource Management
Director, Institute of Space and Earth Information Science

Research Interests:
Microwave Remote Sensing Image Processing and Analysis
Virtual Geographic Environments (VGE) Spatial Database and Data Mining
Spatially Integrated Humanities and Social Science

Keynote Speech Title:
InSAR Remote Sensing for Urban Infrastructure Health Diagnosis

Abstract. The metropolitan area of Hong Kong is characterized by large reclamations with high density skyscrapers and infrastructure. Any inevitable movement of the infrastructure and built environment may pose a threat to infrastructure health and public safety. The development of InSAR remote sensing technology has shown its potential for the diagnosis of the infrastructure health.

Name: Prof. Shuliang Wang

Beijing Institute of Technology, Beijing, China

Shuliang Wang, Ph.D., a scientist in data science and software engineering, is a professor in Beijing Institute of Technology in China. His research interests include spatial data mining, and software engineering. For his innovatory study of spatial data mining, he was awarded the Fifth Annual Info Sci-Journals Excellence in Research Awards of IGI Global, IEEE Outstanding Contribution Award for Granular Computing, and one of China's National Excellent Doctoral Thesis Prizes.

Guest Editor:
International Journal of Systems Science
International Journal of Data Warehousing and Mining
Lecture Notes in Artificial Intelligence

Keynote Speech Title:
Spatial Data Mining Under Big Data

Abstract. It offers a systematic and practical overview of spatial data mining, which combines computer science and geo-spatial information science, allowing each field to profit from the knowledge and techniques of the other. To address the spatiotemporal specialties of spatial data, the authors introduce the key concepts and algorithms of the data field, cloud model, mining view, and Deren Li methods. The data field method captures the interactions between spatial objects by diffusing the data contribution from a universe of samples to a universe of population, thereby bridging the gap between the data model and the recognition model. The cloud model is a qualitative method that utilizes quantitative numerical characters to bridge the gap between pure data and linguistic concepts. The mining view method discriminates the different requirements by using scale, hierarchy, and granularity in order to uncover the anisotropy of spatial data mining. The Deren Li method performs data preprocessing to prepare it for further knowledge discovery by selecting a weight for iteration in order to clean the observed spatial data as much as possible. In addition to the essential algorithms and techniques, the book provides application examples of spatial data mining in geographic information science and remote sensing. The practical projects include spatiotemporal video data mining for protecting public security, serial image mining on nighttime lights for assessing the severity of the Syrian Crisis, and the applications in the government project 'the Belt and Road Initiatives'.

Name: Prof. Yong Wang

University of Electronic Science and Technology of China, Chengdu, China
East Carolina University, Greenville, USA

Current research activities

- Investigation of scale and scale effect on SAR application to urban target Evaluation of water level variations in reservoirs using In SAR technique Thin cloud removal for Landsat 8 imagery
- Submerged aquatic vegetation (SAV) assessment
- Flooding mapping using geo-spatial datasets in rural area

Keynote Speech Title:

Issues in Applying Geoinformatics and Big-Data as Additional Assessment Tools for Macro-Socioeconomic Development

Abstract. Annual socioeconomic datasets released by governmental agencies at the local, state, and national levels portrait socioeconomic statuses within different levels of political boundaries. The data collection costs labor, time, and money. The collected data may consist of errors. Remote sensors provide constant Earth observation. Remotely sensed datasets are multi-temporal and freely available mostly. The datasets are widely used to assess landuse and land cover (LULC) types changes through time, and the changes intuitively reflect the socioeconomic status and development. Thus, the development of additional assessment tools through analyses of remote sensed data is of great interest. Unfortunately, analyzing both types of datasets, one constantly faces analytical and/or statistical challenges. No matter what an approach is applied, following issues must be considered. Otherwise, one will undoubtly concern the results and decisions/actions made based on the outcomes. The issues include data selection, distributions of selected datasets, data transformation, missingness of data, single or multiple independent variables, sensitivity of results to sample sizes, and finally alternative. In this study, we use socioeconomic development of Chengdu City, China between 1978 and 2014 as an example to address above issues. In particular, areas of the impervious surface and agricultural land are derived using spaceborne multi-temporal Landsat data. The domestic gross productivity (GDP) per person released by the statistic department of the municipal government of Chengdu is selected. Between 1978 and 2014, the area of the impervious surfaces and GDP per person increase approximately exponentially. The area of agricultural decreased. Proper transformation is individually applied so that each dataset varies linearly with time. Due to pervasive cloud cover in Chengdu, areas of the impervious surfaces and agricultural lands cannot be derived annually. The multiple imputation method based on the Monte Carlo Markov chain (MCMC) approach is used. Then, GDP per person as the function of the impervious surface area, and as the function of the impervious surface area and agricultural area are statistically established and assessed. The result is satisfactory in regression analysis and crosstab evaluation. It should be noted that the minimum number of required sample size increase rapidly as the number of independent variables increases. Therefore, the use of one or two LULC types as independent variables is recommended.

Name: Prof. Huada Daniel Ruan

Beijing Normal University, Beijing, China
Hong Kong Baptist University, Hong Kong, China
United International College (UIC), Zhuhai, China

Research interest:

- Synthesis, activation, modification and characterization of nanomaterials, their applications as sorbents, catalysts, medications, pigments, additives in environment, agriculture, chemistry and medicine, and their commercialization
- Applications of modified mineral-waste and organic-waste materials for the removals of heavy metals and toxic organic compounds in relation to environmental remediation
- The characteristics of environmental pollutants relating to human health Environmental auditing and assessment relating to environmental management and evaluation of climate change
- Interactions of soil minerals, heavy metals and microbes in contaminated soil materials and bioremediation of contaminated soils
- Environmental chemistry including water quality; air, water and soil pollution; plant nutrition; sediment chemistry; non-point pollution; eutrophication and heavy metal transport, accumulation and contamination
- Renewable energy with emphasis on bio-fuel and solar energy

Keynote Speech Title:

The Application of Environmental GIS

Abstract. Geographic Information System (GIS) generally fulfils the following applications: mapping, monitoring, modelling, measurement and management for a number of fields including political science, education, health care, real estate, business, urban planning and environmental science. The application of a GIS in environmental science can be drawn in environmental monitoring; risk assessment; watershed, floodplain, wetland and aquifer management; groundwater modelling and contamination tracking; hazardous or toxic facility siting; pollutant distribution and remediation; and simulation of process in urban and natural environment. Fundamental investigation of environmental pollution with case studies related to the application of GIS is addressed, and the development of GIS for environmental research and education is discussed in this study.

Name: Prof. Qiang Gao
Beihang University, Beijing, China

Position held:
Professor in School of Electronic and Information Engineering, Beihang University, Beijing, China

Research Interests:
Wireless Communication; Wireless Networks

Keynote Speech Title:
Outage Performance Analysis and Comparison of Two-Way Relaying Systems

Abstract. Cooperative communication has been an effective method for improving system reliability by utilizing the spatial diversity to combat wireless impairments. However, one-way relaying leads to lower spectrum efficiency because it consumes more resources than conventional direct transmission. Recently, two-way relaying (TWR) has drawn much attention since it can provide spectrally efficient transmission with high reliability.

This talk first compares the outage performance differences between amplify-and-forward (AF) and decode-and-forward (DF) in two-way relaying. It is well known that outage performance differences between AF and DF in one-way relaying are apparently related to the average signal-to-noise ratio (SNR). We reveal that it is the target spectral efficiency rather than SNR that determines the superiority in outage performance of different relaying schemes, i.e. DF outperforms AF in the low target spectral efficiency region and the other way around in the high target spectral efficiency region.

Then we investigate the outage performance of two-way amplify-and-forward relaying over block fading channels. Previous research on TWR has been mainly based on the assumption that the channel quality remains constant for one round of data exchange. However, this assumption does not realistically reflect the actual environment as channel conditions fluctuate over time. Our results show that the outage performance of the TWR-AF system deteriorates over block fading channels compared with that over constant-quality channels. Under block fading channels, the TWR system exhibits the outage floor phenomenon, which is not the case for constant-quality channels.

Name: Prof. Tao Gong

Donghua University, Shanghai, China

Prof. Tao Gong received the MS degree in Pattern Recognition and Intelligent Systems and Ph.D. degree in Computer Science from the Central South University respectively in 2003 and 2007. He is an associate professor of immune computation at Donghua University, China, and he was a visiting scholar at Department of Computer Science and CERIAS, Purdue University, USA. He is the General Editors-in-Chief of the first leading journal Immune Computation in its field, and an editorial board member of some international journals. He is a Life Member of Sigma Xi, The Scientific Research Society, a Vice-Chair of IEEE Computer Society Task Force on Artificial Immune Systems, and Chen Guang Scholar of Shanghai. His research has been supported by National Natural Science Foundation of China, Shanghai Natural Science Foundation, Shanghai Educational Development Foundation and Shanghai Education Committee etc. He has published over 100 papers in referred journals and international conferences, and over 20 books such as Artificial Immune System Based on Normal Model and Its Applications, and Advanced Expert Systems: Principles, Design and Applications etc. His current research interests include computational immunology and immune computation. He is also a committee member of intelligent robots committee and natural computing committee in the Association of Artificial Intelligence of China.

Keynote Speech Title:
Cooperative Immune Computation Against Collaborative Attacks in Cyberspace

Abstract. A security problem of cooperative immunization against collaborative attacks such as Blackhole attacks and wormhole attacks, in the mobile ad hoc networks such as the Worldwide Interoperability for Microwave Access (WiMAX) networks, was discussed. Because of the vulnerabilities of the protocol suites, collaborative attacks in the mobile ad hoc networks can cause more damages than individual attacks. In human immune system, nonselfs (i.e., viruses, bacteria and cancers etc.) can attack human body in a collaborative way and cause diseases in the human body. With the inspiration from the human immune system, a tri-tier cooperative immune model was built to detect and eliminate the collaborative attacks (i.e., nonselfs) in the mobile ad hoc networks. ARM-based Network Simulator (NS2) tests and probability analysis were utilized in the prototype for immune model to analyze and detect the attacks. Experimental results demonstrate the validation and effectiveness of the model proposed by minimizing the collaborative attacks and immunizing the mobile ad hoc networks.

Name: Prof. Ji Zhang

University of Southern Queensland, Toowoomba, Queensland

Research Interest:

Prof. Ji Zhang is currently working for the University of Southern Queensland (USQ), Australia. He is an Australian Endeavour Fellow, Queensland Fellow and Izaak Walton Killam Fellow (Canada). He received his degree of Ph.D. from the Faculty of Computer Science at Dalhousie University, Canada. Prof. Zhang's research interests in the area of Computer Science include knowledge discovery and data mining (KDD), Big Data analytics, bioinformatics, information privacy and security, and health informatics. He has published over 90 papers, some appearing in top-tier international journals including IEEE Transactions on Dependable and Secure Computing (TDSC), Information Sciences, WWW Journal, Bioinformatics, Knowledge and Information Systems (KAIS), Soft Computing, Journal of Database Management and Journal of Intelligent Information Systems (JIIS) and international conferences such as VLDB, ACM CIKM, ACM SIGKDD, IEEE ICDE, IEEE ICDM, WWW, DASFAA, DEXA and DaWak. Prof. Zhang is the recipient of a number of prestigious grants and awards including International Science Linkages Grants by Australian Academy of Science (2012 & 2010), Australian Endeavor Award (2011), USQ Research Excellence Award (2011), Head of Department Research Award (2011), Queensland International Fellowship (2010), Izaak Walton Killam Scholarship, Killam Trust, Canada (2007–2008) and IEEE ICDM Student Travel Award by Microsoft and IBM, USA (2006). He was the visiting professor of Michigan State University, USA in 2010 and Nanyang Technological University (NTU), Singapore in 2011.

Keynote Speech Title:

A Parallelized Graph Mining Approach for Efficient Fraudulent Phone Call Detection

Abstract. In recent years, fraud is becoming more rampant internationally with the development of modern technology and global communication. Due to the rapid growth in the volume of call logs, the task of fraudulent phone call detection is confronted with Big Data issues in real-world implementations. In this talk, I will present a highly-efficient parallelized graph-mining-based fraudulent phone call detection framework, namely PFrauDetector, which is able to automatically label fraudulent phone numbers with a "fraud" tag, a crucial prerequisite for distinguishing fraudulent phone call numbers from the normal ones. PFrauDetector generates smaller, more manageable sub-networks from the original graph and performs a parallelized weighted HITS algorithm for significant speed acceleration in the graph learning module. It adopts a novel aggregation approach to generate the trust (or experience) value for each phone number (or user) based on their respective local values. We conduct a comprehensive experimental study based on a real dataset collected through an anti-fraud mobile application, Whoscall. The results demonstrate a significantly improved efficiency of our approach compared to FrauDetector and superior performance against other major classifier-based methods.

Name: Prof. Quan Zou

Tianjin University, Tianjin, China

Editorial Board Member of Scientific Report, PLOS ONE
Special issue guest editor for Neurocomputing, Current Proteomics
Organizing Committee Chair of BIIP2015
Special Session Organizer of IJCNN2016

Program Committee member of the CCIB2011 (Special Session on Computational Collective Intelligence in Bioinformatics, during the 3rd International Conference on Computational Collective Intelligence, ICCCI2011 Gdynia, Poland September 21–23, 2011); WAIM2014,2015,2016 (International conference on Web-Age Information Management); FSDK2014(The 11th International Conference on Fuzzy Systems and Knowledge Discovery); APWeb2016

Outstanding Reviewers for Computers in Biology and Medicine (Elsevier, top 10th percentile in terms of the number of reviews completed within two years, 2015.2)

Reviewer of Bioinformatics, Briefings in Bioinformatics, IEEE/ACM Transactions on Computational Biology and Bioinformatics, IEEE Journal of Biomedical and Health Informatics, Scientific Reports, BMC Bioinformatics, PLOS One, Amino Acids, Gene, Neural Networks, Journal of Theoretical Biology, Computers in Biology and Medicine, Computational Biology and Chemistry, Molecular Biology Reports, BioMed Research International, Current Bioinformatics, Protein & Peptide Letters, Computational and Mathematical Methods in Medicine, Frontiers of Computer Science, etc.

Keynote Speech Title:

Computational Prediction of miRNA and miRNA-Disease Relationship

Abstract. MicroRNA is a kind of "star" molecular, and serves as a "director" since it can regulate the expression of protein. In 2006, related works on gene silence won Nobel price, which made miRNA be the hot topic in molecular genetics and bioinformatics. Mining miRNA and targets prediction are two classic topics in computational miRNAnomics. In this talk, we focus on the miRNA mining problems from machine learning views. We point out that the negative data is the key problem for decreasing the False Positive rather than exploring better features. miRNA-disease relationship prediction is another hot topic in recent years. We introduce some novel network methods on calculating miRNA-miRNA similarity, which is the key issue for miRNA-disease relationship prediction.

Name: Dr. Arun Kumar Saraf
Department of Earth Sciences, Indian Institute of Technology Roorkee, India
Research specialization: Geographic Information System (GIS), Remote Sensing &
Digital Image Processing

Honours and Awards:

a. INSA – Royal Society, UK Fellowship – 2002
b. INSA – Chinese Academy of Sciences Bilateral Fellowship - 2011
c. National Remote Sensing Award-2001
d. GIS Professional of the Year-2001
e. National Scholarship for Study Abroad 1986, Govt. of India
f. Indo-US S&T Fellowship, 1994–1995
g. Khosla Research Award 1996
h. Khosla Research Prize 1996
i. Khosla Research Prize 1997
j. Excellent Performance Recognition by IITR for the years 2001–2002
k. Excellent Performance Recognition by IITR for the years 2002–2003
l. Excellent Performance Recognition by IITR for the years 2003–2004
m. Excellent Performance Recognition by IITR for the years 2004–2005
n. Best Paper Award in Map Asia 2004 (Beijing, China)
o. Nominated as Scientific Board Member of the International Geoscience Programme (IGCP) Scientific Board of UNESCO and IUGS

Keynote Speech Title:
Geoinformatics in Mapping of Fog-Affected Areas over Northern India and Development of Ion Based Fog Dispersion Technique

Abstract. Fog is a phenomenon that affects the Indo-Gangetic Plains every year during winter season (December – January). This fog is sometimes in the form of radiation fog and other also occurs as a mixture with other gases, known as smog (smoke + fog). There are various factors contributing to the formation of fog, that may be either meteorological, topographical or resulting from pollution. Fog has been mapped for the winter seasons of the years 2002–2016. In these winter seasons, fog affected areas were found to be changing significantly. The net cover of fog during a season varies in space, time intensity and frequency of occurrence. Presently, it is now possible to map and to predict fog formation to some extent. However, so far it has not been possible to disperse fog, though theoretically it has been discussed in literature. In the current work, experiments were conducted to find out the possibility and effectiveness of a negative air ionizer for fog dispersion. The experiments were carried out with fog, dhoop smoke and a mixture of both to generate smog. Two different glass chambers of different sizes were used in a closed room and the impact of air ionizer on dispersion was studied by testing the time taken for dispersion with or without the ionizer. The results show a significant performance with air ionizer indicating the effectiveness of the ion generator, which reduced the time taken for dispersion (in comparison to without ionizer) by about half.

Abstracts of Invited Talks

Abstracts of Invited Talks

Name: Dr. Ismail Rakip Karas

Karabuk University, karabük, Turkey

Research Interests:
GeoInformatics, Geographic Information Systems, GIS, Three Dimensional

Geographic Information Systems (3DGIS), Network Analyses, Software Development for GIS, Web based GIS, Geo-Databases, Spatial Data Structures, Computer Graphics, Computational Geometry, Image Processing, Graph Theory, Location Based Services

Speech Title:
3D Network Analyses Based on Smart Evacuation System for Indoor

Abstract. The number of buildings, which are very tall, complex and located on wider areas, has been increasing in today's modern cities. Having dozens of floors, hundreds of corridors, and rooms, and passages, these buildings are almost like a city in terms of their complexity and number of people accommodated. Due to size and complexity of buildings, there are many new problems to be addressed. Evacuation of the buildings quickly and seamlessly is the leading problem in case of emergency. Fire, power outage, terrorism (explosions, bomb threat, hostage-taking incidents), chemical spills, earthquake, flood, etc., are some of the extraordinary occasions that may be encountered or affect indoors. In such kind of cases, formation of panic, crowd, congestion, crush, unable to reach exit, etc. are frequently encountered.

In this talk, 3D Network Analyses and Interactive Human Navigation System for indoor which consists of three components will be presented. The first component is used to extract the geometrical and 3D topological vector data automatically from architectural raster floor plans. The second component is used for network analysis and simulations. It generates and presents the optimum path in a 3D modeled building, and provides 3D visualization and simulation. And the third component is used to carry out the generation of the guiding expressions and it also provides that information for the mobile devices such as PDA's, laptops etc via Internet.

In addition, an Intelligent Evacuation Model for Smart Buildings will be introduced in this presentation. The model dynamically takes into account environmental (smoke, fire, etc.) and human-induced (age, disability, etc.) factors and generates personalized evacuation route by performing network analysis interactively and in real-time. Intelligent Control Techniques (Feed-Forward Artificial Neural Networks) has been used in the design of the model.

Name: Dr. Huan Yu

Chinese Academy of Sciences, Beijing, China

Research Area:
Intelligent Simulation of Landscape Changes; Remote Sensing Application

Education Backgrounds:

> 2013 - Working as Associate Professor at Chengdu University of Technology;
> 2012–2014 Working as post-doctoral scientist at Chengdu University of Technology;
> 2010–2013 Working as lecturer at Chengdu University of Technology;

Speech Title:
The Distribution Characteristics of Halogen Elements in Soil Based on RS and GIS Methods

Abstract. Soil chemical elements are important parameters for soil origin diagnosis, and are sensitive indicators of human disturbance process. The present study attempts to evaluate the influence from human activities on halogen elements (fluoride and iodine). This study also attempts to seek a route to explore the spatial relationships between human disturbances and halogen elements according to geospatial theories and methods. Moreover, the spatial correlations between element anomalies and human disturbed landscapes are calculated to explore the influence from human activities on halogen elements, thereby determining the specific response mechanism. The study results indicate that landscapes influence halogen elements in diverse ways and that element iodine is closely related with road and mine landscapes. Furthermore, strong relationships exist between fluoride and road landscapes, which suggest that this element is affected by road landscapes significantly. Fluoride and iodine are unrelated with city landscapes, and fluoride is unrelated with mine landscapes. These provide a reference for the research on the interaction mechanism between halogen and environment. Therefore, it can be concluded that a response mechanism exploration of soil element aggregation and human disturbance is practicable according to geospatial theories and methods, which provides a new idea for studying the soil element migration.

Name: Prof. Chong-yi Yuan

Peking University, Beijing, China

Graduated from Department of mathematics, Nanjing University, 1964
Graduated from institute of Mathematics, Chinese Academy of Sciences, 1968
Switch from mathematics to the study of computer software, 1975
2 more years in Canada as a visiting scholar, Toronto University and Waterloo University, 1977–1979
3+ years in Germany as a visiting scholar to learn Petri Nets from Prof. Carl Adam Petri, 4 times in the 80s last century
Left Institute of Mathematics and started teaching in Peking University, Dec. 1992, Department of Computer Science at that time, School of Electronics Engineering and Computer Science now
Two master courses were taught: Petri Nets and Parallel Program Design from 1993 to 2009
Retired 2005
Professor and Ph.D. supervisor, named Chong-yi Yuan, born 1941
4 books: Petri Nets (1989), Petri Net Principles (1998), Principles and Applications of Petri Nets (2005), Petri Net Applications (2013)

Speech Title:
OESPA: Semantic Oriented Theory of Programming

Abstract. Testing is now a necessary step before a program is put to use. Formal semantics, including operational semantics, functional semantics etc., do not help in this regard. OESPA is a new theory that combines syntax and semantics together to allow program verification instead of testing. It consists of 3 parts: OE, operation expression, for programming, SP, semantic predicates, for precise semantics description, a semantic axioms. To compute semantics from OE. Examples are included for illustration.

Contents – Part I

Ecological and Environmental Data Processing and Management

Contents – Part II

**Applications of Geo-informatics in Resource Management
and Sustainable Ecosystem**

Smart City in Resource Management and Sustainable Ecosystem

Study of Ecosystem Sensitivity Based on Grid GIS in Leishan County

Shanshan Zhang[1,2], Zhongfa Zhou[1,2(✉)], and Xiaotao Sun[1,2]

[1] School of Karst Science, Guizhou Normal University,
Guiyang 550001, Guizhou, China
zssfly519@sina.com, fa6897@163.com
[2] State Engineering Technology Institute for Karst Desertification Control,
Guiyang 550001, Guizhou, China

Abstract. In order to diagnose the environmental problems and to protect the ecological environment timely, this paper take Leishan County as the study case and take the grid as the evaluation unit to construct evaluation index system of ecosystem sensitivity. Analytic Hierarchy process (AHP), regional synthesis method and the grid GIS technology are used in the evaluation system to the gridding expression of 10 m × 10 m grid scales. Spatial overlay method of raster data and evaluation model of ecosystem sensitivity comprehensive evaluation of the ecological sensitivity of Leishan County. The result shows: Leishan has a higher sensitivity, among which the non-sensitive area and slight-sensitive area are smaller, accounted for only 5% and 13.4% of the total area respectively; Followed are the middle sensitive area and extremely sensitive area, accounted for 26.3% and 17% of the total area; highly sensitive area is the largest, accounted for 38.3% of the total area.

Keywords: Ecosystem sensitivity · Analytic hierarchy process (AHP) · Grid GIS · Key ecological function area · Leishan county

1 Introduction

Ecosystem sensitivity is the sensitivity of ecosystems to human activities, which reflects the possibility of ecological imbalance and the ecological environment [1]. Naturally, interaction between organisms and environment are influence with each other and restrict each other, they are in a relatively stable state of dynamic balance. But when the natural state of the ecosystem is disturbed by the external environment, this balance will be broken, some ecological processes will take the opportunity to expand, so that a series of ecological environment problems occurs. The study of ecosystem sensitivity is the evaluation of the possibility of potential environmental problems under the natural state, which characterize the possible consequences of external interference, and the results will be implemented into the corresponding spatial region [2]. At present, with the contradiction between human and land going deeper and economic unbalance development, ecosystem sensitivity research has been paid more attention by researchers at home and abroad, and some of them has achieved certain results, but most of the researches are aimed at the ecological environment

© Springer Nature Singapore Pte Ltd. 2017
H. Yuan et al. (Eds.): GRMSE 2016, Part I, CCIS 698, pp. 3–11, 2017.
DOI: 10.1007/978-981-10-3966-9_1

problems in Karst region, such as ecological security research [3]; and studies on the sensitivity of ecological environment [4]; ecosystem health change diagnosis [5] and ecological vulnerability assessment and so on, researches on the ecological environment of non-Karst region is relatively less. Study on the sensitivity of non-Karst areas are mostly concentrated on the study of individual ecological environment problem and the provincial level, such as studies on the sensitivity of hydrological system to climate change [6]; studies on the sensitivity of acid rain [7–9]; Wu Jinhua and other's [10] land sensitivity study; Fan Feide and other's [11] studies of the sensitivity evaluation of soil and water loss in Karst region and its spatial distribution characteristics; Wan Zhongcheng [1] and other's evaluation the ecosystem sensitivity of Liaoning province from three aspects: soil erosion, land desertification and soil salinization; Yang Zhifeng [12] and other's sensitivity evaluation of urban ecosystem, but sensitive researches on the comprehensive evaluation of county level or multi factor are few.

Leishan County is the key ecological function areas at the provincial level, it bears the environmental protection, maintaining ecological balance and integrity task in the region, its sensitivity studies can be a scientific diagnosis of the health of the ecosystem. Therefore, this paper take Karst District in Leishan County as an example, through vector quantization the evaluation index data, and with the help of grid GIS technology, take the grid and the natural unit as the evaluation unit, analytic hierarchy process and regional synthesis method are used for the comprehensive evaluation of ecosystem sensitivity from quantitative and qualitative, natural and human factors in Leishan County, so as to provide theoretical method and scientific basis for the environmental remediation, ecological protection and the coupling development between man and nature for the sensitive areas, at the same time, it has important research significance for the ecological environment evaluation of non-Karst district and county scale area.

2 Study Area

The county is located in southwest Qiandongnan Miao and Dong Autonomous Prefecture of Guizhou Province, between 107°55′E–108°22′E and 26°02′N–26°34′N, with a total area of 1204.35 km^2, it has jurisdiction over 5 towns and 3 townships and is in the slope transition zone between Yunnan-Guizhou Plateau and Hunan and Guangxi Basin. It is topographically high in the northeast, low in the southwest with the highest and lowest altitude of 2178.8 m and 480 m separately. It belongs to subtropical monsoon humid climate, the annual sunshine hours are 1225 h, 28% sunshine rate and with an average annual temperature of 14.4–15.4 °C, the total annual accumulated temperature is between 5110–5475 °C, 248–259 d of frost free period, 1250–1500 mm annual precipitation and an average relative humidity about 80% annually. It has 7 main soil types: yellow soil, the mountain yellow brown soil, mountain shrub meadow soil, yellow red soil, purple soil, alluvial soil and paddy soil. With its domestic forest coverage rate over 67.35%, it is a provincial key ecological function area.

3   Materials and Methods

3.1   Sources of Date

Remote sensing image data of Leishan County are processed through ALOS with a spatial resolution of 10 m and 2.5 m and multispectral and panchromatic images mosaic, registration, fusion and cutting. Based on remote sensing image, the derived data of soil erosion, vegetation coverage, land use, the ecological fragility are generated from ENVI5.1 and ArcGIS10.1 software; Slope and altitude data extraction from the County's DEM; the disasters data is calculated by the multi factor regional synthesis method and the maximum factor method; Monitoring data are from hydrological and meteorological station. Population, economy, education and other data collected from the local data.

3.2   Establish Evaluation Index System

In special terrain conditions and under the background of extensive economic growth mode as well as the development of economy, increase in population density, efforts to develop natural resources continue to increase, the problem of ecological environment has become increasingly prominent. Simultaneously, as the key ecological function areas, Leishan's environment is diversity and vulnerability, therefore, in the process of ecosystem sensitivity evaluation, the impact mechanism of the ecological system should be fully understood. Under the principle of representative, measurable, data availability, a comprehensive analysis of the ecological environment in Leishan on natural and man-made factors influence level is conducted in this paper. Meanwhile, based on the natural environment, status of ecological environment, humanities and social aspects established an evaluation index system of ecosystem sensitivity including three secondary level 13 indicators according to the actual situation of the area, and the reference of Li Rongbiao [4] and Wei Xiaodao's [3] research results (Table 1).

3.3   Determination of Weight

The weight value is to evaluate the relative importance of each index factor in the evaluation system, it can objectively show the primary and secondary relationship between the evaluation factors. At present, the method of determining the weights are mainly: grey correlation analysis method, composite index method, analytic hierarchy process (AHP), entropy method and principal component analysis method, among them, the AHP is the most widely used one, it is the most preferred method, so, the analytic hierarchy process is used to calculate the weights in the ecosystem sensitivity evaluation. First, hierarchy the decision problems (target layer, criterion layer and scheme layer), according to the relationship of various factors, gather combination of different levels to form a structure model of multi-level [13]. Then consult relevant experts to determine the weight of indicators, and through one-time test to obtain the weight value of the sensitivity evaluation indicators (Table 1).

Table 1. Index System of ecosystem sensitivity evaluation.

Target layer (A)	Criterion layer (B)	Element layer (C)	Index layer (D)	Data sources	Weight
County ecosystem sensitivity evaluation index system	Foundation of natural environment	Landforms	Altitude	DEM data	0.0321
			Slope	DEM data	0.1127
		Climate	Annual temperature	DEM data	0.0154
			Annual precipitation	DEM data	0.0172
	Status of ecological environment	Vegetation	Vegetation coverage	RS+GIS +monitoring data	0.1216
		Soil	Land use	RS+GIS+DEM data	0.0637
			Soil type	Soil type distribution map	0.0425
		Ecological characterization	Ecological vulnerability	RS+GIS +monitoring data	0.1326
			Soil erosion	RS+GIS +monitoring data	0.1401
		Natural disaster	Natural disaster grade	Natural hazard assessment map	0.0633
	Humanities and social response	Education level	Population quality	Local data collection	0.0535
		Economics	Economic development level	Evaluation of economic development level	0.1004
		Population	Population density	Evaluation map of population aggregation	0.1049

3.4 Evaluation Standard

The classification of each individual evaluation index of the ecosystem sensitivity assessment is reference to the national, industry and local standards, such as: "the Status of Land Use Classification Standards" (GB/T201010-2007); "National Standard General Rules for the Comprehensive Management of Soil and Water Conservation" (GB/T15772-1995); "Classification Standard for the Risk of Water and Soil Loss" (SL718-2015); "Technical Regulation of Soil and Water Conservation Monitoring" (SL277-2002) and Remote Sensing Mapping Standard [14] etc. use the regional synthesis method and maximum factor method, sensitivity degree of single ecological factor will be divided into 5 levels, and was assigned a value to a certain level to form a classification criteria for the ecosystem sensitivity indicator.

3.5   Ecosystem Sensitivity Evaluation Model

Draw on Hui Xiujuan [15] and Xiang Wanli [16]'s research results and according to the index of ecosystem sensitivity and grading standard, combined with the weight value of each evaluation index, consulting previous research scholars, establish 5 levels of ecosystem sensitivity evaluation (Table 2), based on the single factor index, the mathematical model of the comprehensive evaluation of ecosystem sensitivity was constructed, calculate the ecosystem sensitivity index (ESIS) in the study area, comprehensive evaluation of multiple factors on ecosystem sensitivity in Leishan County, the calculation formula is as follows:

$$ESIS = \sum_{i=1}^{n} C_i W_i \tag{1}$$

Table 2. Sensitivity grade classification of ecosystem

Grade	Sensitivity value	Sensitivity level	Total area(km^2)	Proportion (%)
I	≤ 1.4	Non-sensitive	60.06	5.0
II	(1.4,1.8]	Slight-sensitive	162.06	13.4
III	(1.8,2.7]	Middle sensitive	316.39	26.3
IV	(2.7,3.4]	Highly sensitive	461.50	38.3
V	>3.4	Extremely sensitive	204.34	17.0

In the type: C_i is the standard value of evaluation factor; W_i is the weight of evaluation factors; n is the total number of indicators for ecosystem sensitivity assessment, $n = 13$; i is the index factor.

3.6   Evaluation Method in Grid GIS

Combined with previous studies, based on the 10 m × 10 m grid and natural element resolution as the basic unit of evaluation, quantification of each evaluation factor, and assign them a certain value. On the data grid and normalized basis, combining with grid GIS algorithm and taking ArcGIS10.1 as the platform, through clipping, intersection analysis, identification analysis and update analysis module in spatial analysis tools, according to the model of ecosystem sensitivity evaluation, weighted overlay and spatial calculation of each index layer by using the raster calculator, get the evaluation of the ecological sensitivity map (Table 2).

4   Results and Analysis

4.1   The Spatial Pattern of Ecosystem Sensitivity

Grid ecosystem sensitivity evaluation results (Table 2) showed that: Leishan ecosystem is highly sensitivity, its ecological environment security is threatened, and most areas are in the range of highly sensitive. The distribution area of the non-sensitive and the slight-sensitive was scattered in high vegetation cover. Leigong Mountain is one of the three major parks of the Miaoling Mountains National Geological Park, its vegetation coverage is high, the distribution area and the buffer area are non-sensitive and slight-sensitive area, non-sensitive area and slight-sensitive area here are 60.06 km^2 and 162.06 km^2, accounted for 5% and 13.4% of the total area respectively. Middle sensitive area has a wide range, mainly distributed in the southern and southwest of Leishan County. Its total area is 316.39 km^2, a relatively higher proportion which accounted for 26.3% of the total area of the county. Highly sensitive areas are widely distributed in Leishan County, it is concentrated in each township, mainly distributed in southeastern and central of Leishan County. Its total area is 461.5 km^2, the proportion to the study area is as high as 38.3%, mosaic with extremely sensitive areas, equivalent to the total area of the slight-sensitive area and the middle sensitive area. The proportion of extremely sensitive area is small, mainly distributed in the central of Leishan County, the southeast and northeast, with area of 204.34 km^2, accounted for 17% of the total county area. Area of extremely sensitive and highly sensitive in Leishan is greater than 1/2 of the total county area, it is result from the combination of interaction of unreasonable human activities disturbance (illegal cultivation, quarrying, etc.), natural disasters and ecological environment factors.

4.2 The Sensitivity Evaluation of Natural Unit

The results of the sensitivity evaluation of natural element ecosystem (Fig. 1) showed that: Sensitivity level area of each town is not evenly distributed, in the 8 towns, the distribution area of the non-sensitive is the smallest, and the proportion of highly

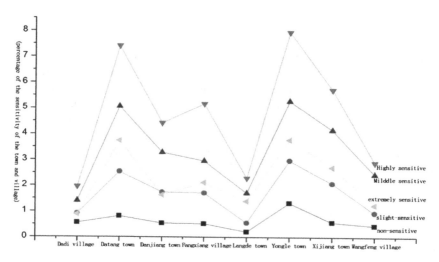

Fig. 1. the Sensitivity of the ecological system in different towns

sensitive area is the largest. For various towns, Yongle and Datang town are the most sensitive ones, Yongle's areas of extremely sensitive and highly sensitive accounted for 3.78% and 7.94% of the total area separately, its strong sensitivity is mainly caused by the large impact of natural disasters, relatively low vegetation coverage; Datang's extremely sensitive areas and highly sensitive areas accounted for 3.74% and 7.41% of the total area, they are caused by the great impact of natural disasters, strong soil erosion and other factors. On the other hand, the common cause of the large area of non-sensitive area of the two towns are the higher level of economic development, lower population density, more money to invest in environmental governance. Dadi village, Langde town and Wangfeng village are relatively low sensitivity. The area of the non-sensitive area and slight-sensitive area in Dadi village accounted for 0.56% and 0.91% of the total area; Langde's and Wangfeng's accounted for 0.21%, 0.56% and 0.44%, 0.94% of the total area respectively. Higher population density, low population quality, low economic development level are the important reasons resulting in the small area in the three areas; but the overall low sensitivity is due to weak human disturbance, the good authenticity of ecological environment and the stability of the ecosystem. Danjiang town is the main area of Leishan County, with higher economic development, concentrated small businesses but strong human activity, extremely impact of natural disasters, high population density, which lead to its enhancement of sensitivity. With frequent commercial activities, strong human activity, higher population density, Xijiang's sensitivity is also high, and so it is with the Fangxiang village under its administer for great impact of natural disasters.

4.3   Recommendations for Ecosystem Sensitivity

The Leishan County is the key ecological function areas in Guizhou province, it undertake the task of protecting the ecological integrity and biodiversity, in view of the environmental problems, the corresponding measures should be taken to reduce the sensitivity of the ecosystem. In Yongle and Datang town with strong ecosystem sensitivity, afforestation should be strengthened and returning farmland to forest, implementation of water and soil conservation and sewage treatment and other construction projects, do well natural disasters early warning and prevention work. For Langde town, Dadi, Wangfeng and Fangxiang village they should focus on the development of the characteristics of green environmental protection industry, speed up economic development, increase investment in education, improve people's quality; and for Dadi village since it has more dry land, planting fruit is preferable, which can enhance the vegetation coverage rate as well as increase economic benefits. Danjiang town's heavy industry should be replaced by high tech industries, and it should also liberate its extensive economic growth mode. The Xijiang town is the biggest tourist town in Leishan County, it should strengthen the protection of the environment while develop its third industry. Besides, in response to the government "Environmental friendly Li County, Tourism County" strategy, it should carried out relocation policy for poverty alleviation to realize the protection and restoration of ecological environment through land reclamation and planting of trees.

5 Conclusions and Discussion

Evaluation results show that Leishan is extremely sensitive and highly sensitive with an area over 55.3% of the total area, they are mainly distributed in the central and southeastern regions; the area of the non-sensitive and slight-sensitive accounted for 18.4%, mainly distributed in the West and northeast; middle sensitive area accounted for 26.3%, distributed in the South and southwest. In the villages and towns Datang and Yongle town are extremely impacted by relatively low vegetation coverage and strong soil erosion, they are the most sensitive area, their extremely sensitive areas and highly sensitive areas accounted for 11.15% and 11.72% of the total area respectively; With good authenticity of ecological environment, Dadi village, Langde town and Wangfeng village are low sensitivity. Danjiang town is the main area of Leishan County, it has a higher level of economic development, most small businesses are concentrated in the region, strong human activity and high population density, all together lead to its sensitivity enhancement; For Xijiang town, its frequent commercial activities or human activity, great impact of natural disasters, lead to its high sensitivity; Fangxiang village under its govern suffers from natural disasters, the sensitivity is also higher. which requires us to strengthen environmental protection while developing the non-sensitive area and slight-sensitive area, in extremely sensitive area and highly sensitive area, we should reduce human disturbance intensity and the development of heavy industry, focus on the protection of the ecological environment so as to the realize the harmonious development of human and nature.

The combination of GIS method and ecological evaluation grid and the evaluation unit of the grid and the natural unit has a certain advancement, however, the research on the humanities and social factors in the evaluation of ecosystem sensitivity is still scarce, and meanwhile, about the determination of the sensitivity index, evaluation criteria and evaluation methods, they still need further study. In addition, "3S" technology, spatial modeling technology should also be strengthened in the application of ecological system to realize systematize and automation of the evaluation and management of ecosystem sensitivity [17].

Acknowledgments. In this paper, the research was sponsored by Karst rocky desertification area of national natural science fund project "regional ecological balance and regional poverty coupling mechanism research" (41661088). Guizhou major application based research project: Karst rocky desertification ecological restoration and ecological economic system optimization control research rock and soil type (Guizhou S&T Contract JZ [2014]200201). Guizhou science and technology projects "The development and application of management system of high efficiency agricultural industrial park intelligent based on the Beidou satellite" (Guizhou S&T Contract GY [2015]3001). Guizhou province soft science research project "the construction of ecological civilization and scientific and technological support in the state key ecological function area"–Taking Guizhou as an example (Guizhou S&T Contract R [2014]2012). The County of Leishan development and Reform Bureau commissioned the project: Town, agriculture and ecological planning of three kinds of space division of Leishan County.

References

1. Wan, Z., Wang, Z., et al.: Sensitivity evaluation of ecosystem in Liaoning province. J. Ecol. **25**(6), 677–681 (2006)
2. Liu, Z., Zhou, Z., Guo, B.: Ecological sensitivity evaluation of key ecological function areas in Guizhou province. Ecol. Sci. **33**(6), 1135–1141 (2014)
3. Wei, X., Zhou, Z., Wang, Y.: Research on the Karst ecological security based on gridding GIS. J. Mt. Sci. **30**(6), 681–687 (2012)
4. Li, R., Hong, H., Qiang, T., et al.: Karst ecological environmental sensitivity evaluation index classification method research: in Duyun city land use types for an example. Chin. Karst **28**(1), 87–94 (2009)
5. Chen, S., Zhou, Z., Yan, L.: Grid GIS based on karst rocky desertification control in the process of ecosystem health change diagnosis: taking Guizhou flower river demonstration area as an example. Chin. Karst **34**(3), 266–273 (2015)
6. Muzik, I.: Sensitivity of hydrologic systems to climate change. Can. Water Res. J. **26**, 233–252 (2001)
7. Yang, X., Chou, R., Cen, H.: To acid deposition in terrestrial ecosystem sensitivity and its influencing factors. Agro Environ. Prot. **18**(2), 92–95 (1999)
8. He, L., Yang, H., Zhou, X.: Evaluation of relative sensitivity of geographic information system and ecosystem to acid deposition. J. Environ. Sci. **18**(2), 177–180 (1998)
9. Zhou, X., Qin, W.: The sensitivity of soil to acid rain in the three provinces of Southern China province (region) and its partition map. J. Environ. Sci. **12**(1), 78–83 (1992)
10. Wu, J., Li, J., Zhu, H.: Evaluation of land ecological sensitivity of Yanan city based on ArcGIS. J. Nat. Resour. **26**(7), 1180–1188 (2011)
11. Fan, F., Wang, K., Xiong, Y., et al.: Sensitivity evaluation and spatial variation characteristics of soil and water loss in southwest Karst region. Chin. J. Ecol. **31**(21), 6353–6362 (2011)
12. Yang, Z., Xu, Q., He, M., et al.: The city ecological sensitivity analysis Chinese. Environ. Sci. **22**(4), 360–364 (2002)
13. Zhang, F., Su, W.: Guizhou province, water resources, economy, ecological environment, social and system coupling coordination evolution characteristics. J. Irrig. Drain. **34**(6), 68–72 (2015)
14. Zhou, Z.: Application of remote sensing and GIS technology in the study of land rocky desertification in Karst area of Guizhou. Bull. Soil Water Conserv. **21**(3), 181–188 (2001)
15. Hui, X., Yang, T., Li, F., et al.: Health assessment of Liaohe River water ecosystem in Liaoning province. J. Appl. Ecol. **22**(1), 181–188 (2011)
16. Xiang, W., Dai, Q.: Rocky desertification area in the watershed ecosystem health quantitative evaluation. China Soil Keep Sci. **9**(4), 11–15 (2011)
17. Hu, B., Wang, S., Li, L., et al.: Karst rocky desertification early warning and risk assessment model system design - in Du'an Yao Autonomous County of Guangxi as an example. Progress Geogr. **24**(2), 122–130 (2005)

The Design and Implementation of Field Patrol Inspection System Based on GPS-Tablet PC

Shengchun Shi[1(✉)] and Yicheng Yin[2]

[1] Oxbridge College, Kunming University of Science and Technology,
Kunming 650106, China
419714078@qq.com
[2] Yunnan Engineering Institute of Surveying and Mapping,
Kunming 650093, China
276198715@qq.com

Abstract. According to the actual demand of land regulation, this paper presents a field patrol inspection system combining GPS and tablet PC through the integration of hardware and software. This paper has overall designed the function and framework of the field patrol inspection system. On the fundamental of solving some key technical problems such as serial communication between PDA and tablet PC and intelligent processing of date, this paper has developed a integrated field patrol inspection system by combining GPS and tablet PC on the basic of GIS. This technology is already verified in practice and proved to be feasible, which will improve the efficiency of land management and can achieve indoor and field integration of field patrol inspection work.

Keywords: GPS · Tablet PC · Land management

1 Introduction

Land utilization is a dynamic project and land regulation is a routine work for land management. The main technique methods of supervising presently are GPS-PDA, aerial survey, satellite remote sensing and field paper checks.

The development of GPS/PDA achieves the transformation from manual labor to intelligent operation, which would improve the work efficiency of workers working in the basic units [1]. Owing to the small storage capacity, slow running of imagines, limitations to the operation of each layer, the GPS/PDA with PDA as the basic hardware platform is mainly used in field location acquisition, which should work with interior work [2].

The aerial survey is typically applicable to the use of investigation updating, which is difficult to be applied in the land change survey due to the high price, complex treatment technology and long cycle; satellite remote sensing is usually used in the dynamic monitor of land use, which can lead field investigation through grasping pattern spots [3]; the method of "direct field annotation" that widely used at present uses the paper figure as work base, measures with conventional methods, writes the changing information by hand with the modification questionnaire form, then processes the interior work data and statistics the area, which has a large workload [4].

© Springer Nature Singapore Pte Ltd. 2017
H. Yuan et al. (Eds.): GRMSE 2016, Part I, CCIS 698, pp. 12–19, 2017.
DOI: 10.1007/978-981-10-3966-9_2

Therefore, establishing an integrated system, which can maintain the daily patrol process in land law enforcement and accuracy requirement, has high work efficiency, can realize the basic geographic information navigation and positioning and the processing of graphs and dates, has an important real-world implication. Thus, the field patrol inspection system based on GPS/tablet computer was discussed, which including the system integration and designing, GPS data receiving and processing, figure spots changing and so on, in this paper. Basing on these studies, the field patrol inspection system based on GPS/tablet computer was designed and applied.

2 System Overall Design and Function Design

After processing the remote sensing data and thematic data under GIS environment, field patrol inspection system is undergone with the assist of the embedded GPS devices, which can exert the fullest potential of all date and sharply improve the efficiency and precision of the field investigation.

2.1 GPS/Tablet Computer System Design

Hardware Design
In the process of hardware design, first of all the work need to be done is demand analysis on the basic of early stage research, next is the overall design of hardware and then is detailed design about each module, finally is the test of hardware system. The flow chart of design is as follows:

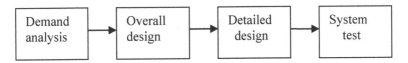

Software Design
The software design of GPS/tablet computer system follows the ideas of software engineering. The design of the system software has experienced some stages, they are the feasibility study, demand analysis, general design, detailed design, coding, testing, iterative development and software test phase, etc. [6].

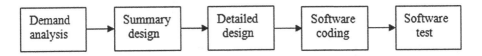

Specific tasks at each stage are as follows:

(1) Demand analysis. During the early stage of the software design, field research is undergone to understand the land survey work process and make the business requirements of the basic land regulation clearly. Based on these information, the results of demand analysis is got, which could determine the business data stream,

the data dictionary of GPS/tablet computer system, and divide the software into base station data service subsystem, field data acquisition subsystem and data processing subsystem.

(2) General design. Based on the object-oriented software design thought, modular analysis and data abstraction of each subsystem are proceed and improved, which leads the general design of GPS/tablet computer system model.

(3) Detailed design. Firstly, dividing each module into specific program flow chart and algorithm, and then defining the communication interface between each module [7], finally forming the detailed design of each subsystem.

(4) Software coding. Technicians code each subsystem of system on account of software development tools. In order to maintain the continuity of development, the compatibility of hardware platforms and operating systems should be paid more attention, and the code quality and the effectiveness of the proposed code comments should be controlled strictly, according to the characteristics of the various subsystems running environment.

(5) Software testing. The testing and coding of GPS/tablet computer system software follows the ideas of iterative development. Every internal distribution of hardware system (internal release versions) is performed under the tester's test in detail, then the test report document is created, after modifying some problem by developers, regression testing is proceeding.

2.2   Functional Framework

The basic function of the field data collection in land use status update survey realized by the integrated GPS/tablet land supervision data acquisition system is shown in Fig. 1.

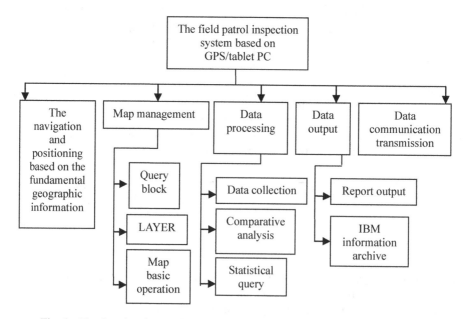

Fig. 1. The function frame of GPS/tablet computer field patrol inspection system

(1) System initialization and setting function: setting the survey area code (provincial, city, county, township and village), associated data file name, setting the file storage mode, geology code system, GPS data receiving parameters and the transformation parameters between different coordinate system;

(2) Base map operation function: realizing the function of importing, opening, displaying, querying, hiding and shut down the base map, and the hierarchical display and management of the base map data;

(3) File operation function: realizing the function of creating, opening, saving and deleting the file, and archiving the search results after patrolling;

(4) View operation function: realizing the function of amplifying, narrowing and roaming the view;

(5) Navigation function: guiding field inspectors arriving the destination rapidly, through setting the identification of the destination and using GPS navigation and positioning;

(6) Recording data function: realizing the receiving of GPS point data in real time, recording the inflection point, line, figure spot, linear features, sporadic feature and the topological relationship between other patterns of the terrain and GPS point data, and recording the attributes including its connected relationship with graphics, which can achieve the modification of graphics connection relationship and attribute data;

(7) Coordinate calculating, coordinate transformation, data exporting functions: realizing the calculating of GPS receiving data, the transformation from latitude and longitude coordinates to land coordinates, and the transmission of original data, recorded data and calculating data;

(8) Data format conversion function: realizing the transmission between recorded data format and land use standard data format;

(9) Analysis function: contrastively analyzing the collected field data that is superimposed on the fundamental data, and confirming the region land use information immediately;

(10) Transmission function: realizing the data communication between the law enforcement supervisory monitoring system and the on-board remote law enforcement supervision system, the collected data was sent to metro monitoring center and law enforcement patrol vehicles through the offline copy, Bluetooth, and 3G technologies, etc.

According to the basic function design, field patrol inspection system based on GPS/tablet PC can be divided into five modules: the navigation and positioning based on the fundamental geographic information, map management, data processing, data output and data communication transmission. Map management mainly includes map basic operation, layer operation and query block. Data processing includes data collection, comparative analysis and statistical query. Data output includes report output and information archiving.

2.3   The Acquirement and Processing of GPS Signals

The technology of reading GPS locating information through the serial communication between tablet personal computers and GPS cards is widely used in many fields. Due to small in size, large capacity, fast running speed and easier to site operation of tablet computer, so the GPS communication and information processing by using tablet computers shows some advantages [5].

The connection between tablet computer and GPS data is achieved by a serial port or parallel port of tablet computer and GPS-OEM mainboard, tablet computer can read the data of GPS-OEM mainboard, control the OEM mainboard and calculate the GPS cable data. Because of the slow reading speed (9600 bit rate), the COM cache interface is needed, so the design of data interface between tablet computer and GPS data is a relatively independent part of system software.

The JGG20 OEM mainboard that produced by LAVAP company was selected in this paper, which configurate with GRIL language (LAVAP company's copyright) that independent with the hardware. The thread of Window CE was designed independently to send GRIL language instruction through the COM interface of OEM, and to control OEM, which can achieve the data communication, controlling, extracting the cable information and saving the period GPS data in real time.

Under the development environment of Windows, the system that with the help of Supermap Objects and visual studio2005 development tool of Microsoft was explored, this system could run on a tablet computer or a computer with Windows operating system.

```
serialPort = (SerialPort) portId.open("GISLAND", 60);
serialPort.setSerialPortParams(4800, SerialPort.DATABITS_8,
SerialPort.STOPBITS_1, SerialPort.PARITY_NONE );

in = new BufferedReader( new
InputStreamReader(serialPort.getInputStream()));

The computer regards the GPS modules as a COM interface,
the code is connected to COM interface, and some code to read
GPS information with correct bit rate as following:

serialPort = (SerialPort) portId.open("GISLAND", 60); //
Opening port, reading one GPS message every 60 seconds

serialPort.setSerialPortParams(4800,
SerialPort.DATABITS_8, SerialPort.STOPBITS_1,
SerialPort.PARITY_NONE ); //Setting port parameters,

in = new BufferedReader( new
InputStreamReader(serialPort.getInputStream()));
//Reading message
```

(1) When starting or stopping the GPS communication, serial port should be opened through opening a file, and related parameters should be configured;

(2) After opening and setting the communication port, a system timer event can be created, and massages can be received and processed by events trigger mode;

(3) GPS receiver can receive, calculate and send GPS navigation and positioning information continuously to the computer through ports when it in the state of work. The fields of information must be extracted from cache byte streams through systems and be transformed into locate information data that has practical significance and could support the figure spot changes;

(4) GPS coordinate transformation: the calculated GPS point data is based on the WGS-84 geocentric coordinate system and the field patrol inspection results is based on the located Gauss plane coordinate system, so the GPS positioning results of geodetic coordinate (L, B) must be transformed to local Gauss plane coordinate (x, y).

There are two steps: the first step is to transform the WGS-84 geodetic coordinate (L, B) to WGS-84 Gauss plane coordinate; the second step is to combine the Gauss plane coordinate with local Gauss plane coordinate through plane coordinate transformation. The flow sheet of GPS communication signal module and analysis module is shown in Fig. 2.

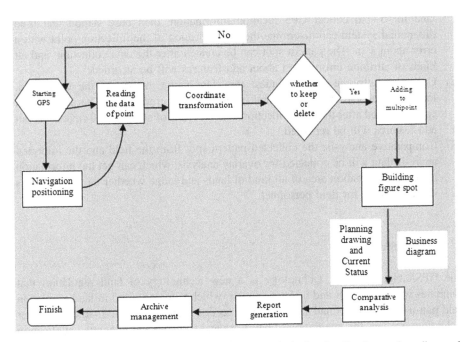

Fig. 2. The comparative analysis flow chart between GPS signal collecting and reading and figure spot

The flow of GPS communication signal module and analysis module: start the receiver GPS, reading the point data from figure inflection point, transform the WGS-84 geodetic coordinate (L, B) of point to plane coordinate, save plane coordinate date, adding to multipoint, building figure spot, obtain area of various types land through comparative analysis between planning drawing and current status map, output area form of various types land, archive save.

2.4   Figure Collection and Analysis

Figure collection and analysis is the key module of a system.

Figure collection mainly includes the graphical data acquisition, data collection and comparative analysis of figure spot, present status and plan. And graphical data acquisition includes the graphic drawing automatically and manually. Graphical data acquisition can achieve the graphic drawing automatically or manually according to the point information obtained by the GPS signal reception and module analysis and acquisition parameters set by the user. As for the attribute data, actual survey information can be inputted to the system by field patrol inspection.

(1) Point messages can be collected through receiving GPS signals and transformed to inflection point coordinates of pattern spot. The module receives these data and stores them into temporarily point sets. The specific method as follows: The inspector hand-held GPS/tablet computer and stand at inflection point of patrolled land for 5 s to collect GPS location information, then base station real time differential system can determine the exact location of the inflection point with a error about 1 m. The pattern spot can be created after the data collection, and all kinds of attribute information about add features will be recorded.

(2) Collecting attribute data includes: the class name, nature of the ownership, location, practical applications and other attribute information. The pattern spot can be created after the data collection, and all kinds of attribute information about add features will be recorded.

(3) Comparative analysis: the collected pattern spot from the field and the imported analysis data will be collected for overlay analysis, which can get the information about the occupation area of all kind of lands and judge whether the using land is illegal or legal for field personnel.

3   Conclusions

The GPS/tablet computer technology is a new technology of land regulation that combines with GPS, GIS and tablet computer, which is mainly used in land regulation field patrol inspections. Land field patrol inspection system with GPS/tablet computer technology realizes the integration of hardware and software and integrated application of GPS and GIS, which represents a development direction of land management applications with GPS [8]. After applied this system in some national project teams, it got a great success and greatly met the land regulation field patrol inspection field investigation work.

Comparing with GPS/PDA that used recently, land field patrol inspection system with GPS/tablet computer technology realizes the navigational positioning that based on basic geographic information and guides the field inspectors reach the destination immediately. This system has a fast running speed and can get specific information about land utilization by comparatively analyzing the collected figure spot, the present status and plan and make a judgment at once.

After the application in the demonstration zone, this system can largely improve the accuracy, efficiency and automation of land regulation field verification, which can achieve the "real-time supervision law enforcement and decision", improve the working efficiency, enhance the land regulation and realize the popularization and application in land supervision business. This is the essential technological base of land regulation, and has great significance for improving the science and technology support system of land and resources.

Acknowledgment. The authors are grateful for the financial support from the scientific research fund project in Yunnan province department of education (Grant no. 2016ZDX240).

References

1. Jia, W., Liu, J., Lina, Yu., et al.: Development and application of field survey technology based GPS and GIS for land consolidation. Trans. Chin. Soc. Agric. Eng. **25**(5), 197–201 (2009)
2. Zhijun, M., Chunjiang, Z., Xiu, Z.: Field multi-source information collection system based on GPS for precision agriculture. Trans. Chin. Soc. Agric. Eng. **19**(4), 13–19 (2003)
3. Chang, Q., Wei, D., Zhu, G., et al.: Study on the Application of GPS/PDA in Land Change Survey. J. Anhui Agric. Sci. **34**(23), 6273–6275 (2006)
4. Yan, C., Yuan, L., Xiaoping, L., et al.: Data collection system designation in changed land usage surveying based on integration of GPS-PDA. J. Nanjing Normal Univ. (Eng. Technol.) **5**(4), 77–81 (2005)
5. Starbird, K.: Digital volunteerism during disater: crowdsourcing information processing. ACM (5), 7–12 (2011)
6. Liu, G.: Development and application of transmission line inspection system in Tianshengqiao Bureau. Power Syst. Prot. Control **38**, 45–51 (2010)
7. Ren, Z.: Obstacle-navigation control of inspection robot for power transmission lines. J. North Univ. China **32**, 33–36 (2011)
8. Yang, G., Wu, W., Liu, Q., et al.: Design and realization of land use change investigation system based on PDA. J. Liaoning Tech. Univ. (Nat. Sci. Ed.) **26**(4), 502–504 (2007)

The Vehicle Route Modeling and Optimization Considering the Dynamic Demands and Traffic Information

Chouyong Chen[✉] and Jun Chen

Management School, Hangzhou Dianzi University,
Hangzhou 310018, Zhejiang, China
cychen@hdu.edu.cn, 514134894@qq.com

Abstract. This paper is aimed to solve this kind of problem and cope with the actual requirements containing time window and dynamic demands. Therefore the smooth and continuous time dependent function is introduced and the two-stage model including the "initial optimization stage" and "real-time optimization stage" is established. At the same time, a hybrid algorithm based on genetic-tabu algorithm and simulated annealing algorithm is designed to solve the model. In the end the effectiveness of the hybrid algorithm and the model is verified by comparing the results of simulation and other algorithms.

Keywords: Route optimization problem · Traffic information · Dynamic demand · Optimization algorithm

1 Introduction

In the late 1980s, the definition of the dynamic vehicle problem was offered by Haghani and other researchers which were followed by dynamic researches on the vehicle route problem. For decades, a great deal of researches has been done domestically and abroad. For instance, Haghani [2] etc. established a dynamic VRP model based on the continuous time dependent function and the Malandraki model. Hu [3] etc. designed a hybrid constraint model to solve the VRP problem in the dynamic network in which the real-time information such as traffic condition and new demands were taken into account. Li [4] etc. has made a lot of researches on the traveling time considering the dynamic demand, but there are some limitations such as his fuzzy definition of traveling time in the research. After summarizing the DVRP researches home and abroad, Zhou [5], considering the effect of dynamic demand and network on the route problem, has mainly studied the dynamic uncertainties of traffic performance. However, the researches on the dynamic vehicle route problem focus on the uncertain demand more than other uncertainties such as the traffic information.

As the vehicle route optimization problem is characterized by complexity and high requirements for the real-time performance, heuristic algorithms such as genetic algorithm has been widely used. While this kind of algorithms still have some limitations, for instance, premature convergence, inefficiency of searching and slow solution speed etc. Therefore in this paper, a two-stage model including the "initial

© Springer Nature Singapore Pte Ltd. 2017
H. Yuan et al. (Eds.): GRMSE 2016, Part I, CCIS 698, pp. 20–33, 2017.
DOI: 10.1007/978-981-10-3966-9_3

optimization stage" and "real-time optimization stage" is established. After studying and comparing genetic algorithm, tabu search algorithm and simulated annealing algorithm, a new hybrid algorithm based on these three algorithms is proposed. Through some examples, this new hybrid algorithm is verified to be stable and reliable and it has obvious advantages in the searching ability and efficiency.

2 The Vehicle Route Problem and Model Considering the Traffic Information and Dynamic Demand

2.1 Description of the Problem

The vehicle route problem considering the traffic information and dynamic demand is characterized by information uncertainties. In this paper, the effect of uncertain traffic network condition and demand uncertainties are taken into account. The details are as following:

A distribution center, with K trucks of the same kind, is in charge of goods distribution for L customers. The quantity that customer i demands is P_i. Each customer can be provided service by any car, but only once every time. During the service process, the dynamic demand of customers will be defined as the new customer and the traffic information updating will lead to the change of traveling time between two points. A suitable route should be developed to achieve the goal of the minimum cost as well as meet the constraints of the premise.

In order to solve the problems mentioned above, a two-stage model of vehicle route problem considering the traffic information and dynamic demand is established and it includes the "initial optimization stage" and "real-time optimization stage".

2.2 Initial Optimization Stage Model

Initial optimization stage model is as following: the code number of distribution center is 0, the code number of customer can be defined as 1, 2, ..., L; the distribution center and any customer will be respectively represented by letter i and j; the vehicle will be represented by letter k, its code number will be 1, 2, ..., K; the vehicle loading limitation is Q; the quantity that customer i demands is $q_i (i = 1, 2, ..., L)$, $q_i < Q$; the unit cost of transportation is c; the time window for customer i is $[E_i, L_i]$, the vehicle arrives at the location of customer i at the time s_i, the service time for i is f_i, the vehicle spends t_{ij} arriving at the location of customer j from i.

Defining the decision variables:

$$x_{ijk} = \begin{cases} 1 & \text{From customer } i \text{ to customer } j \text{ by vehicle } k \\ 0 & \text{others} \end{cases}$$

$$y_{ik} = \begin{cases} 1 & \text{customer } i \text{ served by vehicle } k \\ 0 & \text{others} \end{cases}$$

Initial optimization stage model:

Objective function:

$$\min Z = \sum_{k}^{K} \sum_{i=0}^{L} \sum_{j=0}^{L} cd_{ij} x_{ijk} + c_i(t_i) \tag{1.1}$$

$$\sum_{i=1}^{L} q_i y_{ik} \leq Q \quad i \in L \cup \{0\}, \forall k \tag{1.2}$$

$$\sum_{i=1}^{L} x_{ijk} - \sum_{j=1}^{L} x_{jik} = 0 \quad i,j \in L \cup \{0\}, \forall k \tag{1.3}$$

$$\sum_{k=1}^{K} y_{ik} = 1 \quad i \in L \cup \{0\}, \forall k \tag{1.4}$$

$$\sum_{i=1}^{L} x_{ijk} = y_{jk} \quad j \in L \cup \{0\}, \forall k \tag{1.5}$$

$$\sum_{j=1}^{L} x_{ijk} = y_{jk} \quad i \in L \cup \{0\}, \forall k \tag{1.6}$$

$$\sum x_{ijk} \leq |L| - 1 \quad L \subset \{1, 2, \ldots, L\} \tag{1.7}$$

$$s_j = \max(E_j, s_i + f_i + t_{ij}) \tag{1.8}$$

$$s_j \leq L_j \tag{1.9}$$

$$c_i(t_i) = c_1 \sum_{i=1}^{L} \max\{(a_i - s_i), 0\} + c_2 \sum_{i=1}^{L} \max\{(s_i - b_i), 0\} \tag{1.10}$$

Equation (1.1) is the objective function which represents the minimum cost in total including the vehicles' transport costs and time costs; constraint (1.2) ensures that the total good demand on every route is less than the truck's load limitation; constraint (1.3) ensures that the vehicle arriving at customer i will have to leave form i; constraint (1.4) ensure that each customer can be serviced; constraints (1.5) (1.6) ensures that each customer can only be served by one truck; constraint (1.7) ensures that no truck returns the way it has passed; constraints (1.8) (1.9) meet the soft time windows required by customers; constraint (1.10) meet the soft time windows penalty requirements.

2.3   Real-Time Optimization Stage Model

Based on the research of Ge [9], a smooth continuous time dependent model is introduced. In order to reflect the actual truck speed changes in macro sense, the traveling speed is assumed to change smoothly according to different time segments such as the morning and evening rushing hours to meet the requirements of FIFO model.

At the beginning of real-time optimization stage, many vehicles have left the distribution center and served some customers which means that the loading of each truck varies. It is difficult to reschedule the trucks as some of them are located at the customers' places. Therefore the virtual distribution center is introduced in this paper and is defined as a customer place in which a truck is parked. The model is as following:

Assuming in the first stage that the remaining loading capacity of a truck is b_k ($k = 1, 2, \ldots\ldots, K$), N represents the total quantity of the need-to-served customers in the first stage and the new customers in the second stage. H is the number of virtual distribution center and its code numbers are $N + 1, N + 2, \ldots\ldots, N + H$, the former distribution center's code number is $N + H + 1$, T extra vehicles will be sent out. $M(m = 1, 2, \ldots\ldots, M)$ represents the time segments of a day. $[E_i, L_i]$ is the time window for customer i. t_{ij}^m represents the hours a vehicle travels from customer i to j during the time segment m. s_i represents the time a vehicle reaching customer i and f_i is the service time for customer i.

Objective function:

$$\min\left\{\sum_{k=1}^{H}\sum_{i=1}^{N+H+1}\sum_{j=1}^{N+H+1} cd_{ij}x_{ijk}^m + \sum_{k=H+1}^{H+T}\sum_{i=1}^{N+H+1}\sum_{j=1}^{N+H+1} cd_{ij}x_{ijk}^m + c_i(t_i)\right\} \quad (1.11)$$

$$\sum_{i=0}^{N} q_iy_{ik} \leq b_m \quad k, m = H+1, H+2, \ldots, H+T \quad (1.12)$$

$$\sum_{k=1}^{H+T} Y_{ik} = 1 \quad i = 1, 2, \ldots, N+H+1 \quad (1.13)$$

$$\sum_{i=1}^{N+H+1} x_{ijk}^m = \sum_{j=1}^{N+H+1} x_{jik}^m \quad i = 1, 2, \ldots, N+H+1 \quad (1.14)$$

$$\sum_{i=1}^{N+H+1} x_{ijk}^m = y_{jk} \quad j = 1, 2, \ldots, N+H+1, \ K = 1, 2, \ldots, N+T \quad (1.15)$$

$$\sum_{j=1}^{N+H+1} x_{ijk}^m = y_{jk} \quad i = 1, 2, \ldots, N+H+1, \ k = 1, 2, \ldots, H+T \quad (1.16)$$

$$\sum x_{ijk}^m \leq |L| - 1 \quad L \subset \{1, 2, \ldots, L\}, \ k = 1, 2, \ldots, H+T \quad (1.17)$$

$$s_j = \max(E_j, s_i + f_i + t_{ij}^m) \quad (1.18)$$

$$s_j \leq L_j \quad (1.19)$$

Equation (1.11) is the objective function and it includes three parts: 1 the transportation cost of the trucks sent in the first stage providing service to the need-to-serve customers in the first stage and the new customers in the second stage; 2 the transportation cost of the vehicles sent in the second stage; 3 time window constraint penalty cost; constraint (1.12) ensures that the total good demand on every route is less than the truck's load limitation; constraint (1.13) ensures that each customer can be serviced; constraint (1.14) ensures that the vehicle arriving at customer i will have to leave form i; constraints (1.15) (1.16) ensures that each customer can only be served by one truck; constraint (1.17) ensures that no truck returns the way it has passed; constraints (1.18) and (1.19) meet the customers' time windows requirements.

3   Algorithm Design

3.1   Solution Strategies

To solve the two-stage mathematical programming model for the vehicle route problem, the two-stage solving strategies are introduced which include "the designing of the initial optimization route" and "real-time optimization scheduling". According to the strategy, at first an initial solution is generated based on the acquired information; when the dynamic information occurs, the initial solution will be adjusted partially in some time segments to avoid the interference of the frequent information as well as avoid the trap of getting the sub-optimal solution because of the local optimization strategy. The process mentioned above is shown in Fig. 1.

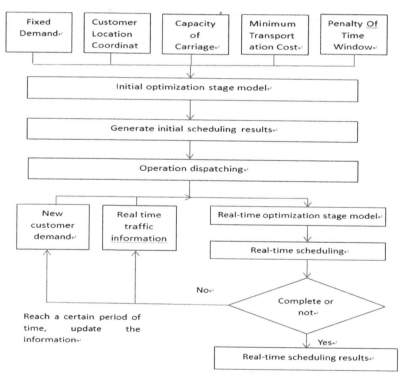

Fig. 1. Solution strategy flow chart

3.2 The Idea of Hybrid Algorithm Based on the Genetic Algorithm, Tabu Search Algorithm and Simulated Annealing Algorithm

Taking into account the characteristics of the designing principle and the parallelism of the genetic algorithm, it can be used as a whole framework of the hybrid optimization algorithm which will take much better of its advantage. The key point is that the taboo factor is introduced during the genetic algorithm's copy operation and the individual's diversity is expanded to avoid the "premature" phenomenon led by the local optimization. After the new species group is generated by the genetic algorithm, the simulated annealing algorithm is used to deal with this kind of new species group. When the simulated annealing algorithm is being used, the changing temperature and the searching range for solutions will be controlled to generate more initial solutions and expand the solutions' diversity. At the same time, the previous operation also can improve search efficiency, keep better solutions, improve the convergence of the hybrid algorithm and accelerate the convergence of the algorithm at later.

3.3 The Solving Process of Hybrid Algorithm

The process is shown in Fig. 2:

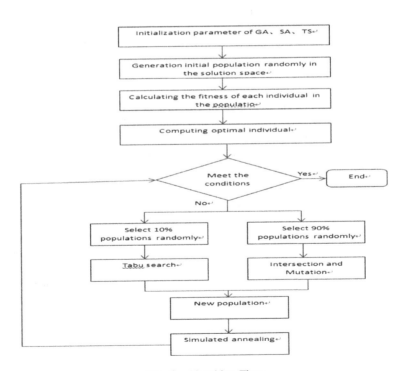

Fig. 2. Algorithm Flow

① Coding method

Due to the particularity of the solution of vehicle routing problem, natural coding method can be used to encode the chromosome, that is, there are n customers and m vehicles in the distribution network. It can be compiled to be a $n + m + 1$ length of chromosome. Taking 9 customer distribution problem as an example, its chromosome code is 0235046017890 which means the nine customers' transportation and path arrangement will be accomplished by three trucks:

Route 1: distribution center $0 \rightarrow$ customer $2 \rightarrow$ customer $3 \rightarrow$ customer $5 \rightarrow$ distribution center 0;
Route 2: distribution center $0 \rightarrow$ customer $4 \rightarrow$ customer $6 \rightarrow$ distribution center 0;
Route 3: distribution center $0 \rightarrow$ customer $1 \rightarrow$ customer $7 \rightarrow$ customer8 $\rightarrow$ customer $9 \rightarrow$ distribution center 0.

② Initial species group

In order to make the hybrid algorithm converge to the global optimum, the optimal species group should be in a large scale. When initializing the chromosome, n customers' full arrays are generated randomly in which 0 is inserted at the beginning and end showing that the vehicles leave from the distribution center and eventually returns to the distribution center. And $m-1$ 0 is inserted randomly into the sequences according to the loading limitation and time window. This can be repeated until enough chromosomes are generated.

③ Fitness function

The fitness function used in this model is consistent with the objective function. Based on the research emphasis and the requirement of loading limitation & time windows, a maximum number is set to the objective function Z for those chromosomes violating the constraints. Therefore the less the fitness is, the better the individual is.

④ Selection operator

The selection operator is also called regenerative operator or replication operator. The purpose of selection operator is to directly copy the optimized individuals to the next generation or to generate new individuals through a pair of mating and then pass the new individuals to the next generation. In this paper, the roulette strategy is used to make the choice and a number of individuals are chosen from the group. The probability of being selected for an individual is inversely proportional to the value of their fitness. Therefore, the smaller the value of fitness is, the greater the probability of being selected for an individual.

⑤ Crossover operator

Crossover operation refers to a process in which two individuals are selected randomly from a group and then exchange some fragments from each other in a certain probability. The main method of generating new individuals in the crossover operation determines the global searching ability of the genetic algorithm. Due to the constraints on vehicle routing problem, a large number of infeasible solutions will be generated if the simple crossover operators are used. Therefore, in this paper, the order crossover OX is introduced and the different sub-arrays are crossover in order. In this way, the convergence speed of the algorithm will be accelerated and excellent sub-strings will be passed to the next generation.

⑥ Mutation operator

Mutation operation is to change the value of some genes of an individual. The introduction of mutation operator in genetic algorithm has two purposes: (1) to enable the genetic algorithm with local random search ability. When the optimal solution domain is reached with the help of crossover operator in genetic algorithm, the mutation operator can be used to accelerate the convergence to the optimal solution. (2) to obtain the diversity of population in genetic algorithm as well as avoid premature convergence. In this paper, the reverse mutation is used which refers to a process in which the sub-strings in the reverse areas-the middle part between two randomly selected points in the process of individual encoding- are put in reverse order and then inserted back to the original position.

⑦ Tabu search operator

In this paper, the two exchange method is used to generate the forbidden field solution. As for the vehicle scheduling problems, high-quality solutions are required when designing the vehicle route. To meet the high requirements, the whole field is searched for the optimal solution.

⑧ Annealing operator

According to the basic principle of simulated annealing algorithm and the Metropolis criteria, the local optimal solution can be obtained after a large number of calculations [12]. In this paper, the whole field is searched with a similar method to the tabu search method. After obtaining the optimal solution, it will be compared to the former solution based on the Metropolis criteria.

⑨ Termination criterion

When the current evolutionary algebra of the algorithm is greater than the predetermined set value N, the algorithm ends.

4 Experimental Verification and Analysis

4.1 Experimental Data

Taking one day's orders in a logistics distribution system as an example, a practical verification is carried out to verify the validity and practicability of the vehicle routing problem with dynamic demand. Among them: (1) initial customer number is 20, the customer point coordinates and demand information are shown in Table 1; (2) all customers are serviced by the same type of vehicles with loading limitation of 20 tons; (3) the average speed of each vehicle is at 50 km/hour; (4) unit cost of transportation is 5 RMB/km; (5) the time window when a distribution is opened and closed during the day is [7.5,17]; (6) soft time windows penalty cost C_1 is 10 RMB/hour, C_2 is 15 RMB/hour.

Table 1. Initial customer information

Customer	Relative position coordinate of customer and distribution center (x, y)	Customer demand	Time window	Service time
1	[−11.2, 3.5]	2.2	[8, 9.5]	0.92
2	[5.6, 10.3]	4.3	[9, 14]	0.75
3	[18.6, −11.9]	1.7	[11, 13.5]	0.5
4	[−1.5, 3.8]	3.3	[12, 14.5]	0.58
5	[7.3, 23.5]	3.5	[13, 15]	0.67
6	[3.1, 8.6]	1.6	[8, 9]	0.33
7	[−5.3, −4.5]	1.9	[10, 14]	0.98
8	[8.4, 2.6]	2.1	[11, 16.5]	0.9
9	[0.8, −5.1]	2.5	[15, 15.8]	0.42
10	[8.6, −12]	2.8	[8.5, 11]	0.33
11	[−2.4, 14.3]	5.6	[9, 10]	0.25
12	[−11.5, −8.8]	3.1	[13, 15]	0.67
13	[12.6, −1.4]	1.6	[8, 16.5]	0.58
14	[−22.5, −15.1]	4.1	[10, 12.5]	0.5
15	[−14.1, 23.5]	2.5	[8, 14.5]	0.4
16	[−1.2, 22.5]	2.3	[9, 11]	0.42
17	[5.5, −6]	3.7	[9.8, 11]	0.8
18	[15.8, −8.6]	1.5	[8, 12]	0.35
19	[−27.5, −9.2]	1.2	[10, 11.5]	0.75
20	[−18.6, −0.3]	2.8	[9, 10]	0.8

4.2   Initial Optimization Scheme Determination

All algorithms are programmed under the development environment of MATLAB, which is a kind of mathematical software, Intel (R) Core i5 (TM) CPU M480 @ 4.00 GB 2.67 GHz memory.

First of all, in the initial optimization model the hybrid algorithm based on genetic algorithm, tabu search method and simulated annealing algorithm is used to deal with the customer points and the results are as following: generating a group of 50, crossover probability and mutation probability respectively 0.95 and 0.05, evolutionary iteration number is 100, initial temperature of SA is 2000, cooling function T = T0 * 0.99 K, K stands for genetic iterations. The initial optimal scheme is shown in Table 2 after calculating randomly for 10 times (Fig. 3).

Table 2. Initial optimal distribution scheme

Distribution vehicle number	Distribution route	Total cost per route distribution	Loading capacity
1	0-6-11-15-16-5-4-0	409.4735	18.8
2	0-1-20-19-14-12-7-9-0	372.2337	17.8
3	0-17-10-3-18-8-2-0	315.8739	17.7

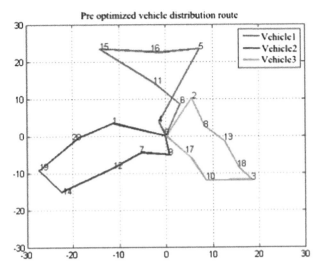

Fig. 3. Pre optimized vehicle distribution route

4.3   Dynamic Optimization Scheme Determination

In this paper, the real effect on the vehicle scheduling in real-time optimization stage is mainly analyzed while considering dynamic demand and traffic information. At the same time, in order to maintain the scientific nature of the modeling, a series of abstraction and simplification of the realistic problems are needed. Including: (1) in the experimental tests, the main road speed dependent function is used and discrete time is divided into 5 segments according to the morning and evening rushing hours, as is shown in Fig. 4; (2) in order to facilitate the research, choose the dynamic information at 10:00 as a basis for scheduling and the initial optimization solution should be implemented before the real-time scheduling.

At 10:00 am, customer demand information has been updated when 4 new customers propose service requests and their point coordinates and detailed information are shown in Table 3.

The results show that the distribution center has issued 3 vehicles, which are respectively located near the points 15, 19, 10 and with respective residual loading of 10.3, 13.8, and 16.3. Customer points 6, 11, 15, 20, 19, 17 and 1 have been served. In the real-time optimization stage, it will be the first priority to meet the customers' demand in the initial distribution scheme under the constraints of time window and vehicle loading. If the residual loading and time window are available, new customers can be added or an extra truck can be sent. To solve the real-time model mentioned above, the first phase algorithm is used in which the iteration occurs for 100 times within a group of 50. The results are as following:

The results can be seen from Table 4: under the condition of traffic information and dynamic demand, three routes are obtained with GTSA algorithm. The penalty cost of

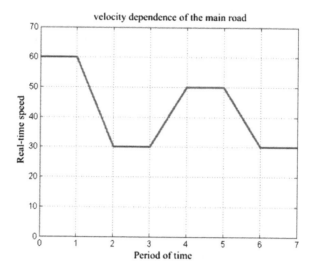

Fig. 4. Velocity dependence of the main road

Table 3. New customer point information

Customer	Relative position coordinate of customer and distribution center (x, y)	Customer demand	Time window	Service time
21	[−5.8, 26]	1.2	[10, 12]	0.92
22	[−25.2, −14.5]	1.8	[12, 16]	0.85
23	[10, −10]	2	[11, 14.5]	0.6
24	[6, −6.1]	8.6	[10, 12]	0.58

Table 4. Optimal distribution scheme in real time

Distribution vehicle number	Distribution route	Total cost per route distribution	Loading capacity
1	0-6-11-15-21-16-5-4-0	434.5675	20
2	0-1-20-19-22-14-12-7-9-0	386.3467	19.6
3	0-17-10-23-3-18-8-2-0	323.7321	19.7
4	0-24-0	42.7814	5.6

each route is zero if the customers' soft time window requirements are met on each route. On the fourth route, the new customers are regarded as the incoming customers at the next real-time stage as there is no truck returning to the distribution center at that moment. At the end of the day, the need-to-serve customers' requirements will be scheduled on the next working day (Fig. 5).

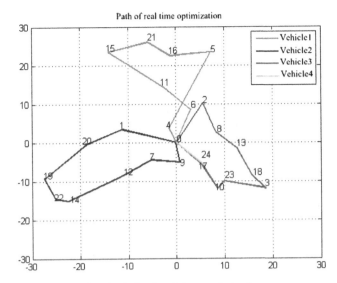

Fig. 5. Path of real time optimization

4.4 Dynamic Optimization Scheme Determination

In order to verify the performance of the hybrid optimization algorithm, genetic algorithm, simulated annealing genetic algorithm and hybrid optimization algorithm are used to obtain the optimal results and optimize the optimization algorithm.

As is shown in Table 5, through GA, GSA and GTSA the optimization of the solution as well as the optimal solution are achieved. But GTSA algorithm has achieved much higher success rate in searching and 16.9% lower average distribution distance than GA algorithm. It shows that the algorithm with best global searching ability is GTSA, GA is with the worst global searching ability and GSA is between them. Comparing to the GSA and GA algorithms, the average computing time of GTSA algorithm is reduced by 71.16% and 75.52% respectively which reflects the GTSA algorithm has obvious advantage in the convergence speed.

Table 5. Comparison and analysis of simulation results

Algorithm	Search success rate	Average distribution distance	The average number of iterations of the final solution for the first time	Average calculating time
GA	30%	465	40	7.56
GSA	48%	386	19	2.18
GTSA	64%	385	8	1.85

Experiments results show that the improved genetic algorithm is reasonable and effective. When solving the vehicle routing problems, the hybrid optimization algorithm performs much better than other algorithms in global search and convergence speed. It can ensure the immediacy and effectiveness when solving the dynamic distribution problems.

5 Conclusions and Suggestions

In this paper, the customers' dynamic demands and the real-time traffic conditions are given fully consideration. And then a two-stage mathematical programming model for the vehicle route problem considering the traffic information and dynamic demand is established which contains two parts: "initial optimization and real-time optimization". The idea of solving the model based on the smooth and continuous time dependent function is proposed as following: considering the traffic information's effect on the distribution route when dealing with the dynamic demand to improve the model's practicability. The experimental results show that this method meets the FIFO criteria and the real-time requirements of the scheduling problems.

To simplify the model, we assume that the traffic condition is comparatively stable between the customers' locations. However this is not the true in the reality. The different road condition, vehicle flows and real-time dynamic customers can have different influence on the optimization process, so the dynamic route optimization will be studied in my further research to reflect the this kind of effect and enrich the dynamic vehicle scheduling theories.

Acknowledgment. Supported by the National Natural Science Foundation of China (71171070, U1509220).

References

1. Psaraftis, H.N.: A dynamic programming solution to the single vehicle many-to-many immediate request dial-a-ride problem. Transp. Sci. **14**(2), 130–154 (1980)
2. Haghani, A., Jung, S.: A dynamic vehicle routing problem with time-dependent travel times. Comput. Oper. Res. **32**(11), 2959–2986 (2005)
3. Hu, T., Liao, T., Lu, Y., et al.: A study on the solution approach for dynamic vehicle routing problems under real-time information. Transp. Res. Board Annu. Meet. **1857**(1), 102–108 (2003)
4. Li, B., Zheng, S.F., Cao, J.D., et al.: Method of solving vehicle routing problem with customers' dynamic requests. J. Traffic Transp. Eng. (2007)
5. Zhou, B.: Research on vehicle routing problem with stochastic travel time. Dalian Maritime University (2005)
6. Lang, M.: Two-phase algorithm for dynamic distribution vehicle scheduling problem. J. Transp. Syst. Eng. Inf. Technol. **9**(4), 140–144 (2009)
7. Mantawy, A.H., Abdelmagid, Y.L., Selim, S.Z.: Integrating genetic algorithms, Tabu search, and simulated annealing for the unit commitment problem. IEEE Trans. Power Syst. **14**(3), 829–836 (1999)
8. Hasan, M., Alkhamis, T., Ali, J.: A comparison between simulated annealing, genetic algorithm and Tabu search methods for the unconstrained quadratic Pseudo-Boolean function. Comput. Ind. Eng. **38**(3), 323–340 (2000)

9. Wang, X., Ge, X., Dai, Y.: Research on dynamic vehicle routing problem based on two-phase algorithm. Contl. Decis. **27**(2), 175–181 (2012)
10. Bin, W.U., Wei-Hong, N.I., Fan, S.H.: Particle swarm optimization for open vehicle routing problem in dynamic network. Comput. Integr. Manuf. Syst. **15**(9), 1788–1794 (2009)
11. Yu, D., Zhang, Q., Yi, H.: New genetic algorithm syncretized the mi proved simulated annealing. J. Comput. Appl. **25**(10), 2392–2394 (2005)
12. Huang, T., Gui, W., Yang, C.: Simulated annealing genetic hybrid algorithm and its applications, pp. 641–645 (2000)

Developing a 3D Routing Instruction Engine for Indoor Environment

Ismail Rakip Karas$^{(\boxtimes)}$, Umit Atila, and Emrullah Demiral

Department of Computer Engineering, Karabuk University, Karabuk, Turkey
{ismail.karas,umitatila,emrullahdemiral}@karabuk.edu.tr

Abstract. The need for 3D visualization and navigation within 3D-GIS environment is increasingly growing and spreading to various fields. When we consider current navigation systems, most of them are still in 2D environment that is insufficient to realize 3D objects and obtain satisfactory solutions for 3D environment. For realizing such a 3D navigation system we need to solve complex 3D network analysis. The objective of this paper is to investigate and implement 3D visualization and navigation techniques and develop 3D routing instruction engine for indoor spaces within 3D-GIS. As an initial step and as for implementation a Graphical User Interface provides 3D visualization based on CityGML data, stores spatial data in a Geo-Database and then performs complex network analysis. By using developed engine, the GUI also provides a routing simulation on a calculated shortest path with voice commands and visualized instructions.

Keywords: 3D-GIS · Network analysis · Visualization · Navigation

1 Introduction

The need for three-dimensional (3D) visualization and navigation within 3D Geographical Information System (GIS) environment is increasingly growing and spreading to various fields. Most of the navigation systems use 2D or 2.5D data (e.g. road layer) to find and simulate the shortest path route which is lacking in building environment [5]. When we consider current navigation systems, most of them are still in 2D environment and there is a need for different approaches based on 3D aspect which realize the 3D objects and eliminate the network analysis limitations on multi-level structures [1, 4, 8, 10].

Passing from 2D-GIS toward 3D-GIS, a great amount of 3D data sets (eg. city models) have become necessary to be produced and satisfied widely. This situation requires a number of specific issues to be researched, e.g. 3D routing accuracy, appropriate means to visualize 3D spatial analysis, tools to effortlessly explore and navigate through large models in real time, with the correct texture and geometry [6].

Unlike some researches that focus on initial requirements of 3D navigation in 3D GIS environment [6], lack of 3D visualization on network analyses [5] or only elaborate on concepts, establishing framework and its application from a bigger scope of view [8] this paper presents how to manage 3D network analyses using Oracle Spatial

H. Yuan et al. (Eds.): GRMSE 2016, Part I, CCIS 698, pp. 34–42, 2017.
DOI: 10.1007/978-981-10-3966-9_4

within a Java based 3D-GIS implementation and developed routing instruction engine for indoor spaces.

2 Visualization of 3D Geometry and Network

Visualization of 3D building model is performed by Java based 3D-GIS implementation. Data in the CityGML format is read and an OpenGL graphic library is used for visualization of 3D spatial objects. CityGML [7] is a common semantic information model and exchange format for the representation of 3D urban objects that can be shared over different applications [3]. Buildings, terrain models, city furniture, vegetation, land use, water bodies, transportation (e.g. streets, railways) are defined in thematic modules which can be extended in the future.

The data model in CityGML is based on the GML3 standard which permits one to define the spatial properties of a model – mainly geometry, but also the topology of a model may be included. CityGML supports five levels of detail (LOD): LOD0 is the coarsest, essentially this is a 2.5D digital terrain model; LOD1 is a block model –

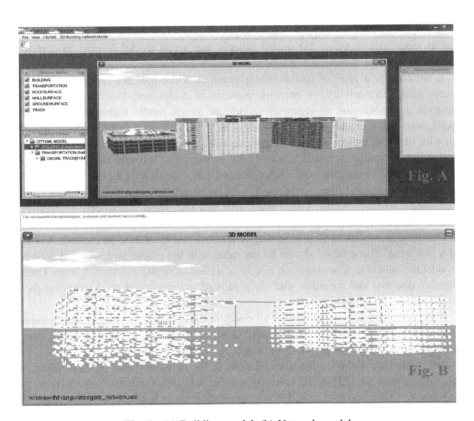

Fig. 1. (a) Building model. (b) Network model.

buildings are represented as blocks with flat roofs; in LOD2 more complex buildings can be modelled – complex roofs, installations like stairs and balconies are available; LOD3 allows for architectural models – detailed walls, roofs, doors windows, etc. are possible; LOD4 completes LOD3 and includes interior structures like rooms, doors, stair, furniture, etc. It is possible to represent the same object simultaneously in different LODs. The prepared 3D-GIS implementation uses *citygml4j* Java class library and API for facilitating work with the CityGML and JOGL Java bindings for OPENGL to carry out visualization. CityGML datasets from LOD0 to LOD2 are supported. Building model is represented in LOD2 described by polygons (Fig. 1a) and network model is represented as a linear network in LOD0 using Transportation Module of CityGML (Fig. 1b).

3   3D Network in Geo-DBMS

Using Geo-DBMS in 3D modelling and spatial analysis has a lot of advantages. Beside the standard advantages of DBMS with respect to centralized control, data independence, data redundancy, data consistency, sharing data, data integrity and improved security, geo-DBMS brings efficient management of large spatial data sets. The management of a 3D network requires usage of graph model in DBMS. While CityGML is used to store and visualize 3D spatial objects, the graph model is used to perform network analysis.

A network is a type of mathematical graph that captures relationships between objects using connectivity. A network consists of nodes and links. Oracle Spatial maintains a combination of geometry model and graph model within Network Data Model. Network elements (links and nodes) may have geometric information associated with them. A logical network contains connectivity information but no geometric information. A spatial network contains both connectivity information and geometric information. In a spatial network, the nodes and links are SDO_GEOMETRY objects representing points and lines, respectively. A spatial network can also use other kinds of geometry representations. One variant lets you use linear referenced geometries. Another lets you use topology objects.

To define a network in Oracle Spatial, at least two tables should be created: A *node* and a *link* table. These tables should be provided with the proper structure and content to model the network. A network can also have a *path table* and a *path link table*. These tables are optional and are filled with the results of analyzes, such as the shortest path between two nodes.

Node table (see Table 1) describes all nodes in the network. Each node has a unique numeric identifier (the NODE_ID column). Other optional columns are geometry, cost, hierarchy_level, parent_node_id, node_name, node_type and active.

Link table (see Table 2) describes all links in the network. Each link has a unique numeric identifier (the LINK_ID column) and contains the identifiers of the two nodes it connects. Other optional columns are geometry, cost, bidirected, parent_link_id, active, link_level, link_name and link_type (Kothuri et al. [3]).

Table 1. Columns of node table in network model.

NODE_ID	230
NODE_NAME	NODE-230
GEOMETRY	MDSYS.SDO_GEOMETRY(3001,NULL,MDSYS.SDO_POINT_TYPE (42.2019449799705,100.382921548946,-3.7),NULL,NULL)
ACTIVE	Y

Table 2. Columns of link table in network model.

LINK_ID	15
START_NODE_ID	452
END_NODE_ID	455
LINK_NAME	Link-452-455-Corridor
GEOMETRY	MDSYS.SDO_GEOMETRY(3002,NULL,NULL, MDSYS.SDO_ELEM_INFO_ARRAY(1,2,1), MDSYS.SDO_ORDINATE_ARRAY (115.306027729301,85.9775129777152,1.8,115.306027729301, 82.9483382781573,1.8))
LINK_LENGTH	3.029174699557899
ACTIVE	Y
LINK_TYPE	Corridor

In this study we use spatial network with SDO_GEOMETRY type for representing points and lines.

Oracle Spatial Network Data Model is composed of a data model to store networks inside the database as a set of network tables, SQL functions to define and maintain networks (SDO_NET), network analysis functions in Java and network analysis functions in PL/SQL (SDO_NET_MEM) which is a "wrapper" over Java API that executes inside database.

There are two ways to define data structures for a network. One is to create network automatically by calling CREATE_SDO_NETWORK procedure defined in SDO_NET package in Oracle Spatial. This procedure creates all the tables and populates the metadata. This procedure is not atomic. If it fails to complete it may cause a half created network. The automatic network creation method gives very little control over the actual structuring of the tables and gives no control at all over their physical storage (table spaces, space management, partitioning and so on). But this procedure is easy to use and makes sure the table structures are consistent with metadata.

The other and more flexible way is creating tables manually. Creating tables is not enough to define a network in Oracle Spatial. The actual naming of the tables that constitute a network and their structure should be defined in a metadata table called USER_SDO_NETWORK_METADATA as shown below by an insert statement ensuring that the table structures are consistent with metadata.

```
INSERT INTO USER_SDO_NETWORK_METADATA
(NETWORK,NETWORK_CATEGORY,GEOMETRY_TYPE,
NO_OF_HIERARCHY_LEVELS,NO_OF_PARTITIONS,LINK_DIRECTION,
NODE_TABLE_NAME,NODE_GEOM_COLUMN,NODE_COST_COLUMN,
LINK_TABLE_NAME,LINK_GEOM_COLUMN,LINK_COST_COLUMN,
PATH_TABLE_NAME,PATH_GEOM_COLUMN,PATH_LINK_TABLE_NAME,
NETWORK_TYPE)
VALUES('CORPORATION_PUTRAJAYA','SPATIAL','SDO_GEOMETRY'
,'1','1','UNDIRECTED','CORP_NETWORK_NODE',
'LOCATION',NULL,'CORP_NETWORK_LINK','GEOMETRY',
'LINK_LENGTH','CORP_NETWORK_PATH','GEOMETRY',
'CORP_NETWORK_PATH_LINK','Corp_Network')
```

Our implementation automates network definition in Oracle Spatial database using manual network creation method presented in this section. As soon as the 3D model of a building in LOD2 and its linear network model in LOD0 opened from CityGML format, the network model creation menu of the implementation gets active to be used (Fig. 2). Network creation tool reads CityGML data, creates tables to define network, inserts proper data into tables and defines network.

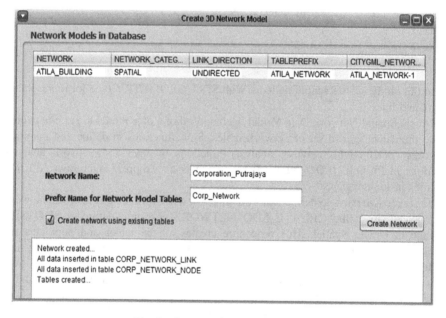

Fig. 2. Automated network creation.

4 Network Analyses for Indoor

The implementation performs network analysis based on Java API provided by Network Data Model of Oracle Spatial. Many kind of network analysis such as shortest path, travelling salesman, given number of nearest neighbors, all possible shortest paths between given nodes, all nodes within given distance, finding reaching nodes to a given node, finding all possible paths between two nodes and finding shortest paths to a node from all other nodes in the network can be performed as well as under some kind of constraints like avoided nodes, links and so on.

In this section some examples for network analysis will be presented. Figure 3a shows a shortest path analysis without any constraint and Fig. 3b shows how the shortest path is updated after links associated with elevators are avoided showed by red lines which means elevator is not in use any more in that part of building.

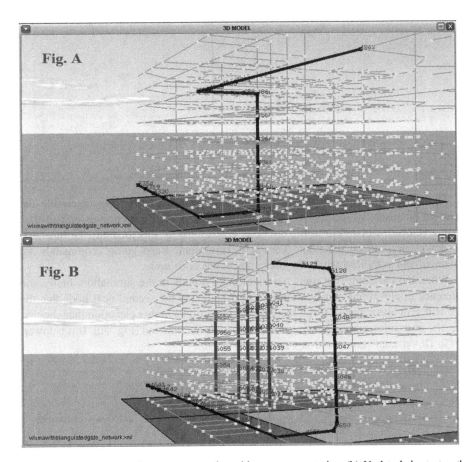

Fig. 3. (a) Shortest path between two nodes without any constraint. (b) Updated shortest path when the elevators are all out of order in a part of building.

5   3D Routing Instruction Engine

One of the most important component of an ideal navigation system is an engine which should produce real time instructions for users to assist them accurately till they arrive destination. Our implementation has such an instruction engine which is integrated into simulation module to produce voice commands and visual instructions for users dynamically on the way to the destination. It is intended to be the infrastructure of a voice enabled mobile navigation system for indoor spaces in our future work (Fig. 4).

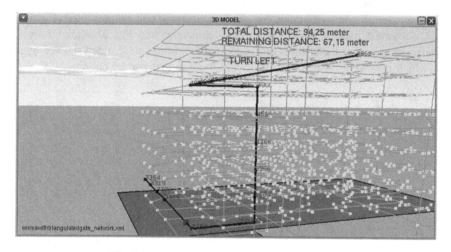

Fig. 4. Simulation process of instruction engine.

The most significant job for producing routing instructions is to determine the direction that users should follow. A method for generating instruction commands has been developed. According to the direction determined by this method, "Go upstairs, Go downstairs, Go on the floor, Turn left, Turn right, Keep going" commands are generated and vocalized while the user approaches each node.

When the red point (the user) passes through a node in the simulation, firstly, the difference of elevations between the first next node and the second next node that the user will visit is compared. If the second next node is at a higher elevation than the first one, the instruction engine generates a "Go Upstairs" command (Fig. 5a). If it is lower, a "Go Downstairs" command is generated (Fig. 5b).

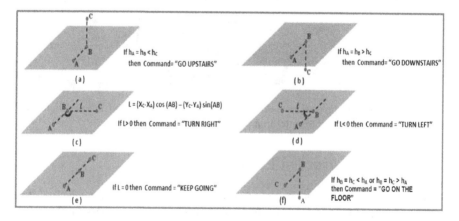

Fig. 5. Determining routing instructions. (Color figure online)

Apart from these, the user needs to walk on the floor after descending or ascending by using an elevator or stairs. In other words, if the elevation of the first next node and the second next node are equal, but the current node is different, a "Go on the Floor" command is generated (Fig. 5f).

If the elevations of the three nodes are equal then the instruction engine decides to go straight or turn right or left. To make this decision, offset (perpendicular distance) calculations should be performed. By using offset calculations of surveying computations, it can be determined if a node is on the right side or on the left side of a line segment. Accordingly this calculation, for a line segment which starts with node A and ends with B, if a node C is on the right side of this line segment then the sign of the perpendicular distance of C is obtained as positive (+), otherwise negative (−) [9]. Assuming A is the node that the user passes through, B is the first subsequent node and C is the second subsequent node that the user will visit, if the length of the perpendicular distance of node C to the line segment AB is calculated, the instruction to the user can be determined by checking the sign of the distance. If it is positive (+), the command should be "Turn Right" (Fig. 5c), if negative the command is "Turn Left" (Fig. 5d). If the calculated distance is zero the instruction engine produces a "Keep going" command (Fig. 5e).

After all these processes, the generated command is vocalized by the simulation while the red point step by step approaches to the first subsequent node. The explanations of the terms of equation in Fig. 5 are as follows:

A: The node that the user is currently passing through.
B: The first next node that the user will visit
C: The second next node that the user will visit
D: The perpendicular distance of the C node to the AB line
E: The elevation of the nodes
(AB): Bearing of the AB direction.

The most significant job for producing routing instructions is to determine the direction that users should follow. According to the determined direction, "go upstairs, go downstairs, go on the floor, turn left, turn right, keep going" commands are produced and vocalized. For producing instruction commands the method proposed by Karas has been used [2]. According to this method, the difference on elevations of the next node and the next node in the two that user will visit is compared. If the next node in the two is in a higher position then the instruction engine produces "go upstairs" command (Fig. 5a), else if opposite then "go downstairs" command is produced (Fig. 5b), else if the user has just used elevator or stairs, command to be produced is "go on the floor", otherwise the elevation of the nodes are equal and instruction engine decides to turn right or left. To make this decision a offset calculation should be performed.

Offset calculation determines if a point is on the right side or on the left side of a line segment. If we suppose the end points of a line segment as A and B, then if a point C is on the right side of this line segment then result is positive (+), otherwise negative (−). Assuming point A is the node user stands, point B is the next node to A and the point C is the next node to B, if we calculate the length and sign of perpendicular distance of point C to line segment AB then the final direction can be determined. If the sign of the result is positive (+) then the command should be "turn right" on the point B (Fig. 5c), otherwise the command is "turn left" (Fig. 5d). If the calculated distance is zero the instruction engine produces "keep going" command (Fig. 5e).

6 Conclusions

This paper presented a Java based 3D-GIS implementation which can visualize 3D building and network models from CityGML format and automate 3D network definition in Oracle Spatial's Network Data Model. We also elaborated the instruction engine developed for producing voice commands and visual instructions for assisting people dynamically on the way to the destination. Our experiments successfully showed that our 3D-GIS implementation could be improved to design an ideal indoor navigation system.

Acknowledgements. This study was supported by Karabuk University BAP Unit and TUBI-TAK - The Scientific and Technological Research Council of Turkey (Project No: 112Y050) research grant. We are indebted for their financial support.

References

1. Cutter, S., Richardson, D.B., Wilbanks, T.J. (eds.): The Geographical Dimensions of Terrorism, pp. 75–117. Routledge, New York, London (2003)
2. Karas, I.R.: Objelerin Topolojik İlişkilerinin 3B CBS ve Ağ Analizi Kapsamında Değerlendirilmesi. Ph.D. thesis, YTÜ FBE Jeodezi ve Fotogrametri Anabilim Dalı Uzaktan Algılama ve CBS, pp. 99–101 (2007)
3. Kolbe, T.H.: Representing and exchanging 3D city models with CityGML. In: Lee, J., Zlatanova, S. (eds.) 3D Geo-Information Sciences, pp. 15–31. Springer, Heidelberg (2009). Kothuri, R., Godfrind, A., Beinat, E.: Pro Oracle Spatial for Oracle Database 11g. Apress, New York
4. Kwan, M.P., Lee, J.: Emergency response after 9/11: the potential of real-time 3D GIS for quick emergency response in micro-spatial environments. Comput. Environ. Urban Syst. **29**, 93–113 (2005)
5. Musliman, I.A., Rahman, A.A.: Implementing 3D network analysis in 3D GIS. In: International Archives of ISPRS, vol. 37, Part B, Comm. 4/4, Beijing, China (2008)
6. Musliman, I.A., Rahman, A.A., Coors, V.: 3D navigation for 3D-GIS — initial requirements. In: Abdul-Rahman, A., Zlatanova, S., Coors, V. (eds.) Innovations in 3D Geo Information Systems, pp. 125–134. Springer, Heidelberg (2006)
7. OGC: City Geography Markup Language (CityGML) Encoding Standard. Open Geospatial Consortium Inc. (2012)
8. Pu, S., Zlatanova, S.: Evacuation route calculation of inner buildings. In: van Oosterom, P.J. M., Zlatanova, S., Fendel, E.M. (eds.) Geo-information for Disaster Management, pp. 1143–1161. Springer, Heidelberg (2005)
9. Serbetci, M., Atasoy, V.: Jeodezik Hesap (Survey Computations), Trabzon, Turkey, pp. 72–78. Karadeniz Technical University Publications, Trabzon (1990)
10. Zlatanova, S., van Oosterom, P., Verbree, E.: 3D technology for improving disaster management: geo-DBMS and positioning. In: Proceedings of the XXth ISPRS Congress, Istanbul, Turkey (2004)

Saliency Detection for High Dynamic Range Images via Global and Local Cues

Dengmei Xie, Gangyi Jiang, Hua Shao, and Mei Yu[✉]

Faculty of Information Science and Engineering,
Ningbo University, Ningbo 315211, China
yumei2@126.com

Abstract. Aiming at the problem that saliency detection algorithms for low dynamic range (LDR) images are unsuitable for high dynamic range (HDR) images, we propose a new saliency detection method for HDR images, where the global and local cues are considered. Firstly, according to human visual perception of high dynamic range content, the luminance and chrominance are processed respectively. Secondly, the bottom-up saliency map (BU-SM) is obtained by the global information. Then, we construct the foreground and the background codebooks based on the BU-SM, and use the sparse coding to get the top-down saliency map (TD-SM). Finally, in order to well account for the global and local factors, BU-SM and TD-SM are combined to get the final saliency map of HDR images. The experimental results show that the proposed method is superior to the state-of-the-art methods.

Keywords: High dynamic range · Saliency detection · Human visual system

1 Instruction

With the development of sensor, digital display and signal processing technology, high dynamic range (HDR) images have widely applications [1]. It reflects the real scene better than the traditional LDR images, giving people an unprecedented visual experience. However, it also brings serious challenges at the same time. Because the pixel value of HDR images is up to one million, how to make it efficient representation and storage is the problem of HDR image processing. On the one hand, we can only store and code the "important" information extracted from HDR images, stimulated by human visual system (HVS) that people allocate the limited system resources to get the "interesting" and "important" information from the mass of information [2]. On the other hand, Bremond et al. [3] show that the traditional method can't extract the salient regions of HDR image effectively. In this case, it is necessary to consider the visual attention problem of HDR images.

Moreover, saliency detection plays an important role in many fields, such as object extraction, image coding, image quality assessment (IQA), and so on. Up to now, there are fewer works for HDR images. Bremon et al. [3] proposed the contrast features (CF) saliency detection model which are applicable to high dynamic range image.

© Springer Nature Singapore Pte Ltd. 2017
H. Yuan et al. (Eds.): GRMSE 2016, Part I, CCIS 698, pp. 43–51, 2017.
DOI: 10.1007/978-981-10-3966-9_5

Comparing to Itti's model, this model has greatly improved which can obtain the saliency map containing more significant regional, but the CF model does not address the intensity and color perception under wide luminance range. Consequently, Dong et al. [4] proposed a saliency detector for HDR content, which considers visual perception to HDR content from the luminance and color channel separately, and gets better results. However, this method only considers the global features of the image without considering the local information, and the saliency map is very vague.

Based on the evidences in visual physiology, that is, HVS typically first produces a global perception, and then gradually focuses on specific local areas when viewing a scene [2], this paper presents a saliency detection method for HDR images via global and local cues. Different from the previous methods, the proposed model make full use of the global BU measure and local TD measure, based on the HVS model that can simulates the human perception of HDR pixel values well. Experimental results show the effectiveness of the proposed method.

2 The Proposed Model

Compared with LDR images, the HDR image has better quality, because of its strong luminance contrast and complex texture. Therefore, in the process of extracting SM, luminance and texture information should be considered, besides the color information. Furthermore, we consider not only global factors but also local factors, for the local factors could find the region with strong contrast, where it maybe not so salient under global factors. Because the HVS's properties under the wide luminance ranges and wide color gamut is different from the LDR's, there are some special characteristics of HDR image saliency map, i.e., the conventional Itti's model can only detect the most salient areas in the image and ignore the others. The main reason for this phenomenon is that the perception of luminance and color for HDR is different from for LDR, and it is not a simple linear or logarithmic relationship [5]. Thus, it is significant to design HVS model to perceive HDR content. Here, the high dynamic image is processed, which mainly consists of two channels: color perception and luminance perception. In view of the particularity of HDR image, we adopt the bottom-up and top-down saliency mechanism, which consider not only the global low level characteristics, but also the local characteristics. Besides, HDR image is processed by superpixel segmentation and HVS model in Fig. 1.

HVS model is designed to simulate the human perception of HDR pixel values, including color and luminance. Because traditional color space such as YUV [6] cannot represent color perception under a wide luminance range, color appearance model (CAM) has been given to predict the color perception in given luminance condition. For luminance, HVS model takes into account the sensitivity-change of the visual perception at different light levels and spatial frequencies using amplitude nonlinearity (AN) process and contrast sensitivity function (CSF).

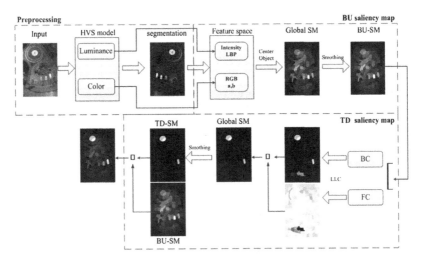

Fig. 1. Method block diagram (Color figure online)

(1) Color appearance model: To simulate the human perception of color, the Hunt-Pointer-Estevez transform is used to transform XYZ to the LMS cone space. The absolute response of cone cells [7], L', M', S', can be modeled. Based on the theory of color opponent process, HVS uses the opposite way to deal with the color information by the cone cells. According to the psychophysical results, two opposing color signals are derived, red/green channel $(a = (11L' - 12M' + S')/11)$, yellow/blue channel $(b = (L' + M' - 2S')/9)$.

(2) Luminance appearance model:

(a) *Amplitude nonlinearity (AN):* The human perception of the HDR luminance is neither linear nor logarithmic, described in AN process of HVS model. According to [8], the process is described as follows, which consists of three different functions.

$$\text{luma}(L_a) = \begin{cases} 769.18L_a & L_a < L_1 \\ 449.12L_a^{0.17} - 232.25 & L_1 \leq L_a < L_2 \\ 181.7\ln(L_a) - 90.16 & L_a \geq L_2 \end{cases} \quad (1)$$

where $L_1 = 0.061843$ cd/m^2 and $L_2 = 164.1$ cd/m^2.

(b) *Contrast sensitivity function (CSF):* CSF describes the relationship between visual sensitivity and spatial frequency under different light conditions, which indicates that the visual sensitivity is a function of spatial frequency. In HVS model, to get the sensitivity of visual perception in different spatial frequency effectively, we use multi-scale CSF [9] to filter the gray scale image. Different luma of the CSF curve is shown in Fig. 2, $L_a = \{0.0001, 0.01, 0.1, 1, 10, 100, 1000, 3000\}$ cd/m2.

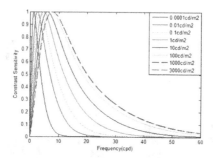

Fig. 2. CSF curves

Recently, some researchers have proposed some coding methods in the study of HDR image, such as the logarithmic transformation, encoding PU [4], but the results are not satisfactory, as shown in Fig. 3. In Fig. 3(c), the whole image is very clear, but it lost contrast effect the original image. Figure 3(d) is not clear overall image but keeps some detail and contrast information. However, compared to the previous two practices, the effect is improved. Figure 3(e) is obtained by the proposed method, where it is better than PU coding effect in detail and contrast information.

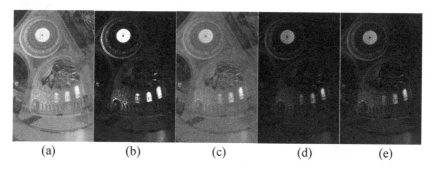

| (a) | (b) | (c) | (d) | (e) |

Fig. 3. The effect of different luminance processing methods, (a) HDR image; (b) The grayscale of (a); (c) The luminance of (a) by using log coding; (d) The luminance of (a) by using PU coding; (e) The luminance of (a) by using the proposed method.

2.1 BU Saliency Map

First of all, the simple linear iterative clustering (SLIC) is used to over-segment the gray scale processed by HVS model into small regions, instead of the original image. After segmentation, N superpixel blocks can be obtained, which are considered as the basic processing units, and then the feature of the image is extracted, where the main features will be divided into three categories: (1) color features: RGB, red - green channel, and yellow - blue channel; (2) intensity characteristics: pixel intensity I, which is processed by HVS model; (3) texture features: local binary pattern (LBP) that is a term used to describe the local texture feature of an image and has rotation invariance and gray scale invariance advantages. Note that the average value of each region is used to represent the color and intensity features, and the quantization histogram of each region is used to represent the texture feature.

Each image is divided into N regions, $\{r_i\}, i = 1, 2, \ldots, N$. Each region r_i has several features (I, a, b, RGB, LBP) that are concatenate into a uniform feature vector of 65 dimensions. And then we can get the saliency value based on the superpixels level, according to the prior theory of object and the center [10]:

$$\hat{P}(r_i) = \left(\frac{1}{M} \sum_{j=1}^{M} d(r_i, c_j) \times Ob(r_i) \times Ce(r_i) \right). \tag{2}$$

where $\{c_j\}, j = 1, 2, \cdots, M$ represents the regions on the image border, $d(r_i, c_j)$ is the difference between the two regions, $Ob(r_i)$ is the average object value of the region r_i, and $Ce(r_i)$ is the center prior of region r_i.

However, the achieved saliency map is superpixel-level, where the blocking artifact is apparent, so we have to generate saliency maps with pixel-level accuracy. In this paper, a kind of effective processing method is utilized to get pixel-level BU-SM P_b, which smooths the edge by the color histogram in the literature [11].

2.2 TD Saliency Map

Top-down visual attention mechanism needs to consider with some prior knowledge in the proposed method, so it is usually necessary to use the method of learning. In the LDR image saliency map researches, some sparse coding methods in [12, 13] are used to obtain salient regions. According to [13], firstly, we obtained the foreground codebook (FC) and the background codebook (BC). Then the function for Locality-constrained Linear Coding (LLC) algorithm is shown as follow:

$$\min_{\mathbf{B}} \sum_{i=1}^{N} \|\mathbf{f}_i - \mathbf{D}_i \mathbf{b}_i\|^2, \quad \text{s.t.} \quad \mathbf{1}^T \mathbf{b}_i = 1, \forall i. \tag{3}$$

where $\mathbf{D}_i$ is the codebook of region $r_i, i = 1, 2, \ldots, N$. By solving this function easily, the solution can be provided analytically using the following equations

$$\mathbf{b}_i = \frac{1}{\mathbf{C}_i + \lambda \times \operatorname{tr}(\mathbf{C}_i)} \quad \text{and} \quad \tilde{\mathbf{b}}_i = \frac{\mathbf{b}_i}{\mathbf{1}^T \mathbf{b}_i}. \tag{4}$$

$\mathbf{C}_i = \left(\mathbf{D}_i - \mathbf{1} \mathbf{f}_i^T \right) \left(\mathbf{D}_i - \mathbf{1} \mathbf{f}_i^T \right)^T$ is the covariance matrix of the feature. λ is a regularization parameter and is set to be 0.1 in this paper. Then, the saliency value of the region r_i is estimated based on the reconstruction error by $\hat{P}_t(r_i) = \|\mathbf{f}_i - \mathbf{D}_i \tilde{\mathbf{b}}_i\|^2$.

On the basis of foreground codebook and background codebook, each input image could produce two saliency maps. Thus synthesis of the two algorithms can improve the effectiveness of the proposed method:

$$P_t(r_i) = \hat{P}_t^b(r_i) \otimes (1 - \hat{P}_t^f(r_i)). \tag{5}$$

where $\hat{P}_t^b(r_i)$ and $\hat{P}_t^f(r_i)$ denote the normalized reconstruction error of region r_i using BC and FC, respectively. $\otimes$ is the fusion method, which means multiplication to better highlight the foreground and restrain the background. Finally, we obtain the pixel-level saliency map P_t as the process of BU-SM.

2.3   Saliency Map Based on Global and Local Cues

So far, we have obtained a BU-SM based on the global feature and the TD-SM based on local features, as shown in Fig. 4. Figure 4(a) is the HDR image; Figs. 4(b) and (c) express BU-SM and TD-SM, Fig. 4(d) show that the combination SM of global and local characteristics. As can be seen, the final SM combines the advantages of the two SMs and expressed the salient region of the image better.

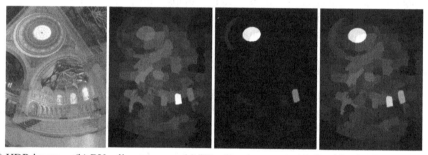

(a) HDR image,   (b) BU saliency map,   (c) TD saliency map,   (d) Combined saliency map

Fig. 4. The results of saliency maps

3   Experimental Results and Discussion

To verify the effectiveness of the proposed method, we compare the performance of the proposed method with some classical methods. In addition, we examine the applications of the proposed method for the task of HDR-IQA.

It is significant that how to consider the human visual system perception in the process of obtaining the HDR image SM. For example, we utilize the gray image processed by the HVS model to do superpixel segmentation, instead of the original HDR image. As shown in Fig. 5, we can easily see that the window position is extracted much better in (a) than in (c). Moreover, because the SM is based on super pixels calculation, the effectiveness of segmentation will affect to the final results, as shown in (b) and (d), where (b) have clearer salient region.

The maps generated by the proposed method have the highest quality compared with state-of-the-arts. The results are shown in Fig. 6. It is obvious that Figs. 6(b), (c) and (d) cannot extract accurate salient regions, so the method for LDR images is not applicable to HDR image. In addition, from Figs. 6(e) and (f), it can be seen that existing calculating HDR saliency map method could obtain SM containing more

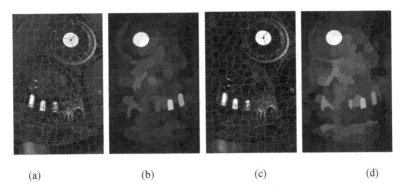

(a)   (b)   (c)   (d)

Fig. 5. The performance of segmentation and the final saliency map. (a) and (c) The superpixel segmentation results of the perception grayscale and HDR images, respectively; (b) and (d) Final saliency map corresponding to (a) and (c), respectively.

significant regional, but the saliency map is not ideal and cannot extract the high brightness and texture complex region accurately. Finally, Fig. 6(h) is corresponds to the method of this paper, which can effectively extract the salient regions in the HDR image and is suitable for shooting from indoor to outdoor scenes especially.

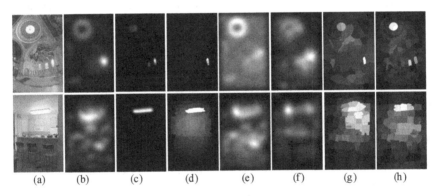

(a)   (b)   (c)   (d)   (e)   (f)   (g)   (h)

Fig. 6. Saliency map obtained by different methods. (a) The original image; (b) Itti et al.'s method [14]; (c) AIM method [15]; (d) BSCA method [16]; (e) CF method; (f) Dong et al.'s method; (g) The method of [13]; (h) The proposed method.

We also apply the proposed method into HDR image quality assessment (HDR-IQA), denoted as Proposed IQA, where visual attention is an important cue in image quality assessment. To demonstrate that the proposed method is more suitable to visual perception, refer to [17], we calculate the performances of HDR-IQA, from Dong's method and the proposed method (see Table 1). It is easy to know that the proposed method has great improvement in HDR-IQA performance.

Table 1. Comparison results for HDR-IQA methods with different saliency map methods

HDR-IQA method	PLCC	SROCC	RMSE
TMQI [18]	0.7491	0.7768	1.2370
Q-AIM [17]	0.7521	0.7850	1.2105
Q-Dong's	0.7711	0.8046	1.1602
Proposed IQA	**0.8090**	**0.8486**	**1.0387**

4 Conclusion

A new saliency map extraction method based on global and local feature is proposed for high dynamic range (HDR) images, where it not only considers the low level features of an image, but also considers the prior experience and the perception of the human eye to HDR luminance and chrominance, and combines with the bottom-up and top-down visual attention mechanism to acquire the final saliency map, where the result is more consistent with human perception than the state-of-the-arts. But there are also some inadequacies. The algorithm is mainly used in the scene shooting from indoor to outdoor with narrow scope of application. In addition, the dark regions of HDR image also has a high texture complexity, and catch our attention in other academic field, such as the image quality evaluation of HDR images, but the proposed method doesn't pay attention to these areas, so in a follow-up study, we will improve the existing algorithm.

Acknowledgement. This work was supported by Natural Science Foundation of China (61271270, 61671258), and Natural Science Foundation of Zhejiang Province, (LY15F010005).

References

1. Bandoh, Y., Qiu, G., Okuda, M., Daly, S.: Recent advances in high dynamic range imaging technology. In: IEEE International Conference on Image Processing, pp. 3125–3128 (2010)
2. Xie, J.: The Principle and Application of Vision Bionics. The Science Publishing Company, Beijing (2013)
3. Brémond, R., Petit, J., Tarel, J.-P.: Saliency maps of high dynamic range images. In: Kutulakos, K.N. (ed.) ECCV 2010. LNCS, vol. 6554, pp. 118–130. Springer, Heidelberg (2012). doi:10.1007/978-3-642-35740-4_10
4. Dong, Y., Pourasad, M., Nasioulos, P.: Human visual system based saliency detection for high dynamic range content. IEEE Trans. Multimedia **18**, 549–562 (2016)
5. Narwaria, M., Silva, M.P.D., Callet, P.L.: HDR-VQM: an objective quality measure for high dynamic range video. Signal Process. Image Commun. **35**, 46–60 (2015)
6. Wiseman, Y.: The still image lossy compression standard – JPEG. In: Encyclopedia of Information and Science Technology, vol. 1, pp. 295–305 (2014). Chapter 28
7. Kim, M., Weyrich, T., Kautz, J.: Modeling human color perception under extended luminance levels. ACM Trans. Graph. **28**(27) (2009)
8. Mantiuk, R., Myszkowski, K., Seidel, H.-P.: Lossy compression of high dynamic range images and video. In: Electronic Imaging (2006). Article no. 60570V

9. Mantiuk, R., Daly, S., Myszkowski, K.: Predicting visible differences in high dynamic range images: model and its calibration. In: Proceedings of SPIE, vol. 5666, pp. 204–214 (2005)
10. Borji, A., Sihite, D.N., Itti, L.: Salient object detection: a benchmark. In: Fitzgibbon, A., Lazebnik, S., Perona, P., Sato, Y., Schmid, C. (eds.) ECCV 2012. LNCS, vol. 7573, pp. 414–429. Springer, Heidelberg (2012). doi:10.1007/978-3-642-33709-3_30
11. Liu, Z., Zhang, X., Luo, S., et al.: Superpixel-based spatiotemporal saliency detection. IEEE Trans. Circuits Syst. Video Technol. **24**, 1522–1540 (2014)
12. Li, X., Li, Y., Shen, C., Dick, A.R., van den Hengel, A.: Contextual hypergraph modeling for salient object detection. In: ICCV, pp. 3328–3335 (2013)
13. Tong, N., Lu, H., Zhang, Y., Ruan, X.: Salient object detection via global and local cues. Pattern Recogn. **48**, 3258–3267 (2014)
14. Itti, L., Dhavale, N., Pighin, F.: Realistic avatar eye and head animation using a neurobiological model of visual attention. In: Proceedings of SPIE, vol. 5200, pp. 64–78 (2004)
15. Bruce, N., Tsotsos, J.: Attention based on information maximization. J. Vis. **7**, 950 (2007)
16. Qin, Y., Lu, H., Xu, Y., Wang, H.: Saliency detection via cellular automata. In: IEEE Conference on Computer Vision and Pattern Recognition 2015, pp. 110–119 (2015)
17. Nasrinpour, H.R., Bruce, N.D.B.: Saliency weighted quality assessment of tone-mapped images. In: IEEE International Conference on Image Processing, pp. 4947–4951 (2015)
18. Yeganeh, H., Wang, Z.: Objective quality assessment of tone-mapped images. IEEE Trans. Image Process. **22**, 657–667 (2013)

Research on Vegetable Growth Monitoring Platform Based on Facility Agricultural IOT

Qingxue Li[1,2,3] and Huarui Wu[1,2,3(✉)]

[1] National Engineering Research Center for Information Technology
in Agriculture, Beijing 100097, China
{liqx, wuhr}@nercita.org.cn
[2] Beijing Research Center for Information Technology in Agriculture,
Beijing Academy of Agriculture and Forestry Sciences,
Beijing 100097, China
[3] Key Laboratory of Agri-Informatics,
Ministry of Agriculture, Beijing 100097, China

Abstract. To monitor the environmental parameters in vegetable greenhouse in real time and reduce the impact of climate disasters on vegetable growth, we develop a technology platform including the environmental data acquisition, transmission, disaster warning, remote control, and information push through using Internet of Things technology. The platform can achieve the greenhouse equipment control through using ZigBee, transmit the data to the cloud service center through GPRS, and be implemented through Java EE. The deployment of field tests in Beijing XiaoTangShan show that the platform is stable and reliable. Moreover, the platform satisfy the need of real time monitoring and early warning, and increase the management level in agricultural park facilities and the ability of coping with disasters.

Keywords: Greenhouse · IOT · Intelligent management system · Wireless sensor network · Environment control · Agent

1 Introduction

In recent years, the planting area of vegetable in China's greenhouse has increased greatly. Due to the differences in management knowledge and the increase of extreme weather, how to ensure the yield and quality has become the bottleneck. Therefore, the constructing of platform to monitor the vegetable growth and prevent the disaster can guarantee the suitable planting environment, further ensuring the production and quality.

Agriculture Internet of Things is to apply the Internet of Things technology in agricultural production, management, and service [1]. That is, collecting filed planting, horticultural facilities, livestock and poultry information through sensors can realize the monitoring and scientific management in agriculture [3].

For the vegetable planting, a basic part of the environmental monitoring equipment has been installed. Generally, the PC is equipped in the local to gathering and controlling the facilities environment, which requires the highly cost and IT skills demand.

© Springer Nature Singapore Pte Ltd. 2017
H. Yuan et al. (Eds.): GRMSE 2016, Part I, CCIS 698, pp. 52–59, 2017.
DOI: 10.1007/978-981-10-3966-9_6

In addition, the production environment cannot ensure the stable operation of data server within 7×24 h.

The paper aims to research the monitoring methods of facilities environment temperature in real time. The data collection, transmission, calibration, and analysis can provide low cost and efficient ways to cope with the meteorological disasters, further realizing the intelligent monitoring and artificial auxiliary management of greenhouse.

2   Research of Key Technology

The key technology in the production environment monitoring including environment information collection, data transmission and filtering, and information service.

2.1   Collection of Environmental Data

The environmental information acquisition module mainly include the external sensors, processing chip, DTU, and ZigBee. The external sensor connect the acquisition board with interface, and can be carried on the dynamic configuration based on the actual monitoring indicators, as shown in Fig. 1. The processing chip built into the environmental information collection procedures can be used to acquire each external sensor real-time data information. The acquisition program supports dynamic parameter configuration, and the configuration information is saved in the ROM. DTU (Data Transfer unit) is specifically designed to convert serial data to IP data or convert IP data to serial data transmission through wireless communication network of wireless terminal equipment. It is a collector with a server-side data transmission channel. Considering the problem such as wiring is not convenient in the vegetable greenhouse, acquisition module and environmental control equipment (fan, shutter, curtain, etc.) communicate each other using ZigBee agreement. The equipment adopts the model of ad-hoc network access, as shown in Fig. 1 [4, 5].

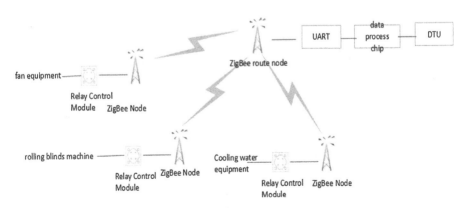

Fig. 1. The topography structure of the system

2.2 Data Transmission

The collected environment data is transmitted to the cloud services platform after dealing with protocol coding. The data parsed and processed by the cloud services platform parse, and then turned into the storage management.

(1) Transport protocol

The data collected by acquisition unit transmit to the cloud service platform through GPRS network. The platform parse the data according to the custom rules of the transport protocol rules, and mapped to the corresponding fields in the relational database, which realizing the logic abstraction of the raw data to the target data. The data transmission integration process is presented in Fig. 2

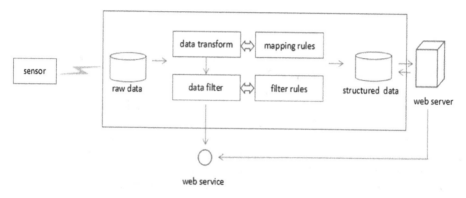

Fig. 2. Data transmission integration process

In this paper, we select the custom message structure, as shown in Table 1.

Table 1. Data structure of network package.

Start symbol	Business type	Head part	Data part	End symbol
<	1 bit	5 bit	variable length	>

The message choose "< >" as the symbol of beginning and ending. Moreover, the message with incomplete start-stop symbol will be abandoned directly.

Business categories
As shown in Table 2, it can indicate the current data message belong to which scope of business. The head of a message and data are different for different type of business.

Table 2. Business categories of data package

Code	Description
C	Heart beat package
D	Data package
G	Warning package
H	Configure package

The values of business categories vary from a-zA to Z0-9, that is, a maximum of 62 kinds of business.

(2) The abnormal data processing

Due to the damage of the sensing devices in harsh conditions, there will be abnormal data. The abnormal data can't reflect the real environmental information, and can disturb the decision support and data mining. Therefore, the data collected by the sensor must be checked before application.

The system regulates a reasonable range for data collected by sensor. When the data beyond the scope, the data is considered to be illegal data, and be discarded.

The system identify the senor data in the form of coding, the sensor category code is shown in Table 3.

Table 3. Reasonable data range for vegetable growth

Code	Description	Reasonable data range
A	Air temperature	−30–50 °C
B	Air humidity	0–100%
C	Soil temperature 20CM	−30–50 °C
D	Soil humidity 20CM	−30–50 °C
E	Illuminance	0–20,000 lx

2.3 Information Service

After the data collection and transmission, the facilities environment perception information has been gathered by cloud services platform. According to the data validation rules and vegetables grow appropriate environmental indicators requirements, as well as production management requirements, information service is designed to include monitoring service, visual statistics, remote control, intelligent decision-making and push service modules. Basing on the vegetable growth models, decision support module can control the greenhouse environment automatically, such as blower, the control of water, rolling machines and other equipment.

(1) Monitoring service
   Display the latest live data from the database of greenhouse online, and label the abnormal data visually.
(2) The visual statistics
   To analyze the sensing data online and dig the correlation between each index in graph form.
(3) Remote control
   The control command control the issuance of the gateway through GPRS network by using GPRS and ZigBee technology. The gateway control the command among the control units, thus realizing remote heat blower, shutter, the water on or off online.

Taking cucumber as an example, we focus on modeling the optimum temperature in every period, and supporting the production personal to control the temperature.

The temperature indicator of cucumber in different growth periods is listed in Table 4 [2, 6].

Table 4. Suitable growth temperature range of cucumber

Period	Reasonable data range
Germination period	28–30 °C
Seeding period	10–22 °C
Flowering period	24–14 °C
Fruit period	Day 28–32 °C, night 18–15 °C
Germination period	28–30 °C

According to the environmental requirements in Table 4, we implement the decision support service and remote control during the process of cucumber growth.

3 Test and Analysis

This test is conducted in greenhouse in XiaoTangShan cucumber plant base. We deploy small stations, fan and shutter devices in every greenhouse, and further validate the collected environmental data.

(1) The platform architecture

The architecture of platform is Java EE, and the development environment is Windows 7 + JDK1.7 + MyEclipse2014. The server is Tomcat 7.0, and the database is Oracle 11g, which mainly includes data perception layer, data layer and data service layer.

Information collection and transmission filter
This experiment mainly collected air temperature, air humidity, soil temperature, soil humidity, radiation (or light) data, and the intervals is 18 min one cycle.

The platform build a uniform data access entrance in the form of Socket service, and the vegetable greenhouses sensor devices push the data to the platform through GPRS network. The platforms parse the complete message, further filter the data according to the filter conditions, and finally warehouse the effective data.

After three months consecutive 7 × 24 h non-stop running, system acquisition success rate reached more than 99%, and the information acquisition module running stability.

(2) Information service platforms

(i) Monitoring display service
The platform provides managers with real-time sensory data checking and service analyzing through visualization technologies such as chart, as shown in Figs. 3 and 4.

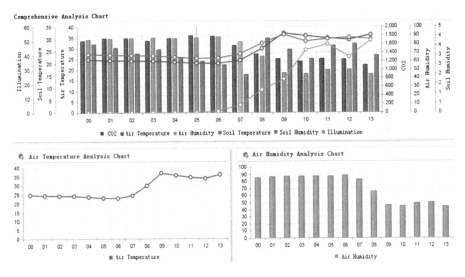

Fig. 3. The analysis page of greenhouse

Data Time	CO2 (ppm)	Air Temp (℃)	Air Humidity (%)	Soil Temp (℃)	Soil Humidity (%)	Illuminance(lux)
2016-09-12 13:31	1109	36.2	44.4	29.7	3	48.6
2016-09-12 13:11	1091	35.8	44.4	30	3	52.1
2016-09-12 12:51	1191	34.8	47.8	30.9	4.5	52.2
2016-09-12 12:32	1260	33.4	51.1	30.6	3.2	48
2016-09-12 12:12	1265	33.3	50.1	30	3.1	16.3
2016-09-12 11:53	1240	35.1	47.4	30.2	2.7	48.4
2016-09-12 11:33	1196	34.8	48.6	30.5	3.6	50.5
2016-09-12 11:14	1296	33.6	50.9	29.1	4.3	44
2016-09-12 10:54	1143	34.5	47	29.4	2.4	46.4
2016-09-12 10:35	1215	34.6	47.2	28.8	3.6	43.9
2016-09-12 10:15	1193	37.8	40.7	28.9	2.5	40.6

Fig. 4. The data list of real-time monitoring

(ii) Remote control

The system provides the online remote control function, managers can remotely control vegetable greenhouse internal blower, shutter and curtain through internet, as shown in Fig. 5.

This system adopts the electric heater as heating device. The electric heater is placed in the middle of the sunlight greenhouse test area, located near the back wall, and the power is 2000 W. The outlet temperature of electric heater is stabilized about

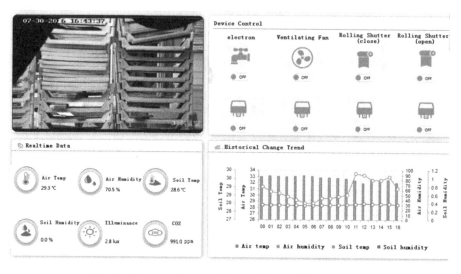

Fig. 5. The page for monitoring and controlling remotely

70 °C, and the wind speed is 3 m/s. To avoid the hot air burning seedlings, the heater outlet must be protected.

The results show that managers can remote control the electric heater, and the heating effect is obvious. Through calculation, it is found that the average temperature in heat greenhouse is 4.2 °C higher than that in contrastive greenhouse. Moreover, the average minimum temperature of heat greenhouse is 4.5 °C higher than that in contrastive greenhouse. The space temperature distribution is more uniform, and it will not affect crop growth in neatly. Especially in the continuous low temperature and extreme weather conditions occur, the effect is more obvious. It is suitable for solar greenhouse in north China to deal with short-term harm use at low temperature.

The system provides the full-automatic control module. The system can remote start and stop according to the present rules to relative equipment, realizing the relative temperature and humidity of two indicators of automatic regulating function.

(iii) Push service

The system offers SMS, WeChat, and telephone push service. Once the environmental indicators beyond a present rules, the system will remind the administrators. If the relative index is in grave danger of state, the system can automatically call administrator.

4   Conclusions

The vegetable growth monitoring and early warning platform can effectively solve the agricultural facilities vegetable greenhouses in low temperature disaster monitoring and early warning. The technology allows user to access and control information through a browser, and further guide the disaster prevention. Different from the original enclosed individual monitoring mode, the platform can realize the greenhouse environment monitoring network, the wireless remote acquisition and analysis technology, intelligent management. It can effectively implement the "WenShiQun" intelligent remote monitoring and management, and change the traditional agriculture management pattern, further improving the management efficiency and ability to cope with disasters.

Acknowledgements. This work was supported by Youth Found of Beijing Academy of Agriculture and Forestry Sciences (Research on Management and Analysis of Internet of Things of Agricultural Facility Data) and Beijing Municipal Natural Science Foundation (Grant number 4151001).

References

1. Li, Z., Wang, T., Gong, Z., Li, N.: Forewarning technology and application for monitoring low temperature disaster in solar greenhouses based on internet of things. Trans. Chin. Soc. Agric. Eng. **29**, 229–236 (2013)
2. Kequn, L., Ming, L.F., Wengang, Y.: The microclimate characteristics and greenhouse climate. Meteorology, 101–107 (2008)
3. Zhihua, D., Liping, C., Gang, W., Chunjiang, Z., Jun, W., Cheng, W.: Design and implementation of wireless monitoring system for facility environment. J. Agric. Eng., 146–150 (2008)
4. Liu, S., Wang, F., Wang, D., Gu, W., Zhu, J.: Study on the key technology of greenhouse environment control. Agric. Inf. Netw., 17–19 (2008)
5. Bao, C.C., Shi, R.Z., Ma, Y.Q., Rongchang, L., Lun, M., Qingzhu, W., Liu, S.U.: Based on ZigBee technology of measuring and controlling system for agricultural facilities design. Chin. Soc. Agric. Eng., 160–164 (2007)
6. Zhao, H., Zhang, Q., Yang, Q., Deng, Z., Wang, R., Ma, P.: Loess Plateau semi-arid rain feed region of solar greenhouse climate analysis. J. Appl. Meteorol., 627–634 (2007)

A Novel Framework for Analyzing Overlapping Community Evolution in Dynamic Social Networks

Hui Jiang, Xiaolong Xu[✉], Jiaying Wu, and Xuewu Zhang

College of Internet of Things Engineering, Hohai University,
Changzhou, Jiangsu, China
jsczjh@163.com, xuxl@hhuc.edu.cn,
wujiaying@hhu.edu.cn, lab_112@126.com

Abstract. Finding overlapping communities from social networks is an important research topic. Previous research mainly focus on static networks, while in real world the dynamic networks are in the majority. Therefore lots of researchers turn to study dynamic social networks. One specific area of increased interest in dynamic social networks is that of identifying the critical events. However these proposed algorithms more or less exist some problems. Here in this paper we propose a novel event-based framework for analyzing overlapping community evolution in dynamic social networks. In addition, we give an index that is community tag to depict the changing process of communities over time intuitively. Moreover five indexes based on events are presented to construct the neural network prediction model, only five indexes make the complexity computation of our prediction model is simpler than the existing algorithms. Experimental results show our framework performs better, and the prediction accuracy is also acceptable.

Keywords: Overlapping community · Community evolution · Dynamic social network · Event-based framework

1 Introduction

Social network analysis has attracted the attention of many researchers, and different methods have been proposed [1–5]. However one specific area of increased interest in social networks is that of finding communities, a community can be regard as the set of nodes, and the connections between nodes in a community are dense, while relationships between communities are sparse. In real world the nodes of social networks always represent individuals and edges mean the relationships. Early research mainly focused on detecting communities from static network, which has produced a large number of community detection algorithms with varied results [6–8]. There are also many excellent methods, such as the CNM algorithm which is proposed by Newman et al. [9]. The algorithm ranks as one of the algorithms with the lowest time complexity as of current research. The only drawback is that it is a non-overlapping community detection method (a node can only belong to a single community). In real world applications, especial in social networks, individuals may belong to multi-communities.

© Springer Nature Singapore Pte Ltd. 2017
H. Yuan et al. (Eds.): GRMSE 2016, Part I, CCIS 698, pp. 60–70, 2017.
DOI: 10.1007/978-981-10-3966-9_7

Palla et al. [2] proposed the first overlapping community in 2005, and it is named as k-clique algorithm. The Speaker-listener Label Propagation algorithm is proposed by Xie et al. [9] which is an extension of the Label Propagation algorithm [10]. The method detects both individual overlapping nodes and the whole overlapping communities by applying the underlying network structure. An overlapping community detection algorithm based on CNM is presented by Zhang et al. [11] which firstly gain the non-overlapping communities by CNM, and then extrapolate the overlapping nodes.

Recently the discovery of community evolution in dynamic network becomes an important research topic [12]. This is because most networks are dynamic changes in the real world, especially for social networks. An important way to study the evolving community is to use the events to characterize the life cycle. The method mainly contains two steps, where dynamic networks are considered as the set of different static networks. The first step is to extract communities from these static networks, and then analyze the change of communities over times. Takaffoli et al. [20] proposed a framework for finding community evolution in social network, and some important events are also defined, where these events are not only between two consecutive timestamps, this will result in a high time complexity.

In this paper we propose a novel framework for analyzing overlapping community evolution in dynamic social networks. We redefine the key events in dynamic network. In order to illustrate the events better and depict the changing process of community intuitively, an index named *community tag* is presented. Moreover five indexes involving events are proposed to construct a neural network model, the model can be applied to predict the number of events. So firstly we use overlapping community detection proposed by Zhang et al. [11] (the method has a high classification accuracy in detecting communities) to obtain communities in static networks, and then calculate the number of all kinds of events in dynamic network. Finally a neutral network model is built based on proposed indexes to predict events. In the experimental section, we compared our framework with other framework, mainly Asur and Takaffoli framework. we apply five indexes [11] (Scale, RNE, Popularity, Popularity, Influence, Sociability) to construct the neural network model. The main mission of the experiments was to test the performance of our framework.

2   Experimental Setup

In this section the metrics and processes in the experiments conducted to measure the performance of the proposed framework is introduced. In order to measure the performance we compared our framework with other framework, such as Takaffoli framework [15] and Asur framework [14]. While the parameter k is used in Asur framework, and according to the experimental setup in their paper we set k to 40. Similarly in Takaffoli framework we also set parameter k (the definition of these two parameters are different) to 0.4. All algorithms proposed in this paper are coded in the Java programming language, while BP neural network is accomplished by MATLAB, and experiments were conducted on a PC with 3.0 GHz processor and 4G memory.

The main mission of the experiments was to test the performance of our framework. In order to achieve it, we applied a co-author network and synthetic networks. These data were used to evaluate the constructed BP neural network model. The co-author network is DBLP Chinese dataset, which is a dataset involving natural language processing field, which can be downloaded from http://www.datatang.com/. Each node in DBLP dataset represents a scientist and an edge means two scientists have worked together. Table 1 shows the node/edge count of the DBLP dataset. The synthetic networks are generated by the tool provided by Derek Greene et al. and the parameter values shown in Table 2 are used to generate three networks, and database 1 will be used to the *community tag* allocation, while database 2 is applied to tune important parameter, and the last database 3 is devoted to test the number of events with three framework.

Table 1. Nodes and edges of DBLP dataset with five years.

Year	Nodes	Edges	Year	Nodes	Edges
2005	1038	2858	2008	1373	3976
2006	1283	3874	2009	1217	3136
2007	1253	3876			

Table 2. Nodes and edges of DBLP dataset with five years.

Parameter	Definition	Dataset1	Dataset2	Dataset3
s	Time steps	10	10	10
n	Number of nodes	100	5000	10000
k	Average degree	10	30	40
$k$max	Max degree	15	50	60
Cmax	Max community size	15	50	75
Cmin	Min community size	10	30	50
u	Mixing parameter	0.2	0.2	0.2

2.1 Parameter Tuning

In this section, three significant parameters t, α and β will be tuned. t is a threshold involving events *remain*, *birth* and *death*, while α can affect most events and control the number of them, and the number of *contract* and *expand* events are determined by β. It is easy to determine the proper value of these two parameters by the following experiments.

First of all, we applied the artificial dataset 2 mentioned in Table 2 to tune parameter t. From Fig. 4 we observe that the average number of *remain* event is enhancive with the increase of the value of t. however for the *birth* and *death* events are the opposite. This kind of situation is easily explained. As the value of t increased, the condition required for the *remain* event is reduced, and thus more communities can satisfy the requirement, yet other two events are reversed. Figure 1 shows these three

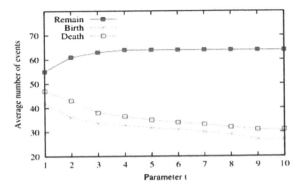

Fig. 1. The average number of events with different t.

events all take more time as *t* increased. Therefore we can set different values according to different demands. In this paper, we set the *t* value to 3, which can obtain enough events and running time is also acceptable.

Figure 2 describes the change trend of events with the increase of parameter α, and the *t* value of *remain*, *birth* and *death* are also 3. The number of *birth* and *death* events are increased as α becomes greater, while they are just the opposite for the remaining three events. The change trend of these events are not the same. Thus if we need more *birth* and *death* events, a high α is a good choice, while for the opposite requirement we can choose a low α. And we select the value of 0.5, which can generate the enough events.

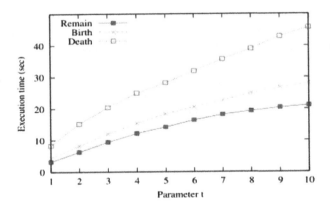

Fig. 2. The execution time of events with different t.

From Fig. 3 we can see the number of *contract* and *expand* events are increased with an increase of β. It seems that a high value of β is better, while if we select a high β, these two events may include other events. Therefore, we set it to 0.3. In the following experiments we will use this value.

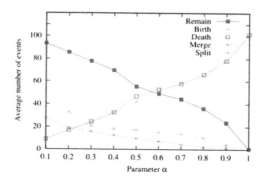

Fig. 3. The average number of events with different α.

2.2 Community Tag Allocation

In this section we applied a simple instance to illustrate how to allocate *community tag*, as the *community tag* is useful for describing the change process of communities in dynamic network, and moreover it is an important step before detecting the events. The dynamic network used in this section is dataset 1, which is a synthetic network shown in Table 2. The dataset consists of ten static networks, and each network has 100 nodes which exist in eight communities.

As we can see from Fig. 4, the nodes with the same *similarity* are sharing with the same *community tag*. These nodes with same *community tag* belong to different timestamps, so it indicates that there happens *remain* events. For the community 3 at timestamp $T = 1$ with tag 3, while at timestamp $T = 2$ it becomes community 1 with the same tag 3. Therefore community 3 still exists at $T = 2$, and a *remain* event just appears. However other communities which do not remain at next timestamp will gain new tags, which leads to the appearance of other new events. In Fig. 4, community 5 with community tag 5 at timestamp $T = 1$ splits in communities 5 and 6 with tags 10 and 11 at timestamp $T = 2$, similarly, community 6 with tag 6 at timestamp $T = 1$ happens a *split* event. Besides, in Fig. 4 we can see that community 1 and community 8 merge into community 4, which is shown by community tag 1 and 8 at timestamp $T = 1$ and community tag 1 at timestamp $T = 2$. Figure 4 also shows the *birth* and

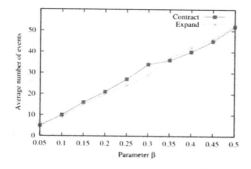

Fig. 4. The average number of events with different β.

death events. In short, through community tags we can observe the variation of communities and the occurrence of different events intuitively.

3 Occurred Events Comparison

3.1 DBLP Dataset

In this section we used the DBLP dataset to test the feasibility of our framework, and is also compared with Takaffoli framework [15] and Asur framework [14]. The DBLP dataset includes the data for five years, in addition the nodes and edges of every year are presented by Table 1.

Results in Table 3 show the number of events found by Asur framework. We can easily observe that the *birth* and *death* events appear at most. This means that the dataset is instable, as many communities are formed or dissolved at most time. However, the *remain*, *merge* and *split* events are almost do not occur in every timestamp, partly because the condition of these events may not very precisely, due to the reason of DBLP dataset. The dataset has only about one thousand nodes, while the detected communities are about three hundred, so most communities have only several nodes, and small communities are usually dramatic changes. From the Table 5 we can see the number of events detected by our framework. Our framework discover more events than Asur framework, even for the *merge* and *split* events. Two extra events *contract* and *expand* in our framework are regarded as a complement of other events. However compared with Table 4 which is the events gained by Takaffoli framework, our framework only detects more *birth* and *death* events. The Takaffoli framework detects more events, yet it takes a lot of time. For *birth*, *merge* events at timestamp T, it will count the similarity between community and any of the following communities, and it is the same with the other events. However, the *merge* and *split* events should be computed between continuous timestamps, which is more accord with reality.

Table 3. Number of events occurred with DBLP using Asur Framework.

Events year	2005	2006	2007	2008	2009
Birth	–	281	318	334	325
Death	241	286	331	338	–
Remain	4	2	3	2	–
Merge	–	0	0	0	0
Split	0	0	0	0	–

Table 4. Number of events occurred with DBLP using Takaffoli Framework.

Events year	2005	2006	2007	2008	2009
Birth	–	288	328	334	335
Death	251	292	331	338	–
Remain	30	29	20	21	–
Merge	–	4	0	2	1
Split	4	2	0	2	–

Table 5. Number of events occurred with DBLP using Our Framework.

Events year	2005	2006	2007	2008	2009
Birth	–	305	345	356	342
Death	264	314	348	353	–
Remain	28	19	16	19	–
Merge	–	2	0	1	0
Split	2	1	0	1	–
Contract	20	9	5	5	–
Expand	7	6	7	8	–

3.2   Synthetic Dataset

The dataset 3 shown in Table 2 were used to evaluate our framework, Takaffoli framework and Asur framework. This dataset is a synthetic dataset consisting of 10000 nodes with ten networks.

Tables 6, 7 and 8 respectively provide the number of events found by Asur, Takaffoli and our framework. Experimental results show that our framework is performed better than Asur, while similar to the above results our framework detects more *birth* and *death* events than Takaffoli framework. There are a small number of *remain*, *merge* and *split* events detected by Asur framework, this is because the requirements of these events are too strict. For our framework we find the *birth* and *death* are almost the same, which is in line with the actual situation. In a dynamic network with the demise of the community there will be the birth of a new community. Furthermore *merge* and *split* events, *contract* and *expand* events are also in the same situation.

Table 6. Number of events occurred with artificial dataset using Asur Framework.

Events time	T = 1	T = 2	T = 3	T = 4	T = 5	T = 6	T = 7	T = 8	T = 9	T = 10
Birth	–	44	49	50	54	56	55	52	57	56
Death	44	49	48	54	56	55	52	57	56	–
Remain	12	10	16	10	7	5	9	10	8	–
Merge	–	2	0	1	5	3	2	1	2	2
Split	2	3	7	4	1	3	3	5	2	–

Table 7. Number of events occurred with artificial dataset using Takaffoli Framework.

Events time	T = 1	T = 2	T = 3	T = 4	T = 5	T = 6	T = 7	T = 8	T = 9	T = 10
Birth	–	54	59	54	64	66	65	63	65	66
Death	54	59	58	64	66	55	62	67	66	–
Remain	130	120	100	101	99	102	112	113	110	–
Merge	–	33	35	32	26	26	24	25	24	24
Split	32	33	32	24	31	33	33	25	22	–

Table 8. Number of events occurred with artificial dataset using Our Framework.

Events time	T = 1	T = 2	T = 3	T = 4	T = 5	T = 6	T = 7	T = 8	T = 9	T = 10
Birth	–	74	69	70	74	76	75	82	87	86
Death	74	69	70	74	76	75	82	87	86	–
Remain	120	110	92	87	87	75	79	65	81	–
Merge	–	22	24	16	11	13	9	11	9	12
Split	30	30	28	26	26	23	22	24	23	–
Contract	80	70	77	78	69	66	77	76	71	–
Expand	64	61	71	75	78	64	69	79	72	–

3.3 Facebook Dataset

The facebook dataset is part of datasets compiled by Viswanath et al. each month of facebook dataset is regarded as a network. The dataset used by us is from January to October in 2007, therefore, there are ten networks in our experimental dataset.

Table 9 shows the number of events detected by Asur Framework. Similarly the number of *birth* and *death* is bigger than other events, which results from the strict definition of the events defined by Asur Framework. The strict definition directly leads to the less number of communities which are able to satisfy the conditions of events. Tables 10 and 11 separately describe the numbers of events using Takaffoli framework and our framework, compared with the number of *birth* and *death* events, our framework can detect much more events in experimental dataset. Although the number of *remain*, *merge* and *split* events using our framework is a little less than that detected by Takaffoli framework, the complexity of computation of our framework is much simpler than Takaffoli framework.

Table 9. Number of events occurred with facebook using Asur Framework.

Events year	2005	2006	2007	2008	2009
Birth	–	321	327	344	336
Death	321	314	345	348	–
Remain	40	42	43	42	–
Merge	22	20	23	25	–
Split	21	19	22	23	–

Table 10. Number of events occurred with facebook using Takaffoli Framework.

Events year	2005	2006	2007	2008	2009
Birth	–	338	345	357	355
Death	331	372	352	354	–
Remain	320	319	321	324	–
Merge	184	182	183	181	–
Split	183	182	181	182	–

Table 11. Number of events occurred with facebook using Our Framework.

Events year	2005	2006	2007	2008	2009
Birth	–	356	355	368	362
Death	384	414	403	394	–
Remain	308	291	301	309	–
Merge	172	170	174	175	–
Merge	171	171	170	172	–
Contract	280	265	275	270	–
Split	225	260	263	258	–

4 Prediction Accuracy Evaluation

In this section we will construct neural network models for different events, and then predict the accuracy of detecting events. Details are shown as the following.

4.1 DBLP Dataset

From Table 12, we can observe the accuracy of detecting events between these three frameworks. As the other two frameworks do not have *contract* and *expand*, thus we do not construct model for them. The performance of Asur framework is the worst, while the accuracy of detecting *birth* and *death* events is acceptable. For our framework, the detected accuracy of events are higher than Takaffoli framework except the *remain* event. The reason is that Takaffoli framework discovers more *remain* event, and this may result in the high accuracy of model, while it takes more time. In addition, Takaffoli framework also detects more *merge* and *split* events, however the accuracy is lower than our framework. This is because these two events should just calculate between continuous timestamps, and this also represents our definitions are more precisely.

Table 12. Number of events occurred with facebook using Our Framework.

Events	Asur framework	Takaffoli framework	Our framework
Birth	73%	75%	76%
Death	72%	74%	77%
Remain	67%	81%	80%
Merge	6%	74%	76%
Split	61%	76%	78%

4.2 Synthetic Dataset

Table 13 shows comparison between the accuracy of detecting events with synthetic dataset. The synthetic dataset is a dynamic network with ten timestamps. The result is similar with the result gained from DBLP dataset. The Asur framework also performs badly, the accuracy of *remain* events detected by our framework is lower than Takaffoli

Table 13. Result comparison between the accuracy of detecting events with synthetic dataset.

Events	Asur framework	Takaffoli framework	Our framework
Birth	74%	79%	79%
Death	73%	76%	79%
Remain	69%	83%	81%
Merge	67%	77%	80%
Split	68%	78%	79%

framework. However our framework performs better on other events, so in general our framework has a better performance.

4.3 Facebook Dataset

From Table 14, it is obvious that Asur framework performs worst in the accuracy of detecting each event compared with Takaffoli framework and our framework. Furthermore our framework also has an advantage over Takaffoli framework, even in predicting *remain* event, our framework performs pretty good. What's more, more than 80% accuracy in predicting the occurrence of each event is a good description of how our framework performs in facebook dataset, especially the *merge* event. The most important thing is that our framework has simple enough computation complexity, which is an important factor in social network research.

Table 14. Result comparison between the accuracy of detecting events with facebook dataset.

Events	Asur framework	Takaffoli framework	Our framework
Birth	73%	78%	81%
Death	72%	75%	80%
Remain	67%	84%	83%
Merge	65%	76%	89%
Split	69%	83%	85%

5 Conclusion

In this paper we have described a framework for analyzing overlapping community evolution in dynamic social networks. Seven events involving communities and two events regarding nodes are proposed. In order to illustrate the events better and depict the changing process of community intuitively, the index *community tag* is defined. Two real datasets and a synthetic data are applied to test our framework with other similar frameworks. The experimental results show that the performance of our framework is better than Asur framework, while for individual events it is poorer than Takaffoli framework, however the Takaffoli framework takes more time. Five indexes based on events are presented for constructing neural network prediction model. The above three datasets are also applied to evaluate the prediction accuracy. From the experimental results, regarding the computation complexity we observe that our framework performs better than the other two frameworks and the results also show the practical value of our framework. Therefore, it is believed that our framework is a more appropriate choice. In terms of future research, we aim to propose a better model for predicting event.

Acknowledgments. National Natural Science Foundation of China (No. 61573128, 61273170, 4130144861573128) Central university basic scientific research business expenses special funds (No. 2015B25214).

References

1. Papagelis, M., Das, G., Koudas, N.: Sampling online social network. IEEE Trans. Knowl. Data Eng. **25**(3), 662–676 (2013)
2. Palla, G., Derényi, I., Farkas, I., Vicsek, T.: Urcovering the overlapping community structure of complex networks in nature and society. Nature **435**(7043), 814–818 (2005)
3. Folino, F., Pizzuti, C.: An evolutionary multiobjective approach for community discovery in dynamic network. IEEE Trans. Knowl. Data Eng. **26**(8), 1838–1852 (2014)
4. Takaffoli, M., Tabbany, R., Zaiane, O.R.: Community evolution prediction in dynamic social networks. In: 2014 IEEE/ACM International Conference on Advances in Social Network Analysis and Mining (ASONAM), pp. 9–16. IEEE (2014)
5. Xu, H., Hu, Y., Wang, Y., Ma, J., Xiao, W.: Core-based dynamic community detection in mobile social networks. Entropy **15**(12), 5419–5438 (2013)
6. Lancichinetti, A., Fortunato, S.: Community detection algorithms: a comparative analysis. Phys. Rev. **80**(5), 05117 (2009)
7. Leskovec, J., Lang, K.J., Mahorey, M.: Empirical comparison of algorithms for network community detection. In: Proceedings of the 19th International Conference on World Wide Web, pp. 613–640. ACM (2010)
8. Gregory, S.: Fuzzy overlapping communities in networks. J. Stat. Mech: Theory Exp. **2011**(02), 02017 (2011)
9. Clauset, A., Newman, M.E., Moore, C.: Finding community structure in very large network. Phys. Rev. E **70**(6), 066111 (2004)
10. Xie, J., Szymanski, B.K., Liu, X.: SLPA: uncovering overlapping communities in social networks via a speaker-listener interaction dynamic process. In: 2011 IEEE 1st International Conference on Data Mining Workshops (ICDMW), pp. 344–349. IEEE (2011)
11. Raghavan, U.N., Albert, R., Kumara, S.: Near linear time algorithm to detect community structures in large-scale network. Phys. Rev. E **76**(3), 036106 (2007)
12. Zhang, X., You, H., Zhu, W., Qiao, S., Li, J., Gutierrez, L.A., Zhang, Z., Fan, X.: Overlapping community identification approach in online social networks. Physica A Stat. Mech. Appl. **421**, 233–248 (2015)
13. Bródka, P., Kazienko, P., Kołoszczyk, B.: Predicting group evolution in the social network. In: Aberer, K., Flache, A., Jager, W., Liu, L., Tang, J., Guéret, C. (eds.) SocInfo 2012. LNCS, vol. 7710, pp. 54–67. Springer, Heidelberg (2012). doi:10.1007/978-3-642-35386-4_5
14. Asur, S., Parthasarathy, S., Ucar, D.: An event-based framework for characterizing the evolutionary behavior of interaction graphs. ACM Trans. Knowl. Discov. Data (TKD-D) **3**(4), 16 (2009)
15. Takaffoli, M., Sangi, F., Fagnan, J., Zäiane, O.R.: Community evolution mining in dynamic social networks. Procedia-Soc. Behav. Sci. **22**, 49–58 (2011)

Developing Mobile Software for Extenics Innovation

Siwei Yan, Rui Fan$^{(\boxtimes)}$, Yuefeng Chen, and Xiaohang Luo

Software School, Guangdong Ocean University, Zhanjiang, China
941090791@qq.com, fanrui@gdou.edu.cn,
yuefengch71@126.com, lou_xiaohang@163.com

Abstract. For using the formal method to deduce human creative thinking, Extenics deals with the contradictory problem by computer. Modeling novel mobile software to simulate flexibility innovation intelligence, and help people to develop creative ideas, to inspire innovation will become key issues to promote crowd innovation progress. According to Extenics innovation methods, new personal innovation mobile software is developed. A case shows that the Extenics innovation mobile software can help people to carry out effectively innovation activities.

Keywords: Extenics innovation · Mobile software · Android

1 Introduction

In the past three decades, Extenics [1] has developed into a new discipline with a relatively mature theoretical framework. It made great progress for creativity and innovation in Extenics theory, formal methods and practical application. For using the formal methods to deduce human creative thinking, Extenics deals with the contradictory problem by computer. Therefore, it becomes a hot spot to model novel software to simulate flexibility innovation intelligence and help people to develop creative ideas, inspire innovation. For example, Zhongfei Wang presents software architecture of ESGS based hownet [2]. ChengXiao Li presents a system framework for extension strategy generating [3]. Guang-zai Ye presents an extension strategy generation system based on components [4]. However, with the reason that software they designed consisted of some specific domain problems, it is insufficient about universality, and it is rare of results in mobile software.

For the need of crowd innovation, we devote to develop universality software model for Extenics innovation [5], and self-adaptive software formal model by Extenics innovation methods [6]. As the preliminary research result, this paper presents a novel Extenics innovation software architecture and develops android-based personal innovation mobile software, which guides innovators to deal with the contradictory problem using Extenics innovation method and to improve the ability of innovation.

H. Yuan et al. (Eds.): GRMSE 2016, Part I, CCIS 698, pp. 71–79, 2017.
DOI: 10.1007/978-981-10-3966-9_8

2   Extenics Innovation Mobile Software

This Extenics innovation mobile software is based on the Extenics innovation process. The Extenics innovation process (see Fig. 1) includes problem defining, core problem modeling, extension analyzing, extension transforming and superiority evaluating and so on. The problem defining put forwards initial problem, then the initial problem is decomposed into many sub-problems and form problem tree, each leaves problem is called core problem that can determine the key aspect of the initial problem. The core problem modeling needs users to construct core problem's basic-elements of goal and condition and dependent functions of the related basic-elements. The extension analyzing and extension transforming generate many new basic-elements, which may be candidate solutions, by adding, modifying deleting and fusing basic-elements of condition and goal for simulating users' innovation deductive process. The superiority evaluating will automatically select the best solution for users.

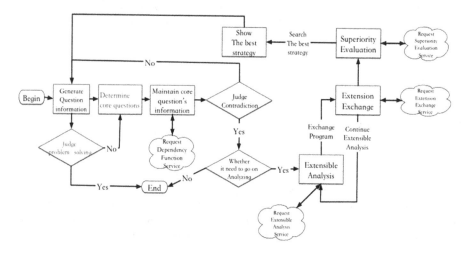

Fig. 1. The extenics innovation process

By the way, the Extenics innovation deductive process needs some local services or remote services help, just like dependency function calculating, extension analyzing, extension transforming and superiority evaluating, to assist innovation activities in carrying out.

Based on the process mentioned above, we must construct universality innovation software architecture, include display layer which is designed to adapt different terminal equipment, for conducting user deducing, control layer which is responsible for the connection between view and control logic, logic layer which adapt innovative reasoning and database layer which mainly accesses various data and information. It can be seen Fig. 2 about Extenics innovation software architecture.

In general, the logic of jumping to the different interface and controlling other thing are provided by the logic layer and the basic-elements' information is administrated by

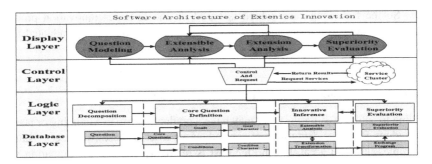

Fig. 2. Extenics innovation software architecture

the database layer. Other layers except the control layer are related closely with the control layer, so the control layer is core. We will spare no effort to explain this layer. It controls contents showing and interfaces jumping, and services requesting.

In problem modeling phase, the control layer needs to control the situation of modeling core problem and to request the dependency function service to calculate the K value, which is the criteria of discrimination to determine whether the basic-element is contradictory.

With the tree structure, the control layer needs to generate the relationship among the basic-elements through searching the database layer, and request extension analyzing services to execute related analysis during of the extension analyzing phase (see Fig. 3, which clearly shows how to guide the user's innovative process in the control layer of the Extenics innovative software structure).

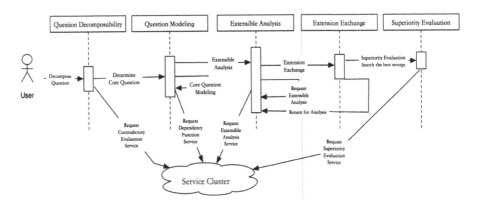

Fig. 3. Interfaces jumping and services requesting control process

3   Case Study

Graduated tourism is indispensable social activities of the University, but graduated tourism and general tourism group is different, because the graduation tourism prefers to customization, and travel group travel may consider the feasibility in different aspects. So the following case, which is based on the Extenics innovation process, represents that how to use our mobile innovative software to solve a graduated tourism's problem.

3.1 Solving Contradiction of Travel Location Selection

According to actual conditions, there may have a graduation trip to Maoming or Hainan in Guangdong Province. But different contradictions exist in these two places. Hainan, for example, although great, but the fee cost is relatively expensive; comparing with Maoming, where can be very convenient to play when on a weekend, the cost is cheap, but just can enjoy in two days and the distance is relatively close. As a result, at a next step, we need to complete the core problem modeling of the location selection contradiction.

With using this mobile innovative software, firstly, we need to initialize the problem's information and take Hainan as an example (see Fig. 4).

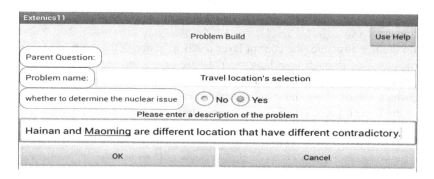

Fig. 4. Initialize the problem defining of Hainan

Then, we need to generate a goal basic-element and a condition basic-element with characters while determining to create a core problem. The detailed of those basic-elements can see Fig. 5. By the way, G represents the goal basic-element and also L represents the condition basic-element, at the meanwhile, these symbols will be used in the same way.

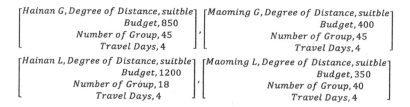

Fig. 5. The detailed information of the core problem

We need to set dependency function for characters like actual cost (see Fig. 6) and set up the relationship between the condition characters and goal characters (see Fig. 7).

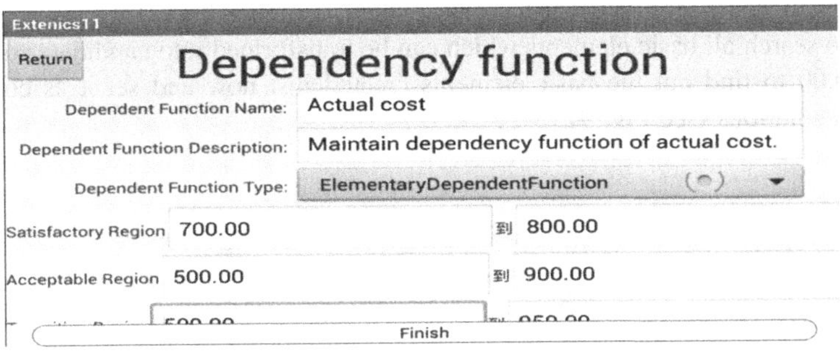

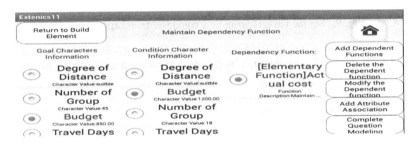

Fig. 6. Set dependency function for actual cost

Fig. 7. Relation of goal and condition characters

Finally, we need to create an evaluation function for the current contradiction information, like P = G*L = G*(actual cost ∧ number of people). The problem modeling for Maoming also do the same things mentioned above. So far, the problem modeling phase has been completed.

In the extension analyzing phase, in order to select a suitably graduated travel location, we need to choose conjugate analysis in extension analyzing's interface, analyzing significance and impact for the graduated travel, which is the imaginary part of these locations. So it is necessary to regenerate new basic-elements of the imaginary part about these travel locations (see Fig. 8).

$$\begin{bmatrix} Hainan\ G, Degree\ of\ Distance, suitable \\ Degree\ of\ meaning, high \end{bmatrix}, \begin{bmatrix} Maoming\ G, Degree\ of\ Distance, suitable \\ Degree\ of\ meaning, high \end{bmatrix}$$
$$\begin{bmatrix} Hainan\ L, Degree\ of\ Distance, suitable \\ Degree\ of\ meaning, high \end{bmatrix}, \begin{bmatrix} Maoming\ G, Degree\ of\ Distance, close \\ Degree\ of\ meaning, low \end{bmatrix}$$

Fig. 8. Regenerate new basic-elements of imaginary part about these travel locations

After conducting the conjugate analysis, it will return to the original analysis interface, and we can see a relationship between the initial extension analyzing basic-element and the new conjugate analysis basic-element. Because we can see the conjugate analysis basic-elements have shown no contradiction. So we can jump to the interface of extension transforming phase.

Simply, it just needs to set a K value, which is higher than 0 means no contradictory, to search all basic elements which can be transformed into candidate solutions. We set 0.00 to find out the basic-elements created just now and set it as conjugate transform solutions (see Fig. 9).

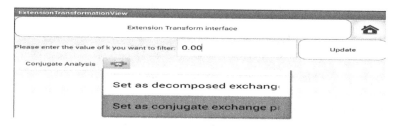

Fig. 9. Set the basic element as a conjugate transform solution

During of superiority evaluating, software can automatically generate the best solution with the superiority value of all basic-elements have been exchanged (see Fig. 10). The result shows that Hainan is the best solution and Maoming has contradictory during the extension analyzing phase.

ChooseElement
Superiority Evaluation interface

Please check the Superiority Evaluation method: (●) Weighted Superiority ( ) Take Max Superiority

Weight:	Superiority Evaluation value
Degree of Distance: 0.7	
Degree of meaning: 0.3	1.00
Original Core	Conjugate Analysis
It save the original core question information.	Conjugate the imaginary part to solve problem

Fig. 10. Show the best solution through Superiority evaluating

3.2 Solve the Contradiction of Tourism

In the problem modeling phase, we can take advantage of the core problem's information of Hainan in see Fig. 5.

According to the initial analysis basic-element, we found that the contradiction of tourism mainly originates from actual cost and the number of people.

First, we analyze the contradiction of the actual cost. With analyzing that the highest composition of contradiction is the actual cost, the actual cost has to reduce.

Second, while jumping to the interface of the extension analyzing phase, we choose the analysis service called OneObjectMoreCharacters to carry out some new characters such as per capita fare(PCF), accommodation fare(AF), ticket fees(TF) and other expenses(OP), and the new basic-elements can see Fig. 11.

$$\begin{bmatrix} Hainan\ G, Degree\ of\ Distance, suitable \\ Budget, 850 \\ Number\ of\ Group, 45 \\ Travel\ Days, 4 \end{bmatrix}, \begin{bmatrix} Hainan\ L, Degree\ of\ Distance, suitable \\ Actual\ Cost, 1200 \\ Number\ of\ people\ , 18 \\ Travel\ Days, 4 \\ Per\ capita\ fare, 455 \\ Accommodation\ fare, 200 \\ Ticket\ fees, 200 \\ Other\ expenses, 355 \end{bmatrix}$$

Fig. 11. One object more characters to create new basic-elements

Then choose the service called TheSameObjectCorrelative to create new relationship for the new characters mentioned above and actual cost (see Fig. 12) like PCF = AF + TF + OP.

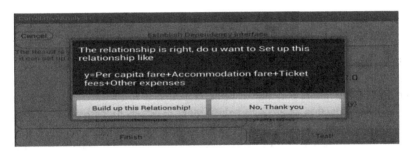

Fig. 12. Choose SameObjectCorrelative service to create a new relationship

The per capita fare is a key character and also it has a relationship with the number of people. So we also can build up the relationship between the per capita fare (PCF) and the number of people (NOP) with the service called SameObjectCorrelative like PCF = 8200/NOP. So we can solve the contradiction of the per capita fare through solving the contradiction of the number of people.

Fourth, choose the service called OneCharacterMoreObject to create a new basic-element representing a class which has a friendship (see Fig. 13).

Then, we choose a service called combined analysis to combine the analysis basic-element created initially with the class basic-element, rising up the number of people to 36 (see Fig. 14), and, for the time being, the contradictory of the number of people is overcome. There may have different ways to solve this contradiction but we just choose one way explained above in this paper.

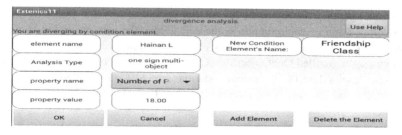

Fig. 13. OneCharacterMoreObject to create basic-element with friendship

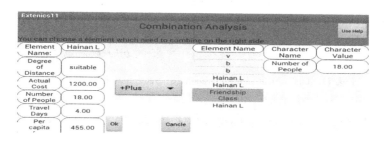

Fig. 14. Combined analysis to combine basic-element

According to the correlation between per capita fare and the number of people, the per capita fare can be updated to 227.5, and also the actual cost to 972.5, but we, unfortunately, find that the actual cost still has a contradiction after calculating dependency function K value.

Therefore, we try to reduce the character named other expenses, decomposing it into the cost of eating and playing. If we eat together, it will reduce the cost per meal, choose scalable reduction analysis to reduce other expenses to increase other expenses by 0.7 times (see Fig. 15) as a result that the other cost is updated to 241.5.

At this time, the actual cost of the dependency function value is 0.09, which means no contradiction if the value is greater than 0. We ought to go on extension transforming.

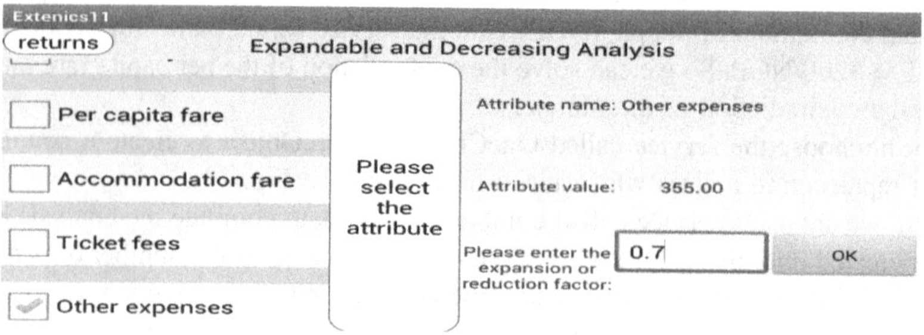

Fig. 15. Choose the scalable analysis to update other expenses by 0.7 times

During of the extension transforming phase, we can set a K value like 0.2 to search non-contradictory all basic-elements, getting some basic-elements which can be transformed into expandable transform solutions.

Finally, in the superiority evaluating phase, we get the best solution by requesting superiority evaluating service.

4 Conclusions

Because of requiring convenient and practical assistance software tool for peoples Innovation, we propose new Extenics innovative software architecture and have developed personal innovation mobile software. A case reveals user-friendly, convenience of the Extenics innovative software. On this basis, further work is to increase intelligence and collaborative innovation ability.

Acknowledgments. This research is supported by the Guangdong Provincial Science and Technology Project (2014A040402010), Guangdong Provincial Science and Technology Project (2016A010101028) and the Guangdong Ocean University Project sail sea (hzfqhjhkjfm2015b14).

References

1. Yang, C., Cai, W.: Extenics. Science Publishing House, Beijing (2014)
2. Wang, Z.: Research and implementation of software architecture of ESGS based HowNet. Guangdong University of Technology, China (2015)
3. Li, C.: Research on design and reuse extension strategy generating system framework. Guangdong University of Technology, China (2011)
4. Ye, G., Li, W., Li, S.: The design and implementation of extension strategy generation system based on components. CAAI Trans. Intell. Syst. **05**(4), 266–271 (2010)
5. Fan, R.: Modelling extenics innovation software by intelligent service components. Open Cybern. Syst. J. **8**, 1–7 (2014)
6. Fan, R., Peng, Y., Chen, Y., Lei, G., Liu, X.: A method for self-adaptive software formal modeling by Extenics. CAAI Trans. Intell. Syst. **10**(6), 901–911 (2015)

Variable Weight Based Clustering Approach for Load Balancing in Wireless Sensor Networks

Xuxun Liu[1,2(✉)] and Hongyan Xin[1]

[1] College of Electronic and Information Engineering,
South China University of Technology, Guangzhou 510641, China
liuxuxun@scut.edu.cn
[2] Key Laboratory of Autonomous Systems and Network Control
Ministry of Education, South China University of Technology,
Guangzhou 510641, China

Abstract. Uneven clustering is one of the feasible methods for energy hole avoidance in a wireless sensor network (WSN). Usually all weight in clustering and routing, such as residual energy of a sensor, distance between different nodes, and so on, are invariable all the time. In fact, with the running of the network, the relative importance of all factors makes the change. A variable weight based clustering approach (VWCA) is composed in this paper for energy hole avoidance in WSNs. The characteristics of the paper are to adjust the weight of residual energy of a sensor and the distance between different nodes. Simulation results show that VWCA does better in energy hole avoidance among all sensor nodes and achieves an obvious improvement on the network lifetime.

Keywords: Wireless sensor networks · Energy hole · Variable weight · Clustering

1 Introduction

Wireless Sensor Networks (WSNs) have wide range of military and civilian applications [1–3]. WSNs, as key components of the Internet of Things [4–6], enable us to achieve the dream of Smart City [7]. In many-to-one pattern wireless sensor networks, sensor nodes closer to the base station have to transmit more packets than those at other places. This easily leads to the phenomenon of energy hole [8, 9]. Much work has been done to avoid energy hole, such as density control, power adjustment, node mobility, clustering techniques, etc. [10]. Energy transfer is a new method emerged in recent years to avoid energy hole [11–13].

Clustering is a good way to achieve high energy efficiency and extend long network lifetime for WSNs [14, 15]. Sensor nodes can be organized hierarchically by grouping them into clusters, where data is collected and processed locally at the cluster head nodes before being transmitted to the sink. In the past, many clustering and routing protocols in the literature have been proposed in order to reduce energy consumption and avoid energy hole. Some performance, such as residual energy of a sensor, distance

© Springer Nature Singapore Pte Ltd. 2017
H. Yuan et al. (Eds.): GRMSE 2016, Part I, CCIS 698, pp. 80–90, 2017.
DOI: 10.1007/978-981-10-3966-9_9

between different nodes have been taken into account in the clustering and routing protocols, and all weight coefficients are invariable all the time. In fact, with the running of the network, the relative importance of all performance makes the change.

Many clustering and routing algorithms have been proposed for WSNs in recent years. LEACH [16] is the representative clustering protocol, in which the network is divided into several clusters. Based on LEACH, there are many variants, such as LEACH-E [17], PEGASIS [18], TEEN [19], HEED [20], and LEACH-M [21]. UCS [22], EEUC [23] and UCR [24] present energy efficient uneven clustering and routing algorithms, which not only consider balancing energy cost among nodes within one cluster, but also take into account energy balance among cluster headers. Tarhani et al. [25] proposes a scalable energy efficient clustering hierarchy (SEECH), which selects CHs and relays separately according to nodes eligibilities. Xu et al. [26] propose a joint clustering and routing (JCR) protocol for reliable and efficient data collection in large scale WSN.

This paper presents a novel uneven clustering and routing protocol based on dynamic weight adjustment for WSNs. The protocol is divided into clustering and routing, and the main ideas are to dynamically adjust the weight coefficients in order to alleviate energy hole and extend network lifetime.

2 System Model

We consider a sensor network consisting of N sensor nodes, denoted by $S = \{s_1, s_2, \cdots, s_N\}$, is distributed uniformly in a planar region. There is an unique base station located in the center of the square sensing field. All nodes have equal initial energy. Each cluster head aggregates the data gathered from its members into a single length-fixed packet, and there is no data aggregation among any cluster heads. The typical model of energy dissipation in [16] is adopted as follows.

$$E_{Tx}(l, d) = \begin{cases} lE_{elec} + l\varepsilon_{fs}d^2, \ d < d_0 \\ lE_{elec} + l\varepsilon_{mp}d^4, \ d \geq d_0 \end{cases} \tag{1}$$

$$E_{Rx}(l) = lE_{elec} \tag{2}$$

The sensing model, such as 0-1 model in [27] or probability model in [28], has not been taken into account.

3 Protocol Description

3.1 General Information

This protocol involves two processes, clustering and routing. For the sake of energy consumption balancing and network lifetime maximizing, the network is unevenly clustered, and each cluster has only one head, which spends its energy on intra-cluster and inter-cluster processing. Clustering involves cluster head election and cluster member election. After clustering, cluster heads act as repeaters from data source to the sink during delivering data.

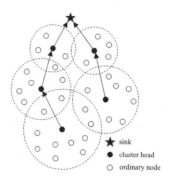

Fig. 1. Topology of the clustered network

An overview of VWCA is shown in Fig. 1, where the circles of unequal size, with broken lines, represent the clusters of unequal size and the traffic among cluster heads illustrates the multi-hop routing method. The detailed descriptions of the uneven clustering and multi-hop routing protocol based on weight adjustment for WSNs are as follows.

3.2   Clustering Description

A WSN is divided into different clusters, each one with a cluster head and some ordinary nodes as its members. Each cluster head is located in the center of a cluster and its role is to gather data from its member and deliver data from data source to the sink. Cluster heads are elected by competitive method based on the residual energy of each node and their distance to the sink. Different from random cluster header election model of EEUC [23], All nodes participate in cluster header election in WACA, which is more impartial to all sensor of the network. Suppose s_i becomes a cluster head candidate and its competition range is $R_c(s_i)$, which, denoted as follows [23], is a function based on its distance to the sink:

$$R_c(s_i) = \left[1 - c\frac{d_{\max} - d(s_i, sink)}{d_{\max} - d_{\min}} \right] R_c^0 \tag{3}$$

where $d_{\max}$ and $d_{\min}$ are the maximum and minimum distance between sensor nodes and the sink, $d(s_i, \sin k)$ is the distance between s_i and the sink, c is a constant coefficient between 0 and 1. According to Eq. (3), the competition radius $R_c(s_i)$ varies from $(1 - c)R_c^0$ to R_c^0. It is distinct that the cluster region closer to the sink is smaller than that farther from the sink, which is the characteristic of uneven clustering. It is prescribed that if s_i becomes a cluster head at the end of the competition, there will not be another cluster head s_j within s_i's competition diameter. Each cluster head candidate s_i maintains a set $SCH(s_i)$ of its adjacent cluster head candidates. Cluster head candidate

s_j belongs to $SCH(s_i)$ if s_j is in s_i's competition diameter or s_i is in s_j's competition diameter, i.e.

$$SCH(s_i) = \{s_j | s_j \in S, d(s_i, s_j) < \max(R_c(s_i), R_c(s_j))\} \qquad (4)$$

When cluster head election has been accomplished, every cluster head may not in the competition diameter of another cluster head.

The clustering process is as follows:

(1) Each cluster head candidate s_i broadcasts a cluster head competition message M1, which contains its competition radius $R_c(s_i)$, residual energy $E_{residual}(s_i)$ and the distance to the sink $d(s_i, sink)$.

(2) Each cluster head candidate s_i checks its $SCH(s_i)$ and decide whether it can become a cluster head based on the residual energy. If sensor s_i finds that its residual energy is more than all nodes of $SCH(s_i)$, it will win the competition and broadcast a election success message M2, otherwise give up the election and broadcast an election failing message M3.

(3) When cluster heads have been elected, each ordinary node s_i chooses an appropriate cluster head s_j, based on the function:

$$f(s_j) = \alpha E_{residual}(s_j) + \beta \frac{1}{d(s_i, s_j)} \qquad (5)$$

$$s_j = \arg\max\{f(s_j)\} \qquad (6)$$

and joins it by sending a joining message. In formula (5), α and β are weight coefficients, which indicate residual energy and distance respectively.

In EEUC [23], when cluster heads have been decided, each ordinary node joins its closest cluster head, but VWCA is different. In the early stage of network running, every cluster head has sufficient residual energy and energy consideration is less important, so ordinary nodes should choose preferentially nearby cluster head to join. In the late stage of network running, most cluster heads' residual energy are few and quite different, so ordinary nodes should choose preferentially cluster head with

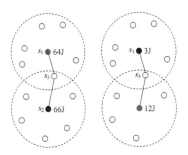

(a) in the early stage   (b) in the late stage

Fig. 2. Different clustering choice in different stages

more residual energy to join. As Fig. 2 shown, in the early stage of network running, residual energy of cluster head s_1 and s_2 respectively is 64J and 66J, therefore node s_3 should select cluster head s_1, which is closer, to join; But in the late stage of network running, residual energy of cluster head s_1 and s_2 respectively is 3J and 12J, node s_3 should select node s_2 to join which has more residual energy. Value α is relatively larger while value β is relatively smaller in the early stage of network running; In the late stage of network running, on the contrary the opposite is true. Evidently, VWCA is characteristic of weight adjustment in different stages.

3.3   Routing Description

After data aggregation is finished by each cluster head from its cluster members, data is delivered by cluster heads to the sink, and this is called routing via multi-hop relay communication. Due to lack of data similarity among different clusters, data aggregation is not adopted between different cluster heads.

At the beginning of the routing, each cluster head broadcasts a message across the network at a certain power which consists of its ID, residual energy, and distance to the sink. The most important is to choose the best next hop node. Cluster head s_i chooses another one s_j as its next hop. Cluster head s_j is in a set $NEXT(s_i)$, which is defined as follows [23].

$$NEXT(s_i) = \{s_j | s_j \in S, d(s_j, sink) < d(s_i, sink), d(s_i, s_j) \leq kR_c(s_i)\} \qquad (7)$$

where k is the minimum integer that let $NEXT(s_i)$ contains at least one item. If there doesn't exist such a k, define $NEXT(s_i)$ as a null set, and s_i will send its data directly to the sink. In order to balance energy consumption and extend the network lifetime, we must choose the relay nodes whose residual energy is as plentiful as possible.

Moreover, energy consumed in the relay process must be taken into account in the same way. In EEUC [23], when s_i selects s_j as its relay node to deliver a l-length packet to the sink, the energy consumed by the two nodes is

$$\begin{aligned} E_{2-hop} &= E_{Tx}(l, d(s_i, s_j)) + E_{Rx}(l) + E_{Tx}(l, d(s_j, sink)) \\ &= l\varepsilon_{fs}[d^2(s_i, s_j) + d^2(s_j, sink)] + 3lE_{elec} \end{aligned} \qquad (8)$$

According to formula (8), an energy consumed function is defined as

$$D_{relay}(s_j) = d^2(s_i, s_j) + d^2(s_j, sink) \qquad (9)$$

by which, energy consumed in the relay process is determined. The smaller the $D_{relay}(s_j)$ is, the less energy will be consumed in the relay process. Obviously, it could save the network energy, when the node s_j is located straight along the way from s_i to the sink.

A tradeoff should be made between the two factors of residual energy and the energy consumed function. Node s_i chooses node s_j as the next hop based on the following formulas:

$$g(s_j) = \rho E_{residual}(s_j) + \lambda \frac{1}{D_{relay}(s_j)} \qquad (10)$$

$$s_j = \arg\max\{g(s_j)\} \qquad (11)$$

In formula (10), ρ and λ are weight coefficients, respectively of residual energy and energy consumed function of a sensor.

In EEUC [23], node s_i chooses node s_j with more residual energy from the two nodes with smallest energy consumed function. However, VWCA is different and more reasonable. In the early stage of network running, every cluster head has sufficient residual energy, so cluster heads should choose preferentially relay with small energy consumed function. In the late stage of network running, most cluster heads' residual energy are few and quite different, so cluster heads should choose preferentially relay with more residual energy.

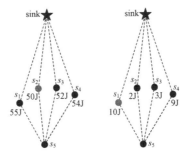

(a) in the early stage   (b) in the late stage

Fig. 3. Different routing in different stages

As shown in Fig. 3, in the early stage of network running, residual energy of cluster head s_1, s_2, s_3 and s_4 respectively is 55J, 50J, 52J, 54J, consequently node s_5 should select relay s_2, with smallest energy consumed function; But in the late stage of network running, residual energy of cluster head s_1, s_2, s_3 and s_4 respectively is 10J, 2J, 3J, 9J, consequently node s_5 should choose relay s_1, with most residual energy. Value ρ is relatively smaller while value λ is relatively larger in the early stage of network running; In the late stage of network running, on the contrary the opposite is true. In the same way, VWCA is also characteristic of weight adjustment here.

4   Experiments and Analysis

The performance of VWCA is evaluated by simulations. Three important performances, network lifetime, number of remaining nodes and residual energy are examined. VWCA is compared with LEACH-M [21] and EEUC [23]. MATLAB is used to finish our simulations. We define the end of the network life when the number of nodes in the network is less than 25% of the total number of them. The network scale is 160 m × 160 m, 200 m × 200 m, 240 m × 240 m, and 280 m × 280 m respectively.

Figure 4 shows the network lifetime in different network scales. It can be seen from the figure that the transmission round of VWCA is the largest, which is followed by that of EEUC. LEACH-M has the smallest transmission round among the three algorithms. Besides, the greater the size of the network is, the shorter the network life is. This is due to large energy consumption when long transmission distance exists.

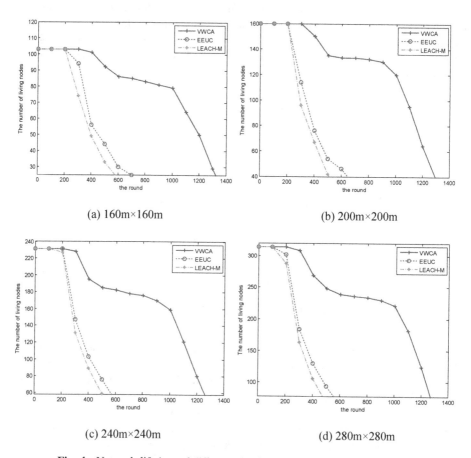

(a) 160m×160m

(b) 200m×200m

(c) 240m×240m

(d) 280m×280m

Fig. 4. Network lifetime of different algorithms in different network scale

It can be seen from Fig. 5 that the number of the living nodes of each algorithms is equal to its initial value and does not reduce when the transmission round is less than 200. But when data is transmitted for more than 300 rounds, all the nodes of VWCA

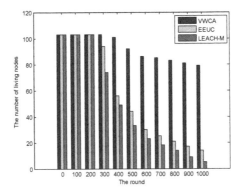

Fig. 5. Number of survival nodes in different transmission rounds

are still survival while the residual nodes of EEUC and LEACH-M have reduced greatly, and the decrease rate of LEACH-M is the fastest. On the other hand, it can be seen clearly that when the transmission round arrives at 1000, the living nodes of EEUC and LEACH-M are very few while the survival nodes of VWCA are more than 80% of its initial level. It is proved that, owing to better energy consumption balancing, the algorithm VWCA has better performance of network life than other algorithms.

Figures 6 and 7 present the performance of different algorithms in terms of the total and the average residual energy of living nodes. Clearly, the total energy consumption of VWCA is more than that of EEUC and LEACH-M. Moreover, the average residual energy of VWCA is less than that of EEUC and LEACH-M with the same transmission round.

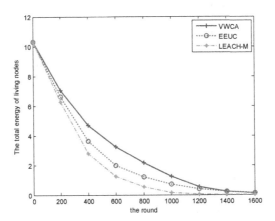

Fig. 6. Changes of total residual energy of nodes

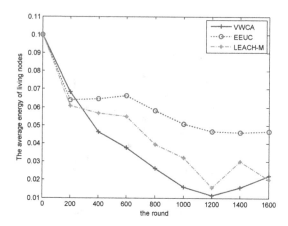

Fig. 7. Changes of average residual energy of nodes

5 Conclusion

In this paper, a novel protocol, VWCA, is introduced for energy hole avoidance in WSNs. An uneven clustering and routing methods is discussed to balance the energy consumption and avoid energy hole among sensor nodes. Clusters closer to the sink have smaller sizes than others, thus cluster heads closer to the sink can preserve some energy for multi-hop communications. The relative importance of all factors is taken into account in VWCA, and the weight of residual energy of a sensor, distance between different nodes, is adjusted based the different stage of the network. Simulation results show that our protocol, VWCA, do better in energy hole avoidance and lifetime prolonging in WSNs.

Acknowledgments. This work was supported in part by the National Natural Science Foundation of China (Grant No. 61001112, 61372082, 61671209, 61271314), the Guangdong Natural Science Foundation, China (Grant No. S2013010012141), the Cultivation Program for Major Projects and Important Achievements of Guangdong Province, China (Grant No. 2014KTSCX012), and the Fundamental Research Funds for the Central Universities, China (Grant No. 2011ZM0030, 2013ZZ0042, 2015ZZ090).

References

1. Liu, X.: A deployment strategy for multiple types of requirements in wireless sensor networks. IEEE Trans. Cybern. **45**(10), 2364–2376 (2015)
2. Kui, X., Sheng, Y., Du, H., Liang, J.: Constructing a CDS-based network backbone for data collection in wireless sensor networks. **2013**, 1–12 (2013)
3. Zhang, Q., Liu, A.: An unequal redundancy level based mechanism for reliable data collection in wireless sensor networks. EURASIP J. Wirel. Commun. Netw. **2016**(258), 1–22 (2016)

4. Liu, Y., Liu, A., Hu, Y., Li, Z., Choi, Y., Sekiya, H., Li, J.: FFSC: an energy efficiency communications approach for delay minimizing in internet of things. IEEE Access **6**, 3775–3793 (2016)

5. He, S., Chen, J., Li, X., Shen, X., Sun, Y.: Mobility and intruder prior information improving the barrier coverage of sparse sensor networks. IEEE Trans. Mobile Comput. **13**(6), 1268–1282 (2014)

6. Liu, X., Wei, T., Liu, A.: Fast program codes dissemination for smart wireless software defined networks. Sci. Program. **2016**, 1–21 (2016)

7. Tang, Z., Liu, A., Huang, C.: Social-aware data collection scheme through opportunistic communication in vehicular mobile networks. IEEE Access (2016). doi:10.1109/ACCESS.2016.2611863

8. Liu, Y., Liu, A., He, S.: A novel joint logging and migrating traceback scheme for achieving low storage requirement and long lifetime in WSNs. AEU-Int. J. Electron. Commun. **69**(10), 1464–1482 (2015)

9. Hu, Y., Liu, A.: An efficient heuristic subtraction deployment strategy to guarantee quality of event detection for WSNs. Comput. J. **58**(8), 1747–1762 (2015)

10. Liu, A., Jin, X., Cui, G., Chen, Z.: Deployment guidelines for achieving maximum lifetime and avoiding energy holes in sensor network. Inf. Sci. **230**, 197–226 (2013)

11. Dai, H., Wu, X., Chen, G., Xu, L., Lin, S.: Minimizing the number of mobile chargers for large-scale wireless rechargeable sensor networks. Comput. Commun. **46**, 54–65 (2014)

12. Dai, H., Wu, X., Xu, L., Wu, F., He, S., Chen, G.: Practical scheduling for stochastic event capture in energy harvesting sensor networks. Int. J. Sens. Netw. **18**(2), 85–100 (2015)

13. Deng, R., Zhang, Y., He, S., Chen, J., Shen, S.: Maximizing network utility of rechargeable sensor networks with spatiotemporally-coupled constraints. IEEE J. Sel. Areas Commun. **34**(6), 1–13 (2016)

14. Liu, X.: A survey on clustering routing protocols in wireless sensor networks. Sensors **12**(8), 11113–11153 (2012)

15. Liu, X., Shi, J.: Clustering routing algorithms in wireless sensor networks: an overview. KSII Trans. Internet Inf. Syst. **6**(7), 1735–1755 (2012)

16. Heinzelman, W., Chanrakasan, A., Balakrishnan, H.: Energy-efficient communication protocol for wireless micro-sensor networks. In: Proceedings of the 33rd Hawaii International Conference on System Sciences, pp. 1–10 (2000)

17. Heinzelman, W., Chanrakasan, A., Balakrishnan, H.: An application specific protocol architecture for wireless microsensor networks. IEEE Trans. Wirel. Commun. **1**(4), 660–670 (2002)

18. Lindsey, S., Raghavendra, C.S.: PEGASIS: power-efficient gathering in sensor information systems. In: Proceedings of IEEE International Conference on Computer Systems, San Francisco, pp. 1125–1130 (2002)

19. Manjeshwar, A., Agarwal, D.P.: TEEN: a routing protocol for enhanced efficiency in wireless sensor networks. In: Proceedings of 15th International Parallel and Distributed Processing Symposium, pp. 2009–2015 (2002)

20. Younis, O., Fahmy, S.: Distributed clustering in ad hoc sensor networks: a hybrid, energy-efficient approach. IEEE Trans. Mob. Comput. **3**(4), 366–379 (2004)

21. Anitha, R.U., Kamalakkannan, P.: Enhanced cluster based routing protocol for mobile nodes in wireless sensor network. In: Proceedings of the 2013 International Conference on Pattern Recognition, Informatics and Mobile Engineering (PRIME), 21–22 February 2013

22. Soro, S., Heinzelman, W.: Prolonging the lifetime of wireless sensor networks via unequal clustering. In: Proceedings of the 5th International Workshop on Algorithms for Wireless, Mobile, Ad hoc and Sensor Networks, pp. 118–129 (2005)

23. Li, C., Ye, M., Chen, G., Wu, J.: An energy-efficient unequal clustering mechanism for wireless sensor networks. In: 2nd IEEE International Conference on Mobile Ad-hoc and Sensor Systems, pp. 597–604 (2005)
24. Chen, G., Li, C., Ye, M., Wu, J.: An unequal cluster-based routing protocol in wireless sensor networks. Wirel. Netw. **15**(12), 193–207 (2009)
25. Tarhani, M., Kavian, Y.S., Siavoshi, S.: SEECH: scalable energy efficient clustering hierarchy protocol in wireless sensor networks. IEEE Sens. J. **14**(11), 3944–3954 (2014)
26. Xu, Z., Chen, L., Chen, C., Guan, X.: Joint clustering and routing design for reliable and efficient data collection in large-scale wireless sensor networks. IEEE Internet Things J. **3**(4), 520–532 (2016)
27. He, S., Shin, D., Zhang, J., Chen, J., Sun, Y.: Full-view area coverage in camera sensor networks: dimension reduction and near-optimal solutions. IEEE Trans. Veh. Technol. **65** (9), 7448–7461 (2016)
28. Yang, L., Cao, J., Zhu, W., Tang, S.: Accurate and efficient object tracking based on passive RFID. IEEE Trans. Mob. Comput. **14**(11), 2188–2200 (2015)

MDPRP: Markov Decision Process Based Routing Protocol for Mobile WSNs

Eric Ke Wang[1], Zhe Nie[2(✉)], Zheng Du[3], and Yuming Ye[1]

[1] Shenzhen Graduate School, Harbin Institute of Technology, Shenzhen, China
{wk_hit,yym}@hitsz.edu.cn
[2] Computer Engineering School, Shenzhen Polytechnic,
Shenzhen 518055, Guangdong, China
niezhe@szpt.edu.cn
[3] National Supercomputing Center in Shenzhen
(Shenzhen Cloud Computing Center), Shenzhen, China
duzheng@nsccsz.gov.cn

Abstract. In this paper we propose a new routing protocol-MDPRP for mobile wireless sensor networks, which adopts Markov Decision Process to make the decision of best next hop to forward the messages. In this scheme, we mainly integrate trust, congestion and distance as the main judgment criterion of the next hop decision. We evaluate the protocol by simulations and the performance results are encouraging.

Keywords: Routing · Congestion · Markov Decision Process

1 Introduction

Security and efficiency are two main goals of routing in mobile wireless sensor networks (WSN). However, it is not easy to achieve them both for routing mobile WSN since it has some special features such as limited resources, openness, and mobility which are quite different from traditional computer networks. Since it is an open network or deployed in an open area, malicious nodes can join in or some nodes may be captured by adversaries. Thus the threat may come from inside networks. Traditional network security measures commonly prevent attacks from outsides by increasing the cryptography ability of authentication, encryption and access control. However, threats come from inside of mobile wireless sensor network should be handled. Besides, since it has limited resources, computation and communication should be reduced. Therefore, lightweight secure routing protocol is required.

Commonly, a mobile wireless sensor network is a typical wireless multi-hop network. As in our previous work, we found that the crucial factor of routing for WSN is the decision of next hop. How to make optimal decision on finding next hop is an important problem to be solved. Therefore, we study routing techniques and propose a new routing algorithm MDPR-Markov Decision Process based Routing Protocol for mobile WSN, which is able to achieve security, low cost and high efficiency by making optimal decision of at each hop, making it suitable for mobile WSN.

© Springer Nature Singapore Pte Ltd. 2017
H. Yuan et al. (Eds.): GRMSE 2016, Part I, CCIS 698, pp. 91–99, 2017.
DOI: 10.1007/978-981-10-3966-9_10

This paper is organized as follows, Sect. 2 introduces related works; the Judgment Criterion to Rationality of Decision Maker is introduced in Sect. 3; Markov Decision Process and routing protocol are described in Sects. 4 and 5. In Sect. 6, we present the simulation result and finally we make a conclusion in Sect. 7.

2 Related Works

A natural way of routing for mobile wireless sensor networks is flooding that the message is broadcasted to the network and any node who receives the message can help forwarding it. However, it results in too much storage, communication overhead. Therefore, later on, researchers have proposed many routing algorithms [1–4] which can be mainly classified into two types, single path routing [5] and multi path routing [6]. Single path routing is to find out one path from source to destination and one message occurs in the network at the same time, while, multi path routing is that employ several paths from source to sink and multiple copied messages occurs in the network. Single path routing has the advantages of small storage and low communication overhead, multi path routing has the robust feature and can choose best one from multiple paths, but with much more communication. Single path routing protocols have some major cases such as LEACH [7] and its improved version [8, 9], Energy aware routing [10] and Rumor routing [11], LEACH assumes that all nodes can directly reach the sink node. Energy aware routing and Rumor routing maintain only single path between source sensor node and sink. Multi path routing protocols have many popular cases such as Braided [12], SPIN (Sensor Protocols for Information via Negotiation) [13]. Braided builds multiple paths for a data delivery, but only one of them is used, while others are maintained as backup paths. In SPIN routing process, nodes name their data meta-data. They remove the transmission of redundant data throughout the network by meta-data negotiations. Moreover, SPIN nodes can decide how to communicate based on both knowledge of the data and knowledge of the available resources. There is another popular routing protocol, data centric routing-directed diffusion [14], all communication is for named data. It provides a mechanism to flood queries based on events or tasks, and then set up the route by establishing reverse gradients to send data back. Directed Diffusion involves several elements: interests, data messages, gradients, and reinforcements. An interest message is a query which describes a user's needs. Each interest contains a sensing task description which is supported by a sensor network for querying data. Directed Diffusion can be single-path or multi-path routing depending on how many paths are reinforced by sink node.

Whatever any routing protocol is employed, security, low power and high efficiency are the main goals for the routing in mobile WSN. However, most of routing protocols lack comprehensive consideration on security, efficiency, low computation and communication overhead of the routing protocol for wireless sensor network. Therefore, in this paper, we propose a new routing protocol for mobile wireless sensor networks, because we found that the main goal of the routing is how to find out the best node as next forwarding node at each hop, so the protocol mainly target the decision of the best next hop.

3 Judgment Criterion to Rationality of Decision Maker

The crucial problem of routing in wireless sensor network is how to find out the best next hop in each hop. As we previous research, we find that "Trust", "Congestion Probability" and "Distance to Destination" are the main impact factors on the decision making of next hop. Trust means that if node N_i have more trust on the other node N_j, then the security and reliability of forwarding data to N_j from Ni are better. Trust has many definitions, in our paper; trust means a comprehensive belief value of honesty, reliability, resource availability, service quality of the target node. From theoretical view, node with biggest trust value should be the best one to be selected to be the next hop. However, in the actual scenario, there would be some problem of efficiency. It includes two sub problems, one is congestion problem, and the other is the distance to the destination. For example, if the strategy of next hop selection is decided to be node k with the highest trust value, node k may have same high trust value in other around nodes. Thus other nodes may also select it as the next hop. Node k would be the bottleneck. Therefore, we need to consider the congestion problem to avoid the congestion, besides; we need to consider which one is closer to the destination to find out the shortest path. Thus it can increase the routing efficiency of routing to destination.

Therefore, the set of impact factors for decision making consist of three elements {trustiness, congestion probability, and distance to destination}. Then the selection criterion of next hop is to find out the one node with high trustiness, low congestion probability, low distance to destination from candidates set.

Suppose at time t a node k has n neighbors, among neighbors the trustiness of node i is T_i, the congestion probability of node i is C_i, the distance to destination of node i is D_i, then impact factors set is $\{T_i, C_i, D_i\}$. The goal of decision is to find out the maximum value of the comprehensive factors (here we call it "decision metric"), as shown in Eq. 3.1:

$$N_{next} = Max\{\alpha T_1 \times \beta \frac{1}{C_1} \times \gamma \frac{1}{D_1}, \ldots \quad \ldots, \alpha T_n \times \beta \frac{1}{C_n} \times \gamma \frac{1}{D_n}\} \qquad (3.1)$$

α, β, γ are the impact weights.

3.1 Trustiness Computation

Commonly, trust in wireless sensor networks can be classified into direct trust Q and indirect trust R. Direct trust $Q_a(b)$ means that node a has previous directed interaction with node b and compute the trustiness value based on direct interaction experience. Indirect trust $R_a(b)$ means that node a has no interaction with node b before, a's trustiness value for b is from other nodes transiting trustiness. As shown in Fig. 1.

Then we combine direct trust and indirect trust, the trustiness value can be computed as Eq. 3.2:

$$T_a(b) = (1 - \tau) \times Q_a(b) + \tau \times R_a(b) \qquad (3.2)$$

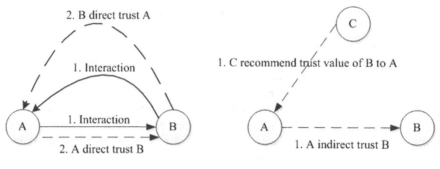

(a).direct trust based on interaction   (b).indirect trust based on Recommendation

Fig. 1. Direct trust and indirect trust

3.2   Congestion Probability Computation

Commonly, the probability of congestion would be bigger if a node has more neighbors than others. Thus the number of neighbors is the main parameter to determine the congestion probability. Besides, number of current forwarding messages is the other parameter. Suppose at time t, node a has n neighbors, and has k messages in the storage to be forwarded. Then the congestion probability should be as following equation:

$$C_t(a) = \frac{n}{n+k} \times 100\% \tag{3.3}$$

3.3   Distance to Destination Computation

Suppose that each sensor node would record the routing hops of historic messages to sink, thus the distance to sink can be computed directly, commonly it uses the latest value of the routing hops. However, there is some special cases that distance to destination is not applicable, since maybe the node has no previews messages for the destination or not any interaction with destination. Thus the factor can be dynamic applicable for the computation. Suppose d_1, d_2, d_k is the distance to sink of historic messages. If d_k is the latest one, then the distance to sink is set to be d_k.

4   Dynamic Markov Decision Process

Selecting the optimal next hop based on current situation is a typical decision making process, so we adopt Markov Decision Process [15] to solve the problem since it is one of the best decision process for random dynamic system. We can look at each hop in the routing as a state; the decision of each hop is to select one of the best next hops. Then after n piece of decisions are made sequentially, the messages can be transferred to the destination efficiently and securely, as shown in Fig. 2.

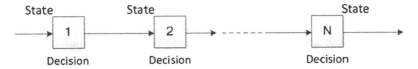

Fig. 2. Dynamic decision process

In each step, the decision making depends on the current situation, and with sequential decisions are made, the whole routing process is the optimal. Since the hops from source to destination are not infinite, so we adopt finite Markov decision process to solve it. The basic idea is: we use maximum of decision metric (as introduced in Sect. 3) in the whole routing process as the goal of the decisions to build a Finite Markov Decision Process model, in order to find out the set of best hops from candidates.

T Phases Markov Decision Process (as shown in following tetrad, Eq. 4.1):

$$\{ S; \quad A(i)|i \in s; \quad P_{ij}(a)|i,j \in s, a \in A(i); \quad r(i,a)|i \in S, \quad a \in A(i) \} \tag{4.1}$$

They are states space, decision set, state transit probability and valued reward.

A. States Space S: we set all possible hops as the states set $S = \{1,2,\ldots,s\}$, S is called the states space in the network. We call that routing is in one state as which hop the message stays currently.

B. Decision Set $A_T(i)$: represent the set of all the possible policies which can be selected at state i, in all T phases.

C. States Transit Probability $p_{ij}(a)$: Suppose in some certain phase, routing is in state i, $i \in S$, the action of selecting next hop is a, $a \in A(i)$, then next phase the probability of the routing is in state j, $j \in S$ is $p_{ij}(a)$.

D. Expected Reward $r(i,a)$: it represents the expected reward $r(i,a)$ when routing is at state i in some phase, the decision is a, $a \in A(i)$.

E. Policy θ: we call each rule of decision as one policy, then a finite Markov policy can be represented by $\theta = (d_1(i), d_2(i), \ldots, d_T(i))$, and $d_t(i)$ is the decision made at state i on time t.

F. Recurrence equation $u_t^*(i)$: it represent the maximum value of total expected reward from t to T (the total expected reward is the accumulate decision metric).

$$u_t^*(i) = max_{d \in A_s}\{r_t(i, d_t(i)) + \sum_{j \in S} p_t(j|i, d_t(i))u_{t+1}^*(j)\} \quad j = T-1, T-2, \ldots, t \tag{4.2}$$

Then we can find out the policies set $\pi = (d_1(i), d_2(i), \ldots, d_T(i))$, that is the final path result of decision.

5 Routing Protocol

Since wireless sensor network is a distributed network, central computing to achieve one route path is not applicable. So each node has the responsibility to compute and make decision in each hop. Therefore, we treat the decision of next hop is a one phase Markov Decision Process, the goal of decision is let the reward of each step is maximum. Then the protocol is as follows:

> *MDPRP-Routing Protocol (suppose node i initialize a routing):*
> *While (j.hasNext ()){*
>   *Node i requests basic information (number of its neighbors, distance to other nodes) of node j from its neighbors set;*
>   *After i received the basic information from neighbors,*
>   *Node i compute the Trustiness, Congestion Probability and Distance To Destination of node j;*
>   *Node   i   compute   the   comprehensive   decision   metric*
>
> $$r_t(i,d_t(i)) + \sum_{j \in S} p_t(j \mid i, d_t(i)) u_{t+1}^*(j) \quad \text{of node } j; \}$$
>
>   *Node i selects the node with maximum of the comprehensive decision metric as the next hop;*
>   *Node i forwards message to Node j.*

6 Simulation

In order to evaluate the efficiency of the MDPRP routing protocol, we simulate the wireless sensor network scenario and to check how the routing protocol can be effective and efficient. We adopt Network Simulator NS2 [16] as the simulation tool. As shown in Fig. 3, we set the area of size 100 m × 100 m, and set the number of nodes is {50, 100, 150, 200, 250, 300}. The data size is 64 bytes, initial power of a node is set to be 8 J, the initial power of a sink is set to be 30 J since commonly the power of a sink is bigger than common sensor nodes. For the judgment criterion computation, we set the weight $\alpha = 1/2$, $\beta = 1/4$, $\gamma = 1/4$. In order to evaluate the efficiency of our protocol, we compare it with other major typical routing protocols (Flooding, Directed diffusion, Energy-aware routing protocol) (Table 1).

The Fig. 3 shows an example of message routing pattern. From source to sink, at each hop the node decides the best next hop as the message forwarding node.

The time delay measures the average time spent to relay data packets from the source node to the sink node. We run the protocol with other comparison protocols for 500 times and compute the average time delay. The Fig. 4 shows the comparison of time delay, from the simulation, we can find out the MDPRP routing protocol is better than three other protocols.

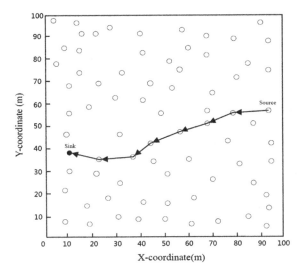

Fig. 3. Message routing pattern

Table 1. Simulation configuration

Parameter	Value
Size of area	100 m × 100 m
Number of nodes	50–300
Data Size	64 bytes
Initial power of a node	8 J
Initial power of a sink	30 J

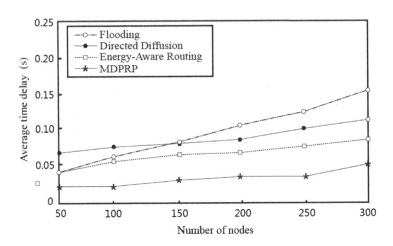

Fig. 4. Average time delay

The average energy consumption (AEC) is compute as following Eq. 6.1:

$$AEC = \frac{\sum_{i=1}^{n} \left(E_{initial}(i) - E_{remain}(i) \right)}{n} \tag{6.1}$$

n is the number of the nodes in the network, $E_{initial}(i)$ is the initial energy of node i, $E_{remain}(i)$ is the remain energy of node i. As shown in Fig. 5, MDPRP has lower energy consumption than other three main routing algorithms.

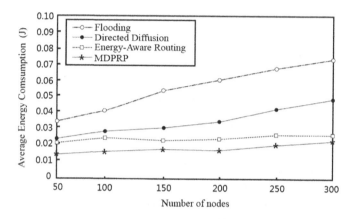

Fig. 5. Comparison of average energy consumption

7 Conclusion

In this paper, we propose a new routing protocol-MDPRP for mobile WSN. We adopt Markov Decision Process to decide best next hop as forwarding node which can forward message efficiently and securely. The simulation result shows encouraging.

Acknowledgments. This research was supported in part by National Natural Science Foundation of China (No. 61572157), grant No. 2016A030313660 from Guangdong Province Natural Science Foundation, No. JCY20150403161923509, JCYJ20150617155357681, JCYJ20160428092427 867, JSGG20141017150830428, JSGG20150512145714247 from Shenzhen Municipal Science and Technology Innovation Project. The authors thank the reviewers for their comments.

References

1. Singh, S.K., Singh, M.P., Singh, D.K.: Routing protocols in wireless sensor networks–a survey. Int. J. Comput. Sci. Eng. Surv. (IJCSES) **1**, 63–83 (2010)
2. Goyal, D., Tripathy, M.R.: Routing protocols in wireless sensor networks: a survey. In: 2012 Second International Conference on Advanced Computing and Communication Technologies (ACCT), pp. 474–480. IEEE (2012)

3. Quang, P.T.A., Kim, D.S.: Enhancing real-time delivery of gradient routing for industrial wireless sensor networks. IEEE Trans. Ind. Inform. **8**(1), 61–68 (2012)
4. Pantazis, N.A., Nikolidakis, S.A., Vergados, D.D.: Energy-efficient routing protocols in wireless sensor networks: a survey. IEEE Commun. Surv. Tutorials **15**(2), 551–591 (2013)
5. Bejerano, Y., Lee, K.T., Han, S.J., et al.: Single-path routing for life time maximization in multi-hop wireless networks. Wirel. Netw. **17**(1), 263–275 (2011)
6. Radi, M., Dezfouli, B., Bakar, K.A., et al.: Multipath routing in wireless sensor networks: survey and research challenges. Sensors **12**(1), 650–685 (2012)
7. Heinzelman, W.R., Chandrakasan, A., Balakrishnan, H.: Energy-efficient communication protocol for wireless microsensor networks. In: 2000 Proceedings of the 33rd Annual Hawaii International Conference on System Sciences, vol. 2, pp. 10. IEEE (2000)
8. Ran, G., Zhang, H., Gong, S.: Improving on LEACH protocol of wireless sensor networks using fuzzy logic. J. Inf. Comput. Sci. **7**(3), 767–775 (2010)
9. Xu, J., Jin, N., Lou, X., et al.: Improvement of LEACH protocol for WSN. In: 2012 9th International Conference on Fuzzy Systems and Knowledge Discovery (FSKD), pp. 2174–2177. IEEE (2012)
10. Shah, R.C., Rabaey, H.M.: Energy aware routing for low energy ad hoc sensor networks. In: IEEE Wireless Communications and Networking Conference (2002)
11. Braginsky, D., Estrin, D.: Rumor routing algorithm for sensor networks. In: Proceedings of the First ACM International Workshop on Wireless Sensor Networks and Applications, Atlanta, September 2002, pp. 22–29 (2002)
12. Ganesan, D., Govindan, R., Shenker, S., Estrin, D.: Highly-resilient, energy-efficient multipath routing in wireless sensor networks. ACM Mob. Comput. Commun. Rev. **5**(4), 11–25 (2001)
13. Kulik, J., Heinzelman, W., Balakrishnan, H.: Negotiation-based protocols for disseminating information in wireless sensor networks. Wirel. Netw. **8**(2/3), 169–185 (2002)
14. Intanagonwiwat, C., Govindan, R., Estrin, D., et al.: Directed diffusion for wireless sensor networking. IEEE/ACM Trans. Netw. **11**(1), 2–16 (2003)
15. Puterman, M.L.: Markov Decision Processes: Discrete Stochastic Dynamic Programming. Wiley, New York (2009)
16. Issariyakul, T., Hossain, E.: Introduction to Network Simulator NS2. Springer, Heidelberg (2011)

Medical Insurance Data Mining Using SPAM Algorithm

Qifeng Cheng and Xiaoqiang Ren[✉]

Qilu University of Technology, Jinan 25000, China
chengqf9116@163.com, Renxq@qlu.edu.cn

Abstract. The sequential pattern data mining technology is widely applied to various fields, and it brings an indispensable value for many areas, especially in the field of medical treatment. But the amount of health-care data is large and the information included is extensive, so some valuable information may have not been found, which needs us to take the further research. By using the Sequential Pattern Mining (SPAM) algorithm to deal with the health-care data, we try to find the user's medical behavior and the doctor diagnosis model or rule. This article first introduces the characteristics and the worth of Medicare data and data mining on it, especially the sequence pattern mining significance. Then discusses the ideas and characteristics of SPAM algorithm and the advantages of high efficiency, we use SPAM algorithm to deal with health care data, and try to find the regularity of visiting doctor, medical treatment characteristics and drug-use mode about insured person in a certain period of time. The experiments reveal the treatment mode and characteristics of the drug of pregnancy, which can provide guidance and reference for the diagnosis and treatment.

Keywords: SPAM (Sequential Pattern Mining) · Sequential pattern mining · Medicare data

1 Introduction

In recent years, the social health insurance has been developed rapidly. The coverage area of social health insurance is larger and larger. It becomes the most important part of the Social Security System of China [1]. Because it's large and wide influence, the social medical insurance is important in our daily life even for the whole country [2–4]. However, the development of domestic insurance industry just gets started. The most of study about risk prevention and control have been stay on the basic statistical methods. Thanks to the development of information technology, the health insurance industry has accumulated vast amounts of data, which contains the disease diagnosis information, treatment information and so on. We could find the valuable information in health care data by means of the technology of data mining, which would be good for the implementation of health work and hospital management. Currently, many domestic and foreign scholars are studying on how to use data mining technology to solve problems of health insurance, such as medical cost projections [5], health insurance fraud detection [6], the medical insurance fund management and others [7].

© Springer Nature Singapore Pte Ltd. 2017
H. Yuan et al. (Eds.): GRMSE 2016, Part I, CCIS 698, pp. 100–108, 2017.
DOI: 10.1007/978-981-10-3966-9_11

This paper attempts to use sequence mode to carry out the Medicare data analysis, and we expect to discover the new knowledge, such as the use of normative issues. Sequential pattern mining is an important branch of data mining [8–10], which has been applied to business intelligence, customer relationship management, electronics access, Web mining and so on. It determines whether the sequence can constitute a sequence mode based on the occurrence rate of sequence or not. The practical application found by sequential pattern mining is also very extensive [11–14]. For example, the retail of mail order business can use this method to cross-Retail, we could find that the sequence mode of events that may lead to a default in the credit card payment transaction. The contributions of the paper can be summarized as follows:

1. Data mining analysis based on the combination of Medicare data and SPAM algorithm.
2. Find the regularity of visiting doctor, medical treatment characteristics and drug-usage mode about insured person in a certain period of time.

The organization of the paper is presented as follows: the Sect. 1 is introduces, the Sect. 2 is sequential pattern SPAM algorithm, the Sect. 3 is experiment and analysis, and the Sect. 4 is conclusions.

2 Sequential Pattern SPAM Algorithm

The common sequential pattern mining algorithms scans the original database repeatedly. However, these would take a lot of time, and even reduces the execution efficiency of algorithm. In addition, the frequent calculation of support has great influence on the efficiency of the algorithm [15, 16]. Ayres puts forward the sequential pattern discovery algorithm based on bitmap (SPAM) [17]. In this section, discusses the ideas and characteristics of SPAM algorithm and the advantages of high efficiency. Subsection 2.1 is the idea of SPAM algorithm, Subsect. 2.2 is advantages of SPAM algorithm.

2.1 The Idea of SPAM Algorithm

The idea of SPAM algorithm is to use the bitmap to represent the database. According to the Depth First Search Strategy, each sequence in the sequence tree is generated in accordance with the expansion of itemsets or sequences in dictionary order. The sequence expansion is that the entry of new transaction is longitudinally added to the original sequence tree. The itemsets extension means that the entry of new transaction is horizontally added to the node of the original sequence tree. After each extension, the length of the sequence adds 1. Then we can get the candidate sequences. The algorithm can calculate the support level efficiently through the calculation of the bitmap, until all frequent sequences. The sequential tree of SPAM algorithm is shown in Fig. 1.

In the first time, SPAM algorithm scans the database [18, 19]. It will build a vertical bitmap for each item in the database. The bitmap would be sorted according to the Cid

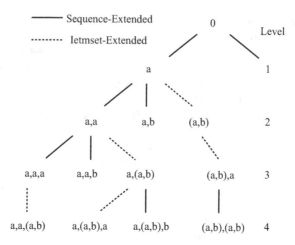

Fig. 1. The sequential tree of SPAM algorithm.

of the patients. In addition, if there is an item in sub-sequence, its corresponding bit is set to 1, otherwise set 0. The definition of sequence size is that the number of items contained in a sequence, that is, the number of transactions with the same user. If the size of sequence in the range of $2^k + 1$ and 2^{k+1}, the corresponding bitmap is identified with the 2^{k+1} bits in the database, meanwhile the minimum value of K is 1. In other words, regardless of the length of the sequence is less than or equal to 4, the bitmap would take up 4 bits. The sample database is shown in Table 1.

Table 1. Sample databases.

Cid	Tid	Item
C001	T001	a(ad)(bc)c
C001	T002	a(ac)c
C001	T003	a(bc)b
C002	T001	a(cb)
C002	T002	a(ac), d

In Table 1, the Cid of example database represents the ID of medical staff, the Tid represents medical records, the Item represents the sequence of drug. We process the Table 1 via the SPAM algorithm, and the result would be shown with the bitmap notation (see Table 2). There are 4 items in the C001 of T001, so it takes up 4 bits. On the other hand, the T002 has 3 itemsets, which means it needs 4 bits. The others are the same.

Table 2. The bitmap data

Cid	Tid	Item			
		a	b	c	d
C001	T001	1	0	0	0
		1	0	0	1
		0	1	1	0
		0	0	0	1
	T002	1	0	0	0
		1	0	1	0
		0	0	1	0
		0	0	0	0
	T003	1	0	0	0
		0	1	1	0
		0	1	0	0
		0	0	0	0
C002	T001	1	0	0	0
		0	1	1	0
		0	0	0	0
		0	0	0	0
	T002	1	0	0	0
		1	0	1	0
		0	0	0	1
		0	0	0	0

2.2 Advantages of SPAM Algorithm

The SPAM algorithm is to use a bitmap represents of the database. The database is only scanned once in the implementation process. In this way, the time of operation is greatly reduced and the efficiency of the algorithm is improved. At the same time, the bitmap method can reduce the occupied memory. We can get the candidate sequences through the extension of Sequences and Item sets. Then according to the principle of Apriori algorithm, we can prune the frequent patterns in a timely manner. What's more, we use the bitmap represent of the database, thus the calculation support is effective. In the dense data set, the advantage of SPAM algorithm is more obvious, and it can be compressed to the data in the first place. Compared with the other two algorithm (AprioriAll algorithm, GSP algorithm), the performance of SPAM algorithm is better.

3 Experiment and Analysis

In this section, we use SPAM algorithm to deal with health care data, and try to find the regularity of visiting doctor, medical treatment characteristics and drug-use mode about insured person in a certain period of time. The Subsect. 3.1 is the treating processes of Medicare data, the Subsect. 3.2 is the Experimental operating results, the Subsect. 3.3 is the Experimental result analysis.

3.1 Medicare Data Processing

Compared with the data stored in hospitals, the health care data has the characteristics of large amount, wide span and long-time accumulation. In our study, we choose the health care data of pregnant woman to be the experimental data with 2 reasons. The first one is that the number of pregnant women are large, and the second one is that the pregnant women would produce a large number of drug bills in the ten months before and after. These bills have obvious characteristics of the sequence. Thus, we use the deliver data as the experimental data, and realize the sequence pattern analysis. In the health care data, we query out the attendance records of the 10 months before pregnant women give birth to child and three months after that.

According to the needs, we choose the ID of pregnant woman that the disease coding is normal within a certain period of time. According to the ID of medical personnel, we can integrate the outpatient information table and medical records in hospital information table into a table. Because each table is composed of medical records of medical personnel, such as cold or typhoid fever during pregnancy test conditions shown in Fig. 2. Each clinic record contains a number of medical bills, each bill contains more than one drug name. We will work up the data into a data set to be analyzed. The data mainly include the ID of medical personnel, medical bills, Drug name and so on. Then, we convert the data to the data format that can be identified by the program of SPAM algorithm. Thus, the SPAM algorithm could be used to deal with the data. At last the data processing would generate the sequential patterns set. Thereafter, we can find the user's medical behave and the doctor diagnosis model or rule. For example, in addition to the normal pregnancy test, the pregnant women may also check cold, typhoid fever and other symptoms that caused by weather condition. We would not sure that whether there were some phenomenon that the drugs were not suitable for pregnant women to use, even the collocation of drugs was not standard when the pregnant women went to the clinic or large hospital for medical treatment.

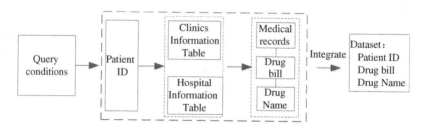

Fig. 2. Medicare data mining structure diagram.

3.2 Experimental Operating Results

The number of patients is 3000 and the sample data is 134272. The minimum support of the experimental set 30%. The numerical results of running SPAM algorithm program are shown in Table 3. The "−1" denotes the end of each items set in the sequence, and the last "−" is the number of patients that had the sequence. For example: $(85, 290, -1, 547, -183)$ on behalf of the sequence $<(85, 290), 547>$ and the number of patients is about 183.

Table 3. SPAM algorithms program results

i_1	i_2	i_3	i_4	i_5
85	−1339			
85	−1	547	−367	
85	547	−939		
85	290	−1	251	−92
85	290	−1	547	−183
85	251	−571		
251	−1	493	−129	
251	−1	85	−233	
251	−1	547	−339	
251	547	−903		
290	547	−1	745	−87
290	547	−973		
290	730	−498		
493	547	−199		

The drug name corresponding to the program running results is shown in Table 4.

Table 4. Chinese sequences of mode.

Chinese sequences of mode	
<(Vitamin C)>	<(Vitamin C, Vitamin K1)>
<Vitamin C, Glucose>	<Vitamin K1, Vitamin B6>
<(Vitamin C, Glucose)>	<Vitamin K1, Vitamin C>
<Vitamin K1, Glucose>	<(Metronidazole, Glucose)>
<(Vitamin K1, Glucose)>	<Vitamin B6, Glucose>
<(Vitamin C, Metronidazole), Vitamin K1>	<Metronidazole, Cefotaxime sodium>
<(Vitamin C, Metronidazole), Vitamin K1>	<(Metronidazole, Glucose), Sodium chloride>

3.3 Experimental Result Analysis

In this paper, the SPAM algorithm is used to mine the data, and the model set of the results is used to analyze the drug data. On the one hand, we try to find the user's medical behave and the doctor diagnosis model from the experimental results. On the other hand, we could find phenomenon that the collocation of drugs was not standard. According to the different support degrees, we analyze the experimental results in two aspects:

(a) Sequential pattern mining analysis

The main sequence of frequent items found by sequential pattern mining is shown in Table 5. The results of the experiment found that about 70% of patients use glucose, more than 60% of the patients replenishes a variety of vitamins during pregnancy in

Table 5. Frequent partial sequence

Frequent partial sequence	
<(Vitamin C)>	<Vitamin K1, Vitamin B6>
<(Glucose)>	<Vitamin K1, Vitamin C>
<Vitamin B6>	<(Vitamin K1, Glucose)>
<(Vitamin C, Vitamin K1)>	<Vitamin B6, Glucose>
<(Vitamin C, Glucose), Vitamin B6>	<Vitamin C, Metronidazole, Glucose>
<(Vitamin C, Vitamin K1, Vitamin B6>	<Metronidazole, Cefotaxime sodium>
<(Vitamin C, Metronidazole), Vitamin K1>	<(Metronidazole, Glucose), Sodium chloride>

order to prevent the occurrence of neonatal birth due to the lack of a vitamin disorder. The number that uses vitamin C is 45%, vitamin B6 is 12%, fat-soluble multivitamins is 11%, and vitamin C and vitamin K1 used together is about 20%. These should be combined the baby's health with the condition of pregnancy to carry out the further study. It was found that 56% patients have used metronidazole and cefotaxime sodium and other anti-infective drugs. This phenomenon is divided into before and after pregnancy and it may also be used for clinical application. The specific conclusion needs to be confirmed in the further.

According to the sequential pattern mining analysis, we found that the doctors have certain characteristics and laws when they use drugs. During the 3 months after pregnancy and pregnancy, the disease code "other neonatal disease" found that vitamin also occupies an important proportion in which the fat-soluble multivitamin accounted for almost 70%. And drug collocation rate also has certain arrangement, so we can get certain diagnosis and treatment mode. We select the drugs data of pregnant women to be the experimental data, and relatively the size of data is small. But some of the regular pattern of medication in pregnancy has a certain guiding significance for some implicit chronic diseases. It is a very complex process to establish the model of diagnosis and treatment. It needs to monitor and record the entire treatment process, and accumulates to a certain extent before the comprehensive analysis. Then we can try to find the basic law model, and establish the relevant path of the content. We also should improve and revise the model of diagnosis and treatment in the late applications.

(b) Problems that may exist

In order to see if there is an irrational phenomenon of drug, the support of experiment is set 3%. The sequence mining has great advantages in data mining. It can found that there were drugs with unreasonable phenomenon. According to the results of the analysis, we can found that there are some unreasonable drug collocations sequences shown in Table 6. The drug used manual suggested that the lidocaine hydrochloride and Mannitol can't be used together. The Furosemide and Mannitol simultaneous used together can lead to kidney failure, and it will also strengthen the role of curare. As to, It is still unknown whether the doctor reminded that the drugs can't take together. Through this analysis, we found that the health clinic in outpatient pharmacy, there are still some problems. On the one hand, it might be caused by unskilled skill of doctors, or the doctor did not understand the drug. On the other hand,

Table 6. The sequence of unreasonable drugs collocation.

The Sequence of unreasonable drugs collocation
<Mannitol, Furosemide>
<Mannitol, Vitamin C, Cimetidine>
<Lidocaine hydrochloride, Furosemide>
<Lidocaine hydrochloride, Glucose, Mannitol>
<Mannitol, Furosemide, Lidocaine hydrochloride>

the pregnant women may forget that she should warn the doctors she is a pregnant woman. But these things still need to draw attention to the relevant personnel, in order to avoid the occurrence of such events.

Not only that, we can also establish the different disease diagnosis and treatment modes according to the information dug out. The content of data mining is still a set of data. The data includes the rules and patterns that is the foundation of treatment mode. But how these data were screened to identify and establish an effective clinical pathway is still a very important problem.

4   Conclusions

Through SPAM algorithm on mining health care data, we can find that the model and rule about the behavior of users and doctors, which provide the reference for future visits behavior. According to the sequence pattern analysis, under certain conditions, there may be not standardized characteristics of the drug usage and it needs to strengthen the supervision of drugs. For example, if a doctor chooses the banned drugs for pregnant in unaware condition, the system can automatically prompt the drug as a "pregnant women" drugs. In this way, it can not only improve the standard of administration for drugs, but also it can strengthen the protection of pregnant women and fetus. And later, by combining the baby's health, we can carry out the association analysis of pregnancy and after the pregnancy. It has a certain guiding significance for the health of baby and pregnant woman. At the same time, how these data were screened to identify and establish an effective clinical pathway is also a very important problem.

Acknowledgments. This work was supported partly by the Jinan youth science and technology star plan (No. 20120104), National Natural Science Foundation of China (71271125) and Scientific Research Development Plan Project of Shandong Provincial Education Department (J12LN10).

References

1. Feng, L.: Review of data mining application in medical insurance in our country. Comput. Knowl. Technol. **21**, 880–882 (2014)
2. Du, Y.: Studying on the design of overall basic medical insurance system. Chin. Health Serv. Manag. **32**, 90–92 (2015)

3. Zhao, Y.: Mechanism of prepaid method of medical insurance in controlling medical expense. Chin. Hosp. Manag. **35**, 45–47 (2015)
4. Lei, Y., Li, M., Hu, W., Song, G.: Efficient methods for rare sequential pattern mining. J. Frontiers Comput. Sci. Technol. **9**, 429–437 (2015)
5. Xu, Y.: Data preparation for risk control of medical insurance fund. Fudan University, Jilin (2010)
6. Liu, S., Yang, H.: Study on medical insurance fraud detection system based on multi-agent. Comput. Technol. Dev. **23**, 171–174 (2013)
7. Qu, G., Cui, S., Tang, J.: Active and passive factors analysis of the medical insurance fund expenditure–taking Dalian as an example. vol. 19, pp. 120–125 (2014)
8. Agrawal, R., Srikant, R.: Mining sequential patterns. In: Proceedings of the 11th International Conference on Data Engineering. IEEE Computer Society, Los Alamitos (1995)
9. Tao, H., Jiang, F.: Application of improved sequential pattern mining in hospital referral. Comput. Syst. Appl. **10**, 253–258 (2015)
10. Xu, T.T., Dong, X.J.: E-MsNFIS: efficient negative frequent itemsets mining based on multiple minimum supports. Appl. Mech. Mater. **411–414**, 386–389 (2013)
11. Cao, L., Dong, X., Zheng, Z.: e-NSP: efficient negative sequential pattern mining. Artif. Intell. **235**, 156–182 (2016)
12. Xiangjun, D.: Mining infrequent itemsets based on extended MMS Model. Commun. Comput. Inf. Sci. **2**, 190–198 (2007)
13. Imran, M.: Authentication user's privacy: an integrating location privacy protection algorithm for secure moving objects in location based services. Wirel. Pers. Commun. **82**, 1585–1600 (2015)
14. Memon, I., Chen, L., Majid, A., Lv, M., Hussain, I., Chen, G.: Travel recommendation using geo-tagged photos in social media for tourist. Wirel. Pers. Commun. **80**, 1347–1362 (2015)
15. Srikant, R., Agrawal, R.: Mining sequential patterns: generalizations and performance improvements. In: Proceedings of the 5th International Conference Extending Database Technology, EDBT 1996, London, UK (1996)
16. Cheng, S., Ma, C., Li, C.: High utility sequential pattern mining algorithm based on MapReduce. Comput. Syst. Appl. **12**, 228–232 (2015)
17. Ayres, J., Flannick, J., Gehrke, J., Yiu, T.: Sequential pattern mining using a bitmap representation. In: Proceeding of the 8th ACM SIGKDD International Conference on Knowledge Discovery and Data Mining, KDD 2002, New York, USA, pp. 429–435 (2002)
18. Chen, J., Weng, Z.: An algorithm of SPAM based on projected database. Comput. Knowl. Technol. **6**, 1537–1539 (2010)
19. Zhang, W., Liu, F., Teng, S.: Improved PrefixSpan algorithm and its application in sequential pattern mining. J. Guangdong Univ. Technol. **30**, 49–54 (2013)

A Genetic-Algorithm-Based Optimized AODV Routing Protocol

Hua Yang[✉] and Zhiyong Liu

Guilin University of Aerospace Technology, Guilin 541004, China
{gl-yh, lzy}@guat.edu.cn

Abstract. The Ad hoc On-demand Distance Vector (AODV) routing protocol is a very important distance vector routing protocol in Mobile Ad hoc Networks (MANET). Due to the mobility of MANET, the performance of routing protocols in many scenarios is not ideal. Based on the consideration of the performance of intermediate nodes, this paper uses genetic algorithm to optimize the routing to find a more suitable route to improve the network performance. The simulation results show that GA-AODV has a significant improvement over AODV in average delay, packet received rate, and routing recovery frequency.

Keywords: AODV · MANET · Genetic algorithm

1 Introduction

For Mobile Ad hoc Networks (MANET), routing problem has been a hot research issue [1–3]. The routing protocol of mobile ad hoc networks requires that a suitable path from the sender to the receiver be found for the data packet in the dynamic network environment, and the path should be as free as possible from the network topology. Therefore, the routing protocol for ordinary networks can't be directly applied to mobile ad hoc networks. As the mobile ad hoc network itself has a dynamic network topology, mobile network nodes, sensitive transmission media and less infrastructure, and so on the routing protocol put forward more requirements.

According to the basic principle of routing protocols, MANET routing protocol can be divided into a proactive routing protocol, reactive routing protocol and hybrid routing protocol. Commonly used routing protocols are AODV [4–8], OLSR [9], TBRPF [10], DSDV [11], DSR [12], TORA [13] and so on. Because it combines the advantages of the DSDV and DSR routing protocols, the AODV routing protocol exhibits good routing performance, making it one of the few routing protocols through the IETF to become RFC documents.

2 AODV Routing Protocol

Ad hoc On-demand Distance Vector Routing (AODV) is an on-demand routing protocol [4–8]. When source node needs to send information to destination nodes in the network, if there is no route to the destination node, it must send the RREQ message in the form of multicast. The RREQ packet records the network layer address of the

H. Yuan et al. (Eds.): GRMSE 2016, Part I, CCIS 698, pp. 109–117, 2017.
DOI: 10.1007/978-981-10-3966-9_12

originating node and the destination node. The neighbor node receives the RREQ and first determines whether the destination node is itself. If yes, the router sends a RREP to the originating node. If not, it searches the routing table for a route to the destination node. If so, it unicasts RREP to the source node. Otherwise, it continues forwarding RREQs.

In MANET, due to the particularity of the network node movement and the performance of the node itself, routing may generate an interrupt at any time. In AODV, node performance and node mobility are not taken into account. Therefore, it may be necessary to frequently start a route lookup procedure in high dynamic network topology, which may cause harm to the whole network. Genetic algorithm is introduced into AODV to find a superior AODV route, which will have a positive effect on improving network usability and survivability.

Paper [5–8] optimizes the AODV from the aspects of route discovery, maintenance, energy saving control and QOS, and obtains a series of results. But it is worth noting that optimization often has some limitations, how to optimize the AODV, is still a hot topic of current research. Paper attempts to optimize the AODV using Genetic Algorithms to achieve an AODV with better overall performance.

3   Genetic-Algorithm-Based AODV

In MANET, a network can be represented as an undirected graph $G(V, E)$, where V is the node set, E is the link set, N_u is the neighbor set, the link (u, x) exists if and only if $x \in N_u$, $D(u, x)$ represents the node u The ability to send data at the link (u, x), $R(u, x)$ represents the ability of the node u to receive data at the link (u, x). According to the ability of nodes to send and receive data, our optimal path should maximize network throughput, so the objective function can be expressed as:

$$\max(E) = \max_{u \in V} \left(\sum_{x \in N_u} R(x, u).g(x, u) + \sum_{x \in N_u} D(u, x).t(u, x) \right) \tag{1}$$

The constraint is $\dfrac{d(\sum\limits_{x \in N_u} R(x, u).g(x, u))}{d_t} \leq F(u), \forall u \in V, N_u \subset V.$

Genetic algorithm (GA) is the first proposed by John Holland [14, 15], which is a stochastic search algorithm based on Darwin's theory of evolution and Mendelian genetic variation theory. It has been widely used in the field of biology. Combinatorial optimization, machine learning, signal processing, adaptive control and artificial life, and achieved a lot of remarkable results. The operation of the genetic algorithm is mainly composed of selection, crossover and mutation.

3.1   Routing Encoding

In solving the AODV optimal routing problem, the binary or real coding in the traditional genetic algorithm can't reflect the nodes passing through the routing well, and the number of nodes in each route is not fixed, so consider using one kind of Variable length encoding. First of all, the nodes in the network ID sequence number, in which the genes in the chromosome node ID to represent the source node to the destination node to form a series of ID routing chromosome. For any of these chromosomes, the start of the gene fragment is the source node, and the end of the gene fragment is the destination node. Because there are usually multiple routes from the source node to the destination node, the number of intermediate nodes in each route is not fixed, so the chromosome length in the algorithm is not fixed either. As shown in Fig. 1, a first-generation chromosome containing a 6-node MANET representation can be expressed as:

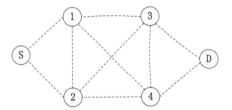

Fig. 1. A MANET with 6 nodes

$$H_1(t) = \{S, 1, 3, D\} \qquad H_2(t) = \{S, 1, 4, D\}$$
$$H_3(t) = \{S, 1, 3, 4, D\} \qquad H_4(t) = \{S, 1, 4, 3, D\}$$
$$H_5(t) = \{S, 1, 2, 3, D\} \qquad H_6(t) = \{S, 1, 2, 4, D\}$$
$$H_7(t) = \{S, 1, 2, 3, 4, D\} \qquad H_8(t) = \{S, 1, 2, 4, 3, D\}$$
$$H_9(t) = \{S, 1, 3, 2, 4, D\} \qquad H_{10}(t) = \{S, 1, 4, 2, 3, D\}$$
$$H_{11}(t) = \{S, 2, 3, D\} \qquad H_{12}(t) = \{S, 2, 4, D\}$$
$$H_{13}(t) = \{S, 21, 3, D\} \qquad H_{14}(t) = \{S, 2, 1, 4, D\}$$
$$H_{15}(t) = \{S, 2, 3, 4, D\} \qquad H_{16}(t) = \{S, 2, 4, 3, D\}$$
$$H_{17}(t) = \{S, 2, 1, 3, 4, D\} \qquad H_{18}(t) = \{S, 2, 1, 4, 3, D\}$$
$$H_{19}(t) = \{S, 2, 3, 1, 4, D\} \qquad H_{20}(t) = \{S, 2, 4, 1, 3, D\}$$

3.2   Design of Fitness Function

Fitness function is used to measure the merits of the individual scale. Genetic algorithm in evolutionary search, only to fitness function as the basis to determine the merits of the individual, so the merits of the function itself directly affects the feasibility of genetic algorithms and convergence speed. The performance of the routing algorithm is mainly measured by network delay and network lifetime. The network delay can be adjusted by node performance. Therefore, to extend the network lifetime, only by optimizing the route to reduce network energy consumption can be achieved, the less energy, the higher the fitness value, so the fitness function can be defined as:

$$f(u) = \frac{1}{\sum_{x \in V} R(x, u).g(x, u) + \sum_{x \in V} D(u, x).t(u, x)} \tag{2}$$

From this function we can see that the ultimate goal of optimizing routing is to maximize the throughput of the network, while the entire network of the least energy.

3.3 Selection

The selection operation is to select the probability of the next generation by the probability of individual fitness. Design selection probability formula is:

$$P(i) = \frac{f_i}{\sum_{i=1}^{n} f_i} \tag{3}$$

In the formula (3), f_i is the fitness value of the $f_i * P(i)$ individual, so the individuals copied directly to the next generation are $f_i * P(i)$. It can be seen from Eq. (3) that there is a direct relationship between the selection probability and the fitness value of the individual, and the probability of selection is larger for individuals with larger fitness, and the probability of the next generation appears in the population. The selection operation is divided into two steps:

(A) First calculate the fitness value of the current individual, in accordance with the size of the order;
(B) According to type (3) to calculate the probability of selection of each individual, according to the corresponding probability of selection to copy to The next generation.

3.4 Cross

Cross-operation refers to the process of reorganizing and replacing part of two parents to generate new individuals. Crossover can cross chromosomes between individuals, and cross-fertilize chromosomes. Because the encoding method of the present invention adopts variable length coding, the single point crossover in the traditional genetic algorithm does not apply here. In the choice of network path, the intersection is to choose the different path between the nodes, so here take between the chromosomes of the gene fragment cross way. The specific operation can be carried out as follows:

(1) Select any two paths in the path between the source node and the destination node to see whether the two paths share the same node. If so, it indicates that there are common genes in the two paths. Cross-operation, or two paths can't cross.
(2) Starting from the first common passing node of the two paths and ending at the next common node, if there is only one common node between the two paths, the destination node is crossed as the next common node.

Suppose that for the two chromosomes of t generation: $C_i(t) = \{S,1,3,D\}$, $C_j(t) = \{S,1,2,3,4,D\}$, Node 1, 3 are the nodes through which path $C_i(t)$ and $C_j(t)$ pass together, after crossing operation, the next generation chromosome is: $C_i(t+1) = \{S,1,2,3,D\}$, $C_j(t+1) = \{S,1,3,4,D\}$.

3.5 Variation

The mutation operation is to ensure the diversity of the population and avoid the local optimization of the genetic algorithm. Mutation can make the individual changes in certain chromosomes, but also can be chromosomal gene changes. The variation of the routing between nodes can be understood as a replacement of the original path with another random path, that is to search for a new path, and thus can be seen as a change in the chromosome gene. Specific implementation process is as follows:

(1) Randomly select two nodes i, j as the starting point of variation;
(2) The genes from the source node S to the node i and from the node j to the destination node D remain unchanged and the node i to node j genes are randomly selected.

4 Simulation and Discussion

4.1 Performance Parameters

We mainly on the following parameters for simulation analysis and comparison.

(1) Packet received rate. The ratio of the number of data packets delivered to the destination node to the number of data packets generated by the CBR source node, or the received throughput of the destination node (in kb/s).
(2) The end-to-end average delay of data packets. Including the latency of buffering, interface queuing delay, MAC layer retransmission delay, airborne propagation delay and transition time during route lookup.
(3) Routing recovery frequency. The total number of route lookup processes initiated per second.

4.2 Experimental Parameter Setting

Use the NS3 emulation system [16]. The MAC has used the IEEE 802.11 DCF protocol. The transfer model uses Lucent's WaveLan. The moving model uses a random point model. Each node moves according to the randomly chosen destination node direction, after a round of the destination node, pauses for a period of time, then randomly selects the next destination node to move toward the new destination node. The maximum moving rate of nodes is variable. The nodes in the network are distributed in a 1000 m × 1000 m area with 80 nodes. Using CBR traffic, the data packet length is 1024 Byte, simulation run duration is 1 h. See Table 1 for details.

Table 1. Parameter setting

Parameters	Value
Simulation times	1 h
Simulation range	1000 m × 1000 m
Number of nodes	Fixed at 80
Motion model	WaveLan model
Node pause time	Random
Data transmision rate	1 Mbps
Wireless channel bandwidth	11 Mbps
Data channel packet size	1024 Bytes
Control channel packet size	128 Bytes
Transmission range	0–80
Max speeds	0–30 m/s

4.3   Simulation Results

Figures 2, 3 and 4 show the performance comparison of AODV, GA-AODV under the change of node moving speed, and the maximum moving speed of node changes in the range of 0–30 m/s. When the node moves at 0, the network is static. The average rate of link disruption is 0–50 times/s, and increases as the moving speed of the node increases. In the static network, AODV and GA-AODV performance is almost the same.

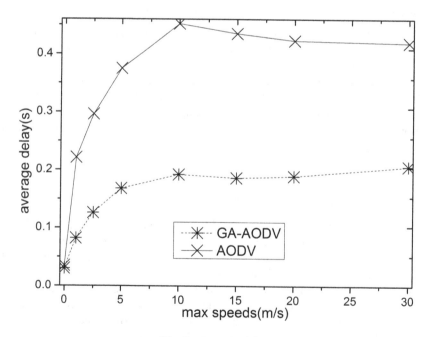

Fig. 2. Average delay

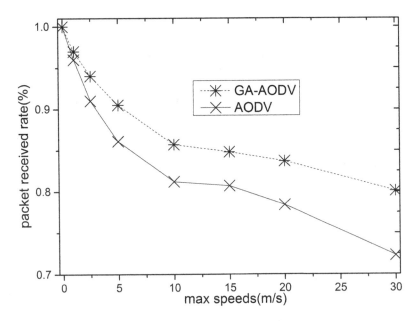

Fig. 3. Packet received rate

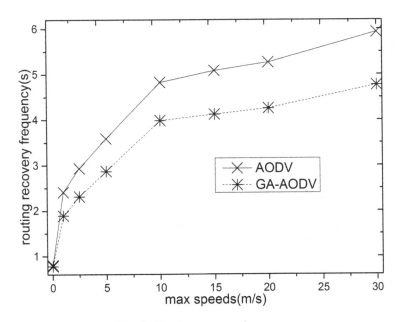

Fig. 4. Routing recovery frequency

The performance advantage of GA-AODV is more obvious with the increase of node's moving speed. Under the condition of increasing mobile speed, the packet loss rate of non-optimized AODV is 3−5% higher than GA-AODV.

As shown in Fig. 2, the GA-AODV significantly reduces the end-to-end average delay because the GA-AODV takes into account the various conditions between the nodes, and the probability of routing outages is significantly smaller than the unoptimized AODV. The end-to-end average delay of AODV and GA-AODV increases with the increase of node moving speed. When the moving speed is more than 10 m/s, the loss of intermediate nodes due to link disruption is mainly caused by the Effect of Routing Transport Packets. That is, the average number of hops of the delivered packet drops rapidly. Thus, when the packets delivered at high speed are almost those packets that are delivered along the shorter path, the end-to-end average delay becomes insensitive to mobility enhancement.

5 Conclusions

As an important routing protocol in MANET, how to further improve the performance of routing protocol and improve the performance of MANET has been a hot topic in MANET. Based on the performance of the intermediate nodes, the genetic algorithm is used to search for a better route to improve the availability of the AODV routing protocol. Simulation results show that GA-AODV is superior to AODV in average network delay, packet acceptance rate, and route rediscovery, and the simulation results show that GA-AODV is better than AODV.

References

1. Xin, M.Z., et al.: Interference-based topology control algorithm for delay-constrained mobile ad hoc networks, in mobile computing. IEEE Trans. **14**(4), 742–754 (2015)
2. Loo, J., Jaime, L.M., Jesús, H.O. (eds.): Mobile Ad Hoc Networks: Current Status and Future Trends. CRC Press, Boca Raton (2016)
3. Pathan, A.S.K. (ed.): Security of Self-Organizing Networks: MANET, WSN, WMN, VANET. CRC Press, Boca Raton (2016)
4. Charles, P., Belding-Royer, E., Das, S.: Ad hoc on-demand distance vector (AODV) routing. No. RFC 3561 (2003)
5. Kazuhiro, Y., et al.: Performance analysis of routing methods based on OLSR and AODV with traffic load balancing and QoS for Wi-Fi mesh network. In: International Conference on Information Networking (ICOIN) 2016. IEEE (2016)
6. Fehnker, A., et al.: Modelling and analysis of AODV in UPPAAL (2015). arXiv Preprint arXiv:1512.07312
7. Tyagi, S., Som, S., Rana, Q.P.: A reliability based variant of AODV in MANETs: proposal, analysis and comparison. Proc. Comput. Sci. **79**, 903–911 (2016)
8. Wang, T., Qiu, R.H.: The AODV routing protocol performance analysis in cognitive ad hoc networks. In: Proceedings of the 2014 International Conference on Control Engineering and Information Systems (ICCEIS 2014, Yueyang, Hunan, China, 20–22 June 2014). CRC Press (2015)
9. Clausen, T., Jacquet, P.: Optimized link state routing protocol (OLSR). No. RFC 3626 (2003)

10. Ogier, R., Templin, F., Lewis, M.: Topology dissemination based on reverse-path forwarding (TBRPF). No. RFC 3684 (2004)
11. Perkins, C.E., Pravin, B.: Highly dynamic destination-sequenced distance-vector routing (DSDV) for mobile computers. In: ACM SIGCOMM Computer Communication Review. vol. 24, no. 4. ACM (1994)
12. Johnson, D.B., Maltz, D.A., Broch, J.: DSR: the dynamic source routing protocol for multi-hop wireless ad hoc networks. Ad Hoc Netw. **5**, 139–172 (2001)
13. Broch, J., et al.: A performance comparison of multi-hop wireless ad hoc network routing protocols. In: Proceedings of the 4th Annual ACM/IEEE International Conference on Mobile Computing and Networking. ACM (1998)
14. Ghamisi, P., Benediktsson, J.A.: Feature selection based on hybridization of genetic algorithm and particle swarm optimization. IEEE Geosci. Remote Sens. Lett. **12**(2), 309–313 (2015)
15. Gao, Q., He, N.-b.: Study on fuzzy classifier based on genetic algorithm optimization. In: Huang, B., Yao, Y. (eds.) Proceedings of the 5th International Conference on Electrical Engineering and Automatic Control. LNEE, vol. 367, pp. 725–731. Springer, Heidelberg (2016). doi:10.1007/978-3-662-48768-6_81
16. http://www.nsnam.org

Performance Analysis of PaaS Cloud Resources Management Model Based on LXC

Xuefei Li and Jing Jiang[✉]

College of Computer Science and Technology, Qingdao University,
Qingdao 266071, China
1028425792@qq.com, jj@qdu.edu.cn

Abstract. LXC is an OS-level virtualization technology supported by the Linux kernel. It can provide a lightweight virtualization technology support for PaaS cloud platform, in order to reach the goals that not only the different tenants are isolated but also the software as well as hardware system resources are shared. On the basis of analyzing the requirements of PaaS, a cloud resources management model of PaaS had been created by this paper based on the Cgroups mechanism of LXC. It conducted performance tests in terms of memory, CPU, disk and network transfer speed, etc., which was respectively to deploy various applications in the LXC-based and KVM-based PaaS. And then these performance results were analyzed and compared. The experimental results show that compared with KVM, the performance advantage of LXC is obvious, and it is very fit to be deployed in PaaS cloud platform of providing the high performance computing to ensure the computing high performance and high availability of PaaS cloud platform.

Keywords: PaaS · Virtualization · Resources management · LXC · KVM

1 Introduction

Cloud computing is a kind of computing paradigm to offer users high throughput services by a large Data Center or Cluster Server. This calculation model can permit users to access the shared resources anytime and anywhere after they have played for them through the interconnection equipments. PaaS cloud mainly focuses on the applications to be hosted and its online usings. And to host applications for a PaaS, the following requirements need to be satisfied: ① isolation of applications on the level of compute, storage and network; ② elasticity, namely better scalability; ③ multiple languages and runtimes support; ④ sharing CPU fairly; ⑤ management of applications lifecycle; ⑥ persistence and migration of applications; ⑦ lower costs. In order to meet the above requirements, the PaaS cloud computing system needs to bring in a new layer namely a middleware, which is a virtualization layer.

Through the virtualization layer, PaaS cloud platform can virtualize a physical computing platform to be several virtual computing platforms. The virtual computing platform has hidden the real software and hardware resources of computer for users, thus the users can use the virtual computing platform just like the physical computing platform. Now there are two kinds of virtualization technologies, one is the traditional Virtual Machine (VM) technology, such as KVM, XEN and VMware, etc. [1]. This

© Springer Nature Singapore Pte Ltd. 2017
H. Yuan et al. (Eds.): GRMSE 2016, Part I, CCIS 698, pp. 118–130, 2017.
DOI: 10.1007/978-981-10-3966-9_13

kind of VM technology is an abstract computing platform of the hardware resources that can run the Guest OS independently, and here the installed Guest OS can be different from the Host OS, as shown in Fig. 1. But the cost of running this kind of VM is much higher, and perhaps thousands of VMs need to be initialized simultaneously in an environment of cloud computing. However, there in fact are a large number of repetitive contents in the image of each VM, so this kind of VM technology's performance is lower. The other kind of virtualization technology is the OS-level virtualization, its idea is to insert a virtualization layer into the OS kernel to partition the physical resources of the machine. So it permits multiple isolated VMs to run simultaneously within a single OS kernel, and this kind of VM is usually called Virtual execution Environment (VE), or container for short [2]. Figure 2 shows the Linux Container (LXC)-based architecture of PaaS platform. In the figure, the container and the physical machine are isolated from each other. And it also can be considered as a complete OS duplicate generated according to users' requirements for the configuration parameters in the container, which is on top of the Host OS namely VE. Compared with

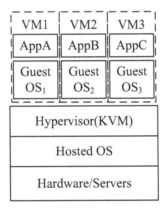

Fig. 1. Architecture of PaaS based on VM

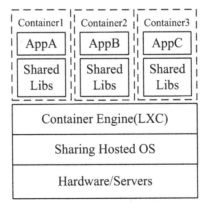

Fig. 2. Architecture of PaaS based on LXC

the virtualization technologies such as KVM and XEN etc., it can avoid the complex system costs that caused by the binary translation of OS request in the full virtualization. There aren't any differences between the container model and the Hypervisor model from a user's perspective. But from the perspective of the principle to realize isolation, they are very different.

The paper mainly aims to study on the cloud resources management mechanism of LXC-based PaaS platform. What's more, compared it with the PaaS platform based on the Hypervisor virtualization solution – KVM to conduct a performance test and the performance analysis about the cloud resources management model of LXC-based PaaS platform is given out.

This paper is organized as follows: In Sect. 2, the LXC-based virtualization PaaS cloud platform is studied and provided. Section 3 gives out the cloud resources management model of LXC-based PaaS. Section 4 is the performance test experiment, and the related experiment parameters and running conditions are set. In Sect. 5, the performance analysis and comparison are conducted for the testing results. Finally, summarizes for all works of the paper.

2 Build LXC-Based PaaS Cloud Platform

A LXC is a process of Linux user space. LXC is implemented through the mechanisms namely Cgroups and Namespaces of the Linux kernel that must be Version 2.6.24 or higher [3, 4]. And the PaaS cloud platform supported by it can meet the requirements of PaaS such as elasticity, low performance overheads, resources sharing, and so on. So it can be used to host the applications exclusively, and the loads to create a new VM for each application can be relieved. Which can meet the demands to maintain each layer of process, network and file system independent, meanwhile it also can reduce the costs of using the traditional VM to run an application of PaaS. But all the VEs of a single container must use the same OS is its disadvantage. For example, a Windows OS cannot run in a Linux container.

For creating a LXC instance, firstly it needs to check whether the current Linux kernel supports the mechanisms namely Cgroups and Namespaces by using the command "lxc-checkconfig". Then the related tools and softwares of LXC are installed. The LXC project provides LXC packages for users, including a variety of tools to manage containers, in order to create and manage containers as well as its hosting applications, and to offer the technical support of LXC-based development to users.

Starting a LXC is done by using the command "lxc-start". Firstly, lxc_start () is called to parse the command line parameters and configuration files, in addition to obtain some necessary information that needed by creating a container. Then lxc_init () is called to execute the initialization operation for the container. And later the lxc_cmd_init () function will be called by lxc_init () to create a socket to listen to so that to receive the external commands. Then the container's state will be set to starting, meanwhile it will be written into another channel file. At last, through the configuration file to set the environment variables of the container, in addition to allocate and create the tty and console device for the container. After initialization, the lxc_spawn () function will be called to execute, it is the core process of starting LXC. The parent process creates child process of the namespace that is related to PID, MNT etc. by calling clone () function. The clone () function executes do_start () function to send message to parent process after returning from child process, and to inform parent process to start to configurate. The parent process begins to create the group Cgroups and adds the created child process into this group. And then to conduct the corresponding configuration for the group Cgroups according to the configuration files. And at last the new process to execute sbin/init is permitted in the new namespace, the new process can execute the hosted applications by calling execvp (). In this way, a container has been started. Figure 3 is a flow chart about creating and using a LXC instance to host the applications.

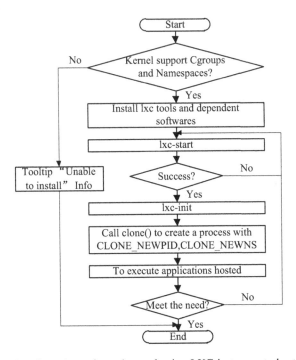

Fig. 3. Execution flow chart of creating and using LXC instances to host applications

3   Cloud Resources Management Model of LXC-Based PaaS

3.1   LXC-based PaaS Cloud Model

LXC is an OS-level kernel virtualization technology [5]. The mechanism Namespaces is a lightweight virtualization solution provided by the Linux OS kernel for the user space, namely to provide a variety of different Views for the process groups, which is referred to as a namespace. In other words, it permits multiple isolated VEs to run in a single OS kernel simultaneously. And from the user's perspective, all the processes running in an independent namespace feel that the software and hardware resources of the system they are using is exclusive. So a namespace is a LXC [6]. As shown in Fig. 4, here the mechanism Namespaces partitions the system resources into several parts, and the processes belonging to the same namespace share the resources of this namespace. The different containers are isolated from each other, in order to reach such a goal that the processes among different containers are noninterferential.

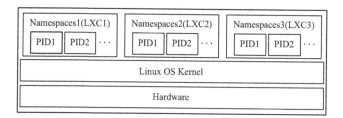

Fig. 4. Resources isolation of Namespaces

Using LXC just like using a real server, it has its own process, file system, user account and network interface with IP address, routing table and other personal settings. What's more, the settings of LXC can be individual customized according to the requirements of different users.

As a controller of the system resources, the container can make its process no longer be the main body of resources. In other words, the process is just the running body of system, while the container is the resource body of system. And after separating, a more accurate and more efficient control for the system resources will be achieved. There usually exists one process or more than one in the container, and the relation between the process with its derived thread and the resource body is not static but dynamically adjustable.

Figure 5 is a cloud resources management model of LXC-based PaaS built by the paper. It establishes many VEs on the basis of sharing a single OS image. And a VE is

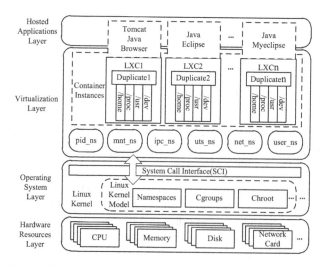

Fig. 5. Cloud Resources management model of LXC-based PaaS

namely a LXC instance, it is similar to the ordinary OS that can do operations such as startup, shutdown and restart. As soon as a VE has been etablished, the system resources, such as disk spaces, CPU usage, and memory space etc. Will be determinately allocated. But if necessary, it can be modified at the running time. Virtualization layer is an important part to implement the deployment architecture of PaaS. What can be seen from Fig. 5 is that the whole architecture of PaaS platform can include n containers. And for the users, every container instance system realized by creating the duplicates of OS is an independent host, it is a View that can be seen by its customer. The applications of every LXC run just like running in a non-container environment, from the users' view, its running environment is the same as the real environment.

3.2   Resources Management Mechanism of LXC Virtualization

The mechanism Cgroups has been designed by Linux kernel to realize the virtualization management for OS-level kernel resources. It aims to implement a unity architecture of the system resources management and control, and its basic unit about resources allocation is the process. Only if a process has been added into the specified instance of cgroups, Cgroups system will point to the address of resources control subsystem information by the domain subsys[] of its core data structure struct css_set {}, and these controllers including CPU controller, Memory controller and Device controller, and so on. A resources controller subsystem must be associated to a hierarchy of a cgroups file system for playing its role. And the hierarchy is implemented by the cgroup structure of the kernel codes. When the customer system in the container conducts a series of resources scheduling and allocation for the process, it will index to the structure css_set of this process and check the quotas setting by its corresponding resources controller. Then to make the right response. For example, when the memory quota requested by customers exceeds the limitation of its memory controller, a message will be sent by the system to warn the customer to make a right processing. Figure 6 provides the overall schematic of Cgroups system hierarchy.

There doesn't directly define the cgroup that is linked to the hierarchy in the task_struct of process. And the process is through the pointer of the member cgroup_subsys_state in struct css_set to access the corresponding resources controller. And css_set is defined as follows:

```
struct css_set{
    atomic_t refcount;          /* retain count */
    struct hlist_node hlist;    /* Hash index */
    struct list_head tasks;     /* process list*/
    struct list_head cg_links;  /* be indexed by the structure cgroup */
    struct cgroup_subsys_state *subsys[CGROUP_SUBSYS_COUNT];   /* pointer
array of the controller structure */
    ...
    }
```

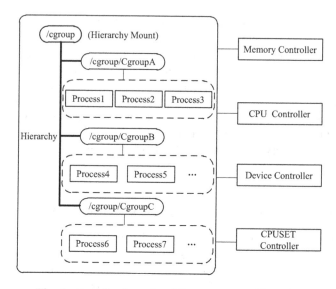

Fig. 6. Overall schematic of Cgroups system hierarchy

There defines the information that a process controls resources through resources controller in this structure. Of which the member cgroup_subsys_state *subsys is an array of pointer, it refers to every subsystem controller information that is associated with the process.

Cgroups is a control group with hierarchy, and every child node inherits the control properties of its parent. Only every subsystem has been attached to the corresponding hierarchy can it play its role It can be seen from Fig. 6 that once the subsystem has been attached to a hierarchy, all the control groups of this hierarchy will be in control of this subsystem [7].

A process can be migrated from one control group to another. The processes of the same group can use the resources allocated by the unit control group, meanwhile it is also under control of this control group. The control group is a tree structure, as Fig. 7. And here three processes, CgroupA, CgroupB and CgroupC are given. Each of the process groups can set different control properties.

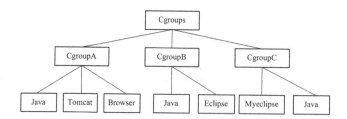

Fig. 7. Control groups of LXC

4   Experiment of Performance Test

The performance test, performance analysis and comparison, including CPU, memory, disk and network transfer speed etc., are conducted by experiment. It focuses on deploying several of applications on the PaaS cloud that respectively based on LXC and KVM.

4.1   Test Environment

The infrastructures used by experiment are shown in Table 1. Ubuntu14.04.1 LTS OS is installed at the host machine in the experiment, it is the Host OS to install LXC and KVM.

Table 1.   Infrastructure List

Physical machine	Processor	Memory	Operating system
Lenovo-G470	Intel(R) Core(TM)i3-2310 M CPU @ 2.10 GHz 2.10 GHz	4 GB	32 bits, Ubuntu14.04.1LTS

Our testing environments are as the following three circumstances:

(1)   The native machine to run Ubuntu14.04.1 LTS OS;
(2)   KVM-based PaaS platform, as shown in Fig. 1, and to install the Guest OS namely Ubuntu14.0.1LTS in KVM;
(3)   LXC-based PaaS platform, as shown in Fig. 2.

4.2   Test Tool

In this paper, the main performance metrics to be chosen are: memory and CPU, disk I/O and network I/O, meanwhile to adopt appropriate benchmarks tools to test them respectively. The descriptions in detail are as follows:

(1)   Memory performance: the performance of memory operation is tested by using the STREAM software [8];
(2)   CPU performance: the computing performance of CPU is tested by using the benchmark tool High Performance Linpack (HPL) [9];
(3)   Disk I/O performance: the benchmark tool IOzone [10] is chosen to make testing on the performance of disk reading and writing;
(4)   Network I/O performance: the Nether [11] is used to test the performance of network load.

In the experiment, 32 times testings have been executed respectively for all above metrics in the three environments, and then the highest and lowest values are excluded.

Then to calculate the average value of the rest 30 results. Based on the results of the native experiment, to compare and analyze the performance test results of LXC and KVM with them.

5 Testing Performance Analysis

In the paper, the results is in scientific notation, and the corresponding evaluation results are obtained by sorting. Here the standard error line is added to show the error degree of the statistical analysis.

5.1 Memory Performance Analysis

The memory throughput performance of PaaS platform plays a more and more important role in improving performance of the whole system. It not only can reduce the system efficiency but also can influence the multi-core and high frequency performances' advantage due to the deficient of memory throughput. Here the STREAM is chosen as it is easy to use for testing the memory throughput. It is realized mainly through four kinds of operations: Copy, Scale, Add and Triad. Among them, Copy refers to access to a memory cell to read its value, then to write it in another memory cell; Scale represents to read the value of a memory cell, then do a multiplication and write the result in another unit; Add refers to read two values from one memory cell, then to write the addition of the two values in another cell; Triad is the combination of above three operations, namely to read two values from one unit, then to do mixed operation about multiplication and addition, and to write the result in another unit.

The testing results of memory throughput about the native, LXC and KVM environments by STREAM are shown in Fig. 8. Note that the throughput is bigger the memory performance is better. What can be seen is that the memory performance of LXC nears to the native one, and the one of KVM is lower than that of LXC.

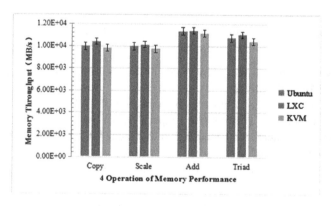

Fig. 8. Memory throughput of STREAM testing

5.2   CPU Performance Analysis

The CPU performance is the indicator to inspect the configured hardware computing performance, thus to use currently the most popular benchmark tool Linpack to calculate the floating peak, so that to realize the comparison of CPU performance between the above two kinds of virtualization technologies. Linpack can be classified into three patterns: Linpack100, Linpack 1000 and HPL. They mainly decide the division standard according to the handle scale namely array size N. There is no limit on the array size in HPL, and it is more suitable for the high performance computing system that develops rapidly in current days. So HPL is chosen to conduct an experiment, and its testing scale N = 10000. Its results are given in GFlops, as shown in Fig. 9.

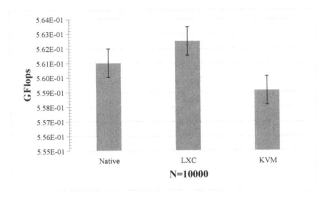

Fig. 9. CPU floating peak of HPL testing

What should be noted is that the relationship between CPU performance and the CFlops values is proportional. And from Fig. 9 it can be shown that LXC performed better than KVM in the aspect of CPU performance.

5.3   Disk I/O Performance Analysis

Disk I/O performance is also an important index to influence the computer performance. In order to analyze the input and output performance of the virtualization file system, respectively to run the IOzone tool in the above three environments. The advantage of IOzone is it can run in the different operation system. And it can make a disk performance test of many models by using IOzone, such as read, write, reread, rewrite, read-backwards, etc., but here only select write and read performances to test, record and analyze. The testing results of disk write and read performance are given in Figs. 10 and 11 respectively.

Choosing 2 Mbytes as the recording size in the experiment because such a testing result is better, and the bigger bandwidth value represents the better disk I/O performance. It can be seen from the analysis results that the write and read bandwidths of LXC are higher than that of KVM.

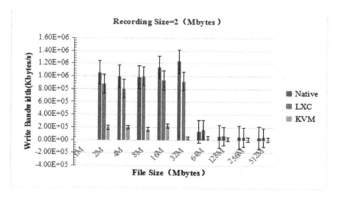

Fig. 10. Disk write bandwidth of IOzone testing

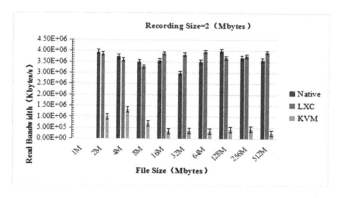

Fig. 11. Disk read bandwidth of IOzone testing

5.4   Network I/O Performance Analysis

The Netperf is used to test the network performance for the two kinds of virtualization technologies. Netperf realizes its function by several predefined testing procedures, it is the typical C/S mode. The main testings include the traffic patterns of TCP and UDP. Because TCP provides the end-to-end reliable transmission, we only conduct testing and analysis of the network I/O performance on TCP. And the concrete testing results are shown in Table 2.

Table 2. Network I/O performance of netperf testing

	Ubuntu	LXC	KVM
TCP_STREAM (Mb/s)	18577.8525	18482.9788	13611.9931
TCP_RR (times/s)	38018.9938	37830.3963	22853.2171
Network Latency (us)	26.8888	27.0038	44.7434

TCP_STREAM in Table 2 is a test for mass transfer rate. It can be seen that the throughput from the native host to the remote server is 18577.8525 Mb/s. Here the value is bigger and the performance is better. It's clear that the network transmission rate of LXC is closer to the native and is better than that of KVM. Another TCP_RR is also a common testing, it is a request/response type that is one end sends a message and another end to reply after receiving. It also can be seen that its performance of LXC is better. And the network latency refers to the average transmission delay of the system in the request/response mode. The smaller of its value represents its performance is better.

All in all, the advantage of LXC performance is obvious compared with KVM. And there is almost no performance overhead for the LXC-based virtualization technology. What's more, starting a LXC instance only needs a few seconds, and it is easy to run in the host and Data Center. LXC can carry more services with less disk space.

6   Conclusions

A cloud resources management model of LXC-based PaaS is presented on the basis of analyzing the PaaS cloud in the paper. And it mainly analyzes two mechanisms namely Cgroups and Namespaces that can be used by LXC to realize the allocation and isolation of its resources. The performance advantage to deploy applications in the LXC-based PaaS cloud platform has been obtained through the performance evaluation experiments about comparing LXC with KVM. Because LXC is an OS-level light-weight virtualization technology, its performance advantage is obvious compared with the virtualization technology – KVM, and it is more suitable for the PaaS cloud platform, which prefers to high performance computing. But the container-based virtualization asks that all containers must use the same OS. While the users' demands is various and ever-changing. All these bring the challenges to the OS-level virtualization technology. Furthermore, the security performance of LXC also needs to be improved currently. Whereas with the development of cloud computing, the LXC-besed virtualization cloud platform will get more and more widely applying. What needs to do next is the adaptive management mechanism research of the container-based PaaS cloud resources, and to provide a solution to ensure the security performance of LXC.

References

1. Bernstei, D.: Containers and cloud: from LXC to Docker to Kubernetes. IEEE Cloud Comput. 1(3), 81–84 (2014)
2. Dua, R., Raja, A.R., Kakadia, D.: Virtualization vs containerization to support PaaS. In: 2014 IEEE International Conference on Cloud Engineering (IC2E), pp. 610–614. IEEE Computer Society (2014)
3. Menage, P., et al.: CGROUPS. https://www.kernel.org/doc/Documentation/cgroup-v1/cgroups.txt. Accessed 20 Sept 2015
4. Linux Programmer's Manual: namespaces-overview of Linux namespaces. http://man7.org/linux/man-pages/man7/namespaces.7.html. Accessed 20 Sept 2015

5. Rosen, R.: Linux containers and the future cloud. Linux J. **240**(4), 86–95 (2014)
6. Yu, C.: Research on the live migration mechanism of Linux containers. University of Electronic Science and Technology of China (2015)
7. Wang, K., Zhang, G., Zhou, X.: Research on virtualization technology based on container. Comput. Technol. Dev. **25**(8), 138–141 (2015)
8. STREAM: Sustainable Memory Bandwidth in High Performance Computers. http://www.cs.virginia.edu/stream/. Accessed 9 Oct 2015
9. HPL-A Portable Implementation of the High-Performance Linpack Benchmark for Distributed-Memory Computers. http://www.netlib.org/benchmark/hpl/. Accessed 25 Oct 2015
10. IOzone Filesystem Benchmark. http://www.iozone.org/. Accessed 15 Nov 2015
11. Welcome to the Netperf Homepage. http://www.netperf.org/netperf/. Accessed 21 Nov 2015

Link Prediction Based on Precision Optimization

Shensheng Gu[⊠] and Ling Chen

Department of Computer Science, Yangzhou University, Yangzhou, China
18252713230@163.com, yzulchen@163.com

Abstract. In complex networks, link prediction involves detecting both unknown links and links that may appear in the future. Recently, various approaches have been proposed to detect potential or future links in temporal social networks. To evaluate the performance of link prediction methods, precision are usually used to measure the accuracy of the predicting results. This paper proposes an algorithm based on the precision optimization. In the method, precision is treated as the objective function, and link prediction is transformed as an optimization problem. A group of topological features are defined for each ordered pair of nodes. Using those features as the attributes of the node pairs, link prediction can be treated as a binary classification where class label of each node pair is determined by whether there exists a directed link between the node pair. Then the binary classification problem can be solved by precision optimization. Empirical results show that our algorithm can achieve higher quality results of prediction than other algorithms.

Keywords: Complex networks · Link prediction · Precision · Optimization

1 Introduction

In recent years, complex network has been witnessed an inconceivable development. Many systems can be described as complex networks, where individuals are denoted as nodes and their relations as links. These links often exhibit patterns that can indicate properties of the nodes such as the importance, rank, or category of the object it represents. In some cases, there may also be some unobserved potential links; therefore, we may be interested in predicting the existence of links between instances. In some real world system, where the links are evolving over time, we need to predict whether a link will occur in the future. Therefore, it is necessary to investigate some effective link prediction methods to predict the missing and potential links according to the known network information [1, 2]. Link prediction problem has a wide range of practical application value in different fields. For example, in online social networks, potential friends of the users can be revealed by link prediction, and can be recommended to the users [3]. By analyzing social relations, we can find potential interpersonal links [4, 5]. Link prediction can also be used in the academic network to predict the type and cooperators of an academic paper [6]. In biological networks, such as protein-protein interaction, metabolic and disease-gene networks, link existing between the nodes

© Springer Nature Singapore Pte Ltd. 2017
H. Yuan et al. (Eds.): GRMSE 2016, Part I, CCIS 698, pp. 131–141, 2017.
DOI: 10.1007/978-981-10-3966-9_14

indicates they have an interaction relationship [7]. Link prediction study not only has a wide range of practical value, but also has important theoretical significance [8]. For example, it helps to understand the mechanism of the evolution of complex network in theory, and can provide a simple and unified platform for a more fair comparison of network evolution mechanisms, so as to promote the theoretical research on complex network evolution model.

In recent years, many methods on link prediction have been reported. For example, In the similarity-based method, typical local indices include Common Neighbors [9], Salton Index [10], Jaccard Index [11], Sorensen Index [12], Hub Depressed Index [13], Hub Promoted Index [14], Leicht-Holme-Newman Index (LHN_I) [14], Preferential Attachment Index [15], Adamic-Adar Index [16] and Resource Allocation Index [17, 20] et al. Global indices require global topological information. Katz Index [18], Leicht-Holme-Newman Index (LHN_II) [14], Matrix Forest Index (MFI) [19] are typical global indices. Quasi-local indices do not require global topological information but make use of more information than local indices. Such indices includes Local Path Index [17, 26], Local Random Walk [21], and Superposed Random Walk [21]. Another group of similarity is based on the random walk, such as Average Commute Time [22], Cos+ [23], random walk with restart [24], and SimRank [25]. Zhou et al. [17] proposed two new local indices, Resource Allocation index and Local Path index.

Liu and Lü [27] studied the link prediction problem based on the local random walk, and found that the limited step may get a better prediction than the result of global random walk. Rao [28] proposed an algorithm based on the MapReduce parallel computation model that can be applied to large complex networks. Dong [29] proposed an algorithm based on the gravitational of the node, which can improve the prediction accuracy while maintaining a low time complexity.

To evaluate the quality of the predicting results, AUC and precision are the most used measurement. Precision is also an important performance measure that has been widely used in many tasks such as cost-sensitive learning, class-imbalance learning, instances ranking, and information retrieval. For link prediction in networks, whatever method we used, our final goal is to optimize the precision score. Since precision is the measurement of link prediction results, we can optimize it directly, instead of calculating other similarity indexes or optimizing other parameters in some supposed models describing the topological features of the network. Inspired by this fact, we present a link prediction method by directly optimizing the precision score. In the method, precision is treated as the objective function, and link prediction is transformed as an optimization problem. A group of topological features are defined for each ordered pair of nodes. Using those features as the attributes of the node pairs, link prediction can be treated as a binary classification where class label of each node pair is determined by whether there exists a directed link between the node pair. Thus the link prediction issue can be reduced to a problem of determining which edges belong to a positive class where the edges do exist, and which belong to a negative class where the edges do not exist. Then the binary classification problem can be solved by precision optimization. Our experiment results on real networks show that the algorithm can achieve higher speed and more accurate results than other methods.

The rest of the paper is organized as follows. Section 2 reviews the problem of link prediction and precision score for evaluating the results. Section 3 illustrates how Sect. 4 proposes the link prediction algorithm based on precision optimization, and describes the implementation details of the algorithm. Section 5 shows and analyzes the experimental results obtained by our algorithm and compares the performance with other similar methods. Section 6 draws conclusions.

2 Link Prediction and Precision

Consider an undirected network $G(V, E)$, where V is the set of nodes and E is the set of links. The total number of nodes and edges in the network are n and M respectively. We use U to denote by the universal set containing all $N(N-1)/2$ possible links. Let $A = [a_{ij}]$ be the adjacent matrix of the network G, where $a_{ij} = 1$ if there is a link between nodes i and j, or $a_{ij} = 0$ otherwise. The task of link prediction is to find out missing links (or the links that will appear in the future) in the set of nonexistent links $U\text{–}E$.

The goal of our link prediction method is to give a score S(x, y) to each ordered pair of nodes (x, y) $\in$ U. This score reflects the existence likelihood of the directed link between the two nodes. For a nodes pair (x, y) in U \ E, the larger S(x, y) is, the higher probability there will exist a link between nodes x and y.

To test the accuracy of the algorithm, the observed links, E, is randomly divided into two parts: the training set E^T, is treated as known information, while the probe set E^P, is used for testing and no information in this set is allowed to be used for prediction. Clearly, $E^T \cup E^P = E$ and $E^T \cap E^P = \Phi$. The non-exist links are in the set $U - E$. Two standard metrics are used to quantify the accuracy of prediction algorithms: area under the receiver operating characteristic curve (AUC) and Precision. A detailed introduction of these two metrics is as follows.

AUC: Provided the rank of all non-observed links, the AUC value can be interpreted as the probability that a randomly chosen missing link (i.e., a link in EP) is given a higher score than a randomly chosen nonexistent link (i.e., a link in U–E). In the algorithmic implementation, we usually calculate the score of each non-observed link instead of giving the ordered list since the latter task is more time consuming. Then, at each time we randomly pick a missing link and a nonexistent link to compare their scores, if among an independent comparisons, n' there are times the missing link having a higher score and n'' times they have the same score, the AUC value is

$$\text{AUC} = \frac{n' + n''}{n}. \tag{1}$$

If all the scores are generated from an independent and identical distribution, the AUC value should be about 0.5. Therefore, the degree to which the value exceeds 0.5 indicates how better the algorithm performs than pure chance.

Precision: Given the ranking of the non-observed links, the precision is defined as the ratio of relevant items selected to the number of items selected. That is to say, if we take the top-L links as the predicted ones, among which m links are right, then the precision equals m/L. That is

$$\text{Precision} = m/L. \tag{2}$$

Clearly, higher precision means higher prediction accuracy.

For link prediction in networks, whatever method we used, AUC and Precision score are the most used measure for evaluating the quality of the predicting results. In essence, the goal of all those link prediction methods is to optimize the AUC or Precision score. Inspired by this fact, we present a link prediction method by directly optimizing the Precision score. In the method, Precision is used to construct the objective function, and link prediction is transformed as an optimization problem.

3   Features of Node Pair for Link Prediction

We treat the link prediction problem as a classification on the node pairs based on their structural property reflected by the features of node pairs. The link prediction task is to find a similarity score s_k for every node pair e_k, where s_k is the probability of the link e_k to appear. In practice, we may have additional information on the network entities. A systematic approach to incorporate those additional data is to add them as covariates. In this paper, for simplicity, we limit our discussion on relational data only but our model can be readily extended to handle covariates.

The key to the successful prediction is to construct the mapping function, such that it generalizes the inherent topological structures underlying the relations of entities.

Node properties such as degree and centrality and relation between nodes such as common neighbors and paths rely on the structure of the entire or significant portion of the network structure, node centrality and shortest distance between nodes often require global knowledge. For large complex networks, having a global knowledge of the network is impossible and at times processing the entire network is impractical. In state, being able to determine existence of links between nodes using only local information becomes very important. Therefore, we use local similarity to determine proximity of two nodes among other node-pair properties. The node pair properties are compiled as a feature vector for node pairs whose connectivity is known. Then a model is constructed by precision optimization to predict the existence of connection from a given feature vector.

For a node pair (v_i, v_j), which represents the potential link $e_{ij} = (v_i, v_j)$, we use node pair properties and node properties to construct a feature vector which consists of the features such as CN, Jaccard, AA, RA, PA, LP and so on.

(1) Common Neighbors (CN) [9]. For a node v_x, let $\Gamma(x)$ denotes the set of neighbors of v_x. If nodes, v_x and v_y, have many common neighbors, they are more likely to have a link. This index is defined as

$$s_{xy} = |\Gamma(x) \cap \Gamma(y)|. \tag{3}$$

(2) Salton Index [10] is also called cosine similarity and is defined as

$$s_{xy} = \frac{|\Gamma(x) \cap \Gamma(y)|}{\sqrt{k_x k_y}},$$

(4)

Where k_x is the degree of node v_x.

(3) Jaccard Index [11]. This index is proposed by Jaccard and is defined as

$$s_{xy} = \frac{|\Gamma(x) \cap \Gamma(y)|}{|\Gamma(x) \cup \Gamma(y)|}.$$

(5)

(4) Sorensen Index [12]. This index is utilized mainly for ecological community data, and is defined as

$$s_{xy} = \frac{2 \times |\Gamma(x) \cap \Gamma(y)|}{k_x + k_y}.$$

(6)

(5) Adamic–Adar Index (AA) [16]. A weight is assigned to each pair of node according to the degree of common neighbors and is defined as

$$s_{xy} = \sum_{z \in \Gamma(x) \cap \Gamma(y)} \frac{1}{\log k_z}.$$

(7)

(6) Resource Allocation Index (RA) [17]. This index is motivated by the resource allocation dynamics on complex networks [30]. Consider a pair of nodes, x and y, and these two nodes are not directly connected. The node x can send some resource to y, with their common neighbors playing the role of transmitters. In the simplest form, we assume that each transmitter has a unit of resource, and will equally distribute it to all its neighbors. Then the similarity between x and y can be defined as the amount of resource y received from x, which is

$$s_{xy} = \sum_{z \in \Gamma(x) \cap \Gamma(y)} \frac{1}{k_z}.$$

(8)

For each node pair, all its features form a vector which can be used in our link prediction algorithm based on precision.

4 The Link Prediction Algorithm Based on Precision

Let d be the number of features on each node pair. The feature vector of node pair $e_k = (v_i, v_j)$ is denoted as $x_k \in R^d$ form an instance space where $x_k = (x_{k1}, x_{k2}, \ldots, x_{kd})^T$ consists of d features of node pair e_k.

To transform the link prediction to a problem of binary classification, we assign a label $y_j \in \{+1, -1\}$ to each instance x_j indicating the existence of link between node pair e_k: $y_i = 1$ Means that there is an edge between node pair e_k and $y_i = -1$ otherwise. A set of training samples consisting of n_+ existing links $e_1^+, e_2^+, \ldots, e_{n_+}^+$ with feature vectors $x_1^+, x_2^+, \ldots, x_{n_+}^+$ and labels $y_i^+ = 1$, $(i = 1, 2, .., n_+)$, and n_- non-existing ones $e_1^-, e_2^-, \ldots, e_{n_-}^-$ with feature vectors $x_1^-, x_2^-, \ldots, x_{n_-}^-$ and labels $y_i^- = -1$, $(i = 1, 2, .., n_-)$. Each node pair e_i is denoted by a tuple (x_i, y_i), and the training set is $D = \left\{ (x_1^+, 1), \ldots, (x_{n_+}^+, 1), (x_1^-, -1), \ldots, (x_{n_-}^-, -1) \right\}$.

To solve this binary classification problem, we denote $S : x \rightarrow R$ as a real value function which indicates the probability of a node pair being connected by a link. Let $w = (w_1, w_2, \ldots, w_d)$ be the weight vector where w_i is the weight of the i-th feature, function S can be defined as:

$$S(x_i) = \sum_{k=1}^{d} w_k x_{ik} = w^T x_i. \tag{9}$$

Let $wx_i^T = \bar{y}_i$ be the result of link prediction using (9). If $\bar{y}_i > 0$ we believe e_i is an existing link, otherwise e_i is a non-existing link. Our goal is to select the optimal weight vector w so that $\bar{y}_i$ can approximate to y_i as close as possible. The main idea of our algorithm is that we use precision metric as the objective function of optimization to link prediction. We define an objective function based on the definition of precision and then optimize the function, aiming to maximize the precision.

According to (2) the precision of the prediction result by (9) is

$$precision = \frac{1}{n_+ + n_-} \left[\sum_{i=1}^{n_+} I[w^T x_i^+ > 0] + \sum_{i=1}^{n_-} I[w^T x_j^- < 0] \right]. \tag{10}$$

Here, $I(b)$ is the indication function:

$$I(b) = \begin{cases} 1 & b \text{ is true} \\ 0 & \text{otherwise} \end{cases}$$

To maximize the value of *precision*, we ought to minimize the loss function:

$$L(w) = \frac{1}{n_+ + n_-} \left[\sum_{i=1}^{n_+} \left(1 - w^T x_i^+\right)^2 + \sum_{i=1}^{n_-} \left(-1 - w^T x_j^-\right)^2 \right] + \frac{\lambda}{2} \|w\|_2^2. \tag{11}$$

In (11), $\frac{\lambda}{2}\|w\|_2^2$ is the regulation term for preventing over-fitting, and λ is a parameter.

We use stochastic gradient descent method to optimize the loss function $L(w)$. Once we get the gradient $\frac{\partial L}{\partial w}$, by theory of stochastic gradient descent, the solution can be obtained by the iteration

$$w_{t+1} = w_t - \eta_t \frac{\partial L}{\partial w} \tag{12}$$

Where η_t is the step length for the t-th iteration.

Let $N = n_+ + n_-$, it is easy to get

$$\frac{\partial L}{\partial w} = \lambda w - \frac{2}{N} \sum_{i=1}^{n_+} x_i^+ + \frac{2w}{N} \sum_{k=1}^{N} x_k^T x_k + \frac{2}{N} \sum_{j=1}^{n_-} x_j^-. \tag{13}$$

Let $m^+ = \sum_{i=1}^{n_+} \frac{x_i^+}{N}, m^- = \frac{1}{m} \sum_{j=1}^{n_-} x_j^-, s = \frac{1}{N} \sum_{k=1}^{N} x_k^T x_k$ then we have

$$\frac{\partial L}{\partial w} = 2(m^- - m^+) + 2w(\lambda + s). \tag{14}$$

After setting a proper initial value w_0, we use formula (12) to update w iteratively until convergence. Framework of the link prediction algorithm based on precision optimization is as follows:

Algorithm   Precision-Optim
Input: A: The adjacency matrix of network;
Output: S: score vector of all the node pairs;
Begin
1. Parameter initialization, set the initial value of w;
2. **For** every node pair $e_{k\,in}\ U$ **do**
                Construct the feature vector x_k and label y_k;
End for
3. **While** not converge **do**
Use formula (12) to update w;
End while;
4. Output the updated w;
5. **For** every node pair $e_{k\,in}\ U$ **do**
Calculate the similarity according to

$$s_i = w^T x_i$$

End for
6. Output the final score vector S;
End

5   Experimental Results and Analysis

In this section, we consider six benchmark data sets [31] representing networks drawn from disparate fields.

5.1   Dataset

In the experiments, we test our algorithm on 6 typical datasets, namely, protein-protein interaction networks (PPI), net of science (NS), American grid (Grid), political blog network (PB), INT and USAir. Their statistical properties are shown in Table 1. N and M denote the number of nodes and edges in the network, and Nc is the number of connected groups and the number of nodes in the largest connected group. For example, 332/1 denotes that there is 1 connected group in USAir and that the largest connected group contains 332 nodes. E denotes the efficiency [32] of the network, C denotes clustering coefficient [33], r is the assortativity coefficient [34], and K is the average degree of the network.

Table 1. Topological properties of the 6 networks tested

Networks	N	M	Nc	e	C	r	K
USAir	332	2126	332/1	0.406	0.749	−0.208	12.807
PB	1224	19090	1222/2	0.397	0.361	−0.079	31.193
NS	1461	2742	379/268	0.016	0.878	0.462	3.754
PPI	2617	11855	2375/92	0.180	0.387	0.462	9.060
Grid	4941	6594	4941/1	0.056	0.107	0.003	2.669
INT	5022	6258	5022/1	0.167	0.033	-0.138	2.492

5.2   Results

We compare the accuracy of the algorithm Precision-Optim (PO for short in Table 2) with that of some classical similarity indexes in the above 6 networks by utilizing the AUC criterion. Table 2 presents the average AUC scores on 10-fold CV tests by different algorithms. In the table, the highest precision scores for each data set by the 11 algorithms is emphasized in bold-face. Experimental results show that the accuracy of Precision-Optim is higher than others shown in the Table 2.

Comparing the results from the 11 methods, we can see that the Precision-Optim algorithm outperform the other ten algorithms on precision. Specifically, compared with that of CN, the AUC score can be improved by 4.6% for PPI, 3.2% for PB, 11% for Grid and 33% for INT. When this comparison is against Salton, Jaccard and LHN_I, the improvement are further enlarged. In particular, the performance of Precision-Optim for disparate networks is very robust; i.e., it is either the best or very close to the best. In contrast, other algorithms can largely fail.

We further move to study the convergence condition of algorithm. In the process of the iteration of weight vector w, we calculate the error of w between adjacent time periods, error formula is $Error^{(t)} = ||w^{(t)} - w^{(t-1)}||_2$, and we plot the relation between

Table 2. The comparison of AUC of 11 algorithms on 6 networks

AUC	PPI	NS	Grid	PB	INT	USAir
CN	0.924	0.996	0.632	0.918	0.641	0.941
Salton	0.918	0.993	0.626	0.880	0.668	0.920
Sorensen	0.918	0.991	0.629	0.873	0.641	0.905
HPI	0.912	0.987	0.635	0.855	0.645	0.867
HDI	0.916	0.992	0.625	0.876	0.647	0.892
LHN-1	0.912	0.991	0.616	0.767	0.643	0.712
AA	0.918	0.991	0.624	0.930	0.666	0.949
RA	0.911	0.991	0.633	0.924	0.620	0.952
LP	0.922	**0.998**	0.702	0.939	0.953	0.912
Katz	0.396	0.319	0.429	0.511	0.535	0.514
PO	**0.968**	0.996	**0.704**	**0.948**	**0.954**	**0.953**

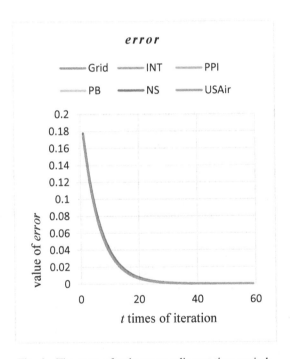

Fig. 1. The error of w between adjacent time periods.

the error and the iteration number of algorithm on different datasets. The result is shown in Fig. 1. From the figure, we can see clearly that the trend of error presents a sharply decrease trend and reaches the state of convergence quickly. For all the data sets, the algorithm converges after about 30 iterations.

6   Conclusions

This paper proposes a link prediction algorithm based on precision optimization. First, we extract some features of the edges according to the similarity indexes to form a feature vector. Then, we define a loss function based on the definition of precision and get the optimal weight vector by iteration. Finally, we calculate the final similarity matrix. Experimental results show that this algorithm can achieve results of higher quality.

Acknowledgements. This research was supported in part by the Chinese National Natural Science Foundation under grant Nos. 61379066, 61070047, 61379064, 61472344, 61402395, Natural Science Foundation of Jiangsu Province under contracts BK20130452, BK2012672, BK2012128, BK20140492 and Natural Science Foundation of Education Department of Jiangsu Province under contract 12KJB520019, 13KJB520026, 09KJB20013. Six talent peaks project in Jiangsu Province (Grant No. 2011-DZXX-032). Foundation of Graduate Student Creative Scientific Research of Jiangsu Province under contract CXZZ13_0172. China Scholarship Council also supported this work.

References

1. Lichtenwalter, R.N.: New precepts and method in link prediction. In: Proceedings of ACM KDD 2010, pp. 243–252 (2010)
2. Lü, L., Zhou, T.: Link prediction in complex networks: a survey. Phys. A: Stat. Mech. Appl. **390**(6), 1150–1170 (2011)
3. Papadimitriou, A., Symeonidis, P., Manolopoulos, Y.: Fast and accurate link prediction in social networking systems. J. Syst. Soft. **85**(9), 2119–2132 (2012)
4. Hossmann, T., Nomikos, G., Spyropoulos, T., et al.: Collection and analysis of multi-dimensional network data for opportunistic networking research. Comput. Commun. **35**(13), 1613–1625 (2012)
5. Jahanbakhsh, K., King, V., Shoja, G.C.: Predicting missing contacts in mobile social networks. Pervasive Mob. Comput. **8**(5), 698–716 (2012)
6. Sun, Y., Barber, R., Gupta, M., et al.: Co-author relationship prediction in heterogeneous bibliographic networks. In: 2011 International Conference on IEEE Advances in Social Networks Analysis and Mining (ASONAM), pp. 121–128 (2011)
7. Li, X., Chen, H.: Recommendation as link prediction in bipartite graphs: a graph kernel-based machine learning approach. Decis. Support Syst. **54**(2), 880–890 (2013)
8. Huang, Z., Lin, D.K.J.: The time-series link prediction problem with applications in communication surveillance. INFORMS J. Comput. **21**(2), 286–303 (2009)
9. Newman, M.E.J.: Clustering and preferential attachment in growing networks. Phys. Rev. E **64**(2), 025102 (2001)
10. Salton, G., McGill, M.J.: Introduction to modern information retrieval. Inf. Process. Manag. **19**(6), 402–403 (1983). ISBN 0-07-054484-0
11. Jaccard, P.: Etude comparative de la distribution florale dans une portion des Alpes ET du Jura. Impr. Corbaz (1901)
12. Sorenson, T.: A method of establishing groups of equal amplitude in plant sociology based on similarity of species content. Kongelige Danske Videnskabernes Selskab **5**(1–34), 4–7 (1948)

13. Ravasz, E., et al.: Hierarchical organization of modularity in metabolic networks. Science **297**(5586), 1551–1555 (2002)
14. Leicht, E.A., Holme, P., Newman, M.E.J.: Vertex similarity in networks. Phys. Rev. E **73**(2), 026120 (2006)
15. Barabási, A.L., Albert, R.: Emergence of scaling in random networks. Science **286**(5439), 509–512 (1999)
16. Adamic, L.A., Adar, E.: Friends and neighbors on the web. Soc. Netw. **25**(3), 211–230 (2003)
17. Zhou, T., Lü, L., Zhang, Y.C.: Predicting missing links via local information. Eur. Phys. J. B **71**(4), 623–630 (2009)
18. Katz, L.: A new status index derived from sociometric analysis. Psychometrika **18**(1), 39–43 (1953)
19. Chebotarev, P., Shamis, E.V.: The matrix-forest theorem and measuring relations in small social groups. Autom. Remote Control **58**, 1505 (1997)
20. Lü, L., Jin, C.-H., Zhou, T.: Similarity index based on local paths for link prediction of complex networks. Phys. Rev. E **80**, 046122 (2009)
21. Liu, W., Lü, L.: Link prediction based on local random walk. Europhys. Lett. **89**, 58007 (2010)
22. Klein, D.J., Randic, M.: Resistance distance. J. Math. Chem. **12**(1), 81–95 (1993)
23. Fouss, F., Pirotte, A., Renders, J.M., et al.: Random-walk computation of similarities between nodes of a graph with application to collaborative recommendation. IEEE Trans. Knowl. Data Eng. **19**(3), 355–369 (2007)
24. Brin, S., Page, L.: Reprint of: the anatomy of a large-scale hypertextual web search engine. Comput. Netw. **56**(18), 3825–3833 (2012)
25. Jeh, G., Widom, J.: SimRank: a measure of structural-context similarity. In: Proceedings of the Eighth ACM SIGKDD International Conference on Knowledge Discovery and Data Mining, pp. 538–543. ACM (2002)
26. Lü, L., Jin, C.H., Zhou, T.: Similarity index based on local paths for link prediction of complex networks. Phys. Rev. E **80**(4), 046122 (2009)
27. Liu, W., Lü, L.: Link prediction based on local random walk. EPL (Europhys. Lett.) **89**(5), 58007 (2010)
28. Rao, J., Wu, B., Dong, Y.X.: Parallel link prediction in complex network using MapReduce. Ruanjian Xuebao/J. Softw. **23**(12), 3175–3186 (2012)
29. Dong, Y.X., Ke, Q., Wu, B.: Link prediction based on node similarity. Comput. Sci. **38**(7), 162 (2011)
30. Ou, Q., Jin, Y.-D., Zhou, T., Wang, B.-H., Yin, B.-Q.: Power-law strength-degree correlation from resource-allocation dynamics on weighted networks. Phys. Rev. E **75**, 021102 (2007)
31. Bhawsar, Y., Thakur, G.S., Thakur, R.S.: Model for link prediction in social network by genetic algorithm approach. Data Min. Knowl. Eng. **7**(5), 191–196 (2015)
32. Verkaria, K., Clack, C.: Biases introduced by adaptive recombination operations. In: Proceedings of Genetic and Evolutionary Computation Conference (GECCO), vol. 1, pp. 670–677 (1999)
33. Poli, R.: Exact schema theory for genetic programming and variable-length genetic algorithms with one-point crossover. Genet. Program Evolvable Mach. **2**(2), 123–163 (2001)
34. Rothlauf, F., Goldberg, David, E.: Pruefer numbers and genetic algorithms: a lesson on how the low locality of an encoding can harm the performance of gas. In: Schoenauer, M., Deb, K., Rudolph, G., Yao, X., Lutton, E., Merelo, J.J., Schwefel, H.-P. (eds.) PPSN 2000. LNCS, vol. 1917, pp. 395–404. Springer, Heidelberg (2000). doi:10.1007/3-540-45356-3_39

Face Feature Points Detection Based on Adaboost and AAM

Xiaoqi Jia$^{(\boxtimes)}$, Qing Zhu, Peng Zhang, and Menglong Chang

School of Software Engineering,
Beijing University of Technology, Beijing, China
{jiaxiaoqi,zhangpeng, changmenglong}@emails.bjut.edu.cn,
ccgszq@bjut.edu.cn

Abstract. Face detection is a classical problem in the field of computer vision. It is widely used in recent years, face detection and face tracking has not only limited to the scope of application of face recognition: in video retrieval, video surveillance, facial expression analysis, gender, race, age discrimination, digital entertainment, and so on. This paper proposed algorithm based on AdaBoost algorithm AAM model of face feature points to identify the improvement in a certain range to solve the present stage AAM algorithm does not consider the grayscale in the exact face of initial position and face face detection problem.

Keywords: Face detection · Detection of side faces · Adaboost · AAM

1 Introduction

With the development of machine learning, more and more people pay attention to the face detection technology. Face detection is a classical problem in the field of computer vision. It is widely used in recent years, face detection and face tracking has not only limited to the scope of application of face recognition: in video retrieval, video surveillance, facial expression analysis, gender, race, age discrimination, digital entertainment, and so on. For example, Beijing Opera Facial Masks attached to the face, so the research on the application of facial features, can not be separated from the facial feature point's detection and tracking. To make some facial projection in faces in motion, the first step is human face detection and tracking. This is a new application aspects of cultural heritage. At present, the face detection and tracking technology is one of the challenging research topic. Currently, there are few methods to detect and track human faces in real time.

Based on the Active Appearance Model (Active the Appearance Model AAM) and Adaboost algorithm as the foundation, this paper proposes a combination of the two algorithms to improve the algorithm. While the human face and the feature points are detected, to improve the efficiency of detection of side faces. First of all using rectangle features (Haar features) and AdaBoost algorithm for face detection for video frames, the detected face area is submitted to AAM and AAM area detects features, multiple iterations to select the optimal solution for increasing weight. In this way, we can improve face recognition and feature robustness and efficiency.

System block diagram is shown in Fig. 1.

© Springer Nature Singapore Pte Ltd. 2017
H. Yuan et al. (Eds.): GRMSE 2016, Part I, CCIS 698, pp. 142–149, 2017.
DOI: 10.1007/978-981-10-3966-9_15

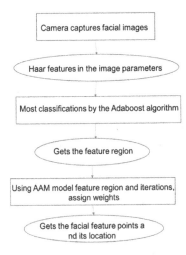

Fig. 1. System block diagram

2   Related Work

2.1   Setted Constriction

The origin of face recognition technology should be traced back to 1960s. Early in the study of typical methods mainly use texture, edge and color of some low level image features [1, 2], with face local organs of geometric constraints [3, 4] to locate the face, or through the image is calculated between each location with the standard face template correlation of face detection. In 1987, Kass et al. [6] proposed known as snake active contour model (active contour model, ACM), which can not detect object contours. In 1995, Cootes et al. [7] proposed a, active shape model (active shape model ASM) model based on the shape feature, ASM uses some research results of ACM, ASM can not only the outer contour of the object is detected and the inner contour of the object is detected. In 1998, Cootes et al. [8] added texture model and put forward the active appearance models (the active appearance model, AAM) based on the ASM shape model, it used the shape information and texture information of the face object, which makes its application more widely. In 2001, Viola [9] proposed AdaBoost based on face detector, not only made high detection rate, but also the detection speed of a substantial increased, which stimulated the further development of the research.

2.2   Based on Appearance

Generally speaking, the early method was based on some heuristic knowledge to detect. It was difficult to adapt to the changes in the actual environment, and real-time detection accuracy was difficult to meet the practical application needs. Aiming at the defects of the early research work, the researchers put forward a face detection method based on the appearance [10–12]. The basic idea of this method was to use statistical learning theory to find the characteristics of face or non-face, and to judge each position in the image. Among them, the cascade Haarlike-Adaboost method had good real-time performance and high detection accuracy, and its excellent detection performance made face detection from theory research to practical application. The proposed method had a milestone significance for the face detection problem. However, Haarlike-Adaboost proposal still suffers from drawbacks: First, the method trained process complex, time-consuming; secondly, the method detected poor robustness, various face global changes, such as occlusion, illumination changes, countenance, posture change etc., may lead to missed. In view of the problems Haarlike-Adaboost method to train the complex and time-consuming, Bourdev et al. [13] put forward a conventional cascade face detection AdaBoost algorithm. The method trained a single classifier, then for each weak classifier trained a rejection threshold. Eventually, there was no loss in accuracy or speed of detection. For Haarlike-Adaboost methods are easily affected by light interference, Liao et al. [14] proposed using LBP (local binary pattern) instead of original haarlike features for face detection, LBP features with some light illumination invariance reduced the light changes interference of human face detection effectively. Although these methods can partially compensate for the defection of Haarlike-Adaboost method, these methods needed a lot of training samples in order to obtain better detection performance. Mairal et al. [15] proposed the use of sparse expression to detect the edge of the target class. Song et al. [16] in the deformable parted model based on the introduction of sparse expression, thus can improve the efficiency of multi class object detection.

3   Methodology

3.1   AdaBoost Algorithm

Haar features can effectively apply the difference between the face and non-face, characterized in Fig. 2 is Haar face in use.

Fig. 2. Haar features in human face images

Haar characteristic feature is the two white rectangle with a black rectangle pixels and pixel difference; each Haar feature is a one to one weak classifier, therefore, weak classifiers eigenvalues are Haar features eigenvalues. If a Haar wherein n, the training sample $N = n$; in the use of Adaboost algorithm n-Haar features to train to enter the n Haar characteristic data set is $\{(\{(x_1, y_1), (x_2, y_2)\ldots(x_n, y_n)\}$, Among them, the number of counterexamples is M, the number of positive cases is L($N = M + L$). x_i represents the characteristic value of the Haar feature, $y_i = 0$ represents counterexamples ($i = 1, 2\ldots n$).

3.2 Active Appearance Model (AAM)

The independent AAM shape model is composed of a set of images on the triangular mesh and the vertex set s of these triangular meshes, and S is denoted as the form of the set of coordinate points: $s = [x_1, y_1, x_2, y_2, \ldots, x_n, y_n]^T$

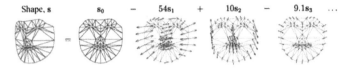

Fig. 3. Shape model

AAM shape model, as shown in the Fig. 3, through the training of such a set of given vertex set, the data alignment, statistical processing, PCA and other practices, and ultimately can get a set of shape models:

$$s = s_0 + \sum_{i=1}^{n} p_i s_i \tag{1}$$

s_0 is the average shape, s_i is the orthogonal base, p_i is S in this group of feature vector projection, that is, a set of shape parameters.

Fig. 4. Texture model

As is shown in the Fig. 4, the data are processed on the basic grid and the principal component analysis is used to obtain a set of texture models:

$$A(x) = A_0(x) + \sum_{i=1}^{m} \lambda_i A_i(x) \tag{2}$$

$A_0(x)$ is the average texture, $A_i(x)$ is a set of orthogonal bases, and the $A_i(x)$ is a set of texture parameters.

Finally, we need to map the texture information in the average shape to the current shape.

4   Experiments

4.1   Face Image Gray

According to the weighted average method, the video frames are processed to form a gray scale image.

$$\hat{x}_{ij} = 0.299x_{ij}^{(R)} + 0.587x_{ij}^{(G)} + 0.114x_{ij}^{(B)} \tag{3}$$

$\hat{x}_{ij}$ of the gray value of the video frame, and $x_{ij}^{(R)}, x_{ij}^{(G)}, x_{ij}^{(B)}$ respectively represent the three component (R, G, B) value of video frames. The upcoming R, G, and B components are mapped onto the R, G, B, and the diagonal of the cube.

4.2   AdaBoost Algorithm Based on Haar Feature

Get all the Haar features of the locations on the gray image, and for each Haar features and its value.

To calculate each Haar features of value.

Initialize the weights of N = n training samples, and give the same weight to the N training samples:

$$\omega_{t,i} = \frac{\omega_{t,i}}{\sum_{j-1}^{n} \omega_{t,j}}, \quad j = 1, 2, \ldots, n \tag{4}$$

Using Adaboost algorithm t = 1, 2... T iteration, for each Haar feature J, training a weak classifier (h_j), and calculate the weighted error rate (ε_j) of all Haar features.

After each iteration to adjust the weights, update the weights $(\omega_{t+1,i})$ of each training sample, and improve the weight of the sample is the wrong classification:

$$\omega_{t+1,i} = \omega_{t,i}\beta_t^{1-e_i} \tag{5}$$

If it is properly classified, then $e_i = 0$, if not, then $e_i = 1$; $\beta_t = \frac{\varepsilon_t}{1-\varepsilon_t}$.

After the T iteration, a weak classifier is obtained, and then weighted constitute a strong classifier (h(x)):

$$h(x) = \begin{cases} 1, & \sum_{t=0}^{T} a_t h_t(x) \geq \frac{1}{2}\sum_{t=0}^{T} a_t, \\ 0, & other \end{cases} \quad 其中 a_t = \log\frac{1}{\beta_t} \tag{6}$$

By classifying the gray images, the face region is obtained.

4.3   AAM Model Generation

Submit the obtained face region to the AAM algorithm. Combine the shape model and texture model of the obtained face region. The p and λ are connected in series to get the apparent vector b:

$$B = (\frac{w_s P}{\lambda}) \tag{7}$$

Using diagonal matrix (w_s) to adjust the difference between p and λ. The B was processed by PCA, and the correlation between shape and texture was eliminated, and the apparent model was obtained:

$$B = B_0 + \sum_{t=0}^{t} c_t B_i \tag{8}$$

B_0 is the average apparent vector, B_i is based on the characteristic values of the first t after the apparent feature vector, C for the control of changes in the parameters of the table.

N (x; q) is defined as the control global deformation function, in which the parameters of q = (a, B), the parameters of the A, B respectively in the global shape change of the scale factor K and the angle of the relevant:

$$N(x;q) = \begin{pmatrix} (1+a) & -b \\ b & (1+a) \end{pmatrix} \begin{pmatrix} x \\ y \end{pmatrix} + \begin{pmatrix} t_x \\ t_y \end{pmatrix} \tag{9}$$

$a = k \cos\theta - 1$, $b = k \sin\theta - 1$, the parameters t_x and t_y respectively represent the displacement.

The global deformation is introduced into the AAM model, face region is obtained in Sect. 4.2, and the minimum value of the following formula is calculated by adjusting the shape parameter P, Q and the value of the texture parameter.

$$\sum_{x \in S_0} [A_0(x) + \sum_{i=1}^{m} \lambda_i A_{i(x)} - I(N(W(x;p);q))]^2 \tag{10}$$

λ is the parameter that controls the change of texture; P is the parameter that controls the local shape change; q is the parameter that controls the global deformation.

Training according to the characteristics of AAM on the profile face, once per iteration, once again, the weights are re assigned, and the parameters are updated with P and λ, and then again, re allocation, re update parameters p and λ. In turn, the minimum mean square error of the face and the AAM model in the image is the minimum mean square error, and the characteristic shape of the face tends to converge.

5 Result

Table 1 compares the method with previous methods herein are matching accuracy in profile cases (Fig. 5):

Table 1. Comparison of the accuracy of various methods on IMM face database

Method	Positive face (%)	Profile face (%)
LBP + SVN	76.38	-
AAM	63.80	-
Multi template AAM	80.40	-
L-AAM	90.96	84.90
Paper method	93.33	86.67

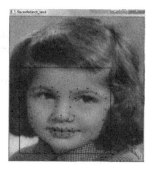

Fig. 5. Demo on Adaboost and AAM.

6 Discussion and Further Work

The proposed algorithm based on AdaBoost algorithm AAM model of face feature points to identify the improvement in a certain range to solve the present stage AAM algorithm does not consider the grayscale in the exact face of initial position and face detection problem. First using the AdaBoost algorithm, the strongest classifier to determine the location of the face in the image region, on the basis of this, using AAM model, prior to the face location area for model fitting, fitting better human face feature points obtained after training and re assigning weights. But this method in face recognition, yet to be improved. It will consider the face recognition and deflection angle more accurate fitting of face recognition.

References

1. Sirohey, S.A.: Human face segmentation and identification. Technical Report, CS-TR-3176, University of Maryland, Maryland (1993)
2. Augusteijn, M.F., Skujca, T.L.: Identification of human faces through texture-based feature recognition and neural network technology. In: Proceeding of the IEEE Conference on Neural Networks, pp. 392–398. IEEE (1993)
3. Yang, G., Huang, T.S.: Human face detection in complex background. Pattern Recogn. **27** (1), 53–63 (1994)

4. Heiseleu, B., Serret, T., Pontils, M., Poggiot, T.: Component-based face detection. In: Proceeding of the IEEE Conference on Computer Vision and Pattern Recognition, pp. 657–662. IEEE Computer Society Press (2001)

5. Lanitis, A., Taylor, C.J., Cootes, T.F.: An automatic face identification system using flexible appearance models. Image Vis. Comput. **13**(5), 393–401 (1995)

6. Kass, M., Witkin, A., Terzopoulous, D.: Snake: active contour models. In: Proceedings of the 1st International Conference on Computer Vision, pp. 259–268. IEEE Computer Society Press, London (1987)

7. Cootes, T.F., Taylar, C.J., Cooper, D.H., et al.: Active shape models their training and application. Comput. Vis. Image Underst. **61**(1), 38–59 (1994)

8. Cootes, T.F., Edwards, G.J., Taylor, C.J.: Active appearance models. In: Burkhardt, H., Neumann, B. (eds.) ECCV 1998. LNCS, vol. 1407, pp. 484–498. Springer, Heidelberg (1998). doi:10.1007/BFb0054760

9. Viola, P., Jones, M.: Rapid object detection using a boosted casecade of simple features. In: Proceedings IEEE on Computer Vision and Pattern Recognition, Kauai, Hawaii, USA, pp. 511–518 (2001)

10. Rowley, H., Baluja, S., Kanade, T.: Neural network-based face detection. IEEE Trans. Pattern Anal. Mach. Intell. **20**(1), 23–38 (1998)

11. Osuna, E., Freund, R., Girosi, F.: Training support vector machines: an application to face detection. In: Proceeding of the IEEE Conference on Computer Vision and Pattern Recognition, pp. 130–136. IEEE Computer Society Press (1997)

12. Viola, P., Jones, M.: Rapid object detection using a boosted cascade of simple features. In: Proceeding of the IEEE Conference on Computer Vision and Pattern Recognition, pp. I-511–I-518. IEEE Computer Society Press (2001)

13. Bourdev, L., Brandt, J.: Robust object detection via soft cascade. In: Proceeding of the IEEE Conference on Computer Vision and Pattern Recognition, pp. 236–243. IEEE Computer Society Press (2005)

14. Liao, SC., Zhu, XX., Lei, Z., Zhang, L., Li, SZ.: Learning multi-scale block local binary patterns for face recognition. In: Proceeding of the International Conference on Biometrics, pp. 828–837. IEEE (2007)

15. Mairal, J., Leordeanu, M., Bach, F., Hebert, M., Ponce, J.: Discriminative sparse image models for class-specific edge detection and image interpretation. In: Forsyth, D., Torr, P., Zisserman, A. (eds.) ECCV 2008. LNCS, vol. 5304, pp. 43–56. Springer, Heidelberg (2008). doi:10.1007/978-3-540-88690-7_4

16. Song, H.O., et al.: Sparselet models for efficient multiclass object detection. In: Fitzgibbon, A., Lazebnik, S., Perona, P., Sato, Y., Schmid, C. (eds.) Computer Vision–ECCV 2012. LNCS, vol. 7573, pp. 802–815. Springer, Heidelberg (2012). doi:10.1007/978-3-642-33709-3_57

Stock Price Manipulation Detection Based on Machine Learning Technology: Evidence in China

Jiangyun Zhang[✉], Shaojie Wang, Shicheng Xu, and Mengxin Yu

School of Information, Renmin University of China, Beijing, China
{zjy1995,w_shaojie,2013202494,onlooker}@ruc.edu.cn

Abstract. Stock price manipulation has become a big concern in stock markets, especially in emerging markets like China. This paper aims to employ machine learning methods to detect the stock price manipulation in China to increase the market fairness and transparency. Based on the information given by China Securities Regulatory Commission, we took the difference of stocks between manipulated time and normal time based on their daily return, trading volume, stock price volatility and market value. We used them as explanatory variables. Then we employed single model, Support Vector Machine (SVM), and ensemble model, Random Forest (RF) for detection. Test performance of classification accuracy, sensitivity and specificity statistics for SVM were compared with the results of RF. As a result, we found that both of them have a meaningful accuracy while RF outperforms SVM. We also found that daily return and market value have a bigger effect on detection than other explanatory variables do.

Keywords: Stock price manipulation · Support Vector Machine · Random Forest

1 Introduction

Manipulation has become an important issue for both developed and emerging stock markets since the very beginning. Stock price manipulation happens when some manipulators use illegal trading strategies or information to manipulate the stock price for earning extra profit from other investors. As we know, price of the stocks must be determined by the market in a healthy market with fairness and transparency. Investors takes all relevant information into account when valuing a stock, so the market price shows stocks common judgment of investors, which means this price is acceptable. One of the most significant stock price distortion is caused by manipulation, so we can regard manipulation as any interference or action to the market mechanism which prevents the stock price come to a fair one. When the manipulated price prevails and investors believe it do not reflect the true value of these stocks, the investors may lose their confidence in stock investment. To be worse, manipulation can cause the loss of common investors. Since manipulators aim at making profit at the expense of other

© Springer Nature Singapore Pte Ltd. 2017
H. Yuan et al. (Eds.): GRMSE 2016, Part I, CCIS 698, pp. 150–158, 2017.
DOI: 10.1007/978-981-10-3966-9_16

investors by several trading strategies, the loss of fairness is another bad consequence of manipulation. Manipulators hurt individual investors and in turn, the stock market.

Recently, data mining has been regarded as an effective tool to analyze economic hotspot, so we use machine learning methods to detect stock price manipulation in China, an emerging stock market. Based on 37 stock price manipulation cases released by China Securities Regulation Commission (CSRC) from 2010 to 2015, we train these machine learning models, Support Vector Machine (SVM) and Random Forest (RF), so that they can classify the manipulated days and the normal days when given out sample datasets.

One of the main contributions of our research work is to apply machine learning models and suitable input data to detect stock price manipulation in real time, which means it can undermine the extent to which individual investors lose money in trading the manipulated stocks. What is more, it offers an option for the market regulator to find those illegal manipulators. Though several approaches have already been taken by China Securities Regulatory Commission, the way to detect manipulation, especially in data mining methods, is still an important and meaningful issue. Our study focuses on developing a detection method in trade-based manipulation in Shanghai Stock Exchange as well as Shenzhen Stock Exchange. An effective and real time method with relative high accuracy can be beneficial for both investors and regulators.

The rest of this paper is organized as follows. Section 2 summarizes the related literature of our research. Section 3 describes our machine learning methods: Support Vector Machine and Random Forest. Empirical analysis and data are presented in Sect. 4. Section 5 has our conclusion and some future works.

2   Related Literature

Many researchers have studied the theory and mechanism of stock price manipulation. The very first researchers who studied stock price manipulation are Allen and Gale [1] and Jarrow [2]. Allen and Gale [1] studied the history of the stock-price manipulation and classified the manipulations as action-based manipulation, information-based manipulation, and trade-based manipulation. Jarrow [2] constructed a model to find out whether a large trader could influence the price of a stock and gain a profit without risk through manipulative trading. Several theoretical models examine trade-based manipulation. Felixon and Pelli [3] studied day end returns manipulation and indicated short-term stock price manipulation might be possible. Kumar and Seppi [4] modeled a manipulator that takes a position in the futures market and then manipulates the price to profit from the futures position. Gerard and Handa [5] investigated possible manipulations around seasoned equity offerings. Chakraborty and Yilmaz [6] developed a model to show how informed insiders should trade so as to manipulate the market. Furthermore, Thoppan and Punniyamoorthy [7] analyzed stock manipulation and surveillance from the securities regulatory commission based on literature and a series of practical implications.

However, studies on detection-based real time data are also meaningful apart from the mechanism. In consideration of the detection data, Aggarwal and Wu [8] developed a model to explain trade-based manipulation and they tested the model based on data from US stock markets. Moreover, attention should not only be paid on the data itself. Detection methods should also be considered. Punniyamoorthy and Thoppan [9] applied quadratic discriminant analysis for detecting stock price manipulation. Kim and Sohn [10], on the other hand, employed peer group analysis method for the detection of stock fraud, namely stock price manipulation.

As for methodologies of stock price manipulation detection, machine learning is a less frequently used but effective tool. Very few works have been done on applying machine learning models for detecting stock price manipulation. The most recent one according to our knowledge is conducted by Öğüt et al. [11]. They used single machine learning methods, Logistic Regression (LR), Artificial Neural Networks (ANN) and Support Vector Machine (SVM) as detecting tools in Turkish stock market. However, when looking at the whole stock market, machine learning has already been a widespread method in researches on stock market. SVM [12, 13] and ANN [14–16] are among the most popular machine learning models in stock market predicting. Also for ensemble models, Rodriguez and Rodriguez [16] explored AdaBoost and Random Forest (RF). In stock price direction prediction literature, both SVM and RF have proven to be top performers [17, 18].

The differences between our study and other literature are at as follows. First, we focus on detecting stock price manipulation in real time based on both single and ensemble machine learning methods in order to protect investors from losing money. Secondly, we use real world data from China stock market while most prior studies mostly used synthetic data in studying stock price manipulation. Last but not least, we apply different explanatory variables and 5-day-before history data as input.

3 Methodology

3.1 Support Vector Machine

Support Vector Machine (SVM) is a popular kernel method in classification. A kernel function $k(x, z)$ is employed to map the dataset into a feature space, where we find an optimized hyperplane separating data of two categories with the margin reaching maximum. Eventually, we need to solve this optimization problem (1)–(3).

$$\max_{\alpha} W(\alpha) = \sum_{i=1}^{m} \alpha_i - \frac{1}{2}\sum_{i,j=1}^{m} y^{(i)} y^{(j)} \alpha_i \alpha_j K(x^{(i)}, x^{(j)}) \tag{1}$$

$$\text{s.t.} \, 0 \le \alpha_i \le c, i = 1, \ldots, m \tag{2}$$

$$\sum_{i=1}^{m} \alpha_i y^i = 0, \tag{3}$$

In (2), c is a regularization parameter controlling the tradeoff between margin and misclassification error.

In this paper, we test four commonly used kernel functions, namely linear kernel, polynomial kernel, RBF kernel and sigmoid kernel. Then, different levels of parameter c is tested for each kernel function. We apply grid search to obtain the best SVM model.

3.2 Random Forest

Random Forest [19] is an ensemble of decision trees. During the training, each tree is trained separately on a subset of features. Due to the randomness of selecting features to split each node, a larger error rate is obtained, but with respect to noise, the model is more robust. Since the stock market is filled with noise, we can certainly expect a good result by applying random forest.

In our study, we need to choose two parameters of the random forest model. One is the number of trees, which is theoretically the bigger the better. However, the computing capability of the computer should be considered, since too many trees will result in a long training time. Therefore, we choose 500 as the number of trees. The other parameter is the features given to each tree. Empirically, we use square root of the number of total features, which is 4 in our study.

4 Empirical Analysis

4.1 Data Description and Evaluation Metrics

In this paper, we use the stock price manipulation cases released by China Securities Regulatory Commission for our study. Once China Securities Regulatory Commission officials find that someone or some investment institution have manipulated one or many stocks in a period of time, they would punish the manipulator and release this case on the website regularly. Based on these cases, we can download the trading data of these stocks, including daily stock price, daily highest price, lowest price, trading volume and market value, in both manipulated and non-manipulated time. Moreover, we can gain some other explanatory variables we need from calculation, such as the difference of the daily return between the whole market and a specific stock and so on.

However, not all the cases in Shanghai or Shenzhen stock exchange would be detected by the China Securities Regulatory Commission and a large part of stock-price manipulation cases could not be a part of our source for sure. As a result, we only use the information released by China Securities Regulatory Commission, which is definitely connected to stock manipulation. If a stock is manipulated since a specific trading day, it means this period of time (until the manipulator sell out the stocks) should all be regarded as the manipulated days. Since our China Securities Regulatory Commission define stock manipulation days in this way, we have to make sure our detection is suitable with the definition.

We select 37 stock-price manipulation cases released by CSRC from 2010 to 2015. Some cases are listed in 0 And we get the data of each manipulated stocks in 250 (or more) trading days. In each trading day, we get all kind of data mentioned above of every stock. For example, according to CSRC, Guoyuan Securities was manipulated from July 2nd, 2015 to July 28th, 2015, we gain the trading data of this stock in all trading days in 2015. Similarly, we get the data of other cases.

The input data for the machine learning models contains trading data such as daily stock price, return rate, trading volume, etc., while the output data is binary, marked by 0 and 1. "One" means that the stock is manipulated in this trading day, while "zero" means the stock is in regular condition. The training set contains 75% of randomly chosen data samples while the test set has the 25% left. We first use training set to train RF and SVM separately, and then test on the trained models with our test set. Finally, we compare the predicted results with real output and make evaluations as follows.

In our study, we employ 6 metrics to evaluate the results of SVM and RF. They are Accuracy (ACC), Positive Predictive Value (PPV), True Positive Rate (TPR, also known as Recall Rate), F1 score (F1), Area Under the Curve (AUC), and Matthews Correlation Coefficient (MCC). The calculation formula of each metric (except AUC) is presented as (4)–(8).

$$ACC = \frac{TP+TN}{TP+FP+TN+FN} \tag{4}$$

$$PPV = \frac{TP}{TP+FP} \tag{5}$$

$$TPR = \frac{TP}{TP+FN} \tag{6}$$

$$F1 = \frac{2TP}{2TP+FP+FN} \tag{7}$$

$$MCC = \frac{TP \times TN - FP \times FN}{\sqrt{(TP+FP)(TP+FN)(TN+FP)(TN+FN)}} \tag{8}$$

TP stands for True Positive, TN for Ture Negative, FP for False Positive, FN for False Negative. AUC is the area under the ROC curve, whose calculation is not easy to be expressed as a formula. Among the 6 metrics, only MCC ranges from −1 to 1, where −1 means a total disagreement between predicting result and true value, 1 means a total agreement, and 0 means a result no better than random prediction. Other 5 metrics all range from 0 to 1 and conform to "the greater the better" rule (Tables 1 and 2).

Table 1. A confusion matrix

		Predicted class	
		Success	Failure
Actual class	Success	TP	FN
	Failure	FP	TN

Table 2. Stock manipulation cases

Stock name	Manipulation information		
	Starting day	Ending day	Stock code
Guoyuan Securities Co., Ltd	2015.7.2	2015.7.28	000728.sz
Jilin Aodong Industry Group Co., Ltd	2015.6.18	2015.7.2	000623.sz
Jiangsu Hua'er Quartz Materials Co., Ltd	2015.6.29	2015.7.31	603688.ss
Quantum Hi-Tech (China) Biological Co., Ltd	2014.4.22	2014.5.22	300149.sz
Digiwin Soft Co., Ltd	2014.8.29	2014.9.3	300378.sz
Jiuzhou Electronic Co., Ltd	2014.4.17	2014.4.29	300040.sz

Sample of some manipulation cases

4.2 Experimental Results

The evaluation of results of our study are illustrated in 0 And 0ACC reflects the general accuracy of the classification, but not of the detection. PPV, on the contrary, perfectly reflect the hit ratio of manipulation detection. TPR, on the other hand, indirectly tells to what extent the models misclassifies the manipulation cases. F1, AUC and MMC are all comprehensive metrics in order to better evaluate SVM and RF classification models (Table 3 and Fig. 1).

Table 3. Results evaluation

Methods	Evaluation metrics		
	ACC	*PPV*	*TPR*
Support vector machine	0.761	0.753	0.765
Random forest	0.812	0.799	0.825
	F1	*AUC*	*MCC*
Support vector machine	0.759	0.761	0.522
Random forest	0.812	0.811	0.624

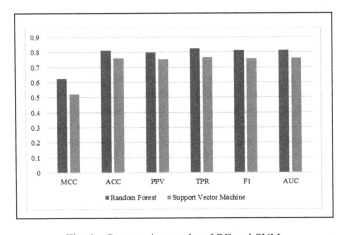

Fig. 1. Comparative results of RF and SVM.

4.3   Detection Analysis

ACC of both models are high, but since we mainly focus on the manipulation detecting, ACC cannot show how well the detection is. A large ACC does not necessarily mean a better detection of manipulation. Therefore, we employed the rest 5 metrics. TPR, also known as recall rate, tells us exactly the rate of correct manipulation detection. The TPR of SVM model is 0.765, while that of RF model is 0.825. Both are rather high, which means most manipulations are successfully detected by our machine learning models and thus the detection is remarkable for the manipulation issues. PPV, on the other hand, presents the accuracy of non-manipulation detection. This is meaningful because we cannot mistake too many non-manipulation cases as manipulation. Apparently, PPV of both models are very high, which means a low rate of mistaking, and hence we can say that the results are quite promising. Apart from these, we also use comprehensive metrics for binary classification problems, F1 score, AUC and MCC, in order to make the results more convincing. All of the three metrics are relatively high according to their own value range. As is illustrated in 0, both machine learning models perform well, and Random Forest (RF) outperforms Support Vector Machine (SVM) in every aspect. One of the reasons is that RF is an ensemble model which deals better with noise and large number of features than SVM.

Apart from the result shown above, we also conduct sensitivity analysis and obtain the conclusion that daily return and market value have a bigger influence on manipulation detection than other variables do.

5   Conclusions and Future Work

5.1   Conclusions

The aim of our research is to develop methods which are capable of detecting the stock-price manipulation in both Shanghai Stock Exchange and Shenzhen Stock Exchange, which can prevent investors from losing money. For this purpose, we use the popular machine learning techniques, Support Vector Machine (SVM) and Random Forest (RF), to estimate a suitable and reasonable model and compare the results of both models. We found that machine learning methods have their own value when detecting stock price manipulation in Chinese stock market. Based on the 6 evaluation metrics, both Support Vector Machine and Random Forest can make meaningful detection based on real time data. As for the comparison, Random Forest has a better performance in all 6 metrics, for ensemble models generally performs better than single models. However, the performance of stock price manipulation detection is not well enough to detect most manipulation cases and make as few mistakes as possible. Therefore, we still have to find some ways to improve the performance.

5.2   Future Work

As for the data, we find the stock trading data from Wind and the number of stock manipulation cases are limited because many manipulation cases might not be found or released by CSRC. A wider range of data might help Support Vector Machine and Random Forest make a better performance in detecting. A more comprehensive and complete list of stock manipulation cases and data can improve the accuracy of detection. What is more, gaining data from some developed stock markets, such as Dow Jones Industrial Average, Standard and Poor's Composite Index and Hang Seng Index, is a reasonable way to enrich our training set and improve the performance of our study.

When it comes to the accuracy, the accuracy 76.1% of SVM and 81.2% of Random Forest are not high enough. There are two ways for us to improve: finding a more efficient method or get data of higher quality. We would like to find more high-quality data in Shenzhen and Shanghai Stock Exchange.

Acknowledgments. To begin with, we would like to extend our sincere gratitude to our supervisor, Dr. Wei Xu, for his instructive advice and priceless suggestions on our thesis. We are deeply grateful of his help in the completion of this thesis.

High tribute shall be paid to the conference, which give us a chance to improve ourselves, and the sponsors who help us improve our manuscript.

Special thanks should go to our friends who have put considerable time and effort into their comments on the draft.

Finally, we are indebted to our parents for their continuous support and encouragement.

References

1. Allen, F., Gale, D.: Stock-price manipulation. Rev. Financ. Stud. **5**, 503–529 (1992)
2. Jarrow, R.A.: Market manipulation, bubble, corners, and short squeezes. J. Financ. Quant. Anal. **27**, 311–336 (1992)
3. Felixon, K., Pelli, A.: Day end returns-manipulation. J. Multinatl. Financ. Manag. **9**, 95–127 (1999)
4. Kumar, P., Seppi, D.J.: Futures manipulation with cash settlement. J. Financ. **47**, 1485–1502 (1993)
5. Gerard, B., Handa, V.: Trading and manipulation around seasoned equity offerings. J. Financ. **48**, 213–245 (1993)
6. Chakraborty, A., Yilmaz, B.: Informed manipulation. J. Econ. Theor. **114**, 132–152 (2004)
7. Thoppan, J.J., Punniyamoorthy, M.: Market manipulation and surveillance: a survey of literature and some practical implications. Int. J. Value Chain Manag. **7**, 55–75 (2013)
8. Aggarwal, R.K., Wu, G.: Stock market manipulations. J. Bus. **79**, 1915–1953 (2006)
9. Punniyamoorthy, M., Thoppan, J.J.: Detection of stock price manipulation using quadratic discriminant analysis. Int. J. Financ. Serv. Manag. **5**, 369–388 (2012)
10. Kim, Y., Sohn, S.Y.: Stock fraud detection using peer group analysis. Expert Syst. Appl. **39**, 8986–8992 (2012)
11. Öğüt, H., Doğanay, M.M., Aktaş, R.: Detecting stock-price manipulation in an emerging market: the case of Turkey. Expert Syst. Appl. **36**, 11944–11949 (2009)
12. Cui, D., Curry, D.: Predictions in marketing using the support vector machine. Mark. Sci. **24**, 595–615 (2005)
13. Lin, Y., Guo, H., Hu, J.: An SVM-based approach for stock market trend prediction. In: The 2013 International Joint Conference on Neural Networks (IJCNN), August 2013

14. Kuo, R.J., Chen, C.H., Hwang, Y.C.: An intelligent stock trading decision support system through integration of genetic algorithm based fuzzy neural network and artificial neural network. Fuzzy Sets Syst. **118**, 21–45 (2001)
15. Chen, M.Y., Chen, D.R., Fan, M.H., Huang, T.Y.: International transmission of stock market movements: an adaptive neuro-fuzzy inference system for analysis of TAIEX forecasting. Neural Comput. Appl. **23**, S369–S378 (2013)
16. Rodriguez, P.N., Rodriguez, A.: Predicting stock market indices movements. In: Brebia, C. (ed.) Computational Finance and Its Applications. Marco Constantino, Wessex Institute of Technology, Southampton (2004). SSRN: http://ssrn.com/abstract=613042
17. Kumar, M., Thenmozhi, M.: Forecasting stock index movement: a comparison of support vector machines and random forest. In: SSRN Scholarly Paper. Social Science Research Network, Rochester, 24 January 2006
18. Patel, J., Shah, S., Thakkar, P., Kotecha, K.: Predicting stock and stock price index movement using trend deterministic data preparation and machine learning techniques. Expert Syst. Appl. **42**, 259–268 (2015)
19. Breiman, L.: Random forests. Mach. Learn. **45**, 5–32 (2001)

Study over Cerebellum Prediction Model During Hand Tracking

Shaobai Zhang[✉] and Qun Chen

College of Computer Science, Nanjing University of Posts
and Telecommunications, Nanjing 210046, China
adzsb@163.com, 443881337@qq.com

Abstract. This paper adopted a new particle filter method to reduce the dimension of particle sampling during hand tracking and describes the posterior probability distribution of state variable with few particles. The manuscript presents three core issues: firstly, we studied the characteristics of relevant kinetics during the hand tracking and the operator's cognitive psychology features under the man-machine interaction condition, and established a cerebellum prediction model by analyzing the operator's behavioral characteristics during hand tracking; secondly, we studied the tracking algorithms related to the cerebellum model built; thirdly, we made a comparison with traditional particle filter algorithm through simulation. As shown in experimental results, the proposed algorithm in this paper can significantly improve both the tracking speed and precision.

Keywords: Cerebellum model · Particle filter · Comparative experiment

1 Introduction

During the recent years human-computer interaction techniques have gained much research focus, the cognitive psychology has gradually become the theoretical basis for human-centered interaction mode and behavioral model has gradually become groundbreaking to solve the high-dimensional problem of hand tracking. From the perspective of cognitive psychology, the pressure, position, visual sense, and other information concerning hand tracking are all related to the interaction channel; which is an important characteristic of hand interaction. However, based on the analysis from computer system's perspective, those concerned by us are closely related to the hand signal's automatic submission of an interaction order in one or more continuous movements and the corresponding degree of freedom etc. This is an important issue during the research on a natural person's hand tracking problem.

Particle filter (PF) is one of the major technical means to realize the moving human hand tracking. The PF algorithm, which is based on Monte Carlo method, signifies the probability with the particle set, so it can be used in the state-space model in any forms; its core thought is to show the particle distribution condition from the random-state particles extracted at the posteriori probability and it is of sequential importance sampling method. The PF method can effectively solve the problems such as, non-Gaussian distribution problem, non-linear problem caused in a shielding condition,

© Springer Nature Singapore Pte Ltd. 2017
H. Yuan et al. (Eds.): GRMSE 2016, Part I, CCIS 698, pp. 159–167, 2017.
DOI: 10.1007/978-981-10-3966-9_17

and complicated background. Thus, it results in an approximate value of the probability distribution of target location at every moment; as a result, in an event of tracking loss, the target can be found again and continuously tracked. Another importance of PF tracking method lies in its strong robustness for the influences of illumination variation, complicated background, hand self-occlusion and other factors. Wei Yucheng uses PF for gesture tracking in his doctoral thesis [1], the particle filter particle mean shift into particle filter process, and avoid particle filter sampling deterioration and the sample impoverishment problem in a certain extent, increasing the number of effective particles. On the basis of real-time gesture tracking, the trajectory analysis method based on time sequence template is proposed, and the two layer classifier is used to recognize the 7 dynamic gestures.

The gesture sampling dimension is one of the major difficulties affecting real-time tracking. However, the PF method needs a large number of samples to show the posteriori distribution probability, therefore to lower the state-space dimension has become a research hotspot. Besides lowering the state dimension at a certain degree by using the relevance between variation scopes of variables determined the human hand's dynamic constraint condition and static constraint condition, people also adopted a variety of methods to study the solutions of high-dimensional states. The general methods are shown below: (1) divide the state space into several subspaces [2–5] (Search Space Decomposition); then, track particles with the PF method in each subspace respectively; (2) obtain the relation between the high-dimensional feature set and hand gestures through machine learning or artificial neural network [5, 6]; (3) use the appearance-based tracking method to avoid a high-dimensional search [7, 8]; (4) lower the dimension of state vector by using intelligent computing and combining various means; (5) reduce the dimension through the Rao-Black wellised particle filter (RBPF); (6) solve the high-dimension problem with an optimization method. Generally speaking, it is widely to transform a high-dimension problem into a low dimension one. However, such methods excessively rely on predicting result of a dynamic model and fail to study Cognitive ability of human behavior. Therefore, even though the expenditure of time and precision can be effectively improved, the difficulties caused by a high-dimensional state space still cannot be solved well through such methods.

The PF method includes two modules, namely prediction and sampling. The Smith prediction method included in the model proposed in this paper is exactly an improvement of prediction model in the traditional PF method. However, the difference from the traditional PF model is as follows: establishes the corresponding behavioral prediction model by combining the cognitive psychology based on the cerebellum mode; finally, it obtains new particles through multichannel regular sampling to reduce the quantity of samplings.

2 Characteristics of Kinetics

The idea of cerebellum intima [9] proposed by Maill et al. in 1998, author suggested that there are two different internal models in the cerebellum: forward model and inverse model. The forward internal model, which can predict the result of an action or behavior, can be used to solve the time delay related to feedback control; the inverse model can provide commands to complete the anticipated actions and behavioral tracks. Besides, the cerebellum can also serve as a delay model, which can exclude the sensorial predictions to match the practical sensorial feedback. The forward and inverse models of cerebellum can copy the current state of hand gesture and an output of motion command generated by the controller as input and to yield state estimation or prediction during hand tracking. In the visual guidance tracking task, the test controls the position of hand by using a certain target and the visual information of hand. Such information is delayed due to visual processing, so they cannot directly change the muscular strength and even the joint angle via the central nervous system to correct a command error. During this rapid hand movement, the feedback of a sense organ can only be used near the end of motion, but the cerebellum model can provide that important missing information. The cerebellum has three basic control structures. The model provides the condition of errors between the anticipated behavior and the practical behavior of the current object; besides, there is also a precise model which can properly delay the output of controller to match with the delayed feedback information. Figure 1 shows the three basic control structures of Smith predictor. In the Fig. 1, the

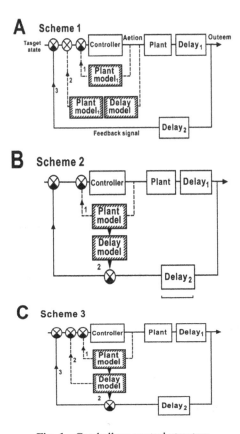

Fig. 1. Cerebellum control structure

controller, plant and (feed forward transmission delay) delay 1 and (feedback transmission delay) delay 2 form a negative feedback loop. There is no transmission delay in loop 1, so it is applicable to the high-gain low-delay negative feedback loop. Thus, when the object model is precise and reliable, loop 1 can offer the optimal control. In loop 2, when there is an error between the practical behavior and anticipated behavior of object, there is a model which can properly delay the output of controller to match with the feedback information in the loop.

Kawato thought that the cerebellum has the same function as of the Smith predictor, therefore, he proposed the Smith predictor cerebellum prediction model (as shown in Fig. 2) [10–13]. The Smith predictor has two forward models: one is used to output the state estimate and the other is used to change and delay the state estimate to form a self-shift estimate. In the cerebrum–cerebellum loop, the forward kinetic model generates a state prediction. Since the physiological sense-motor system has an obvious feedback delay, there is an extra internal delay model, namely forward output model, in the prediction system. The error between the whole predictor (including the forward dynamic model and forward output model) and the true value can be used to correct the corresponding motion during hand tracking and simultaneously maintaining the whole prediction precision.

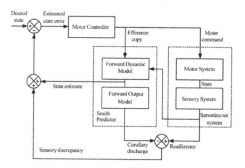

Fig. 2. Smith predictor cerebellum prediction model

3 Characteristics of Hand Motion in the Human-Computer Interaction Scenario

This paper aims to discuss the sampling method of high-dimensional state space from a new perspective so as to improve the time precision and expenditure of traditional PF algorithm. For this reason, we designed a whole set of experimental schemes to study the operator's psychological cognition features and motion physiology features during the operating process.

After analyzing the digital virtual prototyping system and experimental operations on the virtual assembly interaction platform, we divided the experiment into five steps: (1) translation of hand gestures; (2) Grab an object; (3) move to grab the object;

(4) release the object; (5) grab the next object. Divide the tested people into 5 groups with 5 persons in each group; each group of persons completes the same basic operations; then, each tested people is required to wear digital gloves position trackers to complete the same interaction task; during this process, we record the data of each interaction task; their hand gesture change data are stored in the computer's database via the digital gloves. The data acquired are used to analyze and build the prediction model. Finally, tested people are required to describe their psychological cognition processes during the experiment. Figures 3 and 4 show respectively the curve diagram for changes in some velocities and accelerations.

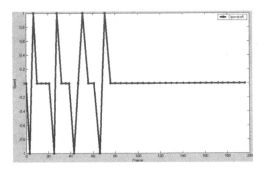

Fig. 3. Velocity change

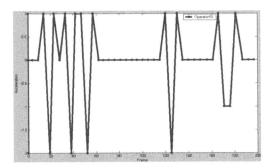

Fig. 4. Acceleration change

The experiment shows that there is almost no change in velocities and accelerations at the first, third and fifth stages, so sampling is not required. In this way, the sampling dimension can be further reduced. However, there are significant changes in velocities and accelerations at the second and fourth stages hence particle sampling can be increased at two stages.

4   Learning Rules of Cerebellum Prediction Model

According to the studies on the Smith predictor cerebellum prediction model in Fig. 2, velocity and acceleration during human hand tracking and human's psychological cognition, we can acquire the learning rules of phased cerebellum prediction model during hand tracking.

The learning rules of cerebellum prediction model, which are subject to the Hebb rules [14], can be signified with the energy function below:

$$E(\omega) = -\psi(\omega^T x) + \frac{\alpha}{2} ||\omega||_2^2 (\alpha \geq 0) \tag{1}$$

ω is the synaptic weight vector; ψ is the differentiable function; x signifies sample vector (representing the velocity and acceleration samples obtained from the experiment); α is Forgetting coefficient. Based on (1), we can obtain the output of cerebellum model as follows:

$$y = \frac{d\psi(v)}{dv} = f(v) \tag{2}$$

$y = \omega^T x$ can derive the learning rules of continuous time via the rapid descending method.

$$\frac{d\omega}{dt} = -u\nabla_\omega E(\omega) \tag{3}$$

u is the learning rate coefficient.

Based on this formula, we can obtain the gradient of (1) below:

$$\nabla_\omega E(\omega) = -f(v)\frac{\partial V}{\partial \omega} + \alpha\omega = -yx + \alpha\omega \tag{4}$$

Thus, we can obtain the learning rules of cerebellum as follows:

$$\frac{d\omega}{dt} = u[yx - \alpha\omega] \tag{5}$$

The learning rules of discrete times are shown below:

$$\omega(t+1) = \omega(t) + u[y(t+1)x(t+1) - \alpha\omega(t)] \tag{6}$$

5   Proposed Algorithm

The proposed algorithm is as follows:

Input: (1) Output status of T-1 time is $X^{(k-1)}$;
      (2) frame image at t moment;
Output: output state $X^{(k)}$ at t moment;

Step 1: predict samples by learning the cerebellum model algorithm to obtain the prediction model;

Compare the image obtained through the cerebellum prediction model and the hand gesture image of current frame to obtain the stage model with minimum error;

Step 2: generate multiple channel samples;

Step 3: calculate the weight of each particle

Step 4: solve the weighted sum of N particles to obtain the output state;

6 Experimental Results

Based on the 10 simulation experiments with the traditional PF algorithm on the interaction test platform, we obtained: Fig. 5 shows the data obtained through the 10 interaction experiments; Fig. 6 shows the corresponding experimental data of proposed algorithm. Through the comparison, it is observed that the time expenditure of the proposed algorithm is significantly improved and the precision is also further enhanced.

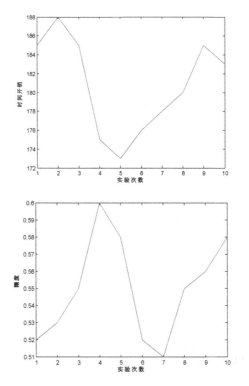

Fig. 5. Time expenditure and precision of traditional PF algorithm

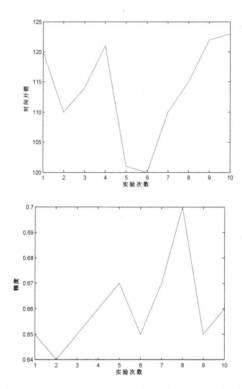

Fig. 6. Time expenditure and precision of algorithm in this paper

7   Conclusion

The proposed algorithm largely improves the precision and time expenditure during hand tracking. The time synchronization and other problems in transmission delay can be effectively handled through the Smith cerebellum prediction model. The combination of Smith cerebellum prediction model and PF classical particle filter algorithm can maximally reduce the quantity of particle samples during tracking and then actually attain the goal of dimension reduction. The posterior probability distribution using fewer particles to describe the state creates favorable conditions for the real-time tracking processing of currently popular particle filter.

There are still many problems to be further studied and explored in the present manuscript. The operator's psychomotor and psychological cognition are required everywhere in the experimental process of human-computer interaction hand tracking based on natural person's hand. Obviously, the study on the operator's psychomotor and psychological cognition in this paper is local and superficial, and the deep relation forms between such psychological features and hand tracking needs further study.

Finally, I would like to thank the national natural fund support,and thank the help of sister Meng-ting SHI and teacher Shao-bai Zhangi.

Acknowledgment. This paper is subsidized by the National Science Foundation (61271334) & (61373065).

References

1. Wei, Y.: Intelligent service robot vision human computer interaction and navigation. Doctoral dissertation. Institute of automation, Chinese Academy of Sciences (2004)
2. MacCormick, J., Isard, M.: Partitioned sampling, articulated objects, and interface-quality hand tracking. In: Vernon, D. (ed.) ECCV 2000. LNCS, vol. 1843, pp. 3–19. Springer, Heidelberg (2000). doi:10.1007/3-540-45053-X_1
3. Deutscher, J., Davidson, A., Reid, I.: Articulated partitioning of high dimensional search spaces associated with articulated body motion capture. In: Proceedings of International Conference on Computer Vision and Pattern Recognition, Hawaii, vol. 2, pp. 669–676 (2001)
4. Zhou, H., Huang, T.S.: Tracking articulated hand motion with eigen dynamics analysis. In: IEEE International Conference on Computer Vision, p. 1102. IEEE Computer Society (2003)
5. Wachs, J.P., Stern, H., Edan, Y.: Real-time hand gesture telerobotic system using the fuzzy C-means clustering algorithm. In: World Automation Congress (WAC), Orlando, Florida, USA, pp. 403–410 (2002)
6. Rosales, R.: The specialized mappings architecture with applications to vision-based estimation of articulated body pose. Ph.D. thesis, University Graduate School of Arts and Sciences, Boston (2002)
7. Shimada, N., Kimura, K., Shirai, Y., Kuno, Y.: Hand posture estimation by combining 2-D appearance-based and 3D model-based approaches. In: Proceedings of 15th International Conference on Pattern Recognition, Barcelona, Spain, vol. 3, pp. 709–712 (2000)
8. Athitsos, V., Sclaroff, S.: An appearance-based framework for 3D hand shape classification and camera viewpoint estimation. In: Proceedings of IEEE International Conference on Automatic Face and Gesture Recognition, pp. 45–50 (2002)
9. Zhang, S., Zhang, Z.: Design of the control model of the cerebellum in the coordination of arm extension and grasping movement time. J. Electron. Inf. Technol. **11**, 2607–2613 (2014)
10. Zhang, S., Zhang, Z., Zhou, N.: A new control model design for the temporal coordination of arm transport and hand preshape applying to two-dimensional space. Neurocomputing **168** (C), 588–598 (2015). (SCI)
11. Mail, R.C., Weir, D., Wolpert, D.M., et al.: Is the cerebellum a smith predictor. J. Motor Behav. **25**(3), 203–216 (1993)
12. Zhang, S., Cheng, W.: An application of cerebellar control model for prehension movements. Neural Comput. Appl. **24**(5), 1059–1066 (2014). (SCI)
13. De Lucia, M., Michel, C.M., Murray, M.M.: Comparing ICA-based and single-trial topographic ERP analyses. Brain Topogr. **23**(2), 119–127 (2010)
14. Tian, D., Liu, Y., Li, B.: Distributed neural network learning algorithm based on Hebb rule. J. Comput. Sci. **30**(8), 1379–1388 (2007)

Forecasting for the Risk of Transmission Line Galloping Trip Based on BP Neural Network

Lichun Zhang$^{(\boxtimes)}$, Bin Liu, Bin Zhao, Xiangze Fei,
and Yongfeng Cheng

China Electric Power Research Institute, CEPRI, Beijing, China
{zhanglc,liubl,fxz,cyf}@epri.sgcc.com.cn,
zb9991987@163.com

Abstract. Due to the strong randomness and nonlinear characteristics of the transmission line galloping, the prediction of the intensity and the characteristics (amplitude, frequency, trip rate, etc.) of the galloping cannot reach a high precision. A BP neural network model is employed to map three main meteorological factors and galloping trip-out risk. Three main meteorological factors, temperature, humidity and wind speed were used as input parameters and the risk of galloping trip as the output parameter of the model. Typical galloping data from State Grid Corporation were used to verify the validity of the model. In order to counteract random factors, the operations were performed for 20 times with the same training and testing data. All of the network results had more than 90% accuracy and the average rate was 92.3%. The results show that it is feasible to use this model to predict the risk of transmission line galloping trip. The research results can provide support for the transmission line galloping prediction and early warning technology, so as to improve the level of intelligent operation and maintenance of power grid.

Keywords: Component · Transmission line · BP neural network · Galloping trip · Risk forecasting

1 Introduction

Galloping is a kind of vibration with low frequency and large amplitude when iced conductor in transmission line is under wind excitation. It would bring large amount of damages with different forms to transmission lines. For example, it always would lead to line trip, brand breakages damaged fittings, etc. Severe cases could lead to conductor's breakage and tower's falling down (as shown in Fig. 1).

Domestic and overseas scholars had carried out a great deal of comprehensive research work on the phenomenon of dancing, and achieved fruitful results [1–6]. However, due to the randomness and strong non-linearity of galloping, to data here there is still no unified control technology to make galloping disasters completely stay away from grid. So under certain weather conditions, galloping still appeared in the lines that had been taken anti-galloping measures. In this premise, it is an effective way to reduce the loss of galloping disaster by using the intelligent control technology to carry out the research on the risk early warning and forecasting. According to the

© Springer Nature Singapore Pte Ltd. 2017
H. Yuan et al. (Eds.): GRMSE 2016, Part I, CCIS 698, pp. 168–175, 2017.
DOI: 10.1007/978-981-10-3966-9_18

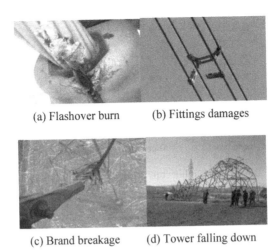

(a) Flashover burn   (b) Fittings damages

(c) Brand breakage   (d) Tower falling down

Fig. 1. Failures brought by galloping

forecast result, operators of power grid could take preventive measures to minimize the loss.

With the rapid development of smart grid technology, artificial intelligence control played an important role in the field of power grid construction, operation and maintenance. Based on the artificial neural network technology, this paper established a risk prediction model of transmission line galloping trip, and made a reasonable assessment of the risk of galloping trip in transmission line. The research results could provide powerful support for the transmission line galloping warning technology, which had not only important academic significance, but also very practical engineering application value.

2 Galloping Trip Model Based on BP Network

Wherever Times is specified, Times Roman or Times New Roman may be used. If neither is available on your word processor, please use the font closest in appearance to Times. Avoid using bit-mapped fonts if possible. True-Type 1 or Open Type fonts are preferred. Please embed symbol fonts, as well, for math, etc.

2.1 Input and Output Parameters Selection

According to the achievements on influencing factors of transmission lines galloping [13], three high related factors, including wind speed, temperature and humidity, are selected as the input parameters of the network. The output is the trip, which contains two kinds of results, "yes" and "no". Hidden layer is only one. The topological structure has been shown in Fig. 2.

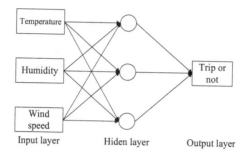

Fig. 2. BP network architecture for galloping trip risk prediction

2.2 Data Preprocessing

(1) Data format conversion

One of the limitations of BP neural network is the strong dependence to samples. Final effect of algorithm is closely related to the typicality of learning samples. If the representative of sample collection was bad and lots of contradictory samples were contained in it, the network would be difficult to achieve the expected performance. Therefore, this paper illustrates the original data pretreatment, and conflicting samples are deleted. At last, 356 historical dancing records appeared in the power grid in China from 1978 to 2010 are selected to train and test the BP neural network.

According to the threshold range of influence factor [13], as input parameters, the value of wind speed, temperature and humidity was represent by code, which could effectively facilitate the establishment of model. Detail convert way is shown in Table 1. Output of trip has two forms, which respectively use "1" instead of "yes", and "0" instead of "no".

According to the above data convert rules, the 356 samples are pretreated. Table 2 lists partial original data and the corresponding data format (due to the limit of length, only the first 10 data are displayed).

Table 1. Threshold convert rules for network input parameters

Parameter	Threshold range	Code
Wind speed (V, m/s)	$0 < V < 4$	1
	$4 \leq V \leq 25$	2
	$V > 25$	3
Temperature (T, °C)	$T < -10$	1
	$-10 \leq T < 5$	2
	$-5 \leq T \leq 4$	3
Humidity (H)	$0 < H < 70\%$	1
	$70\% \leq H \leq 100\%$	2

Table 2. Comparison of data before and after data processing

	Raw data				Processed data			
	Wind speed (V, m/s)	Temperature (T, °C)	Humidity (H)	Tripping or not	Wind speed (V, m/s)	Temperature (T, °C)	Humidity (H)	Tripping or not
1	8–12	−3	80%	Yes	2	3	2	1
2	8–12	−3	80%	Yes	2	3	2	1
3	25–28	−3	80%	Yes	3	3	2	1
4	17–20	−4	80%	Yes	2	3	2	1
5	17–20	−4	80%	Yes	2	3	2	1
6	25–28	−1	80%	Yes	3	3	2	1
7	2–3	−4	70%	Yes	1	3	2	1
8	2–3	−2	70%	Yes	1	3	2	1
9	2–3	−2	70%	Yes	1	3	2	1
10	15	1	60%	No	2	3	1	0

(2) Normalization of data

As the input of artificial neural network, above data should be dimensionless vector. In order to reveal further how much the reaction of output vector is caused by changes of any input vector, both of two vectors is normalized to [−1, 1]. Normalization formula is in the following manner:

$$f(x) = 2(x - x_{\min})/(x_{\max} - x_{\min}) - 1 \qquad (1)$$

Where x_{max} and x_{min} are any given input parameter's maximum and minimum value separately; $f(x)$ is the normalized results, and ranged in [−1, 1]. This normalization method can be implemented by function mapminmax in MATLAB toolbox.

(3) Division of test data and training data

In order to verify the performance of BP neural network, the sample collection was divided into training data and test data. 296 data are selected as training data to predict the risk of galloping trip and 60 data are used as test data to test the prediction accuracy of the neural network model.

3 Risk Prediction of Galloping Trip Based on BP Network

The structure of BP neural network is directly related to the function mapping ability and network performance of the network. The design of network structure includes the choice and determination of the number of input and output nodes, the number of hidden layers and the number of hidden layer nodes. The most important one is the choice of the number of hidden layer and the number of hidden layer nodes.

3.1   Numbers of Hidden Layers and Hidden Layer Nodes

In this paper, we select one implicit layer, which is a typical three layer BP network structure (as shown in Fig. 2).

The determination of the number of nodes in the hidden layer is directly related to both of the requirements of the problem and the numbers of input and output neurons. There is no an ideal analytic formula can be used to determine the reasonable number of the hidden layer neurons, so the usual practice is using the empirical formula to estimate, as shown in (2) [14].

$$\sum_{i=0}^{n} C_M^i > k \tag{2}$$

Where, k is the number of samples; M is the number of hidden layer neurons; n is the number of input layer neurons. If $i > M$, let $C_M^i = 0$.

After training and testing again and again, number of hidden layer neurons is finally determined to be 15. The number of iterations, the required time, the error rate and other indicators could achieve the best.

According to the above network structure selection method, the final galloping trip risk prediction of the BP network structure is determined as 3-15-1, as shown in Fig. 3.

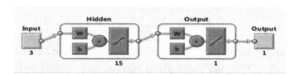

Fig. 3. BP network diagram structure for transmission line galloping trip risk prediction

3.2   Network Function and Weight Determination

Select the default Sigmoid function as transfer function, and LM (Levenberg- Marquardt) algorithm as learning function, this algorithm has been widely used, because it converges fast, and has small mean square error. The initial weights and thresholds are both used default values.

3.3   Processing of Output Data

The output value of the BP network is not limited to 1 or 0, but a real number, so it is needed to convert the output to an integer. Take 0.5 as the threshold, the output that less than the threshold is 0 (no), and the one that is greater than the threshold is 1 (yes).

After all of the above parameters are set up, the network can be trained and tested, and all the processes are shown in Fig. 4.

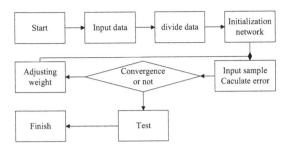

Fig. 4. BP network flow chart for galloping trip risk prediction

3.4   Calculation Result

In order to counteract the effect of random factors, take the same training and test samples for 20 operations, and each output result calculated by network is compared with the actual dancing tripping results among 60 test samples. Accuracy rate and the number of iterations would be counted to verify error performance and convergence speed of the network model.

Statistical results are shown in Table 3. Obviously, accuracy of each calculation is more than 90%. Average accuracy of 20 times operation was 92.3%. Average number

Table 3. The accuracy and the number of iterations of predicting galloping trip risk prediction with BP neural network

	Accuracy	Iteration number
1	90%	4
2	90%	5
3	90%	4
4	90%	5
5	90%	4
6	90%	5
7	95%	4
8	95%	4
9	95%	4
10	95%	4
11	95%	4
12	90%	4
13	90%	4
14	95%	4
15	95%	4
16	95%	5
17	90%	5
18	95%	5
19	95%	5
20	95%	5

of iterations is 4.4 times. In conclusion, the BP network model for prediction dancing trip risk has higher accuracy as well as faster convergence speed.

Select one result with 95% accuracy, the calculation result and the corresponding actual trip have been listed in Table 4 (due to the limit of length, partial results are shown).

Table 4. The predicted value of the trip risk with BP network and the true value

	Calculated value of galloping trip	The true value of galloping trip
1	1	1
2	1	1
3	1	1
4	1	1
5	1	1
6	1	1
7	1	1
8	1	1
9	1	1
10	1	0
11	1	0
12	1	1
13	1	1
14	1	1
15	1	1
16	1	0
17	1	1
18	1	1
19	1	1
20	1	1

4 Conclusion

(1) The results showed that the establishment of galloping trip risk prediction model for transmission lines had higher accuracy, and faster convergence speed, which could bring satisfactory results. Considering three main meteorological elements that have great influence on galloping, this paper made reasonable galloping trip risk assessment. It is obviously that using artificial neural network technology for transmission line galloping trip risk prediction is feasible.

(2) The above model only considered three meteorological elements in predicting galloping trip risk. But in the actual operation of transmission lines, a variety of factors had influence on galloping, which included line's structure parameters, topographic features, installation of anti-galloping device, etc. And this needed further in-depth study. Using the powerful nonlinear mapping ability of neural

network, as well as considering various factors, a more accurate prediction model was set up, which could provide support for the realization of intelligent early warning.

Acknowledgment. This work is supported by State Grid Corporation Project, GC71-16-011.

References

1. Wang, S., Jiang, X., Sun, C.: Study status of conductor galloping on transmission line. High Volt. Eng. **31**(10), 11–14 (2005)
2. Lilien, J.L., Vinogradov, A.: Full-scale tests of TDD anti-galloping device (torsional damper and detuner). IEEE Trans. Power Deliv. **17**(2), 638–643 (2002)
3. Tu, M., Zhang, L., Zhu, K.: Zoning method for galloping of transmission lines. Electr. Power Constr. **32**(4), 26–28 (2011)
4. Zhang, L., Zhu, K.: Research on the law of disaster caused by transmission line galloping. Power Syst. Clean Energy **28**(9), 13–17 (2012)
5. Li, J., Cheng, Y.: Improvement and application on the meteorological geographical method in the drawing of galloping region distribution map. Electr. Power **34**(3), 210–215 (2010)
6. Zhu, K., Liu, C.: Analysis on dynamic tension of conductor under transmission line galloping. Electr. Power **38**(10), 40–44 (2005)
7. Yang, Q., Sima, W.: The building and application of a neural network model for forecasting the flashover voltage of the insulator in complex ambient conditions. In: Proceedings of CSEE, vol. 25, no. 13, pp. 155–159 (2005)
8. Ling, Y., Yan, P.: BP neural network based cost prediction model for transmission projects. Electr. Power **45**(10), 95–99 (2012)
9. Zhu, S., Chen, H.: Forecasting insect appearance areas of dendrolimus punctatus walker by applying artificial neural network (ANN). Chin. J. Agrometeorol. **25**(1), 51–53 (2012)
10. Li, S., Wunsch, D., O'Hair, E.: Using neural networks to estimate wind turbine power generation. In: Power Engineering Society Winter Meeting, p. 977 (2001)
11. Bai, P., Fang, T.J., Ge, Y.J.: Robust neural network compensating control for robot manipulator based on computed torque control. Control Theory Appl. **18**(6), 895–903 (2001)
12. Rong-Long, W.: Robust control for nonlinear motor mechanism coupling system using wavelet neural network. IEEE Trans. Syst. Man Cybern. **33**(3), 489–497 (2003)
13. Li, J., Wang, T.: Rendering method of galloping distribution map based on meteorological-geographical method. Electr. Power constr. **35**(7), 97–103 (2014)
14. Ming, C.: MATLAB Neural Network Principle and Essence of Instance, pp. 1–199. Tsinghua University Press, Beijing (2013)

A Features Fusion Method for Sleep Stage Classification Using EEG and EMG

Tiantian Lv$^{(\boxtimes)}$, Xinzui Wang, Qian Yu, and Yong Yu

Suzhou Institute of Biomedical Engineering and Technology,
Chinese Academy of Science, Suzhou, China
Britney_lv@126.com, {wangxz,yuq,yuy}@sibet.ac.cn

Abstract. To achieve accurate sleep stage classification and improve its generalization ability, we presented a features fusion method to classify sleep stage using Electroencephalogram (EEG) and Electromyography (EMG). We regarded EEG and EMG samples from MIT-BIH Polysomnographic database as analysis objects. First of all, we used the Discrete Wavelet Transform (DWT) to filter noise of signals and extract energy ratio of α, β, θ and δ wave from EEG and the high frequency component from EMG, and used Sample Entropy (SampEn) algorithm to extract nonlinear characteristics of EEG. Then, we compared the accuracy difference of sleep stage classification method between using EEG and using EEG and EMG features fusion by inputting these features to Support Vector Machine (SVM) classifier to train and test. Finally, we used cross-validation method to train and test different samples to verify its generalization ability. The experiment of testing accuracy showed a satisfactory result with accuracy of 91.86%, and the average accuracy raised 4.94% compared to the sleep stage classification method using EEG. The cross-validation results indicated that this method has better generalization ability.

Keywords: EEG · EMG · Energy feature · Sample entropy · Features fusion · Support vector machine classifier · Sleep stage classification

1 Introduction

In this rapid developing society, life pressure is becoming more and more intense thus many people have appeared symptoms of sleep disorder which has huge impacts on people health. Currently, sleep disorder has been identified as a common and harmful disease [1]. Disease related to sleep is diagnosed and treated based on assessment of sleep quality. An objective and effective way of achieving sleep quality assessment is sleep staging through human physiological signals.

The classical sleep staging method is that many characteristic parameters from EEG (Electroencephalogram, EEG) were extracted by using different analytical methods and then classifier can be used to classify sleep stages. Liu [2] used the nonlinear analysis of symbolic dynamics, spectrum analysis and detrended fluctuation analysis to reach the mean of sleep staging accuracy of 92.87%, however, the generalization ability of algorithm is still need to be improved since it only trained and validated separately each sample. Xie [3] improved the generalization ability of model by adopting discrete

© Springer Nature Singapore Pte Ltd. 2017
H. Yuan et al. (Eds.): GRMSE 2016, Part I, CCIS 698, pp. 176–184, 2017.
DOI: 10.1007/978-981-10-3966-9_19

wavelet transform combined nonlinear SVM, but the accuracy rate (81.56%) still could not meet the requirement.

In order to improve the accuracy of sleep staging, several experts paid attention to sleep staging study by using other body physiological signals, such as ECG, EOG, and EMG and so on. Some studies combined EEG and EOG to classify sleep stages [4], and others do sleep classification using EOG and EMG [5], which could perform better performance.

In this study, we proposing a sleep staging method using EEG and EMG features fusion in order to increase its generalization ability and accuracy. Since EEG and EMG have different characteristics wave in various sleep stages, we can extract the time-frequency domain features by calculating energy feature ratio. Besides, we adopted sample entropy algorithm which is suitable for nonlinear analysis of biological signals to extract nonlinear features of EEG because of the EEG signal's non-stationary. We inputted these features into support vector machine classifier to obtain a sleep staging method with higher accuracy and broader generalization.

2 Materials and Method

2.1 Data Source

This article used two signals included EEG and EMG from the MIT-BIH database polysomnography [6] that recorded several physiological signals of 16 test subjects during sleep. This study chose four samples included slp41, slp45 and slp48 as subjects because they had integral sleep stages. And the sampling frequency is 250 Hz. In order to test the accuracy and generalization ability of method, the result was compared to the sleep stage results that recorded by an experienced doctor after every 30 s.

2.2 Method

The flow diagram of this method was shown in Fig. 5. Firstly, the high-frequency noise from raw signals was filtered by using wavelet decomposition. Secondly, the energy ratio of α, β, θ and δ waves and sample entropy from EEG were extracted as the first part of the feature parameters. Thirdly, the energy ratio of high-frequency from EMG was extracted by using wavelet decomposition as the second part of the feature parameters. Finally, the first part of the feature parameters were inputted SVM classification to train and test compared to two parts of the feature parameters.

(1) **Data Preprocess**

EEG and EMG usually contain some unknown frequency components, especially ECG and muscle and eye movements will largely have an effect on EEG. Therefore, wavelet decomposition was used to filter these noises in order to improve the measurement accuracy in this study.

Discrete Wavelet Transform (DWT) is so automatically adapt to analysis time-frequency signal that makes it particularly suitable for non-stationary signals. Generally speaking, the appropriate decomposition level was selected according to the characteristics of signals in practice. For example, the signal was decomposed into i layer, and the frequency range of nodes included A_i and D_i is respectively $0 \sim \frac{fs}{2^{i+1}}$ and $\frac{fs}{2^{i+1}} \sim \frac{fs}{2^i}$, where fs is the sampling frequency.

Generally, researchers are interested in the frequency range of EEG between 0.5 Hz and 35 Hz [7], and the high-frequency components of representing EMG muscle movement are more concentrated in the 30 to 125 Hz [8] in clinical medicine. The signals' sampling frequency is 250 Hz in the database, therefore, the raw signals of EEG and EMG were respectively decomposed into 7 and 3 layer by using "db4" Wavelet. Then a heuristic threshold method was used to filter. The effect was shown in Fig. 1 after data of EEG and EMG from object slp48 for 10 s was filtered.

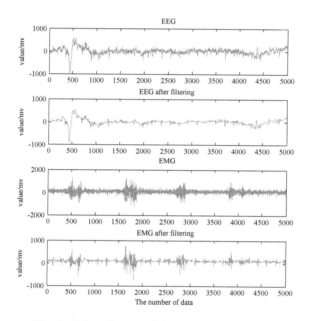

Fig. 1. Effect diagram of EEG and EMG denoise

(2) Feature Extraction

According to the 2007 American Academy of Sleep Medicine (AASM) developed, sleep can be divided into five stages: the Wake phase (W phase), Non-Rapid Eye Movement 1 phase (N1 phase), Non-Rapid Eye Movement 2 phase (N2 phase), Non-Rapid Eye Movement 3 phase (N3 phase) and Rapid Eye Movement phase (REM phase) [9]. It is the crucial step of achieving accurate sleep staging to extract features represented the various sleep stages. Some features of each sleep stage were shown in Table 1.

In order to analysis features, we need to choose some data of sleep subjects' five stages. For example, slp45 sample was chose as analysis object, whose sleep time was 380 min and the length of each data for 30 s was 7500 point. We chose data what each of the stages from EEG and EMG for 25 min.

(a) *Energy Feature*

According to the corresponding relationship of Table 1 between sleep stages and characteristic waves of EEG and EMG, EEG was decomposed into 7 layer by using "db4" Wavelet. Based on the relationship of wavelet decomposition nodes and frequency, D3 represents β wave, D4 represents α wave, D5 represents θ wave and D6 + D7 represents δ wave. The energy ratios of four waves on total energy (0–30 Hz) are calculated. Then EMG was decomposed into 3 layer by using "db4" Wavelet, D1 + D2 represents muscle movement frequency (30–125 Hz), and the energy ratio of that on total energy (0–125 Hz) was calculated. The energy ratio was calculated as in Eqs. (1) and (2):

$$\eta_i = \frac{\sum_{k=1}^{n} |D_i(k)|^2}{E_s} \tag{1}$$

$$E_S = \sum_{k=1}^{N} |D_i(k)|^2 \tag{2}$$

Where, η_i represents energy ratio of the i-th layer frequency band on the total energy. $D_i(k)$ represents the k-th wavelet coefficient of the i-th layer. n represents the data number of i-th layer. E_s represents total energy. N represents the data number of total layers.

The results were shown in Fig. 2, which energy ratio of α, β, θ, δ waves and EMG high-frequency components for each sleep stage were calculated. And the average values of feature energy ratio in different sleep stages were shown in Fig. 3.

We can make conclusion from Figs. 2 and 3: energy ratio of α wave was the most obvious in Wake phase, which began to decrease with the depth of sleep, however, it began to increase when reached REM phase; energy ratio of β wave was smaller than that of α wave, which is similar to α wave in each sleep phase; energy ratio of θ wave was less than other waves in the whole sleep process, but it increased obviously in REM phase; energy ratio of δ wave accounted for a larger proportion in the whole sleep process and achieved maximum in the N3 phase; energy ratio of EMG high frequency components was largest in the Wake phase, followed by N3 phase, and almost had little in the REM phase. These features met the characteristics listed in Table 1, which shown that the method based on discrete wavelet transform to extract feature energy ratio of EEG and EMG can achieve sleep stages classification (Fig. 3).

(b) *Sample Entropy*

Sample Entropy (SampEn) is a measurement method of measuring sequence complexity, which calculation is faster and more accurate.

Sample entropy can be represented with SampEn (m, r, N), where, m is the embedding dimension, which is usually 1 or 2, but we will select 2 in the practical application; r is similar tolerance, He [10], who believed that the result is better when r = 0.2SD (SD standard deviation of the original data); N represents the data length, He [10] proved that the experiment result is ideal when N = 1000. The Sample Entropy of EEG was shown in Fig. 4.

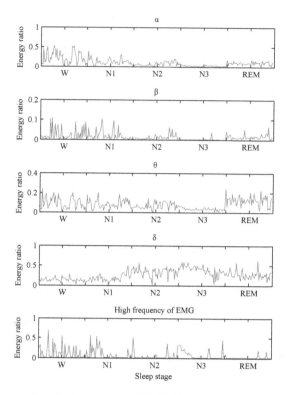

Fig. 2. Energy ratio of four rhythm waves and EMG high frequency components in each sleep stage

Table 1. Features of each sleep stage

Sleep stages	Features
W	α(8–13 Hz), β(13–30 Hz), highest EMG
N1	θ(4–8 Hz), higher EMG
N2	Sleep spindles(12–14 Hz), EMG is weaker than N1
N3	δ(1–4 Hz), lower EMG
REM	α(8–13 Hz), β(13–30 Hz), EMG disappeared

From Fig. 4 we can see that Sample Entropy of EEG was the largest. Because EEG complexity was high due to strong brain activity in Wake phase. Brain activity weakened with the depth of sleep, Sample Entropy of EEG correspondingly decreased. However, EEG activity increased with the brain began to dream in the REM phase. It caused that complexity of EEG increased, so Sample Entropy started to increase. These obvious difference features suggested that Sample Entropy can further show sleep information.

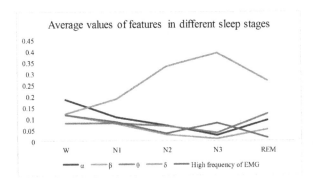

Fig. 3. Average value of different feature parameters in each sleep stage

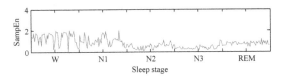

Fig. 4. EEG sample entropy in each sleep stage

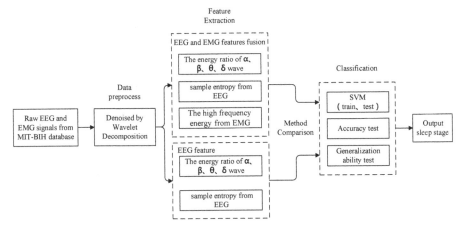

Fig. 5. Flow diagram of sleep staging based on EEG and EMG features fusion

(3) **Support Vector Machine (SVM) Classifier**

SVM can solve many problems such as small sample, nonlinear, high dimension and so on, which is becoming the preferred method of pattern recognition [11]. For nonlinear problems, its basic idea is that function is transformed into linear problem by $x \rightarrow \phi(x)$ mapping into a high-dimensional space in which the optimal classification hyperplane is built. This mapping is achieved by kernel function $K(x_i, x_j) = \phi(x_i) \cdot \phi(x_j)$ [12] to take the optimal classification function, as in:

$$y = sgn\left\{\sum\nolimits_{i=1}^{m} \alpha_i y_i K(x_i, x) + b\right\} \tag{3}$$

This paper selected RBF kernel function, as in:

$$K(x_i, x) = exp(-\gamma * \|x - x_i\|^2) \tag{4}$$

Three steps that this paper utilized SVM to classify were as follows.

Step 1: Six feature parameters of slp41, slp45 and slp48 sample overnight sleep data where slp41 totals 780 groups, slp45 totals 755 groups and slp48 totals 760 groups were extracted.

Step 2: Mixing feature parameters of slp41, slp45 and slp48 samples, we could obtain a total of 2295 groups' sample features, where 70% was used to establish sleep SVM classification model as training samples.

Step 3: The remaining was used to test the accuracy of the classification as testing samples.

3   Results

In order to verify the advantages of this method, we designed an experiment what compared the accuracy between sleep stages classification using EEG and sleep stages classification using EEG and EMG features fusion. Classification results of the two methods were shown in Table 2. The average increase rate of the sleep staging accuracy was shown in Table 3.

Table 2. Comparison of automatic sleep staging results between with EMG and without EMG

Sleep stage	Training samples (70% of slp45 and slp48)			Testing samples (30% of slp45 and slp48)		
	Number	Accuracy (%)		Number	Accuracy (%)	
		No EMG	Add EMG		No EMG	Add EMG
W	392	96.43	100	168	88.10	96.42
N1	368	95.38	99.73	157	84.08	90.45
N2	622	95.18	95.34	267	85.39	88.76
N3	83	96.39	100	35	87.50	91.43
REM	142	95.77	100	61	85.25	96.72
Total	1607	95.64	98.13	688	85.47	91.86

To further verify the generalization capability of the method, this experiment utilized cross-validation to train and test different samples. Firstly, we chose the one of slp41, slp45 and slp48 samples as the training set to build model, and then the remaining two samples were used to test by the model. The generalization capability testing results were shown in Table 4.

4   Discussions

Table 2 shown that the classification result of sleep staging method using EEG and EMG features fusion is better, and the overall accuracy rate can reach 91.86%, which is better than many sleep staging methods mentioned in most articles.

Through Tables 2 and 3 we can see that the accuracy of every sleep stage has increased, especially REM phase is the highest and W phase follows after adding EMG. However, the increase rate of N2 and N3 phase is lower after adding EMG. Therefore, it is a focus of future research directions for us to improve the accuracy of sleep staging. Overall, the average increase rate of sleep staging accuracy is 4.94% after adding EMG. It indicates that adding EMG can effectively improve the accuracy of sleep staging.

Table 3. Average increase rate for each sleep stage accuracy

Sleep stage	W	N1	N2	N3	R	Average
The average increase rate (%)	5.95	5.36	1.77	3.77	7.85	4.94

By testing generalization ability in Table 4, we can see that the effect of cross-validation between different samples was ideal and the average accuracy rate reached 82.68%. It indicates that this sleep staging method based on EEG and EMG features fusion has certain generalization ability compared to the previous algorithm. There are some factors will affect the model accuracy such as artificial calibration errors, the continuous of various sleep stages.

Table 4. Test results of generalization ability of the staging sleep based on EEG and ECG features fusion

Training samples	Testing samples			
	slp41	slp45	slp48	The average accuracy
slp41	——	82.01%	86.43%	84.22%
slp45	81.41%	——	84.45%	82.93%
slp48	82.69%	79.10%	——	80.90%
The average accuracy	——	——	——	82.68%

5   Conclusion

In this study, a features fusion method for sleep stage classification using EEG and EMG was proposed. Through contrastive experiment, we conclude that working with EMG can significantly improve sleep staging accuracy. Cross-validation results show that the method has a certain generalization. The experiment results have high reliability, and the method can complete accurately sleep staging, which provide an effective basis for assessing the quality of sleep, which has great prospects in future clinical application. The results also provide many other interesting resources in classifying sleep stages by using multi-parameter to obtain more accurate results.

Acknowledgment. This work is partially supported by Science and Technology Program of Suzhou-the Medical Devices and New Medicine Program (ZXY201427, ZXY201429).

References

1. Neonates, F.P., Kushida, C.A.: Sleep Deprivation, pp. 121–150. Marcel Dekker, New York (2004)
2. Liu, Z., Zhang, H., Zhao, H.: Study on sleep staging algorithm based on EEG signals. Chin. J. Biomed. Eng. **34**(6), 693–700 (2015)
3. Xie, H., Shi, X.: Research on sleep staging based on discrete wavelet transform for electroencephalogram. Softw. Algorithms **34**(16), 18–20 (2015)
4. Estrada, E., Nazeran, H., Nava, P., Behbehani, K., Burk, J., Lucas, E.: Itakura distance: a useful similarity measure between EEG and EOG signals in computer-aided classification of sleep stages. In: Proceedings of the 2005 IEEE Engineering in Medicine and Biology 27th Annual Conference, Shanghai, China, pp. 1–4, September 2005
5. Li, J., Chen, H., Ye, S.: A self-adaptive threshold method for automatic sleep stage classification using EOG and EMG. MATEC Web Conf. **22** (2015). Article no. 05023
6. Goldberger, A.L., Amaral, L.A.N., Glass, L., Hausdorff, J.M., Ivanov, P.C., et al.: PhysioBank, PhysioToolkit, and PhysioNet: components of a new research resource for complex physiologic signals. Circulation **101**(23), e215–e220 (2000)
7. Lajnef, T., Chaibi, S., Ruby, P., Aguera, P.E., Eichenlaub, J.B.: Learning machines and sleeping brains: automatic sleep stage classification using decision-tree multi-class support vector machines. J. Neurosci. Methods **250**, 94–105 (2015)
8. Constable, R., Thornhill, R.J., Carpenter, D.R.: Time frequency analysis of the surface EMG during maximum height jumps under altered conditions. Biomed. Sci. Instrum. **30**, 69–74 (1994)
9. Wang, H.: The latest interpretation of analytical standard guide on the American Academy of sleep medicine sleep staging. Diagn. Theor. Pract. **8**(6), 575–578 (2009)
10. Weixing, H., Xiaoping, C., Junting, S.: Sleep EEG staging based on sample entropy. J. Jiangsu Univ. **30**(5), 501–504 (2009)
11. Baudat, G., Anouar, F.: Kernel-based methods and function approximation. In: IJCNN 2001 International Joint Conference on proceedings of the Neural Networks, pp. 1244–1249 (2001)
12. Run, J., Luo, D., Luo, H.: Indoor and outdoor scene recognition algorithm based on support vector machine multi-classifier. J. Comput. Appl. **35**(11), 3135–3138 (2015)

Community Detection Algorithm with Membership Function

Dongming Chen, Lulu Jia, Dongfang Sima, Xinyu Huang[(✉)],
and Dongqi Wang

Software College of Northeastern University, Shenyang 110169, China
neuhxy@163.com

Abstract. Most of the community detection algorithms underperform on overlapping community structures. To eliminate the ambiguity, we define a membership function which can compute a node's subordinate level to communities. This paper proposed a heuristic community detection algorithm with membership function (MCDA) utilizing which as a measure for community detection. By computing one node's membership to each community, we can find which community it belongs to. Considering the edge connectivity, information transmission efficiency and other factors, some experiments are taken to demonstrate that the proposed algorithm perform higher accuracy and lower time complexity.

Keywords: Complex network · Community discovery · Membership function · Dynamic network

1 Introduction

In the real network system, the formation of community structure is closely related to many factors, such as connectivity between the nodes, the information transmission efficiency, nodes' attribute similarity and so on. However, most of the existing classical community discovery algorithms [1–3] only consider a single impaction on community detection [4], so these algorithms have minimal scope of application in the real network. In recent years, with the in-depth study of complex network [5–7], people recognized that in real world network system, some nodes are not only members of one group, but activities or actions in a variety of communities, that is to say, there are many overlapping nodes in complex network [8]. This phenomenon fully explained that the definition of the community is fuzzy [9], and make it difficult to get definite results when the boundary of community structure is not clear. So we need to find a solution for vague community definition, and use explicit data to illustrate the membership of a node to each community. It is of great significance to the optimization of the current community detection algorithm.

Aiming at the existing problem of the above mentioned community detection algorithm, we propose a heuristic community discovery algorithm based on membership function [10] (MCDA). Some factors that influence the community detection results are fully taken into consideration when defining the membership function, so we can give a comprehensive subordinate level to each community. First of all, we need to

© Springer Nature Singapore Pte Ltd. 2017
H. Yuan et al. (Eds.): GRMSE 2016, Part I, CCIS 698, pp. 185–195, 2017.
DOI: 10.1007/978-981-10-3966-9_20

define a number of nodes as the initial community. We all know scale-free [11] is a conspicuous feature for complex network, which means the new emerged nodes tend to be connected to nodes with higher degree. As for a node, the larger of the degree and clustering coefficient, the more of the cohesion in a community. After confirming the core node, the rest of the nodes are divided into each community according to the membership function.

The rest of this paper is organized as follows. Section 2 defines community detection algorithm with membership function and analyzes the complexity. Section 3 gives some experiments to demonstrate the good performance on accuracy and efficiency of the proposed algorithm. Finally, Sect. 4 makes the conclusions.

2 MCDA Algorithm

2.1 Related Concepts and Definitions

(1) **Node importance function K**

Function K indicates the importance of a node in a community. Two factors are included in the definition of function K, the degree of the node, and the sum of clustering coefficient and parameter y. Generally speaking, core nodes are often contained in a community, and they are often with bigger degree and clustering coefficient in the community or even the whole network. The importance of a node is proportional to the degree of the node as well as the cluster coefficient [12] of the node, so we define it by using the product of the degree of the node and the clustering coefficient C, and search for the core node of the network according to this function. The node importance degree function is as formula (1):

$$K(i) = d_i * (C_i + y) \tag{1}$$

where i is an arbitrary node in community V, and C_i is the clustering coefficient of node i. In different network, the impact of a node degree on the core level varies, so we select a parameter y which is valued from 0 to 1 to make a modification based on the actual situation. If the node degree of the network has significant impact on the core character, we increase the proportion of d_i in the function $K(i)$, that is, a bigger value should be set to parameter y; and vice versa.

(2) **Membership function with comprehensive factors**

The definition of a comprehensive membership function consists of two parts, one part is the membership function $h_1(ij)$ about the node i to the community j, which is based on the connection of the edge of the network, the other part is the membership function $h_2(ij)$ about the node i to the community j according to the node's information transmission efficiency. The membership is formulated as formulas (2) and (3):

$$h_1(ij) = \frac{E_{ij}}{sumE_{ij}}, j \in 1 \sim com\_num \qquad (2)$$

$$h_2(ij) = \frac{1/L_{ij}}{\sum_{k=1}^{com\_num} 1/L_{ik}}, j \in 1 \sim com\_num \qquad (3)$$

where i represents the node number, and the j represents the community number, the E_{ij} in function h_1 indicates the actual number of edges from node i to community j, $sumE_{ij}$ represents the total number of edges from node i to all of the initial communities, com_num gives the number of communities. The practical significance of this function means the ratio of the number of edges of a node within a community and the total number of the edges of this node to all of the other nodes.

L_{ij} in function h_2 represents the average path length [13] from node i to all nodes in community j, L_{ij} is expressed as formula (4).

$$L_{ij} = (l_{i1} + l_{i2} + \ldots + l_{in})/n_j. \qquad (4)$$

For a pair of nodes, the shorter the shortest path, the faster the information transmits, and the two nodes are more likely to be in the same community. The average path length from a node to the nodes within the same community it belongs to should be calculated to measure the information dissemination efficiency of the node to the community. Here, the smaller the average shortest path of the node to the community, the higher the efficiency of information transmission, and the more probability the node belongs to the community. In addition, this paper combines the definition of membership function in fuzzy mathematics, and makes 1 as the sum of membership degree from a node to each community.

Let symbol H represents the comprehensive membership function, according to the principle of fuzzy comprehensive evaluation model in fuzzy mathematics, we set ω_1 and ω_2 as the weight ratio of the above two factors affecting the division of community. We select (*, +) model of comprehensive evaluation model for calculating, then the final comprehensive membership function H is shown as the formula (5):

$$H(ij) = \omega_1 * h_1(ij) + \omega_2 * h_2(ij). \qquad (5)$$

where ω_1 and ω_2 is the parameters, the range of ω_1 and ω_2 is 0–1, and the sum of ω_1 and ω_2 is 1.

2.2 Implementation of MCDA Algorithm

Input: Adjacency matrix of a static network; the number of initial communities-com_num; the weight ω_1 and ω_2; the parameter y.

Output: The results of community division.

The process of MCDA algorithm is as follows:

Step 1: Network initialization, the main work is as follows:

(1) To calculate the number of nodes- *node_num*, the degree d_i of each node i, and the total number of edges-*edge_num* in the network according to the adjacency matrix.

(2) Find the neighbor node set *NeighborU*[i] of each node i, *NeighborU*[i] = $\{V_j|a_{ij} = 1\}$, and then traverse all the neighbor nodes of node i, to judge any pair of neighbor nodes whether there is an edge between them, if there is an edge, then let $E_i = E_i + 1$;

3. Compute the important degree value of nodes in the network according to C_i (the clustering coefficient of node i) and $K(i)$ (see formula (1)),here the node clustering coefficient C_i is shown in formula (6):

$$C_i = \frac{2 * E_i}{d_i * (d_i - 1)} \qquad (6)$$

where E_i represents the number of edges between all neighbor nodes of node i, d_i indicates the number of adjacent nodes of the node i.

(4) The shortest path length between two nodes is calculated based on the adjacency matrix, and save it to array $l[i][j]$. Here we only consider undirected network, so the shortest path length between node i and node j satisfies $l[i][j] = l[j][i]$ $(i, j \in [0, node\_num])$. If $i = j$, then. If node i is not connected to node j, then according to the definition of empirical value $l[i][j] = INNINITY$, here *INNINITY* is a parameter that indicates the unconnected status between two nodes. We set $INNINITY = 40$ based on plenty of experimental analysis.

(5) Community initialization. Sort the nodes in descending order according to the importance values $K(i)$ of each node of the network, and take the first *COM_-NUM* nodes, as *COM_NUM* initial communities, and mark the initial communities as $V[1], V[2], \cdots, V[COM\_NUM]$, respectively. The remaining unassigned nodes are put into set $V[0]$ in order.

(6) Set iteration number - *flag* to 1.

Step 2: Traversing the nodes in the set $V[0]$, and for each node in $V[0]$, go to Step 3.

Step 3: Calculating the sum of edges connect node i and the nodes in community $V[j]$, and the sum of edges (S_num) which connect node i and the set of community V. If $S\_num \neq 0$, perform Step 4, else if $S\_num = 0$, return to Step 2.

Step 4: Calculating the mean shortest path length from node i to community $V[j]$ and then obtained the membership of node i to each community based on membership function (see formula (4)). Comparing the magnitude of all the membership elements, and finding the corresponding community number j with maximal $H(ij)$, the node i will be added to community j, and remove it from the node set $V[0]$. Judging whether node i is the last one in node set $V[0]$. If so, go to Step 5; if not, return to Step 2.

Step 5: Due to node set $V[0]$ may exist unassigned nodes when $S\_num = 0$ and the loop breaks, it's necessary to determine whether $V[0]$ is empty again. If it is not empty, then $flag = flag + 1$ and go back to Step 2; otherwise, initial community division ends and perform Step 6.

Step 6: Merge communities based on greedy algorithms, the result of the best community division is reached when obtaining the maximum modularity Q [14]. The combined number of iterations is marked as *flag*, and we set the initial value of *flag* to 0. The detailed process is as follows:

(1) Calculating the modularity of the initial community. Among the *com_num* initial communities, the edges matrix between any two communities is E, the element $e[i][i]$ represents the number of edges within community i, according to formula (7) to calculate the initial modularity Q_0.

$$Q = \sum_{i=1}^{m} \left[\frac{E[i][i]}{E} - \left(\frac{\sum_{j=1\sim m}^{j\neq i} E[i][j]}{E} \right)^2 \right]. \tag{7}$$

Formula (7) discloses the difference between the proportion of the number of edges within a community of a network and the total number of edges in the network and the expected value of the proportion of the number of edges between nodes within a community and nodes outside the community and the total edges of the network.

(2) Let $flag = flag + 1$, there are $com\_num * (com\_num - 1)/2$ kinds of possible cases when combining two arbitrary communities. For any combination of two communities, recalculating the symmetric matrix $E[i][j]$ which indicates the number of edges between communities, we can get the $com\_num * (com\_num - 1)/2$ kind of values of Modularity, then combine community i and community j which produce the maximal Q, that is $Q_f = Q_{max}$.

(3) Determine whether the f is less than com_num, if not, the community division stops, and the algorithm ends; otherwise, proceed to (4).

(4) Determine whether $Q_f > Q_{f-1}$ is satisfied, if yes then go back to (2), otherwise the merging stops, and the algorithm ends.

2.3   Complexity Analysis

Given an undirected network G with n nodes, m edges and K initial communities, we analyze the algorithm complexity as the following.

(1) **Time complexity analysis**

Algorithm first calculates the core importance of nodes. According to the adjacency matrix, calculating the degree of the nodes takes time complexity n^2, and obtaining the clustering coefficient takes $\sum_{i=1}^{n} (d_i - 1) < n^2 d_i$, where the degree of node is i. After that, sorting according to the node importance takes time complexity $n(n - 1)/2$, then initiates K nodes to K community, and put $n - K$ nodes into the node set to be divided take time complexity n. The final time complexity of the initialization procedure is $n^2 + n^2 + n + n(n - 1)/2 = 5n^2/2 + n/2$.

Nodes are constantly taken out from community U and are put into initial communities. Among them, the complexity of computing nodes to each community is

$$K + (node\_num[1] + \cdots + node\_num[K]) + K + K + K$$
$$= 4 * K + (node\_num[1] + \cdots + node\_num[K])$$

Here $node\_num[i]$ is the number of nodes in community i. The total number of nodes in each community $node\_num[1] + \cdots + node\_num[K]$ must be less than the total number of nodes n. Therefore seeking a single node to each community takes time complicity no more than $n + 4 * K$. After traversing all the nodes, the maximum time complexity turns to be $n(n + 4 * K)$.

The time complexity analysis of community merging process: The first is about the time complexity of the modularity, the time complexity of calculating the edges between each two communities takes $n(n - 1)/2$. According to the array of the edges of each two communities, the time complexity for computing modularity is K^2. Therefore the modularity computing time of the initial community structure is: $n(n - 1)/2 + K^2$. On the basis of the value of modularity, each of the two communities combined, the process of merging community needs to calculate the modularity $K(K - 1)/2$ times, and each merge changes the attributes of part of the nodes. So the maximum complexity of the process is K, and will merge $K - 1$ times. The overall time complexity of this process is $[K(K - 1)/2] * [n(n - 1)/2 + K] + n * (K - 1)$, the value of the initial community number K is small and can be ignored, the final result can be simplified as: $K^2(n^2 + n)/2$.

In summary, the overall complexity of the algorithm is $7n^2/2 + n/2 + K^2(n^2 + n)/2$, which can be simplified as $O(K^2n^2)$.

2. **Space complexity analysis**

The adjacency matrix of the network occupies space n^2. Storing the degree of all nodes in a community occupies n storage units when initializing the community. The clustering coefficients of nodes and the degree of importance need occupy n storage units respectively. Communities occupy K storage units, each community contains $node\_num[i]$ member nodes as well as a variable indicating the number of community members. And then, when the nodes are divided, $K * n$ spaces are required to store the membership from the nodes to the community. Finally, when calculating the modularity Q, K valuable Q should be stored, which K take storage units. Furthermore, the edges matrix between each of the two communities requires K^2 storage units, which is calculated for $K(K - 1)/2$ times, then it takes K^4 storage units.

After the completion of the algorithm, the total occupied storage space is $n^2 + 3n + K + n + Kn + K + K^4$, simplified as $O(n^2 + K^4)$. In most cases, K is a small value which can be neglect when calculating, so the usual complexity of the proposed algorithm can be $O(n^2)$.

3   Experimental Analysis E

3.1   Zachary's Karate Club Dataset

Zachary's karate club dataset [15] is used to verify the MCDA algorithm, a suitable value y is first selected for node importance function, and when y takes value 0.7, the arrangement result of the nodes importance is most consistent with the truth. We set the number of initial communities from 4 to 9, and we get the approximate same results. The largest modularity is obtained when the network is divided into two communities. Therefore, the number of initial communities is set to 4, which is similar to the final results. Among the 4 communities, two of them are nearly fixed, another two include node 33 and node 2 respectively just because of the initial importance of node 33 and node 2 in the network. Then, according to the value of modularity to merge communities, and we get the maximum value of the modularity 0.247 when the corresponding community detection results are:

Community 1: {9, 10, 15, 16, 19, 21, 23, 24, 25, 26, 27, 28, 29, 30, 31, 32, 33, 34}
Community 2: {1, 2, 3, 4, 5, 6, 7, 8, 11, 12, 13, 14, 17, 18, 20, 22}

The experimental results are shown in Fig. 1. Node 34 and Node 1 are the cores of the two communities, which is essentially consistent with the real network.

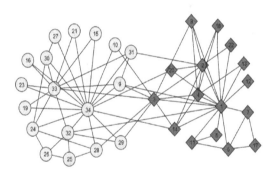

Fig. 1. The final community structure for Zachary club network

3.2   Dolphin Social Network

Following the same experimental process as Zachary's karate club dataset, for Dolphin social network [16] we set the value 0.7 for y, and when the initial number of communities ranges from 15 to 8, we get appropriate the same results, so we choose 8 for initial community number. But when the initial community is set to 7, the final results are dramatically different from before. Experimental results show that the largest modularity 0.287 is reached when two communities are combined, and the corresponding community detection result is:

Community 1: {1, 3, 4, 5, 8, 9, 11, 12, 13, 15, 16, 17, 19, 20, 21, 22, 24, 25, 29, 30, 31, 34, 35, 36, 37, 38, 39, 40, 41, 43, 44, 45, 46, 47, 48, 50, 51, 52, 53, 54, 56, 59, 60, 62}

Community 2: {2, 6, 7, 10, 14, 18, 23, 26, 27, 28, 32, 33, 42, 49, 55, 57, 58, 61}

The final topology of the network community structure is shown in Fig. 2. Node 40, node 8 and node 20 which should be included into community 2 are partitioned to community 1, just because the three nodes are on the boundaries of the two communities, their ownership situation is controversial, the partitioning results are reasonable.

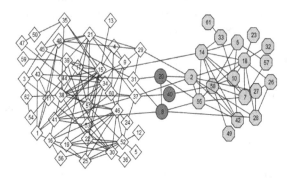

Fig. 2. The partitioned community structure for dolphin network

3.3 Computer-Generated Network Dataset

In this paper, the selected random network composes of 4 sub networks, and the number of nodes and edges of the network is changed every time when the network generation algorithm is executed. The network contains 80 nodes and 496 edges, as shown in Fig. 3, in which four sub-networks (communities) are included and they take nodes {1–20}, {21–40}, {41–60} and {61–80}, respectively. The core nodes of the four communities are nodes 3, node 33, node 53 and node 73.

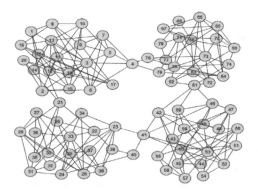

Fig. 3. The original generated network

The results of community division are shown in Fig. 4 when $y = 0.6$ and $K = 8$. The result exactly satisfies the initial network community structure, and we get the value $Q = 0.632$. We experiment on generated networks with other scale, and we get the same correct results.

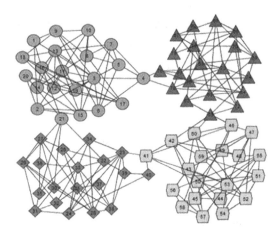

Fig. 4. The partitioned community structure for generated network

Through experiments, the actual running time of each case can be obtained. It can be concluded that the number of initial communities has little influence on running time of the algorithm when given the same network scale. That is to say, the main running time of the algorithm is spent on executing the initial community division. When merging communities, the time complexity is mainly related to K. However, due to the number of initial communities are far less than the number of nodes in the network, it will not take too much time for merging communities following the value of modularity, which is theoretically and practically proved.

It can also be concluded that the running efficiency takes on the power of 2 distribution of the number of nodes, that is, the real complexity of the algorithm is $O(n^2 k^2)$. Compared with the classical Newman fast algorithm and the GN algorithm, the proposed algorithm presents higher efficiency.

The modularity function is a widely accepted indicator for determining whether a community detection algorithm can generate a reasonable community structure, which is proposed by Newman and Girvan [5] according to the ratio of the number of edges in a community to the total number of edges in the network. The modularity is a standard for estimating the quality of complex network partition. In the experiment on Karate club network, Dolphin social network and computer-generated network, we get modularity Q 0.247, 0.287 and 0.623 accordingly. Different modularity values show that the community structure of the Karate club network and Dolphin social network are comparatively fuzzy, while the community structure of the computer-generated network is relatively explicit. It is in conformity with actual conditions.

We also compared with classical clustering algorithms [17] and the experiments results show that the MCDA algorithm has a greater modularity as shown in Table 1.

As shown in Table 1, three algorithm results' modularity are almost consistent and the proposed algorithm can obtain a larger modularity (0.7875 when the community results is 11), which proves the performance of the algorithm.

Table 1. Comparison with classical clustering algorithms

Algorithms	Community number	Q
Betweenness clustering	10	0.6
Hierarchical clustering	23	0.588
The proposed algorithm	11	0.7875

There are several classical algorithms in the research field of community detection. Newman Fast algorithm takes theoretical time complexity $O(n(m+n))$, the GN algorithm takes $O(m^2 n)$, and the fast partitioning algorithm contributes $O(m^2)$. From the above experimental analysis of the static community detection algorithm, we concluded that the time complexity of MCDA algorithm is about $O(K^2 n^2)$, where K is the number of initial communities, n is the number of nodes of the network. In most cases, the number of edges of a network is far more than the number of nodes, that is, $m > n$ always satisfies. So the comparison shows that the proposed algorithm exhibits higher efficiency than the traditional algorithms.

4   Conclusion

In this paper, a heuristic community detection algorithm based on membership function is proposed. The Zachary Karate Club network, dolphin social network and computer generated network are employed to verify the algorithm, experimental results disclose that the proposed algorithm outperforms some typical traditional algorithms in accuracy and efficiency, which indicates that the algorithm in this paper also has a wide range of applicability for the real complex networks.

Acknowledgment. This work is partially supported by the Science Research Project of Liaoning Provincial Department of Education under Grant No. L2015173, the scholarship under China State Scholarship Fund CSC No. 201606085034.

References

1. Kernighan, B.W., Lin, S.: A efficient heuristic procedure for partitioning graphs. Bell Syst. Tech. J. **49**, 291–307 (1970)
2. Radicchi, F., Castellano, C., Cecconi, F., Loreto, V., Parisi, D.: Defining and identifying communities in networks. Proc. Natl. Acad. Sci. USA **101**(9), 2658–2663 (2004)

3. Newman, M.E.J.: Fast algorithm for detecting community structure in networks. Phys. Rev. E **69**(6), 066133 (2004)
4. Wang, X., Liu, Y.: A survey of community structure in complex networks. J. Univ. Electron. Sci. Technol. Univ. Sci. Technol. **38**(5), 537–543 (2009)
5. Wang, X.F., Li, X., Chen, G.: Theory and Application of Complex Network, p. 9. Tsinghua University Press, Beijing (2006)
6. Didwania, A., Narmawala, Z.A.: Comparative study of various community detection algorithms in the mobile social network. In: Nirma University International Conference on Engineering. IEEE (2015)
7. Souza, J.D., Taya, F., Thakor, N.V., et al.: Comparing community detection algorithms on neuroimaging data from multiple subjects. In: International Conference on Signal-Image Technology and Internet-Based Systems, pp. 322–327 (2015)
8. Tanaka, A.: Proposal of alleviative method of community analysis with overlapping nodes. In: IEEE Fourth International Conference on Big Data and Cloud Computing, pp. 371–377. IEEE (2014)
9. Havens, T.C., Bezdek, J.C., Leckie C., et al.: Clustering and visualization of fuzzy communities in social networks, pp. 1–7 (2013)
10. Chen, Z., Zhen, W.: Fuzzy clustering image segmentation algorithm based on particle swarm optimization. J. Fuzhou Univ. (Nat. Sci.) (01), 32–35 (2010)
11. Wu, X., Zhang, M., Han, Y.: Research on centrality of node importance in scale-free complex networks, pp. 1073–1077 (2012)
12. Wang, Y., Wu, X., Zhu, J., et al.: On learning cluster coefficient of private networks. Soc. Netw. Anal. Min. **3**(4), 395–402 (2012)
13. Yen, C.C., Yeh, M.Y., Chen, M.S.: An efficient approach to updating closeness centrality and average path length in dynamic networks. In: IEEE International Conference on Data Mining, pp. 867–876 (2013)
14. Chen, G., Guo, X.: A genetic algorithm based on modularity density for detecting community structure in complex networks. In: International Conference on Computational Intelligence and Security, CIS 2010, Nanning, Guangxi Zhuang Autonomous Region, China, pp. 151–154, December 2010
15. Zachary, W.W.: An information flow model for conflict and fission in small groups. J. Anthropolog. Res. **33**(4), 473 (1977)
16. Lusseau, D., Slooten, L., Currey, R.J.C.: Unsustainable dolphin-watching tourism in Fiordland, New Zealand. Tour. Mar. Environ. **3**(2), 173–178 (2006)
17. Zhang, P., Menghui, L.I., Jinshan, W.U., et al.: The community structure of scientific collaboration network. Complex Syst. Complex. Sci. **2**, 30–34 (2005)

Task Scheduling in Cloud Computing Based on Cross Entropy Method

Ying Ren[1](✉), Lijun Zhou[2], and Huawei Li[2]

[1] Naval Aviation Engineering College, Yantai, China
ren.ying@163.com
[2] Shandong Business Institute, Yantai, Shandong, China
jungle730@163.com, li_hauwei@163.com

Abstract. In order to solve the problem of cloud computing task scheduling in the process of total execution time and the total cost of the conflict, this paper used the cross entropy method to solve the problem of task scheduling in cloud computing. The experimental results showed that using the cross entropy method makes the task of the total execution time is the shortest, lowest cost, customer satisfaction is greatly improved.

Keywords: Cloud computing · Cross entropy · Dispatch · Method

1 Introduction

With the development of cloud computing, the task scheduling problem in cloud computing has been widely concerned and developed. Task scheduling is to find the matching method of n sub tasks and M service resources so that it can meet the task scheduling objectives proposed by users.

There are many ways to solve the problem of task scheduling, for example First Input First Output, capacity Scheduler, Fair Scheduler, these algorithms is relatively simple to achieve but the performance is not good. And also the task scheduling algorithm based on genetic algorithm and particle swarm optimization algorithm and ant colony algorithm, however, these algorithms is a single task scheduling goal, or poor convergence performance and so on. In this paper, the cross entropy method is used to solve the problem of task scheduling in cloud computing, which makes the total execution time of the task is the shortest, and the cost is the lowest, and the user's satisfaction is greatly increased.

2 Task Scheduling in Cloud Computing

Cloud computing platform task scheduling efficiency directly affects the overall performance and service quality of the cloud computing platform, and the performance of the task scheduling strategy has higher requirements. Task scheduling by job scheduler coordinate user service and resource management, rational and effective allocation and scheduling task submitted by users. Therefore, the purpose of cloud computing task

H. Yuan et al. (Eds.): GRMSE 2016, Part I, CCIS 698, pp. 196–202, 2017.
DOI: 10.1007/978-981-10-3966-9_21

scheduling is to find the best scheduling strategy to meet the user's task scheduling objectives (Fig. 1).

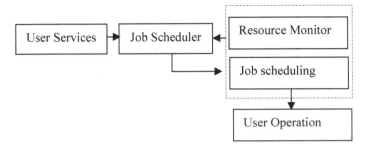

Fig. 1. Task scheduling in cloud computing

3   Cross Entropy Method and Principle

Cross entropy (Entropy Cross, CE) method is through the constant iteration of the occurrence probability of rare events, until the probability of convergence, and this convergence probability is the optimal probability and will generate the optimal solution. Cross entropy is then used to solve combinatorial optimization problems.

Set the minimum probability event $G(x)$, χ is meet the random variables of a distribution $f(x; u)$. If you want to estimate the probability of the event, you must have a large ample of χ, unbiased estimate is:

$$\frac{1}{N}\sum_{i=1}^{N}V_{\{G(X)\geq\gamma\}} \tag{1}$$

In order to achieve the small sample number N, Important sampling for χ, should be "the probability of the original event was estimated by the likelihood":

$$l=\frac{1}{N}\sum_{i=1}^{N}V_{\{G(X)\leq\gamma\}}\frac{f(X_i;u)}{g(X)} \tag{2}$$

The easiest way to get g is to select a g from the distributed cluster $f(X_i; v)$, that is to find the appropriate reference variable v, making the $f(X_i; v)$ and g^* minimum distance. Cross entropy is defined as:

$$D(g,m)=E_g\ln\frac{g(X)}{h(X)}=\int g(x)\ln g(x)dx-\int g(x)\ln m(x)dx \tag{3}$$

we can convert the problem of solving the minimum distance between $f(X_i; v)$ and g^* to the maximize problem

$$\max_{v} \int g^* \ln(f(x; v)dx) \tag{4}$$

The type (3) into the Eq. (4) to:

$$\max_{v} D(v) = \max E_u V_{\{S(X) \geq \gamma\}} \ln f(X; v) \tag{5}$$

Optimization problem is to find the G on the χ of the maximum value of γ^* and to achieve the maximum value of the corresponding state. The solution of cross entropy method is to use a multi-stage iteration of two processes, making the $\hat{\gamma}_i$ close to the optimal value of γ^*.

4 Objective Function Design of Cloud Task Scheduling Problem

The task scheduling problem in cloud computing is how to establish the matching relationship between the scheduling task and service resources by using the scheduling technology, so that it can satisfy the user's scheduling objectives and obtain the scheduling results. The key task is how to put the user's scheduling objectives into the cross entropy algorithm used in the objective function.

Assume that the user submits an N job $J = \{J_1, J_2, \ldots J_N\}$ to the cloud computing platform. Divided into an independent sub tasks $T = \{T_1, T_2, \ldots T_n\}$, and then divided into s class unit sub tasks, and $s \leq k$, cloud platform provides M service resources $W = \{W_1, W_2, \ldots, W_m\}$.

Use the execution time matrix TIME to represent the execution time of each class of tasks on each service resource, as shown in the formula (6):

$$TIME = \begin{pmatrix} time_{11} & \cdots & time_{1m} \\ \cdots & \cdots & \cdots \\ time_{n1} & \cdots & time_{nm} \end{pmatrix} \tag{6}$$

where $time_{ij}$ represents the execution time of the first i unit sub task on the j service resource, $i = 1, 2, \ldots n; j = 1, \ldots m$. Unit service costs can be expressed as $COST = (\cos t_1, \cos t_2, \ldots, \cos t_m)$, of which $\cos t_i$ represents the unit service cost on the first I service resource, $i = 1, 2, \ldots m$. The service cost of the unit task in service resources is represented as a AC matrix.

$$AC = \begin{pmatrix} \cos t_1 \times et_{11} & \cdots & \cos t_m \times et_{1m} \\ \cdot & \cdot & \cdot \\ \cdot & \cdot & \cdot \\ \cos t_1 \times et_{n1} & \cdots & \cos t_m \times et_{nm} \end{pmatrix} \tag{7}$$

Where $ac_{ij} = \cos t_j \times time_{ij}$ represents the service cost of the first I unit sub task on the j service resource, $i = 1, 2, \ldots n; j = 1, \ldots m$.

Using TC represent resource properties $TC = \{TIME, AC\}$, that is $TC_i = \{TIME_i, AC_i\}$ represents a set of properties for all tasks on the first i service resource, the execution time and the execution cost.

The total execution time of all units on the j service resource is:

$$T_1(W_j) = \sum_{i \in (1,n)} time_{ij} \tag{8}$$

The maximum execution time of the resources as shown:

$$\max_j T_1(W_j) = \max_j \sum_{i \in (1,n)} time_{ij} \tag{9}$$

The maximum execution time is smaller, the better so that the total execution time of all tasks is the lowest:

$$\min(\max_j T_1(W_j)) = \min(\max_j \sum_{i \in (1,n)} time_{ij}) \tag{10}$$

The task total service cost on the first j resource is the sum of the service costs of all tasks performed on the j resource, as shown in the formula 11:

$$T_2(W_j) = \sum_{i \in (1,n)} ac_{ij} = \sum_{i \in (1,n)} time_{ij} \times \cos t_{ij} \tag{11}$$

The objective function is to make the lowest total service cost of all tasks:

$$\min \sum_{j=1}^{m} T_2(W_j) = \min \sum_{j=1}^{m} \sum_{i \in (1,n)} ac_{ij} = \min \sum_{j=1}^{m} \sum_{i \in (1,n)} time_{ij} \times \cos t_{ij} \tag{12}$$

Due to the complexity of cloud computing platform, if only considering the task of the total execution time or only consider the task of the total cost of service is very good can't adapt to the need of the users need the cloud task scheduling, the total execution time and the total cost of service combine to consider.

5 Experiment

Experiments are carried out on the CloudSim cloud simulation platform. Assume that the user Xiang Yun platform commit job is divided into 8 independent sub tasks $T = \{T_1, T_2, \ldots T_8\}$, and cloud platform provides 4 service resources $W = \{W_1, W_2, \ldots, W_4\}$. Each sample $X_i = (x_1, x_2, \ldots, x_n)$, x_j indicates the ID number of the service resources allocated to the task T_j, j = 1,2,......8, $x_j \in [1,4]$.

Assume that the length of the 8 tasks are LH = {20365, 37906, 20587, 48635, 19684, 15368, 30147, 36587). The CPU running rate of the 4 service resources are SP = {156, 268, 116, 204), unit for MIPS. The running time of each unit child task on each service resource is $time_{ij} = LH_i/SP_j$, That is:

$$TIME = \begin{pmatrix} 130.54 & 75.99 & 175.56 & 99.83 \\ 242.99 & 141.44 & 326.78 & 185.81 \\ 131.97 & 76.82 & 177.47 & 100.92 \\ 311.76 & 181.47 & 419.27 & 238.41 \\ 126.18 & 73.45 & 169.69 & 96.49 \\ 98.51 & 57.34 & 132.48 & 75.33 \\ 193.25 & 112.49 & 259.89 & 147.78 \\ 234.53 & 136.52 & 315.41 & 179.35 \end{pmatrix}$$

Set the parameters of the cloud computing task scheduling algorithm based on cross entropy method, set the number of iterations and the number of samples, assuming that the unit costs of the 4 service resources are $COST = \{0.23, 0.16, 0.18, 0.25\}$, by formula (7) to get the service cost matrix:

Sequential assignment of the experimental results is shown in Table 1. CloudSim simulation platform to simulate the cloud task scheduling process, so that it can meet the total execution time and the total cost of the implementation of the task is relatively optimal, experimental results as shown in Table 2.

Table 1. Sequential assignment of the experimental results

Task	Server	Execution time	Start time	End time	Service cost
1	1	130.54	0	130.54	30.02
2	2	141.44	0	141.44	22.63
3	3	177.47	0	177.47	31.94
4	4	238.41	0	238.41	59.60
5	1	126.18	130.54	256.72	29.16
6	2	57.34	141.44	198.78	9.17
7	3	259.89	177.47	437.36	46.78
8	4	179.35	238.41	417.76	44.84

It can be seen form the Table 1 that the total execution time of Sequential assignment the task is the latest end time that is 437.36, and the total service cost is the total of service cost that is 274.14. Based on the cross entropy method, the experimental results of the cloud task scheduling strategy in Table 2 shows that the total execution time of the task is 344.10, and the total service cost is the total of service cost that is 247.47. The comparison of the two methods is shown in Table 3.

By comparing the experimental results of the Table 3, it is shown that the total execution time of the cloud task scheduling algorithm based on cross entropy method is the lowest, and the total cost is low.

Table 2. Experimental results on the task of the shortest cloud task scheduling

Task	Server	Execution time	Start time	End time	Service cost
1	4	99.83	0	99.83	24.96
2	2	141.44	0	141.44	22.63
3	1	131.97	0	131.97	30.35
4	2	181.47	141.44	322.91	29.03
5	4	96.49	99.83	196.32	24.12
6	1	98.51	131.97	230.48	22.66
7	4	147.78	196.32	344.10	36.95
8	3	315.41	0	315.41	56.77

Table 3. Comparison of the two methods

Method	The total execution time	The total service cost
Sequential assignment	437.36	274.14
The shortest cloud task scheduling	344.10	247.47

Acknowledgment. This paper uses the cross entropy method to solve the problem of cloud computing environment in the task scheduling problem and is compared with the use of sequential allocation strategy. By comparing the experimental results, the feasibility and effectiveness of using the cross entropy method to solve the problem of task scheduling in cloud computing environments is verified. At present, the research of task scheduling algorithm in cloud computing environment has not yet reached a mature stage, there are still a lot of places to be improved. In the future, we can study from the task scheduling objectives, algorithm parameter setting and algorithm stability.

References

1. Wang, D., Li, Z.: Cloud computing task scheduling algorithm based on particle swarm optimization and ant colony optimization. Comput. Appl. Softw. **30**(1), 290–293 (2013)
2. Wang, Y., Han, R.: Cloud environment task scheduling based on improved ant colony algorithm. Comput. Measur. Control **19**(5), 1203–1204 (2011)
3. Zuo, L., Cao, Z.: Research on scheduling problem in cloud computing. Comput. Appl. Res. **29**(11), 4023–4027 (2012)
4. Xiong, C.-C., Feng, L., Chen, L.: Research on task scheduling algorithm based on Genetic Algorithm in cloud computing. J. Huazhong Univ. Sci. Technol. (Nat. Sci. Edn.) **1**(1), 15–19 (2012)
5. Zhou, W.J., Cao, J.: Cloud computing resource scheduling strategy based on prediction and ant colony algorithm. Comput. Simul. **246**(9), 239–242 (2012)
6. Deng, J.-G., Zhao, Y.-L., Yuan, H.: A cost driven task scheduling strategy for cloud computing. J. Jiangsu Univ. Nat. Sci. Edn. **35**(2), 214–219 (2014)
7. Chun, Y., Ma, Y., Zhao, L.: Cloud computing resource scheduling. Strategy Algorithm **11** (40), 8–13 (2013)

8. Beloglazov, A., Abawajy, J., Buyya, R.: Energy-aware resource allocation heuristics for efficient management of data centers for cloud computing. Future Gener. Comput. Syst. **28** (5), 755–768 (2012)

9. Feng, L., Zhang, T., Jia, Z.: Task scheduling algorithm based on Improved particle swarm optimization in cloud computing environment. Comput. Eng. **39**(5), 183–188 (2013)

10. Ishida, A., Yamazaki, J.Y., Harayama, H., et al.: Photoprotection of evergreen and drought-deciduous tree leaves to overcome the dry season in monsoonal tropical dry forests in Thailand. Tree Physiol. **34**(1), 15–18 (2014)

11. Calheiros, R.N., Ranjan, R., Beloglazov, A., et al.: CloudSim a toolkit for modeling and simulation of cloud computing environments and evaluation of resource provisioning algorithms. Softw. Pract. Exp. **41**(1), 23–50 (2011)

12. Kotzing, T., Neumann, F., Roglin, H., et al.: Theoretical analysis of two ACO approaches for the traveling salesman problem. Swarm Intell. **6**(1), 1–21 (2012)

Bad Data Identification Based on Optimized Local Outlier Detection Algorithm

Jingxian Qi[✉], Yuefeng Cao, and Jianhua Shi

Nari Group Corporation State Grid Electric Power Research Institute,
Nanjing, China
{qijingxian, caoyuefeng, shijianhua}@sgepri.sgcc.com.cn

Abstract. This paper propose an optimized local outlier factor algorithm based on hierarchical clustering over grid bad measurement information, which affect the running safety of power grids phenomenon seriously. The method adopt statistical theory to evaluate the equipment running data and state information. Meanwhile, use clustering algorithm to analyze these data, to achieve the purpose of data reduction. While the relative entropy for data confirm the weight and thus enhance the accuracy of the algorithm. Experimental results show that the algorithm can quickly identify the bad power grid data.

Keywords: Local outlier factor · Hierarchical clustering · Weight · Bad data identification

1 Introduction

Bad data impact equipment operation estimation seriously, it may lead to the state estimation result unavailable or state estimation does not converge [1]. Because of the large number of electrical equipment and uneven data in the state evaluation of the electrical equipment, it is particularly important to deal with the collected data reasonably and effectively [2, 3]. Local outlier factor algorithm is a density anomaly detection algorithm, that is, abnormality factor can be used to identify an object relative to its local neighborhood the degree of abnormality [4]. Meanwhile, the hierarchical clustering algorithm is an important method of data division or packet processing, with the advantage of its simple, fast and efficiently handle large data sets, it has been widely used in many fields [5, 6]. Entropy can be used to measure the degree of clutter and disorder of a system, it can objectively reflect the recognition of the data outliers [7, 8].

This article use hierarchical clustering algorithm for data reduction, and the use of relative entropy values for the data attribute weight in local outlier factor optimization algorithm, used to further improve the efficiency and reliability of the identification results of the algorithm.

2 Modeling

Electrical equipment operating state can be defined by telemetry information to discover the non-normal operation device, at a certain time, the topological state determines the state of this moment before and its conversion to the current state [9, 11].

© Springer Nature Singapore Pte Ltd. 2017
H. Yuan et al. (Eds.): GRMSE 2016, Part I, CCIS 698, pp. 203–212, 2017.
DOI: 10.1007/978-981-10-3966-9_22

Suppose there is no bad telemetry data and topology errors, the power grid operation state can be determined by the reasonableness of the state transition. So you can use the remote metering transition value and remote signal conversion value to define the state transition, using abnormal state transitions to determine whether the presence of the device state error. Suppose the equipment status at the time of t_0 is:

$$[YC_1, YC_2, YC_3, \ldots, YX_1, YX_2, YX_3, \ldots] \tag{1}$$

Where: YC represents remote metering information, YX represents remote signaling information;

The equipment state at the time of t_1 is

$$[YC'_1, YC'_2, YC'_3, \ldots, YX'_1, YX'_2, YX'_3, \ldots] \tag{2}$$

Where: YC' represents remote metering information, YX' represents remote signaling information, and the state transition value is:

$$[YC'_1 - YC_1, YC'_2 - YC_2, YC'_3 - YC_3, \ldots, YX'_1 \oplus YX_1, YX'_2 \oplus YX_2, YX'_3 \oplus YX_3, \ldots] \tag{3}$$

That is the equipment state can be defined by the transition value. Through determining the rationality of the remote metering transition data and remote signaling transition data, to identify the bad data in grid running. Since remote metering and remote signaling information for determining the operating state influence network equipment differently, we need to assign different weights to different attribute during bad data identification.

3 Local Outlier Factor (LOF) Algorithm

3.1 Conversional Local Outlier Factor Algorithm

Local outlier factor algorithm is a density anomaly detection algorithm, that is, abnormality factor can be used to identify an object relative to its local neighborhood the degree of abnormality.

The algorithm process is as follows:

Firstly, compute the k-distance(k_p) of the object p, for any two objects p and o in data set D and any positive integer k, k-distance to the object p is the distance between p and an object o, denoted k_p, object o meet the following criteria:

- at least exist k objects satisfies $d(p, o') \le d(p, o)$, where $o' \in D\backslash\{p\}$;
- at least exist k-1 objects $o' \in D\backslash\{p\}$ satisfies $d(p, o') \le d(p, o)$. The distance between object p and object o is denoted as $d(p, o)$, and the formula is:

$$d(p, o) = \sqrt{\sum_{i=1}^{d} (f(p_i) - f(o_i))^2} \tag{4}$$

Where d is the dimension of data set. $f(p_i)$ and $f(o_i)$ are the value of attribute, i(i = 1,2,...,d) for object p and o.

Secondly, compute the set of k nearest neighbors, which is the data set of q, where the distance p from q is not exceed k_p and the formula is:

$$N_k(p) = \{q|d(p,q) \le k_p\} \tag{5}$$

Thirdly, compute the reachability distance of p and the object in $N_k(p)$, the formula is:

$$R_k(p,o) = \max\{k_o, d(p,o)\}, \; o \in N_k(P) \tag{6}$$

In words, the reachability distance of an object p from o is the true distance of the two objects, but at least the k_o.

Fourthly, compute the local reachability density of an object p is defined by

$$L_k(p) = \frac{1}{\sum R_k(p,o)/|N_k(p)|}, o \in N_k(p), \tag{7}$$

Which is the inverse of the average reachability distance of the object p from its neighbors.

Fifthly, compute the local outlier factor of object p, the formula is:

$$LOF_k(p) = \frac{\sum\limits_{o \in N_k(p)} L_k(o)/L_k(p)}{|N_k(p)|} \tag{8}$$

Which is the average local reachability density of the neighbors divided by the object's own local reachability density.

The larger the local outlier factor, the greater the likelihood the object is abnormal data.

3.2   Optimized Local Outlier Factor Algorithm

Firstly, optimized local outlier factor algorithm realize the data reduction by hierarchical clustering algorithm; secondly, confirm the weight based on the relative entropy; thirdly, use the weighted formula to compute the distance from p to q, then through the local outlier factor detection algorithm to find the abnormal object.

Data Reduction Based on Hierarchical Clustering Algorithm

Hierarchical clustering algorithm decompose the data set hierarchically in accordance with a method until meet certain conditions. In accordance with the principles of classification, it can be divided into two ways: cohesion and division [5].

(a) Cohesion hierarchical clustering algorithm

Cohesion hierarchical clustering algorithm introduced a bottom-up strategy. At first, each object as a cluster, and then merge those clusters until all objects in a cluster, or a termination condition is satisfied. And the vast majority of hierarchical clustering methods fall into this category, but the definition of similarity between the clusters might by different.

(b) Division hierarchical clustering algorithm

Division hierarchical clustering algorithm diametrically opposed compared to cohesion hierarchical clustering algorithm, it introduced the top-down strategy. Firstly, place all objects in the same cluster, and then gradually broken down into smaller clusters until each object in self-contained cluster, or reach a certain termination condition.

Clustering results make the objects in the same cluster as much similar as possible, and to other objects in other clusters as different as possible. By means of setting a suitable termination condition, you can focus on the cluster which contains fewer objects as the abnormal for analysis.

Since cohesion hierarchical clustering algorithm is relatively simple and widely used [6], we use the method in this paper.

Taking into account the Euclidean distance formula does not have transitivity, this paper optimized its formula [10]. Let $\{u_1, u_2, \ldots, u_n\}$ is the union of n objects, and each object has m attributes, that is $\{a_1, a_2, \ldots, a_m\}$. The object u_i equals to $\{x_{i1}, x_{i2}, \cdots, x_{im}\}$ then the distance between the k-th attribute value x_{ik} of u_i and the k-th attribute value x_{jk} of u_j is:

$$d_k(i,j) = \frac{|x_{ik} - x_{jk}|}{a_{kmax} - a_{kmin}} \tag{9}$$

Where: a_{kmax} and a_{kmin} is the maximum value and minimum value of the attribute a_k. Then, the secondary method [11] is used to seek transfer closure T of similarity matrix S, which is $T = t(R)$. Finally, construct the hierarchical clustering diagram referring to T, and get the corresponding clustering results with the stated threshold.

Attribute Weights Assigned

Information theory is a kind of science which taking information for research objects, in order to reveal the essential characteristics and rules of information, and applying the mathematical methods to study the information storage, transmission, processing and control [7]. Entropy can be used to measure the degree of clutter and disorder of a system. The larger the entropy, the more disordered the system. Since the information entropy fully established on the basis of the original data, it is objectively and can be used to identify outlier data, that is to confirm the non-normal state transitions and the bad data. At the same time, based on hierarchical clustering data reduction, it will significantly reduce the amount of computation of local outlier factor algorithm.

An information system is a quadruple $IS = (U, A, V, f)$, where U is a non-empty finite set of objects; A is a non-empty finite set of attributes; V is the union of attribute domains, i.e. $V = \cup_{a \in A} V_a$, where V_a denotes the domain of attribute a: $f : U \times A \rightarrow V$ is an information function which associates a unique value of each attribute with every object belonging to U, such that for any $a \in A, x \in U, f(x, a) \in V_a$.

In a given information system $IS = (U, A, V, f)$ each subset of attributes $B \subseteq A$ determines a binary indiscernibility relation $IND(B)$ as follows:

$$IND(B) = \{(x, y) \in U \times U : \forall a \in B, f(x, a) = f(y, a)\}$$

$U/IND(B) = \{B_1, \cdots, B_m\}$ Denotes the partition of U induced by $IND(B)$. The information entropy $E(B)$ of knowledge $IND(B)$ is defined by

$$E(B) = -\sum_{i=1}^{m} \frac{|B_i|}{|U|} \log_2 \frac{|B_i|}{|U|} \tag{10}$$

Where $\frac{|B_i|}{|U|}$ denotes the probability of any element $x \in U$ being in equivalence class $B_i (1 \le i \le m)$

$$E_x(B) = -\sum_{i=1}^{m-1} \frac{|B_i'|}{|U| - |[x]_B|} \log_2 \frac{|B_i'|}{|U| - |[x]_B|} \tag{11}$$

Denote the information entropy of knowledge $IND(B)$ when removing all objects in $[x]_B$ from U, where $[x]_B$ is the equivalence class of x under relation $IND(B)$, and

$$U/IND(B) - \{[x]_B\} = \{B_1', \cdots, B_{m-1}'\} \tag{12}$$

The relative entropy $RE_B(x)$ of object x under relation $IND(B)$ is defined by

$$RE_B(x) = \begin{cases} 1 - \frac{E_x(B)}{E(B)}, & \text{if } E(B) > E_x(B) \\ 0, & \text{otherwise} \end{cases} \tag{13}$$

In information theory, entropy characterize the uncertainty of information and random variables. According to the definition of relative entropy, given any $B \subseteq A$ and $x \in U$, when we delete all objects in equivalence class $[x]_B$, if the information entropy of knowledge $IND(B)$ decreases greatly, then we may consider the uncertainty of object x under $IND(B)$ is high. Otherwise, if the information entropy of knowledge $IND(B)$ varies little, then we may consider the uncertainty of object x under $IND(B)$ is low. Therefore, the relative entropy of $RE_B(x)$ of x under $IND(B)$ gives a measure for the uncertainty of x. The higher the relative entropy $RE_B(x)$ of x the higher the uncertainty of x.

Each attribute contributes to the abnormality of object is different in the data set, to provide the weights by confirming its properties' relative entropy for distance calculation. So the formula of distance calculation is:

$$d(p, o) = \sqrt{\sum_{i=1}^{d} w_i (f(p_i) - f(o_i))^2} \tag{14}$$

Calculate the Local Outlier Factor

Based on the conventional local outlier factor algorithm, we replaced the formula (4) by formula (14), then calculate the factor.

4   Experimental Results

4.1   Conventional Local Outlier Factor Algorithm

The Table 1 shows that 8 continuous time state data of bus coupler attribute values of some substation, which describe the load, current and remote signaling values.

Table 1. Bus coupler attribute value

Load	Current	Remote signal
72.8	138.9	1
71.2	131.2	1
72.2	128.9	1
0	145.5	0
0	0	0
81.7	154.7	1
82.9	157.2	1
82.9	157.2	1
83.0	162.3	1

The results of lof algorithm are as follows: (Table 2)

Table 2. Object and LOF

Load	Current
U_1	1.225
U_2	0.832
U_3	18.334
U_4	18.138
U_5	1.465
U_6	0.852
U_7	1.214

The results show that u_3 and u_4 are more likely anomalies, in view of the time sequence of these two objects, we need combine all actual operating environment to consider the abnormalities of u_4.

4.2   Optimized Local Outlier Factor Algorithm

Data Reduction Based on Hierarchical Clustering Algorithm

Based on the theory of hierarchical clustering and optimized distance calculation formula, the results as in Fig. 1 show that:

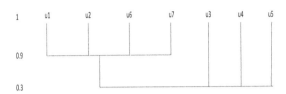

Fig. 1. Hierarchical clustering graph

From Fig. 1, we can see that, in this experiment the 0.9 is the appropriate threshold, the objects of $\{u_1, u_2, u_6, u_7\}$ in a cluster, and the other objects in self-contained cluster, then we can focus on the objects of u_3, u_4 and u_5, which are the suspicious objects.

Object Attribute Weights Assigned

The relative entropy is computed according to the formula 13, the results are as follows: (Table 3)

Table 3. Relative entropy of suspicious objects

Object	Load	Current	Remote signal
U_3	0.5	0.5	1.0
U_4	0.35	0.35	1.0
U_5	0.5	0.5	1.0

Compute the LOF Value

Based on the formula 14 to compute the distance between two objects, the local outlier factor results are as follows: (Table 4)

Table 4. Optimized local outlier factor results

Object	LOF
U_3	21.025
U_4	20.334
U_5	1.267

Comparative results are as follows: (Table 5)

Table 5. Comparative results

Object	Conventional LOF	Optimized LOF
U_1	1.225	0
U_2	0.832	0
U_3	18.334	21.025
U_4	18.138	20.334
U_5	1.465	1.267
U_6	0.852	0
U_7	1.214	0

The result shows that, by using the relative entropy to add the appropriate weight for objects, in practice, it can be more accurately determine the degree of abnormality in the local outlier factor computation.

4.3   Impact Analysis of K

This test system is the running data of some substation, which contains 8 breakers and 29 disconnectors and 3 buses and so on, then according to the optimized local outlier factor algorithm, we changed the k value to find the rules, which can help us confirm the suitable value of k (Fig. 2).

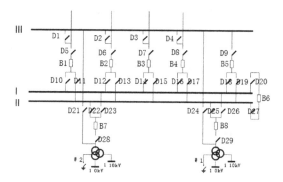

Fig. 2. Connection diagram

The results are as follows: (Table 6)

The results show that the optimized local outlier factor algorithm can be used to find the bad data in electric grid:

Table 6. Comparison results of different K

Nodes	K	Number of outliers found	Number of real outliers	Accuracy
1000	5	3	5	60%
1000	6	5		100%
1000	8	4		80%
1000	10	4		80%
10000	15	9	12	75%
10000	17	10		83.3%
10000	20	12		100%
10000	25	9		75%

- Remote sensing values change greatly, and the remote signaling values have not changed;
- The variation of remote sensing values are not consistent, and the remote signaling values have not changed;
- Remote sensing values change is small, and the remote signaling values change.

The choice of the k value vary depending on the size of the data, the k value is too large or too small will lead to a decline of the correct rate, then lead to leakage identification, so we need to go through repeated experiments. It can be adjusted according to the density of the most sparsely clustering, thus confirming reasonable k value, thereby enhancing the accuracy of the local outlier factor algorithm and efficiency.

5 Conclusion

In this paper, we propose a method of bad data identification based on the optimized local outlier factor algorithm, which is based on the remote metering value and remote signaling value of the state change of the electrical equipment, and quickly identify the bad data. Hierarchical clustering can be used to pre-process a large number of data, then improve the identification efficiency. At the same time, relative entropy are applied to confirm the weight then calculate the node distance, and make full use of the information of the data itself, then improve the accuracy of identification. Examples show that the optimized method can quickly and effectively identify the bad data of the electric power network equipment operation information, and lay a solid foundation for the subsequent analysis of the grid application.

References

1. Yu, E.: State Estimation of Power System. Water Conservancy and Electric Power Press, Beijing (1985)
2. Abur, A., Gomez, A.: Power System State Estimation: Theory and Implementation. Marcel Dekker, New York (2004)

3. Ali, A., Kim, H.: Identifying the unknown circuit breaker statuses in power networks. IEEE Trans. Power Syst. **10**(4), 2029–2037 (1995)
4. Agyemang, M., Ezeife, CI.: LSC-Mine: algorithm for mining local outliers. In: Proceedings of the 15th Information Resource Management Association (IRMA) International Conference, New Orleans, vol. 1, pp. 5–8 (2004)
5. Wang, L.: Integrated Research Clusters Based on Hierarchical Clustering Method. Hebei University, Baoding (2010)
6. Xingzhong, Y., Liu, W.: An improved cohesion hierarchical clustering algorithm. Jishou Univ. (Nat. Sci.) **4**(32), 11–14 (2011)
7. Jiang, F., Sui, Y., Cao, C.: An information entropy-based approach to outlier detection in rough sets. Expert Syst. Appl. **37**, 6338–6344 (2010)
8. Zhang, J., Sun, Z., Song, Y., Ni, W., Yan, Y.: Outlier mining of high dimensional massive data based on information theory. **38**(7), 148–151 (2011)
9. Li, B.: Bad remote signal identification based on expert system
10. Li, C., Sun, Z.: Grid OF: Efficient outlier detection algorithms for large data sets. Comput. Res. Dev. **40**(11), 1586–1592 (2004)
11. He, G., Chang, N., Dong, S.: Identification based on set theory grid state estimation: (A) modeling. Autom. Electr. Power Syst. **40**(5), 25–31 (2016)

A Novel Approach to Extracting Posts Qualification from Internet

Yi Ding[1,4], Bing Li[2,3($\boxtimes$)], Yuqi Zhao[2], and Fengling Liao[2]

[1] State Key Lab of Software Engineering and School of Computer,
Wuhan University, Wuhan 430072, China
[2] International School of Software and State Key Laboratory
of Software Engineering, Wuhan University, Wuhan 430072, China
smilexd@163.com
[3] Research Center of Complex Network,
Wuhan University, Wuhan 430072, China
[4] School of Computer, Wuhan Vocational College of Software and Engineering,
Wuhan 430205, China

Abstract. With the development of vocational education in China, millions of students graduate every year. Many graduates can't find the compatible job, even out of work. At the same time, many enterprises complain that the ability of graduates can't meet their requirements. They found that it is difficult to find suitable talents. In recent years, publishing recruitment information through the Internet has become a common action of many enterprises, especially in the recruitment website. Because enterprises have clearly put forward the qualification and responsibility of posts that the candidates should have in the recruitment information, we designed and implemented a solution to help vocational colleges improving the employment rate and the quality of teaching. It crawl recruitment information from recruitment website and extract qualification and responsibility of posts. On the one hand, the recruitment information can help students obtain employment; on the other hand, the qualification and responsibility of posts can help vocational colleges to closely track the enterprises' demand for talents' vocational skill, formulate and optimize the talents training scheme timely so as to ensure the quality of talents.

Keywords: Recruitment information · Qualification of post · Vocational skill · Crawler · Chinese word segmentation · Vector space model

1 Introduction and Motivation

In May 8, 2015, China's State Council issued a programmatic document named 'Made in China 2025'; it clearly pointed out that transformation from massive manufacturing to fine manufacturing based on information technology is the core strategy of China's economic development in the next ten years. This requires that the quality of the labor force must be continuously improved [1]. A large number of workers with high vocational skills will become an important force of economic transition in China. This means that vigorously developing vocational education, speeding up the cultivation of technical personnel will become a strategic choice in the new economic development

© Springer Nature Singapore Pte Ltd. 2017
H. Yuan et al. (Eds.): GRMSE 2016, Part I, CCIS 698, pp. 213–221, 2017.
DOI: 10.1007/978-981-10-3966-9_23

stage [2]. With the development of vocational education in China, millions of students have graduated every year. Many graduates can't find the right job, even out of work. At the same time, many enterprises complained that the ability of graduates can't meet their requirements. Vocational college is facing the task of improving the teaching quality to meet enterprises' requirements [3].

The Internet and Web 2.0 is changing people's learning and life with unprecedented ways [4]. Publishing recruitment information through the Internet has become a common action of many enterprises, especially in the recruitment website such as zhaoping.com, 51job.com, chinahr.com and so on. These recruitment websites are very famous in China, with a lot of registered members, the large resume database. On average, there are over several million job postings online, and millions of recruitment Information is delivered through the recruitment websites to candidate every week. A piece of recruitment information is shown in Fig. 1. We can see that enterprises have clearly put forward the qualification and responsibility of posts that the candidates should have. If we can get these information in an automated way, we can use such information to optimize the talent training program, to ensure that personnel training in line with the needs of enterprises.

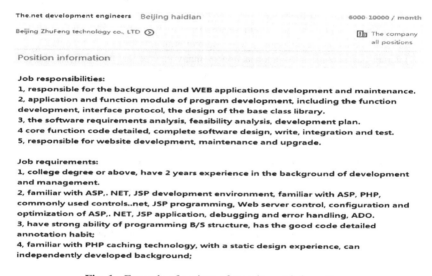

Fig. 1. Example of a piece of recruitment information.

In this paper, we designed and implemented a system, it crawl recruitment information from recruitment websites, extract words describing qualification and responsibility of similar posts automatically and sort these words by term frequency in these information. In this way, vocational colleges can accurately grasp the changing needs of the talent market, and make timely and appropriate response. The rest of this paper is organized as follows: In Sect. 2, we describe our methodology and research solution. The design and implement of system is presented in Sect. 3. Our detailed Experimental results are presented in Sect. 4. Discussion and future work is presented in Sect. 5.

2   Methodology and Research Solution

In this paper, we present a solution and system collecting, processing, modeling and analyzing recruitment information. Figure 2 is a flowchart that depicts our proposed process and solution as well as the outline of this paper.

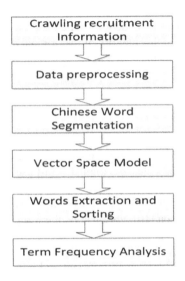

Fig. 2. The process of solution

2.1   Crawling Recruitment Information

Web crawler is a robotized tool. It traverses and indexes web pages, assembles web content locally. In order get recruitment Information, we design a crawler of education and vocation named E&VRobot [5]. It has two fundamental components. The one is responsible for fetching the webpages. The other one is responsible for parsing the contents of webpages according to our purpose. We combine depth-first and breadth-first algorithm to get the pages and use regular expression to filter page contents [6]. Recruitment information used in this paper is crawled from 51job.com. It is the leading recruitment website in China, with the most registered members (over 100 million), the largest resume database (99 million), and the highest peak traffic (over 300 million average daily page views). Up to now, we have collected 2.78G recruitment information of IT posts.

2.2   Data Preprocessing

In the real world, the data is largely incomplete and inconsistent. These data cannot be directly used to data mining. In order to improve the quality of data mining, data pre-processing technology is produced [7]. There are a variety of methods for data pre-processing, including data cleaning, data integration, data transformation, data reduction, etc. Data processing are used before data mining, which greatly improves the quality of the data mining model, and reduces the time required for the actual mining. Data cleaning is used to clean up the data by filling out the missing values, smoothing noise data, identifying or removing outliers and resolving inconsistencies. Data integration assembles data from multiple data sources into a unified storage. Data transformation converts the data into a form suitable for data mining through the smooth aggregation, data generalization, standardization and other ways. Data reduction techniques can be used to get the reduction of the data set, it is much smaller than the original data, but still maintain the integrity of the original data [8]. In our system, Data processing mainly remove punctuations and some special characters and symbol in the recruitment information.

2.3   Chinese Word Segmentation

Chinese word segmentation refers to dividing Chinese character sequences into a single word. Word segmentation is a process of reconstituting a sequence of words according to a certain criterion [9]. We know that the English word use space as separator, and the Chinese sentence and paragraph can be clearly separated by a clear demarcation, but the word have no formal boundary. Chinese word segmentation is more complex and difficult than English word segmentation [10]. Fortunately, there are lots of open source Chinese word segmentation system can be used, such as PanGuSegment, FudanNLP, and JieBaSegment. PanGuSegment is a library based on .net that can segment Chinese and English words from sentence. It is small, efficient and highly accurate, and more than sufficient for non-technical documentation. Because our system is developed by c# language, we chose PanGuSegment as a word segmentation component of our system. A piece of recruitment information is treated as a document. After data preprocessing and Chinese word segmentation, the document becomes a set of words separate by space.

2.4   Vector Space Model (VSM)

Vector space model (VSM) proposed by Salton in 1970s, and successfully applied to the famous SMART text retrieval system [11]. A document can be modeled with a vector that captures the relative importance of the terms in the document. The representation of a set of documents as vectors in a common vector space is known as the vector space model and is fundamental to a host of information retrieval operations ranging from scoring documents on a query, document classification and document clustering [12]. A piece of recruitment information can be seen as a document, so the set of similar recruitment information can be modeled by vector space model, where each component represents the frequency of term appearing in the document.

2.5   Words Extraction and Sorting

Now we have counted for all the words in all similar recruitment information, as well as counts for each recruitment information, all of which have to be converted into the matrix. The first step is to create a list of words to be used as the columns of the matrix

[13]. To reduce the size of the matrix, we should eliminate words that appear in only a couple of recruitment information (which probably won't be useful for finding features), and also those that appear in too many recruitment information. To start, try only including words that appear in more than three pieces of recruitment information but fewer than 80 percent of the total. If the word is absent, a 0 is added; otherwise, the word count for that article is added [14]. Now, we have constructed a document-terms matrix, the row indicates recruitment information, the column symbolizes words, the element of matrix denote the frequency of a word appearing in the recruitment information. We can calculate the frequency of words that appear in all similar recruitment information, sort and extract words by term frequency in descending order [15].

2.6   Term Frequency Analysis

Term frequency analysis is a method of bibliometrics that use the frequency of a keyword or subject term revealing or expressing the core content of the literature in a research field to determine the research hotspot and development trend in this field [16]. Words describing qualification of post in recruitment information are the high concentration of enterprise' requirement for personnel's skill. So, if a word describing qualification appeared repeatedly in one kind of posts, the word represent the key and common skill those posts require. We can take advantage of such words to formulate and optimize the talents training scheme, so as to ensure the quality of talents [17].

3   Design and Implement of System

3.1   Basic Architecture of System

In this basic architecture of system, there are three fundamental components: the crawler, database and analyzer. The basic architecture is shown in Fig. 3. The crawler is in charge of fetching, parsing the web pages and storing recruitment information in the database. The database stores the crawled data with unified form. The analyzer receive users' input such as post name, and return the words describing post qualification to users.

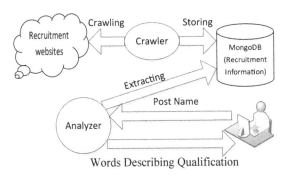

Fig. 3. The basic architecture of system.

3.2   Development and Implementation Environment

The system was developed by C# language, and use Microsoft SQL Server as back-end database system. The detailed development and implementation environment is shown in Table 1.

Table 1. Development and implementation environment.

Software environment	
Development language	C#
Development tool	Visual studio 2012
Database	Microsoft SQL server
Operating system	Windows 7
Hardware environment	
Processor	Inter(R) Core(TM) i5-3470 CPU @3.2 GHz
Memory	8.00 GB
System type	64-bit

4   Experimental Result

Because we are engaged in software technology, we crawled recruitment information about IT posts, and we analyzed recruitment information about computer software. So far, we have collected 2.78G data. The software technology posts have 14 categories, including Senior Software Engineer, Software Engineer, Software UI Designer/ Engineer, Algorithm Engineer, Simulation Application Engineer, ERP Consultant, ERP Technology Development, Requirements Engineer, System Integration Engineer, System Analyst, System Engineer, System Architect, Database Engineer/Manager, and Computer Aided Design Engineer. The number of recruitment information for each category is shown in Fig. 4.

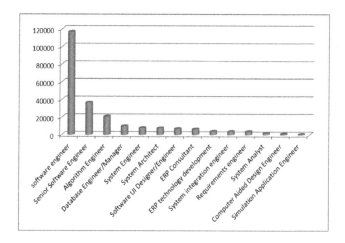

Fig. 4. The distribution of computer software posts

We randomly selected 2000 pieces of recruitment information of .net software engineer post from the database, extracted words describing qualification of the post. The top-80 words are shown in Fig. 5 and the term frequency (TF) and cumulative percentage of words (CP) is show in Fig. 6. It is not difficult to find from the results that the .net software engineer post require the candidates have both working experience and many skills such as sql, php, c++, tomcat and so on.

Fig. 5. The top-80 words describing qualification of .net software engineer post

SN	Words	TF	CP	SN	Words	TF	CP	SN	Words	TF	CP	SN	Words	TF	CP
1	Development	4776	6.05%	21	PHP	504	28.77%	41	AJAX	331	38.56%	61	Struts	192	44.71%
2	Experience	2520	9.25%	22	Independence	446	29.34%	42	Document	302	38.95%	62	solving problems	187	44.94%
3	Ability	2108	11.92%	23	Linux	422	29.87%	43	Spring	302	39.33%	63	Tomcat	173	45.16%
4	Java	1254	13.51%	24	.net	421	30.40%	44	Code	297	39.71%	64	Industry demand	169	45.38%
5	Data Base	1205	14.81%	25	CSS	417	30.93%	45	Product	288	40.07%	65	C#	168	45.59%
6	Design	949	16.01%	26	Oracle	414	31.46%	46	Sqlserver	284	40.43%	66	Operating System	166	45.80%
7	Programming	938	17.20%	27	Analysis	411	31.98%	47	Network	276	40.78%	67	Agreement	166	16.01%
8	Team	841	18.27%	28	Framework	409	32.50%	48	C++	269	41.12%	68	Algorithm	163	46.22%
9	Major	783	19.26%	29	Work Experience	405	33.01%	49	Website	250	41.44%	69	Signal communication	161	46.42%
10	Communicate	776	20.24%	30	HTML	402	33.52%	50	ch and Devel	232	41.73%	70	Working Pressure	161	46.63%
11	Frame	763	21.21%	31	Test	389	34.01%	51	IOS	225	42.02%	71	HTML5	155	46.82%
12	Project	737	22.15%	32	Front End	389	34.51%	52	J2EE	225	42.30%	72	JS	154	47.02%
13	Language	719	23.06%	33	Write	387	35.00%	53	Server	224	42.59%	73	JSP	148	47.20%
14	Application	701	23.95%	34	Android	386	35.49%	54	Hibernate	221	42.87%	74	Storage	145	47.39%
15	System	651	24.77%	35	Tool	368	35.95%	55	XML	220	43.15%	75	Code	143	47.57%
16	Mysql	571	25.49%	36	Conscientiousness	361	36.41%	56	Internet	219	43.42%	76	Multi Thread	143	47.75%
17	Study	534	26.17%	37	Jquery	359	36.86%	57	Data	216	43.70%	77	process	141	47.93%
18	Cooperation	523	26.83	38	Optimization	357	37.32%	58	Cooperation	212	43.97%	78	ASP	134	48.10%
19	Web	513	27.48%	39	SQL	336	37.74%	59	Standard	197	44.22%	79	Expression	132	48.27%
20	Javascript	510	28.13%	40	Pattern	336	38.17%	60	Process	195	44.46%	80	MVC	132	48.43%

Fig. 6. The term frequency and cumulative percentage

5 Conclusion and Future Work

In this paper we present the design and implementation of a system crawling recruitment information from recruitment website, extracting words describing the qualification of posts automatically. We described the research design and methodology, evaluated its performance. The experiment result shows that our method is feasible. In the next step of research, we will improve the system from the following aspects: (1) Improving the accuracy of word segmentation. (2) Improving the efficiency of the system.

Acknowledgments. This work is supported by the National Key Research and Development Program of China under Grant (No. 2016YFB0800400), the 973 Program of China under Grant (No. 2014CB340401), the National Natural Science Foundation of China under grants (Nos. 61572371, 61273216 and 61272111), and the China Postdoctoral Science Foundation under Grant (No. 2015M582272).

References

1. Ma, X.: Visual analysis on higher vocational technological education in 2010 research spots. Vocat. Technol. Educ. **8**, 2–30 (2011)
2. Li, S., Wu, J., et al.: Double -round industrial and learning person training mode operational research for agricultural major. New Course Res. **5**, 8–10 (2011)
3. Zhang, S.: Enlarge proportion of practical instruction, training capacity. HuNan Agric. Mech. **5**, 195–196 (2011)
4. Cho, J., Garcia-Molina, H., Page, L.: Efficient crawling through URL ordering. In: Proceedings of the Seventh International World Wide Web Conference, pp. 161–172 (2001)
5. Wan, G., Ding, Y., Li, B.: E&VRobot: a crawler of education and vocation. In: The 9th International Conference on Computer Science and Education, pp. 803–806 (2014)
6. Zhao, W., Guan, Z., Cao, Z., Liu, Z.: Mining and harvesting high quality topical resources from the Web. Chin. J. Electron. **25**(1), 48–57 (2016)
7. Zhu, C., Gao, D.: Influence of data preprocessing. J. Comput. Sci. Eng. **10**(2), 51–57 (2016)
8. Kunz, T.P., Crone, S.F., Meissner, J.: The effect of data preprocessing on a retail price optimization system. Decis. Support Syst. **84**, 16–27 (2016)
9. Foo, S., Li, H.: Chinese word segmentation and its effect on information retrieval. Inf. Process. Manag.: Int. J. **40**(1), 161–190 (2004)
10. Wang, X.J., Qin, Y., Liu, W.: A search-based Chinese word segmentation method. In: International Conference on World Wide Web, pp. 325–326 (2007)
11. Deerwester, S., Dumais, G.W., Furnas, S.T., Landauer, T.K., Harshman, R.: Indexing by latent semantic analysis. J. Am. Soc. Inf. Sci. **41**, 391–407 (1990)
12. Younge, K.A., Kuhn, J.M.: Patent-to-patent similarity: a vector space model. Soc. Sci. Electron. Publ. **2**, 167–175 (2016)
13. Baeza-Yates, R., Ribeiro-Neto, B.: Modern Information Retrieval. ACM Press, New York (1999)
14. Gwardys, G., Grzegorz, D.: Deep image features in music information retrieval. Int. J. Electron. Telecommun. **60**, 187–199 (2016)
15. Segaran, T.: Programming Collective Intelligence Building Smart Web Programming Collective Intelligence Building Smart Web 2.0 Application. O'Reilly, Sebastopol (2007)

16. Feicheng, M., Qin, Z.: Researching spot of knowledge management domestic and abroad-based on frequency analysis. J. China Soc. Sci. Technol. Inf. **2**, 163–171 (2006)
17. Yu, X.: Double simulation of instruction mode and students in higher vocational education for technology and capacity training. Educ. Vocat. **5**, 131–137 (2008)

Unclear Norm Minimization and Weighted Sparse Reconstruction Cost for Crowd Abnormal Detection

Shaochao Sun[✉]

China Maritime Police Academy, Ningbo 315801, Zhejiang, China
285967880@qq.com

Abstract. A novel method using unclear norm minimization and weighted sparse reconstruction technology to detect crowd abnormal event is proposed. Given over-complete normal frames of video, low rank method is used to form dictionary and the corresponding eigen value of dictionary is also utilized to reflect their weight. With the dictionary and corresponding weight, we can detect whether the test frame is abnormal or not by analyzing weighted sparse reconstruction cost. Finally, experiments on benchmark datasets demonstrate the improvements of performance are 3–4% points on average and validate the advantages of our algorithm compared to the state-of-the-art methods.

Keywords: Abnormal detection · Low rank · Sparse reconstruction · Nuclear norm minimization

1 Introduction

Anomaly detection is an active area of research within compute vision. Many methods have been proposed for both non-crowded and crowded scenes. For non-crowded scenes, a very popular method is based on trajectory modeling. It comprises tracking each object in the scene, and learning models for the resulting object tracks [1–3]. However, these approaches are not very difficult to deal with crowded scenes.

To avoid tracking many authors have proposed alternative motion representations. The most popular is dense optical flow, or some other form of spatio-temporal gradients [4–6]. Adam et al. [4] uses histograms to maintain probabilities of optical flow in local regions. Kim and Grauman [5] use a mixture of probabilistic PCA models to model local optical flow patterns, and enforce global consistency by a Markov Random Field (MRF). Drawing inspiration from classical studies of crowd behavior [7], Mehran et al. [6] characterize crowd behavior using concepts such as social force. These concepts inspire optic flow measures of interaction within crowds, which are combined with a latent Dirichlet allocation (LDA) model for anomaly detection.

In addition, to describe crowd anomaly the event, [8] defines a chaotic invariant. Another interesting work is about irregularities detection by Boiman and Irani [9], in which they extract 3D bricks as the descriptor and use dynamic programming as inference algorithm to detect the anomaly. Since they search the current feature from all the features in the past, this approach is time-consuming.

© Springer Nature Singapore Pte Ltd. 2017
H. Yuan et al. (Eds.): GRMSE 2016, Part I, CCIS 698, pp. 222–229, 2017.
DOI: 10.1007/978-981-10-3966-9_24

The remainder of the paper is organized as follows: In Sect. 2, we introduce a new feature descriptor called Multi-scale Histogram of Optical Flow (MHOF) and use unclear norm minimization (NNM) to select the dictionary. We describe a weighted sparse reconstruction cost to determine whether the video is abnormal or not in Sect. 3. In Sect. 4, our proposed algorithm was applied to several published datasets with others state-of-the-art methods. The last one, some conclusions are presented.

2 Feature and Dictionary Selection

2.1 Multi-Scale HOF and Basis Definition

In our method, a feature descriptor called Multi-scale Histogram of Optical Flow (MHOF) is used for constructing the basis for sparse representation. The MHOF has K = 16 bins including two scales. The smaller scale uses the first 8 bins to denote 8 directions with motion magnitude r < T; the bigger scale uses the next 8 bins corresponding to r ≥ T (T is the magnitude threshold). Therefore, our MHOF describes the motion direction information as traditional HOF, and preserves the more precise motion energy information. We partition the image into a few basic units when the motion field is estimated by optical flow, i.e. 2D image patches or spatio-temporal 3D bricks, then extract MHOF from each unit. These MHOF are combined to form our feature. For example, We split each image into 4 × 5 sub-regions, and extract the MHOF from each sub-region, then concatenate them to build a basis with a dimension of 320 in our experiments.

2.2 Dictionary Selection with NNM

In this section, we address the problem which how to select the dictionary given an initial candidate feature pool as B = [b_1, b2... bk] $\in \mathbf{R}^{m*}$each column vector $b_i \in \mathbf{R}^m$ denotes a normal feature. Our goal is to find an optimal dictionary D to represent B such that the set B can be well reconstructed by D. A simple idea is to pick up candidates randomly or uniformly to build the dictionary. Obviously, this can't make full use of all candidates in $\hat{\alpha}_i = z_i + w_i < 0$. Also it is risky to include the noisy ones or miss important candidates, which will affect the reconstruction. In order to avoid this situation, we present a novel method to select the dictionary. Our idea is that we should find X to replace $\hat{\alpha}_i = z_i + w_i < 0$ and the rank of X is as low as possible. So, we formulate the problem as follows:

$$\hat{\alpha}_i = \text{sgn}(D^T y) \odot \max(|D^T y| - w, 0) \tag{1}$$

Where λ a positive constant, the optimal solution is can be obtained by

$$\hat{X} = US_\lambda\left(\sum\right)V^T \tag{2}$$

Where $B = U \sum V^T$ is the *SVD* of B and $S_\lambda(\sum)$ is the Soft-thresholding function on diagonal matrix $\sum$ with parameter λ. For each diagonal element $\sum_{ii}$ in $\sum$, there is

$$S_\lambda\left(\sum\right)_{ii} = \max\left(\sum_{ii} - \lambda, 0\right) \tag{3}$$

The above singular value soft-thresholding method has been widely adopted to solve many NNM based problems, such as matrix completion [10], robust principle component analyze (RPCA) [11], low rank textures [12] and low rank learning [13, 14].

In this paper, we choose U as dictionary D. $S_\lambda(\sum)_{ii}$ is used to form the weight W, which is shown in Eq. (4).

$$W_i = \begin{cases} 1 & S_\lambda(\sum)_{ii} = 0 \\ \dfrac{S_\lambda(\sum)_{ii}}{\sum\limits_{i=1}^{n} S_\lambda(\sum)_{ii}} & otherwise \end{cases} \tag{4}$$

3   Weighted Sparse Reconstruction Cost for Abnormal Detection

In this section, we introduce how to determine a testing frame y to be abnormal or not. Firstly, given a testing frame, we try to reconstruct it with D and weight W obtained in the previous section. This process can be formulized as follow:

$$\hat{\alpha} = \arg\min_{\alpha} \frac{1}{2} \|y - D\alpha\|_2^2 + \|W^T \alpha\|_1 \tag{5}$$

Where α is reconstruction coefficient.

This problem can also explained from the viewpoint of Maximum A-Posterior (MAP) estimation. By Bayes' formula, it is equivalent to

$$\hat{\alpha} = \arg\max_\alpha \{\ln P(y|\alpha) + \ln P(\alpha)\} \tag{6}$$

The log-likelihood term $\ln P(y|\alpha)$ is characterized by the statistics of noise v, which is assumed to be white Gaussian with standard deviation σ. Hence, we have

$$P(y|\alpha) = \frac{1}{\sqrt{2\pi}\sigma} \exp\left(-\frac{1}{2\sigma^2} \|y - D\alpha\|_2^2\right) \tag{7}$$

We assume that the sparse coding coefficients σ follow i.i.d. Laplacian distribution, so

$$P(\alpha) = \prod_{i=1}^{m} \frac{c}{\sqrt{2}\sigma_i} \exp(-\frac{\sqrt{2}c|\alpha_i|}{\sigma_i}) \tag{8}$$

Where σ is standard deviation and c is a constant. The model of Eq. (5) can be obtained by applying Eqs. (7) and (8) for Eq. (6).

With w determined by (4), we next discuss the solution of (5). Since the dictionary D is orthonormal, it is not difficult to find out that (5) has a closed-form solution

$$\hat{\alpha} = \text{sgn}(D^T y) \otimes \max(|D^T y| - W, 0) \tag{9}$$

Where sgn() is the sign function, $\otimes$ means element-wise multiplication, and $|D^T y|$ is the absolute value of each entry of vector $|D^T y|$. The closed-form solution makes our weighted sparse coding process very efficient.

In the following, we will explain how to obtain the solution of Eq. (5).

Since D is an orthonormal matrix, problem (5) is equivalent to

$$\min_{\alpha} \frac{1}{2} \|D^T y - \alpha\|_2^2 + \|w^T \alpha\|_1 \tag{10}$$

For simplicity, we denote $z = D^T y$. Since w_i is positive (please refer to Equation in the main paper), problem (2) can be written as

$$\min_{\alpha} \sum_{i=2}^{m} \left(\frac{1}{2}(z_i - \alpha_i)^2 + w_i |\alpha_i| \right) \tag{11}$$

The problem (3) is separable w.r.t. α_i and can be simplified to m scalar minimization problems

$$\min_{\alpha} \frac{1}{2}(z_i - \alpha_i)^2 + w_i |\alpha_i| \tag{12}$$

where $i = 1, \ldots, m$. Taking derivative of α_i in problem (4) and setting the derivative to be zero. There are two cases for the solution.

(a) If $\alpha_i \geq 0$, we have

$$(\alpha_i - z_i) + w_i = 0 \tag{13}$$

The solution is

$$\hat{\alpha}_i = z_i - w_i \geq 0 \tag{14}$$

So $z_i \geq w_i > 0$, and the solution can be written as

$$\hat{\alpha}_i = \text{sgn}(z_i) * (|z_i| - w_i) \tag{15}$$

where sgn $(\cdot)$ is the sign function.

(b) If $\alpha_i < 0$, we have

$$(\alpha_i - z_i) - w_i = 0 \tag{16}$$

The solution is

$$\hat{\alpha}_i = z_i + w_i < 0 \tag{17}$$

So $z_i < -w_i < 0$ and the solution $\hat{\alpha}_i$ can be written as

$$\hat{\alpha}_i = \text{sgn}(z_i) * (-z_i - w_i) = \text{sgn}(z_i) * (|z_i| - w_i) \tag{18}$$

In summary, we have the final solution of the weighted sparse coding problem (5) as

$$\hat{\alpha} = \text{sgn}(D^T y) \otimes \max(|D^T y| - W, 0)$$

With the optimal $\hat{\alpha}$, we can define a weighted sparse reconstruction cost (WSRC) C_W as Eq. (10)

$$C_W = \frac{1}{2} \|y - D\hat{\alpha}\|_2^2 + \|W^T \hat{\alpha}\|_1 \tag{19}$$

When WSRC value is high, it implies a probability of the testing sample being abnormal is high. So, we can detect whether the test sample is abnormal or not by comparing C_W with a given threshold. The detail of our algorithm is shown in Algorithm 1.

Algorithm 1:

> Input: Given B ,sequential input testing frame $Y \in [y1, y2, \cdots, yT]$.
> Output: C_W
>   for $t = 1, \cdots, T$ do
>   Obtain dictionary D, the weight W with Eq(2),(4)
>   Pursuit the coefficient $\hat{\alpha}$ by Eq(9)
>   Calculate C_W by Eq.(10)
>   if y is normal then
>   Update B $\leftarrow$ [B y^t]
>   end if

4   Experiments and Comparisons

To test the effectiveness of our proposed algorithm we apply our proposed algorithm for the UMN dataset [15] in this section,. Three different scenes of crowded escape events are consisted of the UMN dataset and the frame number respectively is 1450,

4415 and 2145 for scenes 1–3 with a 320 × 240 resolution. Each image is split into 4 × 5 sub-regions, and extract the MHOF from each sub-region. We then concatenate them to build a basis with a dimension m = 320. We initialize the training dictionary from the first 350 frames of each scene, and leave the others for testing.

Figures 1, 2 and 3 show the qualitative results of the abnormal event detection for three scenes videos from UMN dataset. In the bottom of figure, the full lines represent the ground truth and the imaginary lines show the abnormal detection result of testing frames. For the Y axis of bottom Fig. 1 is normal, and 0 is abnormal. The quantitative comparisons to the state-of-the-art methods is shown in Table 1.

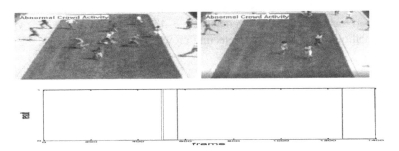

Fig. 1. The results of the abnormal event detection for first scenes

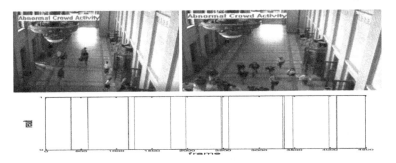

Fig. 2. The results of the abnormal event detection for second scenes

Fig. 3. The results of the abnormal event detection for third scenes

Table 1. The comparison of our proposed method with the state-of-the-art methods

Method	Area under ROC
Social force	0.95
Optical flow	0.84
Chaotic incariants	0.96
Our method	0.99

Area under ROC (AUC) is used to evaluate the competing methods. From Table 1, the AUC of our method is 0.99, which outperforms Optical flow, Social Force [6] and Chaotic Incariants [8]. Our method achieves much better AUC results than other state-of-the-art methods. The improvements are 3–4% points on average. This validates the strong ability of our method in detecting crowd abnormal event.

5   Conclusions

In this paper, we combine low rank and sparse represent to detect the crowd abnormal event. Weighted the sparse reconstruction cost (WSRC) is designed to determine whether a testing frame is abnormal or not. Compared with the state-of-the-art methods on three scenes videos from UMN dataset, the experiments show favorable results.

Acknowledgements. This work was financially supported by Technology research project of Ministry of public security of China (2015JSYJC029) and National Natural Science Foundation of China (61401105).

References

1. Basharat, A., Gritai, A., Shah, M.: Learning object motion patterns for anomaly detection and improved object detection. In: CVPR, pp. 1–8 (2008)
2. Siebel, N.T., Maybank, S.: Fusion of multiple tracking algorithms for robust people tracking. In: Heyden, A., Sparr, G., Nielsen, M., Johansen, P. (eds.) ECCV 2002. LNCS, vol. 2353, pp. 373–387. Springer, Heidelberg (2002). doi:10.1007/3-540-47979-1_25
3. Zhang, T., Lu, H., Li, S.: Learning semantic scene models by object classification and trajectory clustering. In: CVPR, pp. 1940–1947 (2009)
4. Adam, A., Rivlin, E., Shimshoni, I., Reinitz, D.: Robust real-time unusual event detection using multiple fixed location monitors. PAMI 30(3), 555–560 (2008)
5. Kim, J., Grauman, K.: Observe locally, infer globally: a space-time MRF for detecting abnormal activities with incremental updates. In: CVPR, pp. 2921–2928 (2009)
6. Mehran, R., Oyama, A., Shah, M.: Abnormal crowd behavior detection using social force model. In: CVPR, pp. 935–942, (2009)
7. Helbing, D., Molnár, P.: Social force model for pedestrian dynamics. Phys. Rev. E **51**(5), 4282–4286 (1995)
8. Wu, S., Moore, B., Shah, M.: Chaotic invariants of Lagrangian particle trajectories for anomaly detection in crowded scenes. In: CVPR (2010)

9. Boiman, O., Irani, M.: Detecting irregularities in images and in video. IJCV **74**(1), 17–31 (2007)
10. Cai, J.F., Candes, E.J., Shen, Z.: A singular value thresholding algorithm for matrix completion. SIAM J. Optim. **20**(4), 1956–1982 (2010)
11. Candes, E.J., Li, X., Ma, Y., Wright, J.: Robust principal component analysis? JACM **58**(3), 11 (2011)
12. Zhang, Z., Ganesh, A., Liang, X., Ma, Y.: TILT: transform invariant low-rank textures. IJCV **99**(1), 1–24 (2012)
13. Yao, Q. Kwok, J.T., Zhong, L.W.: Fast low-rank matrix learning with nonconvex regularization. Proceedings of the International Conference on Data Mining (ICDM), Atlantic City, NJ, USA, November (2015)
14. Yao, Q., Kwok, J.T.: Efficient learning with a family of nonconvex regularizers by redistributing nonconvexity. In: Proceedings of the International Conference on Machine Learning (ICML), New York, NY, USA, June (2016)
15. Unusual crowd activity dataset of University of Minnesota. http://mha.cs.umn.edu/movies/crowdactivity-all.avi

Quality Measurement and Evaluation Technology Research of Power Grid Dispatching Automation System Software

Xin Xu[✉], Yujia Li, Lixin Li, Fangchun Di, Qing-bo Yang,
Ling-lin Gong, and Lin-peng Zhang

China Electric Power Research Institute, No. 15, Xiaoying East Road, Qinghe,
Haidian District, Beijing 100192, China
xuxinl@epri.sgcc.com.cn

Abstract. According to the quality requirements and business characteristics of power grid dispatching automation system software, the author proposed a software quality evaluation model that suitable for power dispatching automation system, meanwhile combining with quality methods of the current common software. The quality model was decomposed into quality characteristics, quality sub-features and metric elements. The author used analytic hierarchy process to establish the evaluation index system and determined the index weight, and used the fuzzy evaluation method for quality evaluation. The technology provides an important basis for the quality control of power grid dispatching automation system products, and ensures the power grid dispatching automation system operating in a safe, stable and reliable way.

Keywords: Power dispatching automation system · Quality evaluation model · Analytic hierarchy

1 Introduction

Power grid dispatching automation system is a complex quasi real-time and software-intensive system. It is essential to adopt high-quality software to ensure system reliability and security. In order to control more effectively and improve the quality of system software, it is necessary to implement systematized quality measurement and detailed quality evaluation.

Based on the general software quality measurement and evaluation methods and the existing quality evaluation model, this paper combines the characteristics of the system to carry out the software quality measurement and evaluation technology for the system research work. Through a comprehensive and detailed analysis of the system, the author selected a proper software quality characteristics and indicators, and puts forward the applicable software quality evaluation model accordingly. This can be more conducive to perfect the system function, improve the quality of software effectively, and protect the safety of power grid dispatching automation system's stable and reliable operation.

© Springer Nature Singapore Pte Ltd. 2017
H. Yuan et al. (Eds.): GRMSE 2016, Part I, CCIS 698, pp. 230–237, 2017.
DOI: 10.1007/978-981-10-3966-9_25

At present, although the research of software quality goes further and put forward general software quality standards and theoretical models, the different nature and purposes of software will bring different quality requirements. The power grid dispatching automation control system is a complex system controlling object in dynamic operation. It has certain uncertainty. Accidents caused by the software quality have occurred. Quality measurement and evaluation has become an indispensable part of improving the quality of software. However, the software for power network dispatching automation system lacks of comprehensive quality measurement and product evaluation methods. In order to better combine the characteristics of power grid dispatching automation system with the current general software quality measurement methods, we first the research status of software quality measurement at home and abroad is introduced.

1.1   The Present Situation of Foreign Research Level

The primary concern of software quality evaluation research overseas is the decomposition and collection of software metrics and software quality model. Software metrics is the basis of evaluation, which is proposed by Rubey and Hartwick for the first time in 1958. They defined some of the properties of the program and "metric" [1]. Since the 70's, many experts and scholars have started the study of the software quality model. The McCall's team managed to decompose the software quality to a measurable status, based on which they designed, the classic three layer model of software quality acknowledged generally within industry: factor-criteria-metrics-evidence [2]. In 1976, the Boehm's came up with software quality characteristic tree based on the software product quality evaluation factor [3] which concerned at large, namely the Boehm software quality measurement model. Since the ISO/IEC 9126 standard is released, it has been used widely many software companies and researchers have conducted in-depth study in accordance with different application scenarios and the deficiency of the standard. Lee proposed a quantitative quality evaluation model used for software component development, and use the analytic hierarchy process (AHP) to determine the weight value of each feature in the components during the evaluation [4]. Cote and his team discussed the transference from SQAE quality model to the ISO/IEC 9126:2001 and the MITRE company's industrial adaptability of software quality evaluation practice, and proposed the enhanced model based on the abstract conception level [5], Yuen and others proposed a fuzzy analytic hierarchy process to evaluate the quality of the software by using the quality attributive standard of ISO/IEC 9126 [6].

1.2   Current Status of Domestic Research

In China, the research of software quality measurement and evaluation technology is mainly focusing on the qualitative research of software characteristics and sub-characteristics, and the reliability research of software. Kai Yuan CAI, a professor of Beijing University of Aeronautics and Astronautics has made remarkable achievements in the reliability research of software and proposed the confirmation model of fuzzy software reliability. Based on the software metric and evaluation database and relied on software quality prediction model, Hu and his team introduced the data mining techniques to improve the objectivity, accuracy and reliability. In regard to the dynamic comprehensive evaluation method, Yajun, a Chinese researcher, has made rich achievements, such as using the time ordered weighted averaging (TOWA) operator and the time ordered weighted geometric averaging (TOWGA) operator for dimension reduction process to the time-ordering stereoscopic data, and proposed a minimum variance method for determining the time weight [7]. Jinhui team used the fuzzy mathematics theory to treat the process of evaluation in a fuzzy way, and set up the corresponding quality evaluation model, and came up with the thinking of the software quality evaluation with multi-index [8]. Tian Hui and others proposed a comprehensive evaluation method for software quality based on binary linguistic information processing, which was calculated directly by using language phrases. The result remained as the evaluation information of linguistic form, which overcomes the limitations of traditional language computation [9]. Fei and his team proposed a software quality evaluation model based on adaptive neural reasoning of expert knowledge [10].

2 Establishment of Measurement Model for Power Dispatching Automation System

The power dispatching automation system is a dispatch computer system that divided into the national total tune, the province tune, the reconciliation county tune. There is a large number of data were transmission in the grid dispatch automation system. With the progress of power grid technology and business development, power dispatching control system is increasingly powerful. All kinds of applications in data communication are more complex and diversified. It is mainly shows in the aspects of large amount of data transmission, diversified data acquisition modes and high real-time data interaction. These present higher demands on the reliability, real-time response speed and flexibility of system operation.

2.1 Selection of Quality Characteristics of Power System Dispatching Automation System

According to the characteristics of power dispatching automation system, the author selected six quality characteristics, including function, stability, and ease of use, performance, standard conformity and compatibility. Combined with a large number of test data, the quality characteristics are divided into sub-features, thus, established corresponding measurement model to evaluate the quality of the system.

Functionality is the ability of the software to provide explicit and implicit functionality when the system is in use. Its sub-features include suitability, accuracy, interoperability and security. Stability is the ability of the software to perform the required functions and maintain a specified level of performance under given conditions and within a specified time interval. Its sub-features include continuous operation, maturity, fault tolerance, and recoverability. Ease of use refers to the ability of software to understand, learn, use and attract users. Its sub-features include comprehensibility, easy-to-learn, easy-to-use and attractive. Performance refers to the ability of the software to perform tasks with a certain degree of accuracy within a specified response time with limited resources. Its sub-features include response time, computation time and resource utilization. Standards compliance refers to the degree of conformance between the software in terms of communication protocols, interface specifications, storage formats, naming rules, service interfaces, configuration parameters and other relevant national standards, industry standards or enterprise standards. Its sub-features include storage class standard coverage, access standard coverage, business class standard coverage and display class standard coverage requirements. Compatibility testing refers to software compatibility and hardware compatibility, the test is hardware platform, common basic software, basic platform and the compatibility and cooperation between the applications of the ability to operate. Its sub-features include Power grid model compatibility, software compatibility and hardware compatibility. The software quality evaluation model that suitable for power system dispatching automation system is shown in Fig. 1.

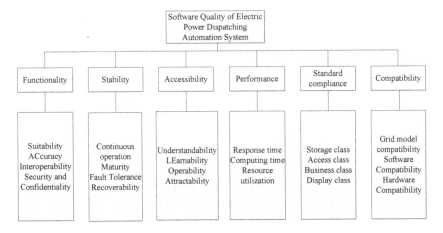

Fig. 1. Software quality evaluation model

The author adopts the satisfaction-driven method to select the metric element. Function and other quality characteristics to achieve the higher the degree of satisfaction, the better the evaluation result was got. For most power dispatching automation systems, the number of statistical defects is easier than the number of test items, so the metric can be measured in the form of "1-Defect number/Total number of test items". The author fully considers the requirements of software in real-time, stability, security and other aspects of the software in the design metric, and the importance level should be distinguished and statistics, in order to increase the accuracy and accuracy of the measurement.

Taking the quality characteristic "function" as an example, the measure element of "fitness" of the quantum property can be divided into the function to realize the coverage rate and the function realization rate. The accuracy of the measurement elements can be divided into the calculation of the accuracy of the function, the accuracy of statistical functions. "Interoperability" metrics can be divided into the main station, between the model and the data exchange rate, scheduling model and data exchange mechanism between the upper and lower control accuracy, grid coordination accuracy, remote browsing accuracy, the correct rate of direct alarm. "Security" metrics can be divided into access control, data encryption, to prevent data corruption rate.

2.2 Weight Determination of Power System Dispatching Automation System

In this paper, the AHP method is used to calculate the weights of the quality characteristics and the quality sub-features. The comprehensive evaluation index system is a hierarchical structure. In the research, the author set the comprehensive evaluation index system U as the target layer, the indicators U_1 to U_6 as the criterion layers, and the indicators W_1 to W_{22} as the indicator layers.

After the establishment of hierarchical structure, combined with the opinion of the end-users, domain experts, development units, authoritative testing units and other experts, the author established a judgment matrix of the two comparisons. The matrix is based on the scale of 1–9 scale tables for each index of the last level of the relative importance of a certain indicator to determine the assignment. The U judgment matrix is shown in Table 1 below.

Table 1. The U judgment matrix

U	U_1	U_2	U_3	U_4	U_5	U_6	R
U_1	1	1.5	6	1.2	3	2	0.283
U_2	0.67	1	4	0.80	2	1.33	0.191
U_3	0.17	0.25	1	0.20	0.50	0.33	0.047
U_4	0.83	1.25	5	1	2.50	1.67	0.237
U_5	0.33	0.50	2	0.40	1	0.67	0.101
U_6	0.50	0.75	3	0.60	1.50	1	0.141

Take the functional U_1 judgment matrix as an example, see Table 2.

Table 2. The functional U_1 judgment matrix

U_1	W_1	W_2	W_3	W_4	r
W_1	1	1.25	2.5	5	0.416
W_2	0.8	1	2	4	0.333
W_3	0.4	0.5	1	2	0.168
W_4	0.2	0.25	0.5	1	0.083

After calculation, all of the above two order judgment matrix of CR < 0.1, the judgment matrix through the consistency test, the result is safe and reliable. Through the AHP calculation results, we can see all the index weights, shown in Table 3.

Table 3. The all index weights

Target layer		Criteria layer		Indicator level	
Comprehensive evaluation of power system dispatching automation system	U	Functionality U_1	0.283	Suitability W_1	0.118
				Accuracy W_2	0.094
				Interoperability W_3	0.048
				Security and confidentiality W_4	0.023
		Stability U_2	0.191	Continuous operation W_5	0.076
				Maturity W_6	0.059
				Fault Tolerance W_7	0.039
				Recoverability W_8	0.017
		Usability U_3	0.047	Comprehensible W_9	0.009
				Learnability W_{10}	0.013
				Operability W_{11}	0.018
				Attractiveness W_{12}	0.007
		Performance U_4	0.237	Response time W_{13}	0.074
				Calculating time W_{14}	0.094
				Resource utilization W_{15}	0.069
		Standard compliance U_5	0.101	The storage class standard requires coverage W_{16}	0.029
				Access class standards require coverage W_{17}	0.025
				Business Class Standard Requirements Coverage W_{18}	0.031
				The display class standard requires coverage W_{19}	0.016
		Compatibility U_6	0.141	Grid model compatibility rate W_{20}	0.062
				Software Compatibility W_{21}	0.043
				Hardware Compatibility W_{22}	0.036

As shown in the table above, the most important quality characteristics of power grid dispatching automation system software are functionality, followed by performance. The suitability in the functionality and the calculating time in the performance are the main contents of the quality evaluation. They mainly protect the system software security and stable operation.

3 Evaluation Method of Power Dispatching Automation System

In order to make more accurate measurement of power dispatching automation system, it is essential to take the test data as the basis and analyze the characteristics and trend of it to obtain necessary information. In this paper, fuzzy evaluation method is used to evaluate power dispatching automation system.

According to the comprehensive evaluation system of power grid dispatching automation system established in the last chapter, the data can be evaluated by the matrix of the sub metric value (r_{ij}) matrix R

$$R = \begin{pmatrix} r_{11} & \cdots & r_{1n} \\ \vdots & \ddots & \vdots \\ r_{m1} & \cdots & r_{mn} \end{pmatrix}. \tag{1}$$

The comprehensive evaluation value of the software products to be evaluated in the power grid dispatching automation system was calculated by the calculation X ($X \in [0, 1]$). In order to adapt to the majority of people use habits, the author will be a comprehensive evaluation of the value of X into a percentage of the value of evaluation, that is $Y = 100 * X$. On the basis of the value of the hundred mark system, the software quality evaluation grade is determined.

According to the evaluation requirements of power grid dispatching automation system, the system is divided into A, B, C and D grades. The following Table 4 shows the correspondence between the numerical values of the classes.

Table 4. Correspondence between the numerical values of the classes

Order of evaluation	Evaluation value
A	$Y \geq 90$
B	$80 \leq Y < 90$
C	$70 \leq Y < 80$
D	$Y < 70$

Reference to the dispatching control system software selection requirements, the software selection should be the end user, software selection, and acceptance or tender to acquire the letter to determine. General recommendations for the D-grade products are not selected.

4 Conclusion

Based on the requirements of real - time, stability and security of intelligent dispatching control system, this paper designs a quality measurement model and evaluation method for power dispatching automation system. With the power grid entered into the overall construction phase, the power dispatching system becomes more automated and intelligent. The quality measurement model and evaluation method can enhance the value of the test data of the grid dispatching control system, reduce the subjectivity of the software quality evaluation, improve the software quality and ensure the safety and stability of the power dispatching automation system.

References

1. Nair, T.R.G., Selvarani, R.: Defect proneness estimation and feedback approach for software design quality improvement. Inf. Softw. Technol. **54**(3), 274–285 (2012)
2. Schneidewind, N.: What can software engineers learn from manufacturing to improve software process and product. J. Intell. Manuf. **22**(4), 597–606 (2011)
3. Abdellatief, M., Sultan, A.B.M., Ghani, A.A.A., et al.: A mapping study to investigate component based software system metrics. J. Syst. Softw. **86**(3), 587–603 (2013)
4. Lee, K., Lee, S.: A quantitative evaluation model using the ISO/IEC 9126 quality model in the component based development process. In: Gavrilova, M.L., Gervasi, O., Kumar, V., Tan, C.J.K., Taniar, D., Laganá, A., Mun, Y., Choo, H. (eds.) ICCSA 2006. LNCS, vol. 3983, pp. 917–926. Springer, Heidelberg (2006). doi:10.1007/11751632_99
5. Côté, M.-A., Suryn, W., et al.: The evolution path for industrial software quality evaluation methods applying ISO/IEC 9126:2001 quality model: example of MITRE's SQAE method. Softw. Qual. J. **13**(1), 17–30 (2005)
6. Yuen, K.K.F., Lau, H.C.W.: A fuzzy group analytical hierarchy process approach for software quality assurance management: fuzzy logarithmic least squares method. Expert Syst. Appl. **38**(8), 10292–10302 (2011)
7. Guo, Y., Tang, H.Y., Qu, D.G.: Dynamic comprehensive evaluation method and application based on minimum variance. Syst. Eng. Electron. **32**(6), 1225–1228 (2010)
8. Zhou, J.H., Wang, Z., Yang, Z.: Research on software quality evaluation based on fuzzy evaluation method. Syst. Eng. Electron. **26**(7), 987–991 (2004)
9. You, T.H., Fan, Z., Li, H.Y.: A synthetic evaluation method of software quality based on two-tuple linguistic information processing. Syst. Eng. Electron. **27**(3), 545–549 (2005)
10. Guo, F., Hou, Z., Dai, Z.: Modeling and simulation of ANFIS software quality evaluation based on expert knowledge. Syst. Eng. Electron. **28**(2), 317–320 (2006)

Identification of Certain Shrapnel's Air Resistance Coefficient in Plateau Environment Based on CK Method

Ming Jiang$^{(\boxtimes)}$, Yuwen Liu, Lijing Cao, and Zhiyuan Zhang

New Star Research Institute of Applied Technology, Hefei 230031, China
389555317@qq.com

Abstract. A calculation program of projectile PRC (Plateau Resistance Coefficient) identification was wrote under Matlab platform based on CK (Chapman-Kirk) method of aerodynamic force parameter identification and shooting range test data of a certain type of shrapnel in this paper. And the difference of resistance coefficient between plateau and plain was analyzed, which showed clearly that, the former was 10–40% less than the latter. The reducing of the PRC would bring 4.24–5.93% farther more range to the field of fire. It was consistent with the plateau test results. It indicated that the change of the PRC and its effect must be considered and calculated particularly when deciding plateau shooting table of the certain type of shrapnel, studying of projectile plateau ballistic characteristics, or designing the plateau module of the artillery information system.

Keywords: Aerodynamic force parameter identification · Ballistic · CK method

1 Introduction

When the artillery projectile was flying, air force was one of the main influence factors which affect its flight path, speed, rotate speed and so on. Its characteristics is described by aerodynamic coefficient [1], obtained by the theoretical calculation method, wind tunnel test, or firing experiment etc. Theory and practice indicated that the resistance coefficient was related to Reynolds number, which is influenced by the air density and viscosity coefficient. Reynolds number has small changes on plain, so its influence can be ignored. However, the air density and viscosity coefficient have great change on plateau, even more obvious according to altitude. Because of the complexity, there isn't an accurate Reynolds number revised method until now, and the empirical formula is the most widely used. But it can't reflect the real impact of plateau environment on the resistance coefficient. Hence, a calculation program of projectile PRC (Plateau Resistance Coefficient) identification was wrote under Matlab platform according to CK (Chapman-Kirk) method of aerodynamic force parameter identification and shooting range test data of a certain type of shrapnel in this paper. And the difference of resistance coefficient between plateau and plain was analyzed, that can provided

© Springer Nature Singapore Pte Ltd. 2017
H. Yuan et al. (Eds.): GRMSE 2016, Part I, CCIS 698, pp. 238–244, 2017.
DOI: 10.1007/978-981-10-3966-9_26

reference for deciding plateau shooting table of the certain type of shrapnel and studying of projectile plateau ballistic characteristics.

2 The Idea and Model of CK Method for PRC Identification

In shooting range test, there are a lot of methods for the aerodynamic coefficient identification, such as maximum likelihood, filter and CK method etc. The CK method, proposed by Chapman-Kirk in the late 60s, was widely and well used in data identification of weapon experiment, which was based on least square procedure.

2.1 The Basic Idea of PRC Identification

The basic idea of CK method for PRC identification was that the test data of speed measuring radar was divided into several small ranges. Assume that the PRC C_D on every small range was constant, and the estimates of the PRC C_D and the initial speed u_0 were given. Ballistic model was calculated, the error between u_{cal} and speed measured value u_{exp} was minimum. After C_D was identified by every small range, a smooth curve was made corresponding Mach number M and every C_D, and it's the curve of C_D-M which to be identified.

In the process of identification, seven to eleven data points were taken on every small range in iterative calculation, and the time interval was 0.01 s between every two data points, and the second and the first range could overlap, the third and the second range could overlap too, in sequence, until to the last range.

2.2 Ballistic Model of PRC Identification

Ballistic model comes in many forms, 2D, 3D, 4D, 5D, 6D etc, according to the external ballistics [2]. To simplify the research problem, the 2D model was taken.

However, the 2D model also comes in many different forms, and the 2D model [3] in natural coordinate system had been taken in the paper, as formula (1).

$$\begin{cases} \frac{du}{dt} = -\frac{\rho S}{2m} C_D(M) u^2 - g_0 \sin \theta \\ \frac{d\theta}{dt} = -\frac{g_0 \cos \theta}{u} \\ \frac{dx}{dt} = u \cos \theta \\ \frac{dy}{dt} = u \sin \theta \end{cases} \tag{1}$$

Integral initial conditions: t = 0, $u = u_0$, $\theta = \theta_0$, x = y = 0.

2.3 Sensitive Model of PRC Identification

In every small range, there was a goal function ε according to the CK method:

$$\varepsilon = \sum_{i=1}^{n} [u_{\exp}(t) - u_{cal}(t)]^2 = \sum_{i=1}^{n} [u_{\exp}(t) - u_{cal}(t)_0 - \frac{\partial u}{\partial u_0} \Delta u_0 - \frac{\partial u}{\partial C_D} C_D]^2$$

$u_{cal}(t)_0$ is the calculated value of $u(t)$ after solving the ballistic mode based on the estimate value of C_D and u_0.

In order to reach the minimum of ε, there could be the flowing the type

$$\begin{cases} \frac{\partial \varepsilon}{\partial \Delta u_0} = 0 \\ \frac{\partial \varepsilon}{\partial \Delta C_D} = 0 \end{cases} \tag{2}$$

If order $P_1 = \frac{\partial u}{\partial u_0}, P_2 = \frac{\partial u}{\partial C_D} P_3 = \frac{\partial y}{\partial u_0}, P_4 = \frac{\partial \theta}{\partial u_0} P_5 = \frac{\partial y}{\partial C_D}, P_6 = \frac{\partial \theta}{\partial C_D}$
Then, we can export the sensitivity equations as follows:

$$\begin{cases} \frac{dP_1}{dt} = -\frac{\rho' S}{2m} C_D u^2 P_3 - \frac{\rho S u}{m} C_D P_1 - g_0 \cos \theta P_4 \\ \frac{dP_2}{dt} = -\frac{\rho' S}{2m} C_D u^2 P_5 - \frac{\rho S}{2m} C_D u^2 - \frac{\rho S u}{m} C_D P_2 - g_0 \cos \theta P_6 \\ \frac{dP_3}{dt} = P_1 \sin \theta + P_4 u \cos \theta \\ \frac{dP_4}{dt} = P_4 \frac{g}{u} \sin \theta + P_1 \frac{g}{u^2} \cos \theta \\ \frac{dP_5}{dt} = P_2 \sin \theta + P_6 u \cos \theta \\ \frac{dP_6}{dt} = P_6 \frac{g}{u} \sin \theta + P_2 \frac{g}{u^2} \cos \theta \end{cases} \tag{3}$$

When the initial value was t = 0, $\begin{cases} P_1 = 1 \\ P_j = 0 \end{cases}$ $j = 2, \cdots 6$

Integral formula (3) needs to be solved with formula (1).
After sorting formula (2), the differential correction equations were:

$$AB = C \tag{4}$$

In the formula,

$$A = \begin{bmatrix} \sum_{i=1}^{n} (P_1)_i (P_1)_i & \sum_{i=1}^{n} (P_2)_i (P_1)_i \\ \sum_{i=1}^{n} (P_1)_i (P_2)_i & \sum_{i=1}^{n} (P_2)_i (P_2)_i \end{bmatrix} \quad B = \begin{bmatrix} \Delta u_0 \\ \Delta C_D \end{bmatrix}$$

$$C = \begin{bmatrix} \sum_{i=1}^{n} [u_{\exp}(t_i) - u_{\exp}(t_i)_0] (P_1)_i \\ \sum_{i=1}^{n} [u_{\exp}(t_i) - u_{\exp}(t_i)_0] (P_2)_i \end{bmatrix}$$

n−the number of radar measured data in small range.
The C_D on small range could be iterative calculated by formula (4).

3   The Selection and Processing of Test Data

The test proceeded in a plateau shooting range, whose altitude is 4302 m. There was weather radar for meteorological guarantee, about 40 Shrapnel of a certain type were shot in group. The ballistic data was tested by radar, containing speed and X, Y, Z coordinates etc. Because of plateau conditions, test data displayed that the ballistic path data did not match the ballistic obviously in several shells (the problem is not in this study), that was to say, the ballistic data had obvious difference. In order to accurately identify PRC, a better test data of ballistic consistency needed to be selected for analysis.

All the test data was statistically analyzed, and two groups were selected for identification, filtered for the sake of obtaining a smooth curve of test result data.

The test result of first group was shown in the Figs. 1 to 2.

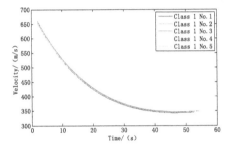

Fig. 1. The curve of T-V   **Fig. 2.** The curve of X-Y

4   The PRC Identification of Certain Type Shrapnel

4.1   Calculation Process

Take certain type shrapnel for instance, a calculation program was written, and the calculation process was shown in the Fig. 3.

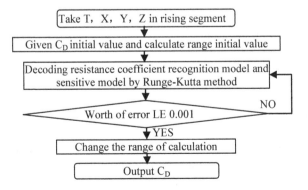

Fig. 3. Calculation process chart

When identifying the PRC, the PRC of different speed range were identified and then compounding, because each group shells in the test had different speed range.

4.2   Result and Analysis of PRC

The first group PRC was shown in Fig. 4, and the second in Fig. 5. In the figure, "I" stands for "identification", which was the result of identification, and "S" stands for "smoothing", which was the result after smoothing.

Figure 4 illustrated that the PRC was identified with the Mach number from 1.15 to 2.05, and in Fig. 5, the Mach number of this result changed from 0.78 to 1.25.

The identification curves were not smooth, for the existence of attack angle. The reason is that the identification model was based on the 2D ballistic model, which paid no attention on attack angle. So, smoothing was also shown in the Figs. 4 and 5. Combining the two results, the final curve was got.

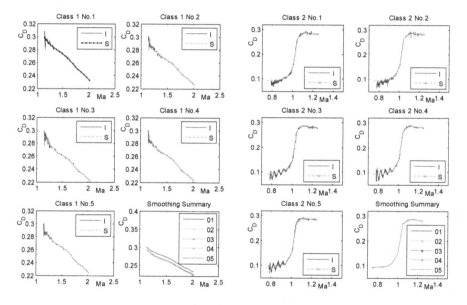

Fig. 4. Class 1 identification   **Fig. 5.** Class 2 identification

In order to practically find out the difference of PRC from the value in wind tunnel experiment, the results of the two stations were compared, and shown in Fig. 6.

It can be seen from Fig. 6 that, the PRC identified is totally less than got from the wind tunnel experiment. The compared calculation shows clearly that, the former is 38% less than the latter in the subsonic region, 13–40% in the transonic region, and 10% in the supersonic region. Consequently, field of fire becomes big in the plateau condition, which is consistent with experiment. Field of fire in plateau condition, is

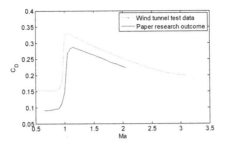

Fig. 6. Test value and identification value

bigger than in plain condition, and is also than the calculated range using the resistance coefficient by the wind tunnel experiment.

As an example of certain shrapnel, whole number tamping, positional elevation 4302 m, the difference is shown in Table 1. X_1 meant range calculated through the wind tunnel test, X_2 meant range under the plateau identification.

Table 1. Range error

Firing angle	X_1/m	X_2/m	Range error/m	Percentage
20°	28293	29665	1372	4.85%
25°	32175	33882	1707	5.31%
30°	35784	37809	2025	5.66%
35°	39187	41488	2301	5.87%
40°	42142	44641	2499	5.93%
45°	44291	46891	2600	5.87%

From the data, we can find that the reducing of the PRC will bring 4.24–5.93% farther more range to the field of fire. It is consistent with the plateau test results.

5 Conclusion

A calculation program of projectile PRC identification is wrote under Matlab platform based on CK method of aerodynamic force parameter identification and shooting range test data of a certain type of shrapnel in this paper. And the difference of resistance coefficient between plateau and plain is analyzed, which shows clearly that, the former is 38% less than the latter in total, 13–40% in the transonic region, and 10% in the supersonic region. The difference between the field of fire calculated through the wind tunnel test and the plateau identification indicates that the reducing of the PRC will bring 4.24–5.93% farther more range to the field of fire. It is consistent with the plateau test results. It indicates that the change of the PRC and its affection must be considered and calculated particularly when deciding plateau shooting table of the certain type of shrapnel, studying of projectile plateau ballistic characteristics, or designing the plateau module of the artillery information system.

References

1. Ji, Y.: Optimization method study of projectile aerodynamic parameters identification. North University of China (2013)
2. Han, Z., et al.: Projectile External Ballistics. Institute of Technology Press, Beijing (2014)
3. Guo, X., Zhao, Z.: Theory and Application of Fire Control Ballistic Model. Weapon Industry Press, Beijing (1997)

Image Semantic Segmentation Based on Fully Convolutional Neural Network and CRF

Huiyun Li[✉], Xin Qian, and Wei Li

State Key Laboratory of Virtual Reality Technology and System,
Beihang University, Beijing 100191, China
li_huiyun1991@163.com, xqianxin@gmail.com,
lwei@buaa.edu.cn

Abstract. Image semantic segmentation is a popular research direction in the computer vision field. Semantic segmentation algorithms based on deep learning outperforms the traditional methods. Fully convolutional neural network (FCN) whose fully connected layers are transformed into convolution layers is a kind of convolutional neural network (CNN). In this paper, FCN is used to operate the image semantic segmentation, which could take input of arbitrary size image and implement end-to-end segmentation task. Due to the limited number of training images, some layers are fine-tuned from AlexNet and the dataset is enlarged by mirroring. The hierarchical feature maps from FCN are combined to improve the segmentation effect. Conditional random fields (CRF) is used on the segmentation result of FCN, which takes into account the positional relationship and color features between any two pixels. Experiments show that our method could refine the segmentation result of FCN, especially using CRF as post-processing.

Keywords: Image semantic segmentation · CNN · FCN · CRF

1 Introduction

Image semantic segmentation is one of the most popular research directions in the computer vision field. In segmentation task, it is necessary to make a prediction for every pixel, divide the image into several segments according to the semantic feature and identify the class that each segment belongs to, which is more challenging than image classification and object recognition. Semantic segmentation has a wide application prospect in the fields of semantic search, medicine and so on.

In recent years, deep learning has achieved great success in the computer vision field. The application of CNN not only improves the accuracy of image classification and object detection task, but also makes progress on semantic segmentation. George Papandreou *et al.* [1] employed a weakly supervised learning method to locate the bounding boxes around the targets and then used CRF to fulfil the segmentation. Jonathan Long *et al.* [2] applied FCN to image semantic segmentation task, so that the network could accept arbitrary size, retain the position information of target objects and implement end-to-end segmentation task. Hyeonwoo Noh *et al.* [3] divided the process of segmentation into two steps, object recognition and semantic segmentation. In the

© Springer Nature Singapore Pte Ltd. 2017
H. Yuan et al. (Eds.): GRMSE 2016, Part I, CCIS 698, pp. 245–250, 2017.
DOI: 10.1007/978-981-10-3966-9_27

first step, RNN was used to detect the objects. Then the objects were resized to the uniform size and input into FCN, which used the deconvolution layers and unpooling layers in alternation to restore the size of feature maps.

In this work, we fine-tune layers from AlexNet using Jonathan Long's method basically. Furthermore, we explore a better combination of shallow layers and deep layers, expand dataset by mirroring and post-process segmentation result by CRF.

2   Algorithm

2.1   Model Transformation Based on Fine-Tune

At the very first, deep learning was used to solve the problem on image classification task. On the one hand, the classification problem is relatively simple. On the other hand, there are more available images and corresponding labels for training set. Consequently, there exist many excellent CNNs for classification, such as AlexNet [4], VGG [5] and GoogleNet [6]. However, it will take many resources to acquire the ground truths of images for segmentation. In the currently public datasets for segmentation, the number of images is often too small to support the training process of a large and deep CNN.

In CNN, convolution layers could extract hierarchical features, from shallow location information to deep semantic information. The feature extraction process is also applicable to semantic segmentation task. Therefore, to avoid the over-fitting caused by a small number of training set but a large number of parameters, it is a good idea to inherit parameters from classification models and fine-tune [7] them on segmentation training set. We have no choice but to fine-tune the parameters of AlexNet model due to the memory limitation (NVIDIA GTX 780, 3G). Alex has trained a CNN model on the ImageNet dataset, which is used to classify 1000 categories. In order to reuse the parameters learned from ImageNet dataset, we fine-tuned the parameters through segmentation training set.

In AlexNet, the output will lose the spatial information because of the existence of the fully connected layers [4–6, 8]. While in the image segmentation task, we not only need to know what the image is, but also where the object is. In order to preserve spatial information, it is necessary to replace the fully connected layers with convolution layers and construct a so-called FCN [2]. Due to the different dimensions of parameters, the modified convolution layers cannot inherit the parameters directly from the classification model. Therefore, the parameter dimensions of the fully connected layers in the AlexNet model need to be transformed. The parameter dimensions of layers fc6 and fc7 are transformed from (4096, 9216) and (4096, 4096) to (4096, 256, 6, 6) and (4096, 4096, 1, 1) by setting the values of parameters Num_output and Kernel_size.

The pooling layers are connected behind the convolution layers to reduce the size of feature maps, thereby reducing the computation. We connect deconvolution layers at the end of modified AlexNet to expand the final feature maps 32 times, acquiring the result having the same size with input image.

2.2   Combining Outputs of Deep Layers and Shallow Layers

A rough segmentation result is obtained by fine-tuning the parameters of AlexNet model. To optimize the segmentation result, we combine the feature maps of a shallow layer and a deep layer. The feature maps acquired by the deep layers of networks are abstract and coarse; contrarily, the feature maps produced by the shallow layers are concrete and fine, which could reflect the location and edge information of objects. Thus, the combination could refine the spatial precision; meanwhile, it turns a line topology into a DAG [2].

We perform several experiments to confirm combining which shallow layers will achieve a better segmentation result. Experiments show that combining the feature maps of the last layer with them of conv4 and conv5 layers could raise the mean IU from 48.0% to 49.5%. Mean IU is the index of measuring the accuracy rate of image semantic segmentation, the formula is,

$$(1/n_c) \sum_i n_{ii}/(\sum_j n_{ij} + \sum_i n_{ji} - n_{ii}) \tag{1}$$

n_{ij} represents the number of pixels of class i predicted to belong to class j, where there are n_c different classes.

The latter part of FCN is shown in Fig. 1, which combines conv4 and conv5 layers. The fc7 layer is a convolution layer transformed from a fully connected layer. The score layer is a convolution layer containing a parameter Num_output, whose value is 21 (the number of classes in training set). The score layer could produce 21 feature maps, which show the possibility of all the pixels belonging to all the categories. The upsample layer is a deconvolution layer. Firstly, the feature maps are expanded by 2 times and fused with the feature maps produced by conv5. Then, the first fusion results are fused with the feature maps obtained by conv4 again. Finally, the second fusion results are expanded by 16 times by deconvolution layer and cut into the expected size by crop layer.

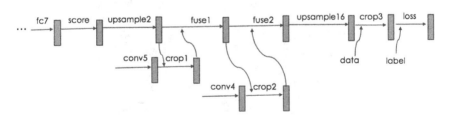

Fig. 1. The network indicates the combination of feature maps produced by deep layers and feature maps produced by shallow layers.

2.3   Expanding Training Set by Mirroring

The simplest and most effective method to prevent over-fitting is expanding training set. In order to expand training set, we generate the vertical reflections of the whole training set. This method would not produce new pixels, but it creates new position relationship between pixels and has a certain effect on improving the accuracy. The mean IU increase 0.5% by using this method.

2.4   Post-processing Based on CRF

Image semantic segmentation based on CRF model is one of the most classic and traditional segmentation algorithm [9]. The CRF model is often used to smooth the rough segmentation result, the energy term in which will connect different pixels in the image. It is usually considered that the spatially adjacent pixels belong to the same class. Short-range CRF model, a kind of CRF model that only connects the spatially adjacent pixels, is mainly used to remove the noise in the coarse segmentation result. However, the segmentation result of FCN is smooth enough and has no noise. Thus, our goal is to obtain a refined segmentation image with clear edges, rather than further smoothing. A fully connected CRF model is adopted for post-processing in this paper, which connects any two pixels in the image and refines the details considering the positional relationship and similarity of color features between two pixels. The fully connected CRF model we used is proposed by Philipp *et al.* [10], which improves the prediction speed by using the average field approximation theory and the high-dimensional Gaussian filters. Experiment shows that mean IU can be increased 3% by using the fully connected CRF model as post-processing. Figure 2 shows the results.

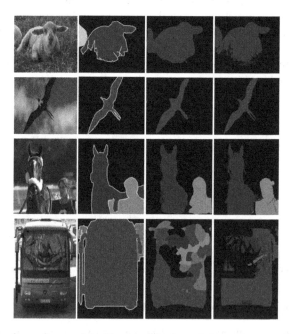

Fig. 2. The first column shows original images. The second shows the ground truths (annotated images). The third shows the results without using CRF. The fourth shows the results processing by CRF based on the results of the third column.

2.5 Experimental Framework

Caffe framework [11] and SGD algorithm are chosen for our experiments. We use a batch size of 20 images, a fixed learning rate of 10^{-4}, momentum 0.9 and weight decay 5^{-4}. We adopt bilinear interpolation kernels for the deconvolution layers, the parameters are fixed and not adjusted dynamically during training phase.

The training set used PASCAL VOC [12], which contains 8498 training images and a corresponding number of annotated images. The dataset contains 20 classes and the annotated images have only a single channel, the pixel value ranges 0 to 20 and each pixel value represents a class of object, 0 indicates the background. To distinguish different categories, a white boundary is added to the edges of objects in the annotated images. Thus, the pixel value 255 should be ignored. PASCAL VOC 2011 val is used as validation set, which contains 736 images.

We use NVIDIA GTX 780 for our experiments.

3 Results

The results of our experiments are shown in Table 1. FCN-AlexNet is a model provided by Jonathan Long *et al*. We improve the segmentation result based on FCN-AlexNet step by step and make comparisons with it. FCN-AlexNet-16S corresponds the network combining deep layer outputs with shallow layer outputs. FCN-Mirror uses the augmented training set by mirroring based on FCN-AlexNet-16S. The FCN-CRF applies CRF on the result of FCN-Mirror. Due to the limitation of our GPU, we chose AlexNet rather than VGG or other deeper networks. While the related works generally adopts deeper networks than AlexNet and lacks comparability with ours. So we only make comparisons with FCN-AlexNet.

Table 1. The results on PASCAL VOC 2011 val. Our models and results are in bold.

Model	Mean IU
FCN-AlexNet	48.0%
FCN-AlexNet-16S	**49.5%**
FCN-Mirror	**50.0%**
FCN-CRF	**52.8%**

4 Conclusion

The FCN can realize the image semantic segmentation effectively. The deep layers of FCN could extract abstract and semantic features; the shallow layers could extract features containing location and boundary information. A combination of both could refine the segmentation results. Enlarging the training set by mirroring improves the segmentation accuracy rate, but the effect is not obvious. The post-processing based on fully connected CRF refines the segmentation boundary and improved the accuracy rate significantly.

References

1. Papandreou, G., Chen, L.C., Murphy, K.P., et al.: Weakly and semi-supervised learning of a deep convolutional network for semantic image segmentation. In: IEEE International Conference on Computer Vision. IEEE, pp. 1742–1750 (2015)
2. Long, J., Shelhamer, E., Darrell, T.: Fully convolutional networks for semantic segmentation. eprint arXiv:1411.4038 (2014)
3. Noh, H., Hong, S., Han, B.: Learning deconvolution network for semantic segmentation. CoRR, vol. abs/1505.04366 (2015)
4. Krizhevsky, A., Sutskever, I., Hinton, G.E.: ImageNet classification with deep convolutional neural networks. In: International Conference on Neural Information Processing Systems, pp. 1097–1105. Curran Associates Inc. (2012)
5. Simonyan, K., Zisserman, A.: Very deep convolutional networks for large-scale image recognition. arXiv:1409.1556 (2014)
6. Szegedy, C., Liu, W., Jia, Y., et al.: Going deeper with convolutions. In: Computer Vision and Pattern Recognition, pp. 1–9. IEEE (2014)
7. Tajbakhsh, N., Shin, J.Y., Gurudu, S.R., et al.: Convolutional neural networks for medical image analysis: fine tuning or full training. IEEE Trans. Med. Imaging **35**(5), 1 (2016)
8. Lecun, Y., Boser, B., Denker, J.S., et al.: Backpropagation applied to handwritten zip code recognition. Neural Comput. **1**(4), 541–551 (1989)
9. Shotton, J., Winn, J., Rother, C., et al.: Textonboost for image understanding: multi-class object recognition and segmentation by jointly modeling texture, layout, and context. Int. J. Comput. Vis. **81**(1), 2–23 (2009)
10. Philipp, K., Koltun, V.: Efficient inference in fully connected CRFs with Gaussian edge potentials. In: Advances in Neural Information Processing Systems, pp. 109–117 (2012)
11. Jia, Y.Q., Evan Shelhamer, E., Jeff, D., et al.: Caffe: convolutional architecture for fast feature embedding. arXiv preprint arXiv:1408.5093 (2014)
12. Everingham, M., Gool, L.V., Williams, C.K.I., et al.: The pascal visual object classes (VOC) challenge. Int. J. Comput. Vis. **88**(2), 303–338 (2010)

Car-Based Laser Scanning System of Ancient Architecture Visual Modeling

Kunyang Wang and Jing Zhang$^{(\boxtimes)}$

3D Information Collection and Application Key Lab of Education Ministry,
Capital Normal University, The West Third Ring Road North 105th,
Beijing 100048, China
dhcampanella@live.cn, zhangjing5946@sina.com

Abstract. Laser point cloud is currently a hot three-dimensional study. *Equipment used in this article are self-developed by Capital Normal University.* This thesis mainly focuses on scan the Zhongshan Park by mobile horizontal pushing type and obtain the distance and angle information by 360° lancer scanner to figure out the 3D laser point cloud relative coordinate, and make color bleeding with image information obtained by linear CCD camera. Then, we get color point cloud information. The scattered point cloud is integrated *based on the reaction of echo date to different target objects,* and the visual modeling towards historic building will be realized. The 3D laser image scan data and 3D modeling and historic building will play an important role in protecting of historic building.

Keywords: Vehicular · Historic building · Laser point cloud · 3D visualization

1 Introduction

In recent years, vehicle-mounted laser scanning system developed rapidly, laser scanning technology matures every year, it plays an increasingly important role in road information collection, the building of digital city, three-dimensional modeling. This paper based on the platform vehicles, integrated GPS, laser scanner, IMU, CCD camera and other sensors of vehicle-mounted laser scanning system. Ancient architecture of the Zhongshan Park gets three-dimensional point cloud data and its application, *and consolidation point cloud by echo data innovatively,* achieved good results. Mobile scans of ancient building in Zhongshan Park and detailed modeling of it. For Web publishing, and visual comparison provides some more complete directions.

Application of three-dimensional laser scanning technology is widely, in the protection of ancient architecture, cultural heritage, historic cities, historic blocks, building of construction quality evaluation, deformation monitoring, record the scene and so is particularly applicable. Famous National Palace Museum, the Terra-Cotta Warriors of XI' an, the statue of liberty, statue of David and other historical buildings or heritage has been scanning by three-dimensional laser point and built digital model [1].

In China, Di using laser scanning system, access the three-dimensional spatial information of migrant villages due to the south-north water diversion project in Zhengzhou Henan. For the village of three-dimensional modeling and visualization

© Springer Nature Singapore Pte Ltd. 2017
H. Yuan et al. (Eds.): GRMSE 2016, Part I, CCIS 698, pp. 251–256, 2017.
DOI: 10.1007/978-981-10-3966-9_28

provides a more complete reference in the future [2]. Li, Mapping Bureau of Hebei, apply three dimensional laser scanning system to the great wall in surveying and mapping, established the real three dimensional model of Shanhaiguan great wall. Realize the great wall ancient measurement function of length, area and volume [3]. Academician Xianlin Liu is trying to use this mobile laser scanning system used by this paper for road extraction of concave on both sides.

In Netherlands, Measurement Department since 1988 the ground fixed lasers laser scanning technology study on extraction of topographic information [4]. The Tokyo University in Japan had attempt the ground fixed laser scanning systems and achieved very good results. J-Angelo Beraldin using laser scanner data to establish the real world of three dimensional virtual environments. Sabry El-Hakim took laser scan data and CAD data, digital photogrammetry and aerial image data combining for complex building reconstruction. Delft University of technology has also been using three-dimensional laser scanning data of vegetation and automatic building recognition and classification research, and the semi-automatic extraction of roads and so on.

2 The Working Principle of the Vehicular Laser Scanning System (SSW)

Laser technology has a rapid development in recent decades and used widely as measurement tool. It can get the surface point coordinate information directly and reflects the spectral geometry of structures [5], thus often used in spatial information extraction and ecological projects in urban construction planning.

Airborne laser scanning data generated in the building technology to rebuild the roof has a relatively mature [6–9]. In the research on forest inventory and analysis, it can analyze a potential volume, an area of woodland and planting plans [10], but airborne-laser-scanning can only obtain the data at the top of the building, lack of facade details. And vehicle-laser-scanning system can make up for these deficiencies.

CNU independent research and development Vehicular Laser Scanning System (SSW) includes 360° laser scanner after calibration, CCD line camera, navigation POS (IMU and DGPS) and Electronic turntable parts. When staff collecting the data, GPS and IMU will be observed and recorded the position and orientation of the sensor data, linear array CCD access images on the side of a building, and Laser scanner will record the road on both sides of the building the laser point cloud data in the same time (Fig. 1).

In vehicles moving in the process, GPS logger you will record measurements of the car in the world coordinate system's absolute coordinates, the laser scanner records the measurement to target buildings on both sides of the road distance and the point of measurement. This distance is the launch point to the target object location on the surface of the laser pulse, based on geometric relationships, we can be calculated the distance in each direction to the target architecture with pulse laser scan points in every time easily [11]. So, we can get the coordinates of the building sites of real-time information in the case of no reference point.

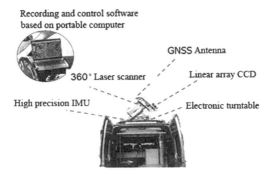

Fig. 1. Sketch map of SSW

Linear array CCD is similar with laser scanner, can be match and laser data much better, gets the same shape and texture information, virtual reproduction of a scene on computer.

3 Landscape Modeling Using Laser Data

The data acquisition of Zhongshan Hall, due to the terrain in the Park is a small, we must be designed to be laser scanner and CCD camera can maximize the angle and position of the data, to ensure every detail can be collected. Then collection the point cloud data in the planned route by SSW, and ensure CCD camera faced on the ancient architecture (Fig. 2).

Fig. 2. Laser point clouds of Pseudo-color

4 Noise Reduction and Integration of Point Cloud Data

During the process of noise reduction of point cloud data, filter out non-terrain echo noise by setting the tolerance value of absolute elevation firstly. Set the maximum building size, topography angles, distances and iterative to classification the ground point [12]. Filter out those points below the surface.

Separate lower is isolated lower level points in the adjacent point cloud. Low points algorithm often used search and significantly the wrong point which lower than the ground. Ground points classification is the most critical step in the point cloud classification. Ground algorithms in isolation on the basis of separate low, search for ground point in a certain range to build a model. According to Max building size, Tin angle and Iteration distance to excluding noise of model.

This paper ingenious use of the laser point cloud echo data. Depending on the characteristics of echo data: *Buildings have only one echo and Vegetation contain multiple return.* Make the scattered point cloud data get integration, *take the point forming a surface* (Fig. 3).

Fig. 3. Comparison between before and after of the noise reduction and point cloud integration

5 Fusion with Laser Point Clouds and Image Information

This paper reference Imaging equation of photo measured, mapping in three dimensional space directly. The photography center S form CCD camera coordinates (xs, ys, zs) while shooting posture angles for (ψ, ω, κ) in WGS-84 rectangular coordinate system, Camera orientation element is (f, x0, y0), as Fig. 4.

Through the imaging equation calculating the mapping relationships of point between cloud data and pixels of the image:

$$\begin{cases} x_p - x_o = -f\dfrac{a_1(x_{84} - x_S) + b_1(y_{84} - y_S) + c_1(z_{84} - z_S)}{a_3(x_{84} - x_S) + b_3(y_{84} - y_S) + c_3(z_{84} - z_S)} \\ y_p - y_o = -f\dfrac{a_2(x_{84} - x_S) + b_2(y_{84} - y_S) + c_2(z_{84} - z_S)}{a_3(x_{84} - x_S) + b_3(y_{84} - y_S) + c_3(z_{84} - z_S)} \end{cases} \quad (1)$$

Take the pixel information has been corrected to point cloud data, can be get a ture-color three-dimensional landscape diagram, as shown in Fig. 5.

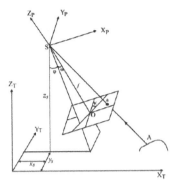

Fig. 4. Principle of CCD camera imaging

Fig. 5. Landscape figure after true-color fusion

6 Summary

Traditional methods of measuring results are often expressed through contour lines or a floorplan method. Cannot be objectively evident to reflect real situations when do the three-dimensional recovery. Fail to reflect the characteristics of ancient architecture.

Fig. 6. Digital modeling effect diagram

By SSW, staff can check point cloud data visually. And details had been fully displayed. Can be Landscape modeling directly. Play a vital role at three-dimensional recovery in the future (Fig. 6).

Overall, SSW for restoration and protection of Ancient architecture provides more reliable methods and ideas. It can reproduce three dimensional model efficiently. In the protection of ancient architecture and other fields, will certainly have a lot of development.

References

1. Fei, W.Y.X.: Demand for date process software of 3D laser imaging scanning technology in ancient architecture mapping and protection field. Archit. Hist. **26**(4), 130–132 (2008)
2. Di, Z.: 3D date acquisition and application based on the vehicle-borne laser scanning system. Geospatial Inf (2), **10**(1), 20–24 (2012)
3. Li, C.: The application of multi 3-dimension laser scanning in the surveying of Shanhaiguan part of the great wall. Bull. Surv. Mapp. (3), 31–40 (2013)
4. Gomes Huising, E.J., Pereira, L.M.: Errors and accuracy estimates of laser data acquired by various laser scanning system for topographic applications. ISPRS J. Photogram. Remote Sens. **53**, 245–261 (1998)
5. Lovell, J.L., Jupp, D.L.B., Culvenor, D.S., et al.: Using airborne and ground-based ranging lidar to measure canopy structure in Australian forests. Can. J. Remote Sens. **29**, 607–622 (2003)
6. Sithole, G., Vosselman, G.: Experimental comparison of filter algorithms for bare-earth extraction from airborne laser scanning point cloud. ISPRS J. Photogram. Remote Sens. **59** (1–2), 85–101 (2001)
7. Axelsson, P.: DEM generation from laser scanner data using adaptive TIN models. Int. Arch. Photogram. Remote Sens. Spat. Inf. Sci. **33**(B4), 110–117 (2000). ISPRS, Amsterdam
8. Halla, N., Brenner, C.: Extraction of buildings and trees in urban environments. ISPRS J. Photogram. Remote Sens. **54**(2-3), 130–137 (1999)
9. Overby, J., Bodum, L., Kjems, E., et al.: Automatic 3D building reconstruction from airborne laser scanning and cadastral data using Hough transform. Int. Arch. Photogram. Remote Sens. Spat. Inf. Sci. **35**(B3), 296–301 (2004). ISPRS, Istanbul
10. Smith, J.: Use of airborne LiDAR and aerial photography in the estimation of individual tree heights in forestry. Comput. Geosci. **31**, 253–262 (2005)
11. Dai, B., Zhong, R., Hu, J.: Research on 3D reconstruction of urban features from data based on vehicle-borne laser scanning. J. Cap. Norm. Univ. (Nat. Sci. Ed.) (6) **32**(3), 89–95 (2011)
12. Schwalbe, E., Maas, H.-G., Seidel, F.: 3D building model generation from airborne laser scanner data using 2D GIS data and orthogonal point cloud projections. In: ISPRS WG III/3, III/4, V/3 Workshop Laser scanning 2005, Enschede, The Netherlands, 12–14 September 2005

Research on Fractal Characteristics of Road Network in Chengdu City

Bowen Qiao and Jing Zhang[(⊠)]

3D Information Collection and Application Key Lab of Education Ministry,
Capital Normal University, Beijing, China
116qiaobowen@163.com, zhangjing5946@sina.com

Abstract. For the importance of intersection in the traffic network, we introduce the concept of node degree into the traffic network fractal calculation. A weighted node degree-radius dimension is designed to better characterize the variation of network connectivity from center to periphery. The road networks of Chengdu city have been analyzed. The research shows that: the node degree-radius dimension can reflect the network characteristics, the weighted dimension will better reflect the spatial connectivity structure of urban system; the truck road is ringy and radial, it is more balanced than overall network, the contact between the second layer and the third layer in Chengdu city is not close enough, peripheral road network has a lot of space for development.

Keywords: Weighted node degree-radius dimension · Fractal

1 Introduction

Traffic network and urban development are closely related and reinforce each other. Research on the traffic network form is helpful to understand the development level of the traffic network scientifically, and make it better match the urban space form and promote the orderly development of the city. In 1967, the famous article: How Long Is the Coast of Britain? Statistical Self-Similarity and Fractional Dimension is published on the authoritative Science Magazine by Mandelbrot [1]. Then he put forward the fractal geometry. After that fractal has been widely used in various fields and becomes a powerful tool for describing the spatial distribution of complex phenomena. Traffic network, as a line collection with irregular spatial structure and distribution, is significant to study the complexity of the distribution by the fractal theory. Fractal Study abroad on the transport network began in the 1990s. Frankhouser [2] studied the suburban railway network in Stuttgart Germany. It is found that the network length L(r) and radius r had a power exponent relationship. Benguigui and Daoud [3] studied the railway in Paris suburb. The results showed that the number of the railway station and the radius r had a power exponent relation [3]. Benguigui [4] in the Paris public transport system of the railway network, proposed the 2 different function of the network (metropolitan area and suburbs) have different fractal dimension [4]. In China, Chen Yanguang systematically study the fractal of transport network. He proposed a new network fractal dimension calculation method in 1999 called Dendrite-radius

© Springer Nature Singapore Pte Ltd. 2017
H. Yuan et al. (Eds.): GRMSE 2016, Part I, CCIS 698, pp. 257–264, 2017.
DOI: 10.1007/978-981-10-3966-9_29

dimension [5]. At the same year, he with Liu Jisheng summarized three types of fractal dimensions to characterize the spatial structure of transport networks. The geographical meanings of these dimensions were illuminated [6]. Many scholars have carried out an empirical research on the traffic network, which indicates that there is a fractal phenomenon in the traffic network [7–23]. There are three main dimensions in fractal calculation of transport network: box-counting dimension, radial dimension, correlation dimension. Box-counting dimension focus on road cover condition; correlation dimension mirrors the urban spatial system; radial dimension reflects the variation of traffic network form center to the hinterland. As an important node of road network, road intersection is the key of the whole city traffic. If the two roads do not intersect, even the roads are very long, there will be disconnected. But in the study of fractal, more attention is paid on the line than the intersection. In this paper, the degree is introduced to fractal calculation. According to different road classes, a weighted node degree radius dimension is designed, which reflects the configuration of traffic network from the perspective of transport nodes rather than the lines. Finally, an empirical study is conducted in Chengdu city.

2 Weighted Node Degree-Radius Dimension

2.1 Node Degree-Radius Dimension

According to the construction method of the radius dimension, we choose the traffic hub as the calculation center with radius r, where r takes 1, 2, 3...n in order. d(r) is the total degrees in the area of πr^2. If $d(r) \propto r^D$, the value D is the fractal dimension called node degree-radius dimension.

2.2 Theoretical Proof

D has fractal properties will be proved from three kinds of road network.

Single Intersection Road. If there are only 2 straight roads in the region forming a cross over and the intersection is taken as the concentric circle center, we can find $d(r) \equiv 4$ and $D = \lim_{n \to \infty} \frac{\ln d(n)}{\ln n} = 0$. Only one point in the region, the point dimension is 0, which is true.

Ringy and Radial Road. Suppose there are 2 straight lines in the region forming a cross over and each r distance is a ring road intersecting with the straight road. We can get $d(n) = 4 + 16n$, so $D = \lim_{n \to \infty} \frac{\ln d(n)}{\ln n} = 1$. Similarly, if the number of radiation road is m, that is, $d(n) = m + 4mn$, we can also get D = 1. All the intersections of ringy and radial road are at radiation lines. The dimension of the straight line is 1. Result is conform to the actual situation.

2.3   Grid Road

Suppose there is a grid network in the area, the distance between each road is r. Because the road network is square, it is not convenient to calculate using circle, so we use square to verify, but the two are essentially the same. Use r as step size to count, $d(n) = 4 * (2n + 1)2$, so $D = \lim_{n \to \infty} \frac{\ln d(n)}{\ln n} = 2$. Checkerboard intersections are showing a planar distribution, the dimension of surface is 2, which is in conformity with the actual situation.

Extend it to the general form, when the road between a single intersection road and fill the whole area, $0 < D < 2$. The node degree-radius dimension reflects the variation of traffic network form center to the hinterland. When $D < 2$, the complexity of the traffic network is decreased from the center to the periphery; when $D = 2$, it means the traffic network balanced everywhere; if $D > 2$, illustrate the complexity of traffic from the center to the periphery increases, this is not a normal phenomenon, that measure center selection is not correct.

2.4   Weighted Node Degree-Radius Dimension

Different level of road capacity is different, so the importance of each intersection node should also be different. For example, some intersections are the connection points of major roads, traffic flow is clearly higher than the intersection formed by local road. The former in the road network structure should be more importance than the latter. Therefore, in order to more accurately reflect the importance of the intersection node, need to assign weight. According to different road types, assign weights to W1, W2... Wn. The number of different types of roads connected by the intersection node is N1, N2... Nn. So the weighted node degree dw = W1 * N1 + W2 * N2 + ... + Wn * Nn. We choose the traffic hub as the calculation center with radius r, where r takes 1, 2, 3... n in order. dw(r) is the total weighted node degrees in the area of πr2. If dw(r) ∝ rD, the value D is the fractal dimension called weighted node degree-radius dimension.

3   Empirical Research

3.1   Data Source and Research Area

This paper chooses Chengdu network to verity. By the end of 2013, the total road mileage reached 22514 km, of which 592 km highway. The data comes from the traffic system map in Chengdu city by Chengdu Planning Administration in 2013. The road network is divided into highway, major road, county road and general urban road. Since highway is closed, only the entrances and exits are considered. According the road type to assign weight, general urban road is 1, county road is 1.2, major road is 1.5 and the highway entrances and exits are 8. The distribution of intersections is shown on Fig. 1.

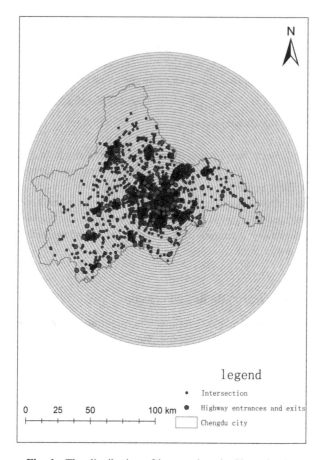

Fig. 1. The distribution of intersections in Chengdu city

3.2   Result and Analysis

Fractal Analysis of the Whole Road Network in Chengdu City. By experience, radius dimension should better choose transportation hubs or urban center as the calculation center. So this article takes the Tianfu Square as the measure center and 2 km as the step size. Based on the method above, we plot on a double logarithmic scale ln $d_w(r)$ as a function of lnr, shown on Fig. 2.

It shows that the road network in Chengdu can be divided into 3 scale-free interval. The first scale-free interval is from 2 km to 26 km, the fractal dimension is 1.64, between 1 and 2, reflecting the network style is a combination of grid road with ringy and radial road. The towns in Chengdu can be divided into 3 layers: downtown, suburb and outer suburb. 26 kilometers are roughly corresponding to the second circle range. It reflects an integration of the first and the second layer. The connectivity of suburb and downtown is good. The second scale-free interval is about 32 km to 42 km, where is just the inner town of the third layer. The fractal dimension is about 0.86, slightly

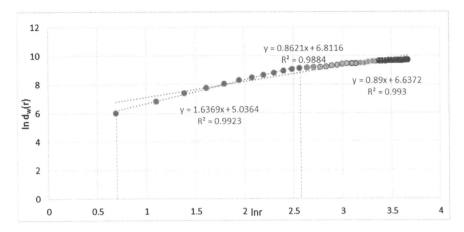

Fig. 2. Log-log plot of the weighted node degree as a function of distance of whole network

less than 1. It means the road network has a weak circle structure, radioactive network take a more important position. Between 26 km and 32 km, the slope of the curve significantly slowed down. This region is located in the middle of the second and third layer, reflecting the area between the two layers have a bad connectivity. The links between the suburb and outer suburb are not very closed. The towns in the third layer are not fully integrated into the downtown. The third scale-free interval is about 48 km to 56 km, where is the farther town of the third layer. The fractal dimension to 0.89, slightly larger than the second scale free interval. According to the general law of city development, it should conform to the core-periphery trend; the slope of double log-arithmic curve should gradually decline. But this range does not decline very much, even slightly improved. This shows the road connectivity in most county center of the third layer are at the same level. What should we take attention is that the slope of the curve also decreases significantly between 42 km and 48 km, where is the intermediate zone from the inner town of the third layer to the farther town. It also shows that the road network in the region is also poor. The links between center towns and the other villages are not closed enough. Other road network generally drops out of the scale-free area, the fractal characteristic is not obvious, need to be developed.

Fractal Analysis of the Trunk Road Network in Chengdu City. High grade road is the skeleton of the entire road network. So we select the intersect node of county road, major road and highway to construct weighted node degree-radius dimension, in order to analyze the shape and connectivity of high grade network. Since the density of high grade network is much lower, we choose 5 km as the step size. The measure center is also Tianfu Square and its double logarithmic curve is shown in Fig. 3.

Figure 3 shows the weighted node degree-radius dimension of high grade network in Chengdu is 1.07, very close to 1. That indicates the structure of skeleton road network is ringy and radial. Actually, Chengdu is a typical ringy and radial road network, and basically formed the "Three ring sixteen shot" fast road network by the end of 2015. The calculation result is valid. Scale-free interval is located roughly 5 km

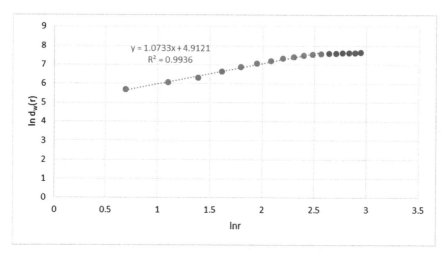

Fig. 3. Log-log plot of the weighted node degree as a function of distance of high grade network

to 65 km, covered the majority towns of Chengdu. It doesn't present the multi-fractal characteristics like the whole road network, showing the connectivity is good, basically throughout the surrounding counties. Scale-free interval of 65 km is also greater than the whole road network of 56 km, showing it is more balanced. Outer suburb is relatively small and the development of local road is not enough, has a large gap with the downtown.

4   Conclusion and Discussion

4.1   Conclusion

For the importance of intersection in the traffic network, we introduce the concept of node degree into the traffic network fractal calculation. Because different levels have different capacity, so the importance of intersection node crossing by different road should be different. In order to better reflect the capacity of crossing, we assign weight to different road. A weighted node degree-radius dimension is designed to better characterize the variation of network connectivity from center to periphery. Through theoretical and empirical studies, the following conclusions are obtained:

- There exist fractal dimension between the intersection node degree and the radius of the circle, and it can be used to characterize the road network. The fractal dimension of ringy and radial road is 1 and the grid road is 2.
- The weighted node degree-radius dimension can reflect the structure of urban system communication.
- The weighted node degree-radius dimension of the whole road network in Chengdu presents multi-fractal characteristics. Suburbs and the central city have almost become whole, suburbs and outer suburbs are not closely linked, mainly connected

by radioactive road. The structure of skeleton road network is ringy and radial and it's more balanced than the whole network. The local road in outer suburb need to be improved.

4.2 Discussion

The weighted node degree-radius dimension is a simple and feasible dimension parameter. It not only can reflect the network structure, but also closely related to the spatial structure of urban system. Of course, this dimension only describes one aspect of the fractal characteristics. In practice it also needs to combine other parameters, to characterize the structure of the road network together.

Because the roads have different level so we assign weight in this paper. Although the results better explain the road intersections fractal characteristics in Chengdu, the weight value looks arbitrary. The same urban roads may have different number of lanes, leading different connectivity level. In the future study, it should assign weight by the lane number, speed and so on, not only by the road grade. We can also assign weight by the vehicle flowrate so as to more truly reflect the interaction level of crossing.

In this paper, the fractal character of the node degree-radius dimension is theoretically demonstrated, and the relationship between the shape of road network and the fractal dimension is obtained. But the actual road network is not so ideal, for example: the interval of loop line may not be the same; the side length of the grid road may not be equal. That lead the fractal dimension and the ideal form are different. It must exist a range, and how to determine this range interval is a problem. Besides, the fractal dimensions are weighted, what extent can affect the range remains to be further studied?

References

1. Mandelbrot, B.: How long is the coast of britain? Statistical self-similarity and fractional dimension. Science **156**(3775), 636–638 (1967)
2. Frankhouser, P.: Aspects fractals des structures urbaines. L'Espace Geogr. **19**(1), 45–69 (1990)
3. Benguigui, L., Daoud, M.: Is the suburban railway system a fractal. Geogr. Anal. **23**(4), 362–368 (1991)
4. Benguigui, L.: A fractal analysis of the public transportation system of Paris. Environ. Plan. A **27**(7), 1147–1161 (1995)
5. Chen, Y.: A new fractal dimension on transport networks and the method of its determination. J. Xinyang Teach. Coll. (Nat. Sci. Ed.) **12**(4), 426–429 (1999)
6. Chen, Y.: A study on fractal dimensions of spatial structure of transport networks and the methods of their determination. Acta Geogr. Sin. **54**(5), 471–478 (1999)
7. Haung, P., Liu, M.: GIS-based study on fractal feature of urban traffic network in Shanghai. J. Tongji Univ. **30**(11), 1370–1374 (2002)

8. Zhang, P., Hang, Z.: A fractal-theory-based analysis on Liaoning's highway network. J. Transp. Syst. Eng. Inf. Technol. **6**(1), 123–127 (2006)

9. Guo, J., Han, Z., Xu, Y.: The evolution of urban traffic network of Dalian based on the gather fractal feature. J. Transp. Syst. Eng. Inf. Technol. **7**(5), 121–126 (2007)

10. Sun, Z.: The study of fractal approach to measure urban rail transit network morphology. J. Transp. Syst. Eng. Inf. Technol. **7**(1), 29–37 (2007)

11. Feng, Y., Liu, M., Tong, X.: Fractal dimension of transportation networks based on weighted length. Complex Syst. Complex. Sci. **4**(4), 32–37 (2007)

12. Bai, C., Cai, X.: Fractal characteristics of transportation network of Nanjing city. Geogr. Res. **27**(6), 1419–1426 (2008)

13. Zhu, H.: Study on Fractal Properties of Highway Network in Henan Province Based on GIS. Henan University, Kaifeng (2008)

14. Cao, F., Wu, J., Xu, M.: A fractal study on the urban spatial structure of Nantong city in Jiangsu province. Hum. Geogr. **25**(5), 69–74 (2010)

15. Chen, B., Wu, Z., Hu, W.: An analysis of urban transport network based on the fractal theory: a case study of Guangzhou. Trop. Geogr. **31**(1), 46–51 (2011)

16. Li, Y., Pan, S., Miao, C.: Study on fractal properties of highway transportation networks in Henan province based on GIS. Areal Res. Dev. **31**(5), 148–153 (2012)

17. Shen, J., Lu, Y., Lan, X.: Relationship between the road network and regional economic development based on the fractal theory. Sci. Geogr. Sin. **32**(6), 658–665 (2012)

18. Fang, D., Yang, Y.: Study on transportation spatial structure fractal characteristics of Yangtze river delta city group in high-speed rail era. Areal Res. Dev. **32**(2), 52–56 (2013)

19. Liu, C.: Spatial complex of urban-rural road network: a case of Wuhan metropolitan area. Central China Normal University, Wuhan (2011)

20. Liu, C., Duan, D., Yu, R., Luo, J.: Multi-scale analysis about urban-rural road network of four major metropolitan area in China based on fractal theory. Econ. Geogr. **33**(3), 52–58 (2013)

21. Liu, C., Yu, R., Duan, D.: Spatial-temporal structure of capacity fractal about urban-rural road network in Wuhan metropolitan area. Geogr. Res. **33**(4), 777–788 (2014)

22. Zhang, X., Wang, Z.: Application of weighted model of fractal dimension based on the theory fractal geometry. J. Wut (Inf. Manag. Eng.) **36**(6), 768–772 (2014)

23. Zhu, S., et al.: A fractal dimension study of road networks in the ancient city of Xi'an. Geogr. Res. **35**(3), 561–571 (2016)

WIFI-Based Indoor Positioning System with Twice Clustering and Multi-user Topology Approximation Algorithm

Xiaofeng Lu$^{(\boxtimes)}$, Jianlin Wang, Zibo Zhang, Haibin Bian, and Erzhou Yang

State Key Laboratory of Integrated Service Networks,
Xidian University, Xi'an, China
luxf@xidian.edu.cn, wangjianlin13@126.com,
zhangzibo612@gmail.com

Abstract. In recent years, indoor positioning technology based on WIFI has been widely researched. However, traditional WIFI-based indoor positioning method can't achieve high localization accuracy due to the clustering errors at some locations. In this paper, RSS and location based twice clustering (RLTC) and Multi-user Topology Approximation Algorithm is proposed. The algorithm is divided into two stages. RLTC method is proposed during offline stage to correct clustering results. During online stage, multi-user topology approximation method is proposed to reduce positioning error on some particular location. Experiments show that the proposed algorithms can effectively improve the positioning accuracy compared to traditional positioning method.

Keywords: Indoor positioning · WIFI · Clustering · Multi-user topology

1 Introduction

Nowadays, with the development of information technology and social networking, there are extensive demands on location-based service (LBS) [1, 2] in recent years. The indoor positioning system (IPS) has shown great promise and researchers have put forward a variety of solutions, such as UWB [3], RFID [4, 5], Bluetooth [6], WIFI [7] and so on. WIFI-based IPS has been widely researched.

The indoor positioning method based on WIFI is mainly divided into two broad categories, Received-Signal-Strength-based (RSS-based) localization method and Geometric-Model-based (GM-based) localization method. Compared with RSS-based method, GM-based method (e.g., time-of-arrival (TOA) [8] or angle-of-arrival (AOA)) needs to measure the time of arrival or the angle of arrival using additional hardware. On the contrary, RSS can be easily obtained by a WIFI module in smart phones without any additional hardware. There are two methods to realize indoor positioning system with RSS. One is using a propagation model to describe the relationship between RSS and position, the other is fingerprinting method using a prebuilt radio map.

In this paper, we use fingerprinting method because it is usually difficult to provide satisfactory positioning accuracy when using propagation model due to the complexity

© Springer Nature Singapore Pte Ltd. 2017
H. Yuan et al. (Eds.): GRMSE 2016, Part I, CCIS 698, pp. 265–272, 2017.
DOI: 10.1007/978-981-10-3966-9_30

of the indoor environment. However, due to the variable wireless environment (e.g., the movement of people and the closing or opening of doors), traditional WIFI fingerprinting method can't achieve high localization accuracy. In order to improve localization accuracy, we have proposed RSS and location based twice clustering algorithm (RLTC) and multi-user topology approximation localization algorithm.

The remainder of this paper is organized as follows: Sect. 2 describes the traditional WIFI based fingerprinting localization method. Section 3 elaborates the proposed RLTC algorithm and multi-user topology approximation algorithm. The experiments and localization results are discussed in Sect. 4. Finally, Sect. 5 concludes the paper.

2   Traditional Fingerprinting Localization Method

The fingerprinting localization method includes two stages [9]. The first stage is online stage, whose main task is to establish a WIFI fingerprinting database. The second stage is online stage. In this stage, we will implement real-time indoor positioning.

2.1   Offline Stage

WIFI fingerprinting database is set up in the offline stage with the WIFI access points existing in buildings. This stage contains collecting RSS data and clustering these RPs.

Collecting RSS Data. During this step, equi-spaced reference points (RPs) at known locations are selected in the region of interest. For each RP, the RSS readings of the nearby WIFI APs are recorded into a database. In order to reduce the error lead from direction, RSS data in four directions are measured.

Clustering these RPs. For the sake of reducing computation complexity in the online stage, the RPs collected in the offline stage are divided into different clusters.

In this paper, the affinity propagation (AP) [10, 11] algorithm is used to create the clusters. AP algorithm classifies data according to their similarity, such as Euclidean distance. In this paper, the similarity between RP i and RP j is defined as

$$s(i,j)^{(d)} = -\left\|\psi_i^{(d)} - \psi_j^{(d)}\right\|^2, \forall i,j \in \{1,2,\ldots,N\}, j \neq i, d \in D = \{0°,90°,180°,270°\}$$

(1)

where $\psi_i^{(d)}$, $\psi_j^{(d)}$ is the RSS vector of RP i and RP j respectively [9].

AP algorithm considers every data as a potential clustering center, and then responsibility message $r(i,j)$ and availability message $a(i,j)$ [10] are iteratively transmitted between pairs of RPs according to similarity until clustering is finished.

2.2   Online Stage

Real-time positioning process is completed during the Online Stage. This stage contains 2 steps, coarse localization by cluster matching and accurate localization by KNN.

Coarse Localization Step. This step can reduce the area the anchor point may be located in. First, an RSS vector from the APs near the anchor point is measured. Then the online RSS measurement is compared with each cluster to identify which cluster the anchor point belongs to.

Accurate Localization Step. The user's accurate position can be obtained in this step by KNN algorithm. The similarity between the online RSS vector and each RPs' RSS vector in the clusters chosen in coarse localization step is calculated. Then find the RPs corresponding to K largest similarity as the Nearest Neighbor. In the end, these K RPs' average coordinates is calculated as the final positioning result.

3 Proposed Localization Algorithms

When using the above fingerprinting database method to implement experiment, we find that large positioning errors exist on some anchor points. In this paper, we propose RLTC and multi-user topology approximation algorithm to reduce the outliers and improve localization accuracy.

3.1 RSS and Location Based Twice Clustering Algorithm

After analysis, we find some RPs far away from each other geographically may have similar WIFI signatures. So these RPs may be divided into a same cluster, which leads to large localization error.

In order to solve this problem, the RSS and location based twice clustering algorithm is proposed in this paper. During the offline stage, the RPs are clustered twice.

First Clustering Process. The first clustering process using affinity propagation algorithm is based on the Euclidean distance in the signal space of RPs. The clusters can be denoted as:

$$C^B = \{C_1, C_2, \ldots, C_M\}, i = 1, 2, \ldots, M, \tag{2}$$

where C_i is one of the clusters, and M is the total number of clusters generated during the first clustering process. These clusters are called big clusters.

Second Clustering Process. After the first clustering process is finished, all the RPs are divided into a number of clusters. However, some RPs far away from each other geographically may be classified into a same cluster.

Note that all the RPs' locations are known, so the second clustering process can be carried out based on RPs' locations. For each cluster $C_i \in C^B$, this clustering process can be divided into the following three steps.

Step 1: Affinity propagation algorithm is used to carry out the second clustering process to generate small clusters according to RPs' locations. These small clusters can be described as:

$$C_i^S = \{C_{i1}, C_{i2}, \ldots, C_{ij}\}, i = 1, 2, \ldots, M, j = 1, 2, \ldots, Q, \qquad (3)$$

where C_{ij} is a small cluster from C_i, and Q is the number of small clusters C_i generating.

Step 2: If $Q \geq 2$, calculate the distance between each of these small clusters. The distance between C_{ip} and C_{iq} is defined as the average distance between each RP belongs to C_{ip} and each RP belongs to C_{iq}.

Step 3: Merge the clusters. Some clusters shouldn't have been split into small clusters need to be spliced into one cluster if their distance is less than a certain threshold. In this paper, the threshold is 4 meters experimentally.

After above steps, the fingerprinting database is established, then the online localization can be carried out as 2.2 describes.

3.2   Multi-user Topology Approximation Localization Algorithm

In order to further improve the positioning accuracy, we can take advantage of multiple mobile phones to implement indoor localization cooperatively. In this paper, multi-user topology approximation localization algorithm is proposed, and the distance between users is applied. There are two challenges to realize the proposed multi-user localization algorithm:

(a) How to measure the distance between two users without additional equipment.
(b) How to use the distance to reduce the localization error.

Distance Measurement with Chirp Acoustic Signal. The users participating in the localization cooperatively are not far from each other, so it is suitable to use acoustic signal to measure the distance [12].

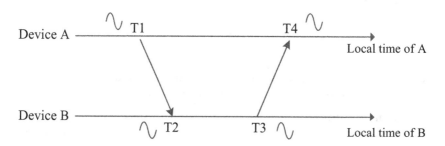

Fig. 1.   The principle of distance measurement.

The speed that acoustic wave spreads in the air is around 340 m/s, which is not very fast. We can measure the time the acoustic wave spreads between two devices, and then the distance can be calculated [13]. The distance measurement process is shown in Fig. 1.

In this method, device A first emits an acoustic signal S1. The signal will be recorded by its own microphone at time T1 and by device B's microphone at time T2. Then device B emits another signal S2 same as S1. S2 is recorded by device B at time T3 and by device A at time T4. So the time it takes for the acoustic signal to spread from device A to device B can be calculated as:

$$\frac{(T4 - T1) - (T3 - T2)}{2} \tag{4}$$

In this method, T1 and T4 are the local time of device A, while T2 and T3 are the local time of device B, so it is not necessary to think about the synchronization between two devices and the software processing delay.

The key problem is how to obtain the accurate time the acoustic wave arrives. In this paper, the acoustic wave is designed as linear chirp signal because of its great autocorrelation property. In order not to be interfered by the voice of the people speak, we choose the frequency range to be between 8–12 kHz.

The error of distance measurement is mainly determined by the sound sampling rate. In this paper, the sound sampling rate is 48 kHz. The theoretical error under this sampling rate is 0.71 cm with the assumption that velocity of sound in the air is 340 m/s. We carry out 200 times ranging experiment, and the average error is less than 2 cm. In addition, the probability of absolute error within 3 cm is 95%.

Multi-user Localization with Known Distance. The distance between users has been obtained, so the next job is how to improve the positioning accuracy with known distance. This multi-user topology approximation localization algorithm can be divided into four stages.

(1) Get Initial Polygon G by WIFI Location Estimations. During this stage, each user gets initial WIFI location estimations by traditional WIFI fingerprinting database method independently. Then the initial polygon G can be obtained by the initial location estimations $p_1(x_1, y_1), p_2(x_2, y_2), \ldots, p_n(x_n, y_n)$.

(2) Get Polygon G' by the Distance Between Users. The distances can be used as edges and constitute a polygon G', which can be obtained by the following steps.

Step 1: Determine the position of the first edge. Assume that the vertices of G' are $p'_1, p'_2, \ldots, p'_n$, where n is the number of users. The distance between any two vertices is known. A plane rectangular coordinate system is established. Let the point p'_1 is located in the origin of coordinates. Then let the point p'_2 on the positive half axis of x. The first edge $p'_1 p'_2$ can be determined.

Step 2: Determine the position of other edges. We can get other edges one by one. The third vertices p'_k along with p'_1 and p'_2 can make up a certain triangle. $\angle p'_k p'_1 p'_2$ can be calculated according to the law of cosines. Then the abscissa and the absolute value of the ordinate can be obtained. However, there are two cases of the position of p'_k, shown as Fig. 2.

It is impossible to obtain the right orientation by rotating the G' with wrong

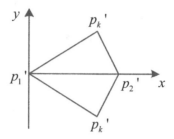

Fig. 2 Two cases of the position of p_k'.

ordinate of p_k'. We can confirm the position of p_k' through the corresponding WIFI Location Estimations by the following equation.

$$S = (x_1 - x_k)(y_2 - y_k) - (y_1 - y_k)(x_2 - x_k) \tag{5}$$

If $S > 0$, the points p_1, p_2 and p_k is anticlockwise as well as p_1', p_2' and p_k', then the ordinate of p_k' is positive. Otherwise, the ordinate of p_k' is negative.

(3) Estimate the Orientation of Polygon G'. The inner product summation between the corresponding edge of G and G' is used to obtain the optimal polygon orientation Φ by rotating the polygon G'.

$$\Phi = \arg\max \sum_{e=1}^{n} v_e' v_e^T \tag{6}$$

where $v_e' = \frac{p_i' - p_j'}{|p_i' - p_j'|}, (i < j)$ is an edge vector of polygon G', and $v_e = \frac{p_i - p_j}{|p_i - p_j|}, (i < j)$ is an edge vector of polygon G.

(4) Estimate the Position of Polygon G'. Finally, the position of polygon G' is determined by the center of gravity of G and G'. Translate polygon G', and make its center of gravity overlap the center of gravity of G. The vertices of G' are the final localization results of multi-user.

4   Experiments and Localization Results

We deliver our indoor positioning system in the Main Building in Xidian University. The CDF of localization errors are shown in Fig. 3.

First we deliver our positioning system with and without RLTC. Figure 3(a) shows that compared with traditional WIFI fingerprinting database method, proposed RLTC can significantly improve positioning accuracy.

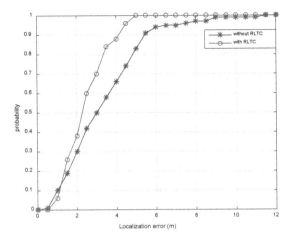

(a) With and without RLTC

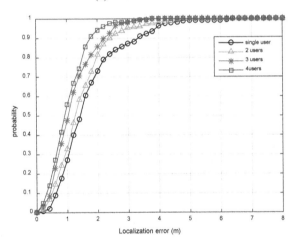

(b) Different number of users.

Fig. 3. Localization errors. (a) With and without RLTC (b) Different number of users.

We also carry out another experiment with different number of users using RLTC. From the localization results shown in Fig. 3(b), we can find that with the increase of number of users, positioning performance is getting better and better. In this experiment, when the user number reaches four, the positioning errors probabilities within 1 m, 2 m and 3 m are 0.56, 0.94 and 0.98, respectively, while the mean localization error is 1.02 m. We reduce the mean localization error by 38.9% when compared to single user.

5 Conclusions

In this work, we have proposed RLTC algorithm and multi-user topology approximation localization algorithm. During offline stage, RPs are divided into several clusters after twice clustering using affinity propagation algorithm. During online stage, the distances between users are measured first, and then their final localization results are determined through topology approximation algorithm.

The experiment results show that the localization accuracy is significant improved.

Acknowledgments. This work was supported by the National Natural Science Foundation of China (61371127 & 61471361 & 61401330 & 61572389), the EU FP7 CROWN Project under Grant Number PIRSES-GA-2013-610524, the National High-Tech R&D Program (863 Program 2015AA01A705) and the 111 Project in Xidian University of China (B08038).

References

1. Chen, Z., Zou, H., Jiang, H., et al.: Fusion of WIFI, smartphone sensors and landmarks using the Kalman filter for indoor localization. Sensors **15**(1), 715–732 (2015)
2. Zou, H., Lu, X., Jiang, H., et al.: A fast and precise indoor localization algorithm based on an online sequential extreme learning machine. Sensors **15**(1), 1804–1824 (2015)
3. Gezici, S., Tian, Z., Giannakis, G.B., et al.: Localization via ultra-wideband radios: a look at positioning aspects for future sensor networks. IEEE Signal Process. Mag. **22**(4), 70–84 (2005)
4. Zou, H., Wang, H., Xie, L., et al.: An RFID indoor positioning system by using weighted path loss and extreme learning machine. In: IEEE International Conference on Cyber-Physical Systems, Networks, and Applications, pp. 66–71. IEEE (2013)
5. Zou, H., Xie, L., Jia, Q.S., et al.: Platform and algorithm development for a RFID-based indoor positioning system. Unmanned Syst. **2**(03), 279–291 (2014)
6. Viswanathan, S., Srinivasan, S.: Improved path loss prediction model for short range indoor positioning using bluetooth low energy. In: Sensors. IEEE (2015)
7. Lim, C.H., Wan, Y., Ng, B.P., See, C.: A real-time indoor WIFI localization system utilizing smart antennas. IEEE Trans. Consum. Electron. **53**, 618–622 (2007)
8. Youssef, M., Youssef, A., Rieger, C., et al.: Pinpoint: an asynchronous time-based location determination system. In: International Conference on Mobile Systems, Applications and Services, 59593-195. ACM (2006)
9. Feng, C., Au, W.S.A., Valaee, S., et al.: Received-signal-strength-based indoor positioning using compressive sensing. IEEE Trans. Mob. Comput. **11**(12), 1983–1993 (2012)
10. Pfeiffer, M., Pfeiffer, M.: Clustering by passing messages between data points. Science **315** (5814), 972–976 (2007)
11. Frey, B.J., Dueck, D.: Response to comment on "clustering by passing messages between data points". Science **319**(5864), 726–730 (2008)
12. Yang, Z., Feng, X., Zhang, Q.: Adometer: push the limit of pedestrian indoor localization through cooperation. IEEE Trans. Mob. Comput. **13**(11), 2473–2483 (2014)
13. Peng, C., Shen, G., Zhang, Y., et al.: BeepBeep: a high accuracy acoustic ranging system using COTS mobile devices. In: International Conference on Embedded Networked Sensor Systems, pp. 1–14. ACM (2007)

Surveillance Camera-Based Monitoring
of Plant Flowering Phenology

Lijun Deng[1], Wei Shen[1], Yi Lin[2(✉)], Wei Gao[3], and Jiayuan Lin[4]

[1] School of Earth Sciences and Resources,
China University of Geosciences, Beijing 100083, China
{denglijun, shenwei}@cugb.edu.cn
[2] School of Earth and Space Sciences, Institute of Remote Sensing and GIS,
Peking University, Beijing 100871, China
yi.lin@pku.edu.cn
[3] School of Geology and Geomatics, Tianjin Chengjian University,
Tianjin 300384, China
gaowei@tcu.edu.cn
[4] Institute of Mountain Hazards and Environment,
Chinese Academy of Sciences, Chengdu 610041, China
linjy@imde.ac.cn

Abstract. Phenology plays an important role in understanding the feedbacks of plants to climate change. However, phenology observation is not a trivial task, particularly for large covers with huge diversities of plants. To handle this issue, this study attempted to apply wide-spread surveillance cameras (SCs) for plant phenology monitoring. In the case of flowering phenology, multiple phenological indices were proposed and derived from SC image series to identify the starting and ending dates. Test showed that the derived flowering phases for *Robinia pseudoacacia* (Fabaceae), *Prunus cerasifera* (Prunus) and *Malus micromalus* (Malus) agreed well with the ground-truth data. The feasibility of assuming SCs for plant flowering phenology monitoring was validated. Furthermore, this study alludes to a potential way of using SCs to compose regional- to continental-scale networks for phenology monitoring.

Keywords: Plant flowering phenology · Surveillance camera · Phenological index

1 Introduction

Phenology plays an essential role in deciding the primary productivity and net exchange of ecosystems [1]. Phenology is also an important indicator of the feedbacks of ecosystems to climate change [2]. Given that global change has become a topic of wide interest in recent decades, increasing numbers of researchers have paid close attention to plant phenology and increasing numbers of countries have made/are making plans to organize their phenological observation networks for understanding global or regional changes [3].

The conventional methods of plant phenology monitoring are based on manual observation. Such means can obtain the key dates of plant phenological events and have already been used for centuries. The in-field phenological recordings have been successfully associated with the environmental factors such as climatic and geographic

© Springer Nature Singapore Pte Ltd. 2017
H. Yuan et al. (Eds.): GRMSE 2016, Part I, CCIS 698, pp. 273–283, 2017.
DOI: 10.1007/978-981-10-3966-9_31

variations [2–5]. However, this kind of methods relies on huge labor-costs, particularly for the phenological events at global and regional scales.

The progress of satellite remote sensing (RS) implies the possibility of learning the continuous changes of vegetation canopies at regional or global scales. Many studies have been conducted to derive phenological information from satellite RS data [6]. However, it is hard to obtain accurate, abundant and valuable transition dates of plant activities from satellite RS data due to their inadequate spatial resolutions and revisit frequencies [7, 8]. The unfavorable factors also include the difficulty of collecting adequate ground-truth samples for the calibrations of phenological information retrievals [9].

A number of endeavors recently attempted to bridge the gap between satellite-derived phenological information and ground-truth data [10]. Digital cameras proved to be able to serve as a practical substitute of manual observation [8]. This technology commonly with higher sampling frequencies can acquire spectral information and objectively record the day-to-day variations of plant growths via continuously shooting of images; thus, it is recognized as the core of the technology of "near-surface remote sensing" [8]. Actually, camera-based solutions have been used to automatically monitor phenology in various ecosystems, including forests [11–13], grasslands [14], wildlife habitats [15] and croplands [16]. Some studies also have used camera images to evaluate the uncertainties in satellite RS measurements [17].

However, installing common digital cameras to compose a phenology monitoring network may still mean a huge cost, particularly for large covers. For this problem, surveillance cameras (SCs), initially installed for security purposes rather than for scientific applications, can be assumed as a solution. SCs have the same resolutions as those cameras specifically designed for phenological monitoring [8], and they need no modifications to adapt to the planted areas of interest. They also have the strength of recording plant dynamics at higher sampling frequencies. Hence, the first objective of this study was to determine if SCs can be applied for plant phenology monitoring. Besides, previous endeavors mainly focused on the dates of leaf unfolding or budburst during the growing seasons [7, 13, 18], whereas few endeavors paid attention to RS-based plant flowering phenology monitoring. Thus, the second objective of this study was to check if plant flowering phenology can be monitored in terms of SC pixel colors.

2 Materials and Methods

2.1 Study Site

The study site was located in the campus (the west side of Lee Shao Kee Humanities Building) of Peking University, Beijing, China (39°54′20″N, 116°25′29″E). Beijing relates to a typical region with the kind of warm temperature and semi-humid continental monsoon climate. The field views of the SCs are predominantly covered by some ornamental plants. The plants for testing in this study mainly include *Robinia pseudoacacia* (Fabaceae), *Prunus cerasifera* (Prunus) and *Malus micromalus* (Malus). The distances from the plants of these three species to the SC are about 22 m, 25 m, and 40 m, respectively. The flowers of both *R. pseudoacacia* and *M. micromalus* appear after their green leaves flush (namely, foliation-first), but for *P. cerasifera*, the pink flowers emerge earlier than purple leaves (i.e., flowering-first).

2.2 SC Image Collection

A series of digital images (form March 27th 2015 to May 10th 2015) were continuously collected using a SC (DS-2AF1-784DS, Hikvision, China), which was installed at a height of ~4 m above the ground and whose field of view was set to face northwest. The imaging module of the SC camera is a 1/3″ progressive scan CMOS sensor, enclosed in a commercial waterproof housing. The mode of camera imaging was set to be automatic exposure and white balance adjustment in an "automatic mode" that once was used by Zhao et al. [19]. Although the position of the SC was fixed, its field of view can be slightly changed by the remotely-controlled settings. This renders the images with differences on horizontal and vertical views during the observations, which can simulate the influences of winds and rains [12]. The three plants lying in the field of view of the SC were recorded in the manner of multiple images per day.

Specifically for the SC images, they were automatically captured and saved in the form of Joint Photographic Experts Group (JPEG, 1920 × 1080 pixels with three channels of 24-bit RGB color information). In favor of subsequent processes of image matching and plant phenological analysis, the JPEG format was transferred into Tagged Image File Format (TIFF, 1920 × 1080 pixels with three channels of 24-bit RGB color information). Compared to the JPEG image format, TIFF is a lossless compression format that can maintain the original information about colors and can make sure a better image quality. Different technical issues such as electronic outage, battery leakage as well as technical failure resulted in that some images were not recorded during the process of image collection (day-of-year (DOY) 93, 101, 106 and 118 with no data). To minimize the influence of different solar angles [11], the images recorded at 7:00 am per day were used.

2.3 Image-Based Phenological Analysis

Since the fields of view of the SC changed occasionally, image registration was conducted in the ENVI (Environment for Visualizing Images) environment for the sake of target representation in consistency. Then, the registered images were sequentially read, without any other pre-processings for image enhancement processing (e.g., histogram equalization or contrast enhancement) so that objective inferences can be obtained [18]. Specifically for analysis of the phenological characteristics of the three trees of interest, three regions of interest (ROIs) were drawn to indicate the crowns of the *P. cerasifera* (ROI A, see Fig. 1a), *M. micromalus* (ROI B, see Fig. 1a) and *R. pseudoacacia* (ROI C, see Fig. 1b) trees. The polygonal ROI covering the full crown of the *M. micromalus* tree was determined to avoid ignoring any trivial changes; but for the *P. cerasifera* and *R. pseudoacacia* trees, analyses were executed on regular rectangular ROIs (ROI A and ROI C). The ROIs were selected to keep a compromise between maximizing crown covers and avoiding influences from the other background species as far as possible. The program to process the digital images was coded in MATLAB (R2015a; The Mathworks, Natick, USA).

In order to determine the onset and ending dates of the flowering phenology, three plant phenological indices based on the digital numbers of the three color channels of the SC camera were proposed as

$$BF = B_{3*3}/\text{Total} \tag{1}$$

$$GF = G_{3*3}/\text{Total} \tag{2}$$

$$RF = R_{3*3}/\text{Total} \tag{3}$$

$$\text{Total} = R_{3*3} + G_{3*3} + B_{3*3} \tag{4}$$

where R_{3*3}, G_{3*3} and B_{3*3} indicate the windows of extracting and averaging each group of 3×3 pixels within the ROI of each tree at daily interval. The variable of Total was defined to calculate the normalized colors of the three color channels, and this can somehow reduce the influences such as different sunlight and rains. The derived BF, GF and RF were used as the phenological indices to determine the onset and ending dates of the flowering phenology.

2.4   Performance Assessment

To validate the feasibility of applying SCs for flowering phenology monitoring and evaluate the performance of the proposed phenological indices for reflecting the onset and ending dates of plant flowering phenology, the phenological dates of the three plants were determined using the common approach of expert knowledge-based verification [10, 12, 20]. Specifically, three independent observers utilized a common protocol to define the following dates for each species:

(1) When the crowns started to show flowers;
(2) When the flowers could not be distinguished.

The transitional dates of flowering were classified to the start date of the flowering period (SFP) and the ending date of the flowering period (EFP). The two parameters identified manually based on the SC images are listed in Table 1 (Obs.), and they were used to assess the indices-based derivations.

3   Results

3.1   SC Image Series

For *P. cerasifera*, its related ROI (ROI A) in the field of the view of the SC was dominated by pink flowers during its blooming; simultaneously, a little portion of the flowers of *P. cerasifera* showed up from among the no-leaves crowns of *R. pseudoacacia* that was at its earlier foliation stage (ROI C) (Fig. 1b). Along with plants growing, *M. micromalus* leaves increased and became greener, and the sizes of its leaves also gradually increased so that rear species no longer could be observed. Besides, the crown of *M. micromalus* (ROI B) (Fig. 1a) appeared obviously with plenty of white flowers but was still dominated by green leaves. Next, the pink flowers of *P. cerasifera* gradually disappeared, and the full crown of *P. cerasifera* was filled with purple leaves and brown branches. Since then the green crowns of the other plants, such as *R. pseudoacacia* lying closer to the SC, more or less emerged in the ROI for *P. cerasifera* and disturbed its observation.

Fig. 1. Two registered sample images showing the flowering dynamics, with the three plants marked out by ROIs A, B and C: (a) Image recorded on 1 April 2015 and (b) Image recorded on 1 May 2015 (The Chinese characters at the bottom-right corner mean the road at the west side, Humanities School of Peking University). (Color figure online)

Because of *R. pseudoacacia* standing adjacent to the SC, there are some overlaps between ROI C (*R. pseudoacacia*) and ROI A (*P. cerasifera*). The plant of *R. pseudoacacia* was in the flourishing stage, with white flowers and green leaves (Fig. 1c). The beginning of its foliation was earlier than the onset of flowering for *R. pseudoacacia*; so, leaves occupied the most proportion of ROI C within the field of view of the SC. The development of crown flowering was spatially in a patchy mode, as different plant species differ at the rates of leaf expansions [18].

3.2   Flowering Phenology Identification

The proposed phenological indices assisted determining the key dates of the flowering stages for the three plants for testing. First, compared to the original digital numbers (DNs) recorded by the three R, G and B channels, the derived RF, GF and BF proved to overcome the influences by rains and sunlight to some extent. Further, as shown in

Figs. 2 and 3, the series of the phenological indices for the three plants exhibit different trends, which demonstrate the possibilities of reflecting the flowering and foliation phenological events. After a comparison of the temporal distributions for all of the phenological indices, RF proved to be more sensitive to the flowering phenology and

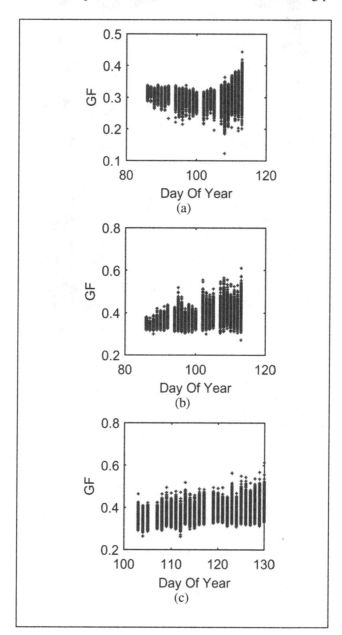

Fig. 2. Phenological patterns indicated by the GF indices for (a) *P. cerasifera*, (b) *M. micromalus* and (c) *R. pseudoacacia*, respectively.

GF can better characterize the state of canopy foliation. Finally, the onset and ending dates of the flowering events for the three plants were determined based on the overall modes of the phenological index series.

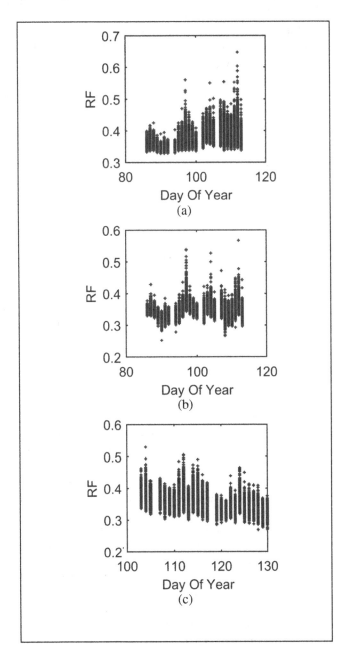

Fig. 3. Phenological patterns indicated by the RF indices for (a) *P. cerasifera*, (b) *M. micromalus* and (c) *R. pseudoacacia*, respectively.

3.3 Performance Assessment

The key dates of the flowering phenological events for *P. cerasifera*, *M. micromalus* and *R. pseudoacacia* were also determined based on visual interpretation and were listed in Table 1. After a comparison of the estimated and observed results, it can be realized that the RF-based derivations about SFP and EFP have good consistencies with their respective ground-truth data (see Table 1). The RF-based dates for SFP showed better agreements with the validation data than the opposites for EFP (the absolute difference is form −1 to 0 VS. from −2 to 1). The results of the three species estimated based on RF are earlier than the derivations by visual interpretation, and this indicates that the index-based method can work with higher accuracies on the task of determining plant flowering phenological events. As regards to the EFP for the flowering-first species, the estimated dates delayed 1 day relative to the observed result, while two foliation-first species advanced 2 day and 1 day.

Table 1. Transitional dates (DOY) of the 3 studied species were showed. The observed dates judged by eye from images (Obs), estimated dates (Est) derived from color index. SFP: the start of the flowering period; EFP: the end of the flowering period. Diff: absolute difference.

Event	*P. cerasifera*		*M. micromalus*		*R. pseudoacacia*	
	SFP	EFP	SFP	EFP	SFP	EFP
Obs	88	99	99	109	114	125
Est	87	100	98	107	114	124
Diff	−1	1	−1	−2	0	−1

4 Discussion

It is a difficult task to interpret the uncertainties lying in the proposed method due to the complexity of tree structures. Although the phenological indices were proposed with the possible influences considered and they proved to be able to work for monitoring the phenological changes of trees, some uncertainties still exist and need to be addressed in order to improve the whole reliability. Noises lying in the DN values have not been fully explored and various weather conditions and illuminations may make image quality low. Unfavorable meteorological conditions and different illuminations may disturb the depiction of crown variations. The greenings of plant understory and saplings may also add additional errors into plant phenological information derivations. A promising approach was to reconstruct the hypothetical horizontal RGB reference that might help overcome illumination influences [21].

The SC cameras shall be fixed so as to capture the same ROIs of the targets of interest, and good resolutions need to be ensured by setting proper distances between the cameras and sample trees in favor of phenological analysis. After all, large distances may affect the quality of the SC images under fog or ill atmospheric conditions. Because the image-based visual interpretation may be impacted by the color sensitivity of human eyes, it is still suggested that the conventional field observation methods need to be used to verify the phenology results as the additional references. SC cameras are often set to be in an auto white balance way, and their color balance was automatically optimized. It was also suggested that data filtering is necessitated to select the optimal data from the similar weather conditions [20]; but for special processes of plant flowering, all images in the given periods are selected regardless of weather conditions.

Overall, although there were still a lot of uncertainties and limitations, the key transitional dates of plant flowering phenological events for the three plants were obtained from the SC digital camera images and were well consistent with the ground-truth data. It can be concluded that SCs can be considered as a promising tool to quantitatively analyze the flowering durations in various categories of plants and RF can be used to accurately extract the transitional dates of the flowering event. The contribution of this study can do favor to improving the urban environments, enhancing the level of urban landscape and provide certain theoretical and practical value for planning rational layout structure.

This work has also laid the foundation for establishing SC-based plant phenology observatory networks at different scales. Future works can focus on reducing the uncertainties and long-term monitoring of various phenological events at the community scale with varying illuminations and weather conditions. Besides, the SC images in this study were only aimed at an individual local environment. In order to ensure that sufficient spatial covers are available, further researches shall focus on different locations, species and multiple years. Moreover, SCs can be installed at flux towers for evaluating the relations between local carbon uptakes and water cycling during the flowering phases. It is necessary to show whether the phenological changes related to biochemical changes in vegetation growths, and consequently, SCs can also be used to calculate gross primary productivity.

Acknowledgment. This work was supported in part by the National Natural Science Foundation of China (Grant No. 41471281), in part by the Research Fund for Doctoral Program of Higher Education of China (Grant No. 20130001120016), and in part by the SRF for ROCS, SEM, China.

References

1. Churkina, G., Schimel, D., Braswell, B.H., Xiao, X.: Spatial analysis of growing season length control over net ecosystem exchange. Glob. Change Biol. **11**, 1777–1787 (2005). doi:10.1111/j.1365-2486.2005.01012.x
2. Wolkovich, E.M., Cook, B.L., Allen, J., Crimmins, T., Betancourt, J., Travers, S., Pau, S., Regetz, J., et al.: Warming experiments underpredict plant phenological responses to climate change. Nature **485**, 494–497 (2012). doi:10.1038/nature11014
3. Dai, J., Wang, H., Ge, Q.: Multiple phenological responses to climate change among 42 plant species in Xi'an, China. Int. J. Biometeorol. **57**, 49–758 (2013). doi:10.1007/s00484-012-0602-2
4. Wu, X., Liu, H.: Consistent shifts in spring vegetation green-up date across temperate biomes in China, 1982–2006. Glob. Change Biol. **19**, 870–880 (2013). doi:10.1111/gcb.12086

5. Aono, Y., Saito, S.: Clarifying springtime temperature reconstructions of the medieval period by gap-filling the cherry blossom phenological data series at Kyoto, Japan. Int. J. Biometeorol. **54**, 211–219 (2010). doi:10.1007/s00484-009-0272-x

6. Ganguly, S., Friedl, M.A., Tan, B., Zhang, X., Verma, M.: Land surface phenology from MODIS: characterization of the collection 5 global land cover dynamics product. Remote Sens. Environ. **114**, 1805–1816 (2010). doi:10.1016/j.rse.2010.04.005

7. Graham, E.A., Yuen, E.M., Robertson, G.F., Kaiser, W.J., Hamilton, M.P., Rundel, P.W.: Budburst and leaf area expansion measured with a novel mobile camera system and simple color thresholding. Environ. Exp. Bot. **65**, 238–244 (2009). doi:10.1016/j.envexpbot.2008.09.01

8. Richardson, A.D., Braswell, B.H., Hollinger, D.Y., Jenkins, J.P., Ollinger, S.V.: Near-surface remote sensing of spatial and temporal variation in canopy phenology. Ecol. Appl. **19**(6), 1417–1428 (2009). doi:10.1890/08-2022.1

9. Studer, S., Stockli, R., Appenzeller, C., Vidale, P.L.: A comparative study of satellite and ground-based phenology. Int. J. Biometeorol. **51**, 405–414 (2007). doi:10.1007/s00484-006-0080-5

10. Fisher, J.I., Mustard, J.F., Vadeboncoeur, M.A.: Green leaf phenology at landsat resolution: scaling from the field to the satellite. Remote Sens. Environ. **100**, 265–279 (2006). doi:10.1016/j.rse.2005.10.022

11. Ahrends, H.E., Brügger, R., Stöckli, R., Schenk, J., Michna, P., Jeanneret, F., Wanner, H., Eugster, W.R.: Quantitative phenological observations of a mixed beech forest in Northern Switzerland with digital photography. J. Geophys. Res. Biogeosciences (2005–2012) **113**, G04004 (2008). doi:10.1029/2007JG000650

12. Ahrends, H.E., Etzold, S., Kutsch, W.L., Stoeckli, R., Bruegger, R., Jeanneret, F., Wanner, H., Buchmann, N., Eugster, W.: Tree phenology and carbon dioxide fluxes: use of digital photography at for process-based interpretation the ecosystem scale. Clim. Res. **39**, 261–274 (2009). doi:10.3354/cr00811

13. Nagai, S., Maeda, T., Gamo, M., Muraoka, H., Suzuki, R., Nasahara, K.N.: Using digital camera images to detect canopy condition of deciduous broad-leaved trees. Plant Ecol. Divers. **4**(1), 79–89 (2011). doi:10.1080/17550874.2011.579188

14. Migliavacca, M., Galvagno, M., Cremonese, E., Rossini, M., Meroni, M., Sonnentag, O., Cogliati, S., Manca, G., Diotri, F., Busetto, L.: Using digital repeat photography and Eddy covariance data to model grassland phenology and photosynthetic CO_2 uptake. Agric. For. Meteorol. **151**, 1325–1337 (2011). doi:10.1016/j.agrformet.2011.05.012

15. Bater, C.W., Coops, N.C., Wulder, M.A., Hilker, T., Nielsen, S.E., McDermid, G., Stenhouse, G.B.: Using digital time-lapse cameras to monitor species-specific understorey and overstorey phenology in support of wildlife habitat assessment. Environ. Monit. Assess. **180**, 1–13 (2011). doi:10.1007/s10661-010-1768-x

16. Zhou, L., He, H.L., Sun, X.M., Zhang, L., Yu, G.R., Ren, X.L., Min, C.C., Zhao, F.H.: Using digital repeat photography to model winter wheat phenology and photosynthetic CO_2 uptake. Acta Ecologica Sinica **32**(16), 5146–5153 (2012). doi:10.5846/stxb201110271606

17. Hufkens, K., Friedl, M., Sonnentag, O., Braswell, B.H., Milliman, T., Richardson, A.D.: Linking near-surface and satellite remote sensing measurements of deciduous broadleaf forest phenology. Remote Sens. Environ. **117**, 307–321 (2012). doi:10.1016/j.rse.2011.10.006

18. Richardson, A.D., Jenkins, J.P., Braswell, B.H., Hollinger, D.Y., Ollinger, S.V., Smith, M.-L.: Use of digital webcam images to track spring green-up in a deciduous broadleaf forest. Oecologia **152**, 323–334 (2007). doi:10.1007/s00442-006-0657-z

19. Zhao, J., Zhang, Y., Tan, Z., Song, Q., Liang, N., Yu, L., Zhao, J.: Using digital cameras for comparative phenological monitoring in an evergreen broad-leaved forest and a seasonal rain forest. Ecol. Inform. **10**, 65–72 (2012). doi:10.1016/j.ecoinf.2012.03.001

20. Ide, R., Oguma, H.: Use of digital cameras for phenological observations. Ecol. Inform. **5**, 339–347 (2010). doi:10.1016/j.ecoinf.2010.07.002

21. Sonnentag, O., Hufkens, K., Teshera-Sterne, C., Young, A.M., Friedl, M., Braswell, B.H., Milliman, T., O'Keefe, J., Richardson, A.D.: Digital repeat photography for phenological research in forest ecosystems. Agric. For. Meteorol. **152**, 159–177 (2012). doi:10.1016/j.agrformet.2011.09.009

Visual Analysis Research of Traffic Jam Based on Flow Data

Wei Tian[1,2], Jinming Zhang[1,2(✉)], and Jialin Ma[1,2]

[1] Information Engineering University, Zhengzhou 450052, China
1158529851@qq.com
[2] Institute of Remote Sensing and Digital Earth,
Chinese Academy of Sciences, Beijing 100101, China

Abstract. With the acceleration of urbanization, road congestion is getting worse. To quickly determine the traffic jam, taking the example of Jiashan traffic data collected by ground sense coil. The paper firstly cleans raw data, eliminates error data, supplements missing data and processes redundant data, which provides reliable data for visual analysis; then using different visual components select congested roads, and collaboratively interacts three visual component which including heat map, bar graph and chord diagram, visually expresses road congestion from macroscopic to microcosmic and judges the position of the traffic jam, and provides reliable route for public trip.

Keywords: Ground sense coil traffic data · Road congestion · Visual analysis · Heat map · Chord diagram

1 Introduction

With the rapid development of society, the acceleration of urbanization, increasing the population and scale of cities, there are different levels of traffic congestion in most large and medium-sized cities, urban problem has became one of the bottlenecks of economic development [1]. Traffic congestion caused huge economic losses, including time loss, fuel loss and environmental costs. The data released by Chinese Ministry of Transportation shows that economic losses caused by traffic congestion are in 20% of the disposable income of the urban population, equaling 5–8% of annual gross domestic product (GDP) loss about up to 250 billion RMB a year, whereas loss in Beijing is 700 billion RMB one year (which more than 80% is the congestion time loss) [2]. Thus, the demand of community traffic control, vehicle navigation and personal travel path planning is growing, an urgent need for real-time traffic information to guide travelers, vehicles and transportation planning needs of individuals travel routes also have the accelerating growth trend.

Due to traffic congestion complexity, dynamic, and irrelevance reasons, it is difficult to predict traffic jams. Using prior knowledge and experience, combined with computer visualization technology, as well as the prediction model to analyze traffic congestion [3]. The paper uses traffic flow data (induction coil, microwave sensors) to analyze the traffic jam by visibility analysis, using the visual components judges congestion roads, which makes users acquire traffic congestion through the visual

© Springer Nature Singapore Pte Ltd. 2017
H. Yuan et al. (Eds.): GRMSE 2016, Part I, CCIS 698, pp. 284–292, 2017.
DOI: 10.1007/978-981-10-3966-9_32

interface and make decisions according to their own needs. For example, taxi drivers want to know the current traffic conditions, or which way is saving time and nearest from the current position to destination; travelers need to choose transport routes considering traffic congestion.

In this paper, we mainly study the visual expression and visual analysis by the ground sensing coil flow data. Using visual component selects congested roads. Heat map is used to display the distribution of ground sense coil flow, from which can directly observe what roads congested and which road unimpeded; histogram, specific display according to flow value recorded by various ground sense coil; chord diagram, show the gridlocked traffic, which makes users understand the interrelation of each section. Visualizing the flow data of the ground sensing coil to analyze the road congestion in Jiashan, judge the congested roads, provide reliable information for public travel.

2 Related Work

In the process of traffic flow data visual analysis, including traffic data cleaning, data visual analysis two parts.

2.1 Traffic Flow Data Cleaning

Traffic flow data use toroidal coil detector to collect the data, the loop is a wide range of traffic sensor in urban roads, can acquire real-time of flow, speed, time share, travel direction and so on, are applied in traffic information collection and traffic state identification, traffic event monitoring system. Traffic flow data cleaning mainly includes data filtering, data recovery, outliers elimination, etc. Chu et al., used dynamic traffic data of Shanghai outer ring to clean noise data by the maximum possible value filling the missing data and moving average method [4]; Lu et al. came up with anomaly detection method based on curve fitting according to traffic/time occupancy rate of the inverted "V" shaped curve model, which can effectively identify the traffic flow data of the abnormal data [5]; Zhang et al. according to the relativity traffic flow theory of flaw detector intersection traffic of prediction [6], is a kind of effective data recovery method.

2.2 Visual Analysis of Traffic Flow Data

The visual analysis of road traffic data can be used to demonstrate the characteristics of spatial and temporal changes of traffic data, assist managers to understand and analyze the operation situation of urban traffic, and provide decision support for managers and users. In recent years, the research on the visual analysis of road traffic is increasing. Jiang et al. using taxi origin and destination point data, designed a Web version's taxi O/D visual analysis system data based on B/S architecture, and a set of explore cognitive rules of taxi O/D data analysis method [7]; Wang et al. make a thorough study about visual analysis of trajectory data, and divide the visual analysis of trajectory data into three types: direct visualization, aggregation visualization, characteristics visualization [8]; Li et al. made massive data and statistical results highly graphical based on the traditional simple chart, showed information and rules more in a shorter period of time [9]; He used the arrow diagram and flow chart to express visualization of road traffic data [10]; Andrienko et al. [11] established a framework of interactive visual interface, which can effectively support the understanding of mobile behavior and mobile mode analysis of human; TripVista interactive visual analysis system researched by the Beijing University of visual development team where Wang [12] and others in, explore abnormal behavior through the research on the microcosmic traffic rules, and show traffic track in a variety of drawing way (direction icon theme River, with a timeline of the scatter diagram, show a high dimensional attribute information of the parallel coordinate plot).

3 The Data Processing

3.1 The Working Process

Visualization of flow data mainly includes two parts. The first part is data cleaning (processing redundant data, eliminating the error data, supplementing the missing data); the second part is visual analysis of traffic flow data, using different visualization techniques on the expression and analysis of Jiashan road congestion (Fig. 1).

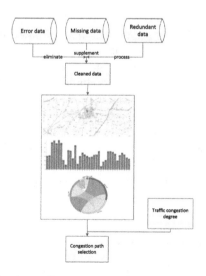

Fig. 1. Traffic data visual analysis work flow.

3.2 Select Data Sets

This paper chooses traffic data collected by the sense coil detector not floating car data, because sense coil detector technology is mature, easy to grasp; detection precision is high, can all-weather work; and low equipment prices. However, due to the limit of coverage and sampling frequency floating car data is not every moment to meet the statistical requirements; in the city buildings, trees, and the viaduct caused blocking of inaccurate GPS positioning; GPS positioning error affects the floating car reliability and availability [1].

Take Jiashan sense coil flow data for sample, total 20 days data from 2015-9-20 to 2015-10-10, the sampling interval is about 30 s, and 36 sense coil detector which distributed in the city main roads, can monitor the city road better, and reflect the real-time traffic congestion. The main attributes of the flow data record of each ground sensor coil include equipment number, direction, number of the pile, average speed, time, average occupation rate, etc., and its properties are listed in Table 1. In this experiment used the device ID to represent selected sections (such as equipment number SBBH14 is Jinyang Road), use direction to judge the uplink and downlink (1 said uplink, 2 said downward), pile flow display number of vehicles through the sense of coil, the time is an important attribute of data traffic, which can determine the vehicle traffic laws through the different time periods and the existence of periodic.

Table 1. Attribute of flow data.

Data types	ID	Connotation
Flow data of ground sensor coil	*SBBH*	Equipment number
	FX	Orientation
	ZHLL	Number of the pile
	PJCS	Average speed
	QSSJ	Time
	PJZYL	Average occupation rate

3.3 Data Cleaning

Original data have many mistakes due to equipment, external environment and other factors, including three kinds of noise data: outliers, missing data, repeat the value in general [13]. By the analysis of the flow data, outliers accounted for 8% of these data, the data loss is 82%, the repeated number is 10%, so the data cleaning mainly for the lack of data processing.

Many reasons lead to data loss, such as the detector scan frequency is not fixed, the transmission line fault and the vehicle is too dense when the detector can't detect vehicles, etc. The processing methods aiming at data missing are removing the data object, interpolation computing missing value, neglecting missing value, using the probability model to estimate the missing value, using the historical data and so on [14]. In this paper, we first make a visual expression of the data (Fig. 2), judge whether the data is missing, and recover the missing data by using historical data [15]. Because people travel is similar in time, the traffic flow data is periodic in different days, using the same time history data to recovery missing data. As follows:

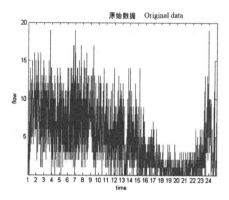

Fig. 2. The original data visualization of line chart.

$$\hat{x}_i = Hx_i \tag{1}$$

In this formula: x_i—estimated value

Hx_i—history data

Using historical data to recovery the missing data, 95% of the missing data can be recovered (Fig. 3), which meets the data requirements better, reflects trends in data change, provides reliable data for subsequent visual analysis.

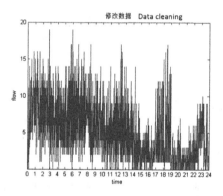

Fig. 3. The visual line chart after data cleaning.

4   Visual Design and Analysis

4.1   Congestion Path Selection

When analyzing Jiashan road congestion data, we need to choose the congestion path, determine which path is clear, which road is congested firstly, then make a visual analysis of the congested road. In this paper, the method of multi view collaborative visualization is used to choose the congested road section.

Multi view collaborative visual expression is used to choose the congestion section, and the congestion section is judged by different visual expression methods. The interactive analysis of the visual components is completed as shown in Fig. 4. Using the flow data acquired by ground sense coil to describe, the visual component 1 which is flow data measured by induction coil is used for the heat map expression, display the flow data of each sense coil collection to highlight form, red is represent for vehicle intensive, green is vehicle through the darker sparse, said the deeper the color the larger the flow, the color is more shallow said the flux is small; the visual component 2 histogram which display specific flow of the sense coil in a time, make up the visual component 1. Component 1 can only judge the range, but cannot determine the specific road congestion, where the component 2 can clearly observe the specific traffic flow at a time and the specific number of the flow. Comprehensive visual components 1 and 2, the congested road is Tangong north road and wood Avenue direction; visual component 3 is displaying the interrelationship diagram of Jiashan County main road congestion by the chordal graph, through the visual component 1 and 2 to determine the main road congestion is wood Avenue, Tangong north road, outer ring road, by the visual component 3 observation to the trend of congested road traffic, such as most traffic of Tangong north road towards the direction of wood Avenue.

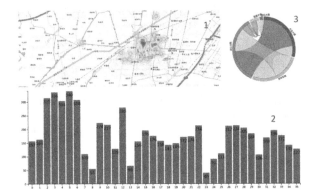

Fig. 4. Traffic data visualization synergy. (Color figure online)

4.2 Traffic Jam Analysis Validation

From above visual components analyzes Jiashan main congestion line is Muye Avenue, outer ring Road, Tangong north road. Due to the existed research on the data flow processing traffic congestion rarely exist, so this paper calculates road congestion degree according to the flow, which judges whether the road congested, and makes traffic data quantify, in order to verify the correctness of the above conclusion of visual analysis.

In the process of deciding the traffic congestion threshold, we cannot to measure congestion just by a simple flow for that the number of vehicles can't discern the road congestion and recording of small flow may be less in the period of driving and is not caused by congestion. Zheng Shujian and others' article described the traffic congestion measured by congestion degree in Japan, defined as the ratio of the evaluation benchmark amount among actual traffic volume and 24 h or 12 h a day in a road [16], such as the formula (2).

$$DC = (\omega \times Q)/C \tag{2}$$

Type: ω as the weight coefficient, $\omega = 1-\alpha/100 + \alpha \times \beta/100$; α as large vehicles mixed rate; β for the standard vehicle equivalent coefficients of large vehicles; Q for 12 h day traffic volume; C as the evaluation base 12 h traffic volume, benchmark traffic can be gained by the level of planning and design traffic capacity, peak rate, in the same direction rate. According to the actual situation of Jiashan (no large vehicle traffic in the urban area), the paper improves the formula (3), as follows:

$$DC = Q/C \tag{3}$$

Type: Q as 1 h traffic volume; C as evaluation benchmark 1 h traffic.

The relationship between the congestion degree and operation level of road network, is that congestion degree by how much to open or congestion, as shown in Table 2:

Table 2. Relationship between the congestion degree and road network.

Congestion degree DC	1 h operation level of road network
DC < 1	Unimpeded
$1 \le DC < 1.75$	Congestion gradually increasing
$DC \ge 1.75$	Chronic congestion

Statistical calculation by formula (3), traffic situation of Jiashan Muye Avenue, Zhujiang Road – Lingshan Road in 2015-9-28 9:00–10:00 as shown in Fig. 5, of which Tangong north Road to Muye Avenue is 75% to the congestion state, where Tangong

Fig. 5. Traffic route choice.

north Road - Zhujiang Road - Lingshan Road is smooth. The result of verifying the traffic flow data visualization by congestion degree quantization is correct.

5   Conclusion

The paper directs at the road congestion matter, taking flow data collected from Jiashan for example, cleans the original flow data, determines the time scale, visualizes and analyzes the flow data. Three interactive visualization components including heat map, bar histogram and chordal graphs, makes users analyze road congestion from macroscopic to microscopic, determine whether the road congested, and whether need to change the travel route. And quantifying the flow by calculating the traffic congestion degree, then judging the rode congestion and what extent of the congestion, which is a serious congestion or slight congestion. The visual analysis results are verified by calculating the congestion degree, which provides a reliable route for public travel.

Acknowledgment. This work is funded by the National Natural Science Foundation of China (No. 41371383).

References

1. Li, Q.: Traffic Geographic Information System Technology in the Development of Frontier. Science Press, Beijing (2012)
2. Beijing Traffic Development Research Center: The 2015 Beijing traffic development report. Traffic Transp. (3), 3–7 (2016)
3. Wang, Z., Lu, M., Yuan, X., et al.: Visual traffic jam analysis based on trajectory data. IEEE Trans. Vis. Comput. Graph. 19(12), 2159–2168 (2013)
4. Chu, H., Yang, X., Wu, Z.: Method analyzing and application of dynamic traffic data pre-processing. In: China Intelligent Transportation Annual Meeting (2005)
5. Lu, W., Shang, N., Qin, M.: A traffic data preprocessing method based on curve fitting exception detection. J. Comput. Res. Dev. 43(s3), 642–646 (2006)
6. Zhang, H., Yang, Z., Li, Y.: Study on forecast of traffic volume at non-detector intersections. J. Highw. Transp. Res. Dev. 19(1), 91–95 (2002)
7. Jiang, X., Zheng, C., Jiang, L.: Visual analysis of large taxi origin-destination data. J. Comput. Aided Des. Comput. Graph. 27(10), 1907–1917 (2015)
8. Wang, Z., Yuan, X.: Visual analysis of trajectory data. J. Comput. Aided Des. Comput. Graph. 27(1), 9–25 (2015)
9. Li, W., Zhou, F., Zhu, W.: The rail transit network traffic data visualization research. Chin. Railw. 05(2), 94–98 (2015)
10. He, X.: Visual analytics of road traffic with large scale taxi GPS data. Zhejiang University of Technology (2014)
11. Andrienko, G., Andrienko, N., Wrobel, S.: Visual analytics tools for analysis of movement data. ACM Sigkdd Explor. Newsl. 9(2), 38–46 (2007)
12. Guo, H., Wang, Z., Yu, B., et al.: TripVista: triple perspective visual trajectory analytics and its application on microscopic traffic data at a road intersection. In: IEEE Pacific Visualization Symposium, pp. 163–170. IEEE Computer Society (2011)

13. Chen, W., Shen, Z., Tao, Y.: Data Visualization. Electronic Industry Press, Beijing (2013)
14. Wang, M.J., Pan, Q.M., Chen, W.: Survey of visualization data cleaning. J. Image Graph. **20**(4), 0468–0482 (2015)
15. Jin, S.: Research on preprocessing methods of loop detector data. Jilin University (2007)
16. Zheng, S., Yang, J.: Traffic congestion evaluation index calculation methods at home and abroad. Highw. Automot. Appl. **04**(1), 57–61 (2014)

A Design of UAV Multi-lens Camera System for 3D Reconstruction During Emergency Response

Junhui Wu, Fei Wang$^{(\boxtimes)}$, and Xiaocui Zheng

Graduate School at Shenzhen, Tsinghua University, Shenzhen 518055,
Guangdong, People's Republic of China
sunwujh@foxmail.com, wang.fei@sz.tsinghua.edu.cn,
2293733306@qq.com

Abstract. Public safety problems increase in recent years and bring a big challenge in the field of fast data acquisition for decision making. UAV (unmanned aerial vehicle) system as an effective way to take images of the emergency scene is highly demanded in many public safety application fields, and the three-dimensional (3D) reconstruction of the scene based on the captured images is also increasingly required. However, multi-lens camera system for taking titled images has not yet maturely applied to small UAVs. In this paper, a novel multi-lens camera system adapting for small UAVs is designed to capture the needed 3D reconstruction images. This system is also able to provide the possibility of transmitting the images to the ground station in real time for further consequent 3D reconstruction processing.

Keywords: Multi-lens camera · Oblique photogrammetry · Emergency response · UAV

1 Introduction

In recent years, the rapid expansion of urbanization, industrialization and globalization in China brings more public safety problems [1]. In some extraordinary situations such as earthquake area and toxic, it is difficult or dangerous for human beings to access the scene of the disaster to get data, which is necessary at emergency response process. However, UAVs can be used in such situations because of its fast reaction, flexibility, and no casualties [2, 3]. UAVs carried a camera can obtain the photos or videos of the emergency scene.

3D reconstruction of the disaster scene based on the images from the UAVs can show the scene more visually to help the manager making decisions. At the recovery stage, it can also help researchers analyze the disaster. In order to get the data used in

Funded by Technical Innovation Project of Shenzhen (CXZZ20140416155802785), Fundamental Research Project of Shenzhen (JCYJ20150331151358139), and Guangdong Emergency Platform Technology Research Center.

© Springer Nature Singapore Pte Ltd. 2017
H. Yuan et al. (Eds.): GRMSE 2016, Part I, CCIS 698, pp. 293–298, 2017.
DOI: 10.1007/978-981-10-3966-9_33

3D reconstruction, UAVs have to carry a special camera system, multi-lens camera system, which is composed of several cameras facing different directions and taking photos at the same time [4].

There are some multi-lens camera systems in the marketplace, but they are not suitable for normal UAVs. For example, Pictometry from America and SWDC-5 from China are so large and heavy that they can only be carried by large-scale UAVs. Compared to large-scale UAVs, small ones are more suitable for emergency situations because of their low cost, simple operation and no requirement for airfield.

As for multi-lens camera system used in small UAVs, there are not mature products at the present stage, because of the strict requirements from small UAVs. First, the system must be small and light enough to be loaded on small UAVs. Secondly, in order to recover the instability of small UAVs and get clear images, the system must have good adaptability and stability. Thirdly, in order to satisfy the requirements of emergency response, the data need to be transferred to the ground station as soon as possible.

A multi-lens camera system that is suitably loaded in small UAVs to get 3D reconstruction images specifically in emergency situations is designed in this paper. It takes advantages of the UAVs' flexibility to get into emergency scene and take images, which can be transferred back to ground immediately to do 3D reconstruction afterwards and help emergency management more efficient in a low-cost way.

This paper first describes the current research state of multi-lens camera system. Then our design of UAV multi-lens camera system for 3D reconstruction at emergency response is introduced. Discussions and further research work are given in the last section.

2 Research Status

2.1 Oblique Photogrammetry

Oblique photogrammetry is rapidly developing in the world in recent years. Traditional aerial photography with a vertical camera can only get roofs of objects on the ground so that it is difficult to build 3D models. As for Oblique photogrammetry, several cameras are loaded on one air vehicle and obtain images of both top and side surfaces of buildings. With the help of high-accuracy position system, images contain 3D coordinate position data and therefore the height of the building can be calculated [5–7].

Nowadays, Oblique photogrammetry technology is widely applied in 3D city modeling. Many companies have launched the projects of developing airborne oblique photogrammetry equipment and 3D reconstruction software.

2.2 Multi-lens Camera

Multi-lens camera system is an important part of oblique photogrammetry. One of the earliest multi-lens camera system is ADS40 produced by Leica in 2000, which is consist of three lenses. Similar camera systems have been developed by many other companies in the world, such as Microsoft in USA, DigiCAM in Germany, Track Air in Netherlands, Visonmap in Israel, Geo-vision in China and so on. There are four major types of multi-lens cameras on the market today, three-line array camera, three-lens camera, four-lens camera, and five-lens camera, among which five-lens camera is the most common type in recent years.

The most popular five-lens camera systems are Pictometry, RCD30 Obliqueby Leica, SWDC-5 by Geo-vision, AMC by Shanghai Hangyao, TopDC5 by TOPRS, and Midas, DigiCAM, A3, etc. Usually, one of the lenses is placed vertically, the other four are facing different directions, and the tilt angle is among 40 to 60 degree. The five-lens camera system shoots multi-view images of the ground which are widely used in basic geography information of digital city and digital earth building. There is also profes-sional software to process the data for five-lens camera systems.

But the multi-lens camera mentioned above are all aimed to large aerial vehicle and they are too large and heavy to be loaded on small UAVs. For example, AMC by Shanghai Hangyao weighs 45 kg. Furthermore, small UAVs are more easily to vibrate in airflow and make images breezing, so that the camera has to have good self-stability.

2.3   UAV Communication System

Communication system is an essential part of UAV, as it transmits not only flight control information including signal of UAV status and instructions from the operator on the ground, but also remote sensing (RS) information including images and videos taken by camera system. Traditionally, flight control information is transmitted by radio communication system or satellite link data transmission system.

As for RS information transmission, it is more complex than the flight control information, because high resolution RS information contains huge data and is vul-nerable to environmental interference. Some research institution have tried methods like microwave transmission, radio transmission, and image compression, but they are still not very stable and mature. The normal method to transmit RS information to the ground is to locally store the data in hard disk of TB storage at UAV and take back the hard disk after the flight.

In order to realize real-time transmission of emergency scene images from UAV, wireless data transmission technique is taken into account [8, 9], which includes infrared transmission, blue tooth transmission, Ultra-wideband (UWB), Zigbee, mobile communication technique and WiFi wireless transmission. Infrared, blue tooth and UWB can only be used in a very short distance, and by Zigbee transmission is very slow. WiFi is a fast and economic way, but the transmission distance is still short within 300 m at outdoor. According to the demands of emergency data transmission distance and velocity, these techniques are not suitable.

Mobile communication technique is developing rapidly in recent years. The fourth generation (4G) mobile communication technique is promoted [10], which is marked by LTE, a universal standard consisted of TD-LTE and FDD-LTE. In China, the three carriers, China Mobile, China Unicom and China Telecom have already get TD-LTE license from Ministry of Industry and Information Technology (MIIT). China Unicom and China Telecom got FDD-LTE license in 2015. Many cities in China have been covered by 4G network. There's a good tendency that 4G is covering more areas and becoming cheaper. The fast and stable transmission and good economic efficiency prove that 4G is a good choice to transmit emergency scene data [11].

3 Design Contents

Our design of multi-lens camera system is aimed at emergency situations and the system is loaded on small UAVs, transmitting multi-view images of the disaster scene in real time to help build 3D models rapidly. According to the principle of oblique photogrammetry, the structural features of UAVs and the requirement of rapid data transmission, the design is divided into three main parts: camera module, camera holder, data processing and transmission module. The general route map of the project is like Fig. 1.

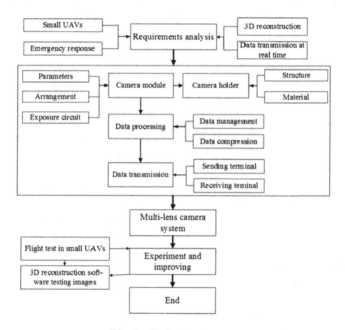

Fig. 1. Project route map

3.1 Maintaining the Integrity of the Specifications

The core part of the multi-lens camera system is camera module, of which the performance parameters have direct influence on quality of emergency scene data. The camera module used in large multi-lens camera system cannot be directly applied to small UAVs, and the parameter selection and lens arrangement have to be redesigned.

Considering about the mature application and software support of five-lens camera, this design uses five-lens camera to get images of the ground from different angles of view. The size of elements must be strictly controlled. The design process concludes:

(1) According to the requirement of images used in 3D reconstruction, choose the parameters of lenses, including pixel resolution, angle of view, focal length, pixel physical size, radiation resolution, light sensitivity, exposure interval, exposure time, etc. Different lenses may have different parameters.

(2) According to the requirement of image overlapping ratio for 3D reconstruction, design the arrangement of lens, including arrangement interval and inclination angle.

(3) In order to avoid errors caused by mechanical deformation and exposure synchronization, design a high accuracy exposure circuit, aiming at the multi-lens camera.

3.2 Camera Holder

Since small UAVs are easy to vibrate in the flight, making the sharpness and continuity of images decreased and even camera equipment destroyed, it is important to design a stable and firm camera holder to protect camera equipment as well as damp vibration.

The holder is designed to load five-lens camera mentioned above. It should be high-strength and waterproof to cope well with the adverse weather at complex emergency situations. The vibration damping function can be realized through two ways, designing a self-stable structure, or using materials that can absorb vibration. Furthermore, the holder should be easy to assemble considering about the urgency. The whole holder should be light and small enough to be assembled in small UAVs.

3.3 Data Processing and Transmission Module

Fast response at emergency requires images to be transmitted to the ground in real time, instead of being stored locally and processed after the flight. With the development of mobile communication technique and the mature of 3D reconstruction algorithm, it becomes possible to realize real-time transmitting.

While the multi-lens means multiple image data, it is necessary to manage and pre-process the data in the air and use a suitable compression format.

The workflow of 4G mobile communication is shown in Fig. 2. Flight control information is transmitted via radio, while image data is transmitted via 4G network to the ground receiving equipment, indirectly through the 4G base station.

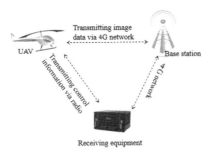

Fig. 2. Data transmission

4 Conclusions

A multi-lens camera system loaded on UAV for 3D reconstruction images taking during emergency response is designed in this paper. The system concludes camera module, camera holder, data processing and transmission module. Based on oblique photogrammetry, 3D reconstruction, and 4G mobile communication technique, it is stably loaded in small UAVs to obtain emergency scene images which is transmitted to the ground instantly and used to build 3D reconstruction. Our future work will be focused on studying parameters, structures and materials of the system in details, realizing a demo and applying it to a small UAV. Although this is a conceptual design with few details, it is proposed to solve emergency problems in actual situations and it can be optimized along with the advancement of scientific research work.

References

1. Fan, W.: Advisement and suggestion to scientific problems of emergency management for public incidents. Sci. Found. China **2**, 2 (2007)
2. Lu, B., Meng, D., et al.: UAV applications and explorations in major natural disasters. J. Catastr. **26**(4), 122–126 (2011)
3. Lei, T., Li, C., He, X.: Application of aerial reomote sensing of aircraft to disaster emergency. J. Nat. Disasters **20**(1), 178–183 (2011)
4. Zhu, Q., Xu, G., Du, Z., Yu, J., Wang, J.: Review of Oblique Photogrammetric Technology (2012)
5. Wang, W., Huang, W., Zhen, J.: Pictometry oblique photography technique and its application in 3D city modeling. Geomat. Spat. Inf. Technol. **34**(3), 181–183 (2011)
6. Gerke, M., Kerle, N.: Automatic structural seismic damage assessment with airborne oblique Pictometry© imagery. Photogramm. Eng. Remote Sens. **77**(9), 885–898 (2011)
7. Gui, D., Lin, Z., Zhang, C.: Research on construction of 3D building based on oblique images from UAV. Sci. Surv. Mapp. **178**(4), 144–146 (2012)
8. Wang, P., Luo, X., Zhou, Z., et al.: Key technology for remote sensing information acquisition based on micro UAV. Trans. Chin. Soc. Agric. Eng. **30**(18), 1–12 (2014)
9. Gao, Z., Deng, J., Sun, J., et al.: Scheme and Key Technologies of Wireless Image Transmission System for Micro Unmanned Air Vehicles, Transactions of Beijing Institute of Technology
10. Liu, X.: Image acquisition and processing system based on UAV remote sensing platform research Zhejiang University (2013)
11. Huang, J., Zhu, X.: Study on the key technology and prospects of mobile communication. China New Telecommun. **16**(3), 119–120 (2014)

Spatial Data Acquisition through RS and GIS in Resource Management and Sustainable Ecosystem

Winter Wheat Leaf Area
Index (LAI) Inversion Combining
with HJ-1/CCD1 and GF-1/WFV1 Data

Dan Li[1,2], Jie Lv[3], Chongyang Wang[2], Wei Liu[4], Hao Jiang[2],
and Shuisen Chen[2(✉)]

[1] CMA/Henan Key Laboratory of Agrometeorological Support and Applied
Technique, Zhengzhou, China
[2] Guangzhou Institute of Geography, Guangzhou, China
css@gdas.ac.cn
[3] Xi'an University of Science and Technology, Xi'an, China
[4] Guangdong Climate Center, Guangzhou, China

Abstract. The LAI is the key factor which has an important influence on crop growth. LAI inversion from remote sensing is an important work in crop management. While, the accuracy of LAI inversion from remote sensing data is restricted by the limited number of observation. Multiple-sensor method has been proposed by the researchers. In this study, two sensor remote sensing data (HJ-1A/CCD1 and GF-1/WFV1) were collected in the study area. The random forest regression (RFR) was adopted in LAI inversion. The MODIS LAI product and the measured wheat LAI were used to calibrate and validate the LAI inversion model. The four spectral indices (DVI, SR, EVI, and SAVI) based on remote sensing data were calculated to develop the LAI inversion model. The accuracy of inversion of wheat LAI by remote sensing image can be improved by adding observations of angle data. Our data analysis resulted in an accuracy of R2 = 0.36, MAE = 0.467, and RMSE = 0.613 for the measured LAI. And in the validation by MODIS LAI product, an accuracy of R2 = 0.48, MAE = 1.05, and RMSE = 2.72 was found, which was a little greater than the average accuracy of mono-angle data for inversion of LAI. The result indicates that the reasonable combination of multi-sensor data can improve the accuracy of LAI estimation.

Keywords: Random forest · HJ-1A/CCD1 · GF-1/WFV1 · Winter wheat LAI

1 Introduction

LAI is an important indicator of terrestrial ecosystem, which is the input parameter of many climate, ecology and recycling models [1]. The fast and wide acquisition of crop LAI is of great significance [2–4]. The inversion of crop LAI is the hotspot and difficult

Work is supported by Open Fund of Agricultural meteorological support and application technology key laboratory in Henan Province/China Meteorological administration (No. AMF201202) and the Guangdong Natural Science Foundation (No. 2015A030313805).

© Springer Nature Singapore Pte Ltd. 2017
H. Yuan et al. (Eds.): GRMSE 2016, Part I, CCIS 698, pp. 301–309, 2017.
DOI: 10.1007/978-981-10-3966-9_34

point in quantitative remote sensing field [5, 6]. The vertical distribution of leaf area index in the canopy is quite complicated, which has an ignorable effect on canopy reflectance characteristics [7]. Multi-angle remote sensing can provide directional information of crop canopy. Recently, multi-angle remote sensing has been used effectively to improve the accuracy of LAI inversion [7, 8]. However, the multi-angle sensor in orbit is little [9, 10]. The combination dataset of the multi-source remote sensing data of different sensor contained the angle information has an important prospect in LAI inversion [4, 11, 12].

Scholars at home and abroad have carried out many researches on multi-angular and multi-source remote sensing data. Zhao et al. [5] combined HJ-1CCD and Landsat 8 OLI data to develop the LAI inversion model by looking-up table. The RMSE between LAI inversion model and measured LAI was 0.71. And the result indicated that the combining of two types of remote sensing data can improve the temporal resolution of LAI product.

Yang et al. [13] proposed the utilization of BJ-1 small satellite and Landsat TM data to construct multi-source and four-angle datasets for inversion of forest LAI, and found that the accuracy of inversion of forest LAI can be improved by adding observations of angle data with the R2 of 0.713 and RMSE of 0.957, which was 20% greater than the average accuracy of mono-angle data for inversion of LAI. Li et al. [14] explored the image features from Chinese HJ-1A/B CCD images for estimating winter wheat LAI in Beijing. Image features were extracted from such images over four seasons of winter wheat growth, including five vegetation indices, principal components, tasseled cap transformations and texture parameters. The LAI was significantly correlated with the near infrared reflectance band. However, the image features cannot significantly improve the estimation accuracy of winter wheat LAI. He et al. [15] proposed the leaf area nitrogen estimation model by multi-angular hyperspectral reflectance in winter wheat across different growth stages, plant types, N rates, planting density, ecological sites and years. They constructed the angular insensitivity vegetation index based on red-edge, blue and green bands. The new model showed the highest association with LNC (R^2 = 0.73–0.87) compared to traditional vegetation indices.

Hall et al. [12] used the spatial variation of multi-angular sensor CHRIS/PROBA images to quantify successfully absorbed photo-synthetically active radiation. Pisek et al. [9] retrieved the clumping index of MISR image over a set of sites representing diverse biomes and different canopy structures, and found that MISR data with 275 m allow clumping index estimation at much pertinent scales (both spatial and temporal). Liu et al. [8] proved that the angular information from MISR bidirectional reflectance factor data can provide a more accurate LAI estimation than previously available. Mousivand et al. [10] used a multi-angle remote sensing approach to determine changes in vegetation structure from differences in directional scattering observed by the MODIS with data atmospherically corrected by the multi-angle implementation atmospheric correction algorithm (MAIAC). The multi-angle approach used in this work may help quantify drought tolerance and seasonality in the Amazonian forests.

The above researches indicated that the multi-source and multi-angular images had the good performance in application of quantitative remote sensing inversion. And the combination of multi-satellites is the good choice for the multi-source and multi-angular images. China HJ-1A/B satellites and GF-1 satellites provide surface spectral information at 30 m and 16 m spatial resolution with a fine revisit frequency, and offer an opportunity to monitor crop LAI. Thus, the objective of the current study is to fully exploit the image of Chinese HJ-1A/CCD1 images and GF-1/WFV1 images using vegetation indices and evaluate the potential of the two types of image for LAI estimation.

2 Materials and Methods

2.1 Study Site

The study was conducted in Kaifeng city (Fig. 1). This area is located in E113°52′–115°02′, N34°12′–35°01′, which is in the middle east of Henan Province. It's in the tip of the Yellow River alluvial plain. The winter wheat is the main grain crop in the area. It belongs to a temperate continental climate with an annual mean temperature of 14.7 °C, extreme high temperature of 43 °C, extreme low temperature of −16 °C, ≥ 10 °C accumulated temperature of 4600 °C, frost-free season of 223–242 d, and the annual precipitation of 634 nm, which can satisfy the demand of winter wheat growth needs.

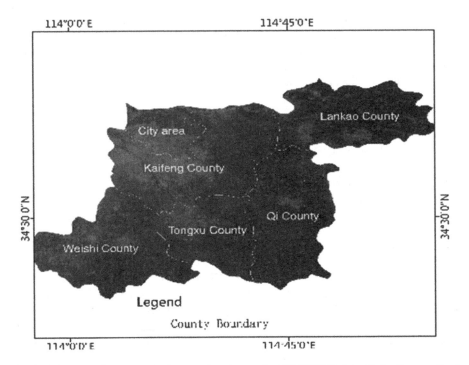

Fig. 1. The true color composite image of study area by GF-1/WFV1 data (Color figure online)

2.2 Field Measurements of LAI

Field experiments were conducted in the April 3–5, 2015. It was the growth season on the tillering of winter wheat. During experimental point, all winter wheat was harvested from 0.5 m × 0.5 m plots that were randomly selected from the central 16 m × 16 m field. The leaf area index of winter wheat was measured by a leaf area meter (LAI-2000, LICOR, Inc., Lincoln, NE, USA) [5]. The LAI measurements of each experiment are listed in Table 1.

Table 1. The details of measured winter wheat LAI

Growth stage	Numbers	Mean value	Value range	Time
Tillering	24	3.04	1.48–4.84	3–5, April, 2015

2.3 Remote Sensing Data and Pre-processing

Two small environment- and disaster-reduction satellites (HJ-1A/B) were launched by the China Center for Resources Satellite Data and Applications (CRESDA) on September 6, 2008. The spatial resolution (30 m) is similar to the first four bands of landsat-8 OLI. The short period facilities temporal analysis at key growth stages for crop monitoring [17]. In our study, one corresponding cloud-free image was acquired (Table 2). The landsat-8 OLI image on April, 2015 was selected to make geographic correction. Four wide-field-of view (WFV) instruments are onboard the Gaogen-1 (or GF-1) satellite. The spatial resolution is 16 m. MODIS LAI product was also selected to compare the performance for the calibration of HJ-1A/CCD1 bands, GF-1/WFV1 bands, and the combination of two type sensors.

Table 2. The information of multi-source and multi-angle data

Data source	Acquisition time	Satellite zenith	Satellite azimuth	Sun elevation	Sun azimuth elevation	Resolution (m)
HJ-1A/CCD1	2015-04-09	27.16	-78.17	52.66	311.64	30
GF-1/WFV1	2015-04-10	63.41	101.37	60.37	151.75	16

Remote sensing images were processed with ENVI 5.1 software through radiometric calibration, atmospheric correction, and geometric correction. The coefficients for radiometric calibration of HJ-1/CCD1and GF-WFV1 data were provided by China Centre for Resources Satellite Data and Application. Land surface reflectance was then retrieved using Fast Line-of-sight atmospheric analysis of spectral hypercube (FLAASH). Each HJ-1A/CCD1 image and GF-1/WFV1 image was co-registered with a geo-referenced Landsat 8 OLI image at a geometric accuracy of <0.5 pixel. Four spectral indices were calculated in LAI retrieval (Table 3).

Table 3. References and **equations** of VIs used in this study.

Vegetation index	Formulas	Reference
Difference vegetation index (DVI)	$DVI = NIR - R$	Xiao et al. [16]
Enhances vegetation index (EVI)	$EVI = 2.5 \frac{NIR - R}{NIR + 6R - 7.5B + 1}$	Liang et al. [3]
Soil-adjusted vegetation index (SAVI)	$SAVI = (1 + 0.16) \frac{NIR - R}{NIR + R + 0.16}$	Liang et al. [3]
Simple ration index (SRI)	$SRI = \frac{NIR}{R}$	Liang et al. [3]

2.4 Methods and Inversion Strategy

Random forest (RF) algorithm was used in modeling because it has presented the good performance in classification and parameter retrieval [3, 18, 19]. Three validation measurements were calculated to evaluate the performance of RF models: mean absolute prediction error (MAE), root mean squared error (RMSE), and squared Pearson correlation (R^2). These indicators are calculated as follows:

$$MAE = \frac{1}{n} \sum_{i=1}^{n} |Pi - Oi| \qquad (1)$$

$$RMSE = \sqrt{\frac{1}{n} \sum_{i=1}^{n} (Pi - Oi)^2} \qquad (2)$$

$$R^2 = \frac{\sum_{i=1}^{n} (p_{i-\bar{o}})^2}{\sum_{i=1}^{n} (o_i - \bar{o})^2} \qquad (3)$$

Where p_i and O_i are the predicted and observed site i, respectively; n is the number of samples; σ_p variances of predicted and observed values; and R2 correlation coefficient between the predicted and observed values. MAE measures the average prediction bias, and RMSE represents the overall quality of the prediction. Prediction becomes increasingly optimal as MAE and RMSE approach zero.

3 Results and Discussion

3.1 Comparison Between HJ-1A/CCD1 and GF-1/WFV1 Data

The LAI estimation results by HJ-1A/CCD1 image and GF-1/WFV1 image were compared in study area. Relevant field LAI measurements (n = 24) were used to validate the inversion result by the two sensors. GF-1/WFV1 data was more sensitive to

Table 4. The validation of RF **LAI** estimation models by field experiment data

Data	r	R^2	MAE	RMSE
GF-1/WFV1 + HJ-1A/CCD1	0.60	0.36	0.496	0.613
GF-1/WFV1	0.27	0.07	0.59	0.740
HJ-1A/CCD1	0.05	0.00	0.639	0.758

LAI than HJ-1A/CCD1 data in winter wheat LAI estimation (Table 4). In validation process, the Pearson correlation coefficient between field LAI and predicted LAI by RF model by GF-1/WFV1 data was 0.27. While which calculated by HJ-1A/CCD1 data was 0.05. The combination of GF-1/WFV1 and HJ-1A/CCD1 data can improve the LAI estimation accuracy. The Pearson correlation coefficient between filed LAI and predicted LAI by RF model was 0.60, with the determination coefficient of 0.36, MAE of 0.37, and RMSE of 0.61.

The validation results by MODIS LAI product was provided in Table 5. For HJ-1A/CCD1 data, the Pearson correlation coefficient was 0.39 between MODIS LAI and the estimation data by RF model (number of sample pairs = 481) with the MAE of 1.01, and RMSE of 2.65. For GF-1/WFV1, the validation result between MODIS LAI

Table 5. The validation of RF LAI estimation models by MODIS LAI product

Data	r	R^2	MAE	RMSE
HJ-1A/CCD1	0.39	0.15	1.01	2.65
GF-1/WFV1	0.67	0.45	0.82	2.16
GF-1/WFV1 + HJ-1A/CCD1	0.69	0.48	1.05	2.72

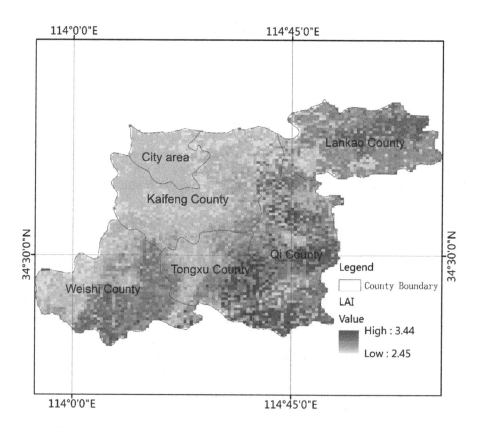

Fig. 2. The inversion LAI **value** by the combination of GF-1/WFV1 and HJ-1A/CCD1 data

and the estimation data by RF model presented the better performance than HJ-1A/CCD1 data with the r of 0.67, R2 of 0.45, MAE of 0.82, and RMSE of 2.16 (number of sample pairs = 490). For the combination of GF-1/WFV1 and HJ-1A/CCD1 data, the r, R2, MAE, and RMSE were 0.69, 0.48, 1.05, and 2.72, respectively (number of sample pairs = 6124, Fig. 2).

3.2 Discussion

The current study focused on retrieving winter wheat LAI using the multi-source and multi-angular image dataset combined by HJ-1A/CCD1 image and GF-1/WFV1 data. The validation results by the field experiment data and MODIS LAI product proved that GF-1/WFV1 can produce a better LAI retrieval result than HJ-1A/CCD1 data. And the combination of two sensors provided the better LAI estimation than that by the single-source data. The accuracy of LAI retrieval was limited by the following reasons. Firstly, the quality of remote sensing data was an important factor which influenced the estimation accuracy in quantitative retrieval [13, 20]. The accuracy in geometric calibration of GF-1/WFV1 data was higher than HJ-1A/CCD1 data because the former had the finer spatial resolution. Meanwhile, the HJ-1A/CCD1 image with larger width is more easily affected by the cloud. Secondly, the angle information of remote sensing data was not optimal. More different angular and source images can be used to improve the accuracy of LAI estimation [5, 13, 15]. In this study, only two different images were collected due to the cloud weather. Thirdly, during the LAI validation, we compared the LAI values between the predicted result and measured result or MODIS LAI product at the same geographic coordinate. The inconsistency of spatial scale between the measured LAI and image retrieval result also made the effect on LAI validation [21]. Last, the modeling method was also an important part on the LAI inversion. The applicability will be discussed in the further study.

4 Conclusions

The current study investigated the applicability of remote sensing image in estimating the winter wheat LAI by Chinese Satellite data. The following conclusions can be drawn:

(1) GF-1/WFV1 is more effective in the LAI inversion of Winter wheat than HJ-1A/CCD1.
(2) The combination of GF-1/WFV1 and HJ-1A/CCD1 images can improve the performance of RF LAI estimation model based on the four spectral indices in winter wheat LAI estimation.

References

1. Tillack, A., Clasen, A., Kleinschmit, B., Förster, M.: Estimation of the seasonal leaf area index in an alluvial forest using high-resolution satellite-based vegetation indices. Remote Sens. Environ. **141**(141), 52–63 (2014)

2. Atzberger, C., Darvishzadeh, R., Immitzer, M., Schlerf, M., Skidmore, A., Maire, G.L.: Comparative analysis of different retrieval methods for mapping grassland leaf area index using airborne imaging spectroscopy. Int. J. Appl. Earth Obs. Geoinf. **43**, 19–31 (2015)
3. Liang, L., Di, L., Zhang, L., Deng, M., Qin, Z., Zhao, S., Lin, H.: Estimation of crop LAI using hyperspectral vegetation indices and a hybrid inversion method. Remote Sens. Environ. **165**, 123–134 (2015)
4. Gray, J., Song, C.: Mapping leaf area index using spatial, spectral, and temporal information from multiple sensors. Remote Sens. Environ. **119**, 173–183 (2012)
5. Zhao, J., Li, J., Liu, Q.H., Fan, W.J., Zeng, Y.L., Xu, B.D., Yin, G.F.: Leaf area index inversion combining with HJ-1/CCD and Landsat 8/OLI data in the middle reach of the Heihe river basin. J. Remote Sens. **19**(5), 733–749 (2015)
6. Viña, A., Gitelson, A.A., Nguy-Robertson, A.L., Peng, Y.: Comparison of different vegetation indices for the remote assessment of green leaf area index of crops. Remote Sens. Environ. **115**(12), 3468–3478 (2011)
7. Yingshi, Z.: Principles and Methods of Remote Sensing Application Analysis. Science Press, Beijing (2013)
8. Liu, Q., Liang, S., Xiao, Z., Fang, H.: Retrieval of leaf area index using temporal, spectral, and angular information from multiple satellite data. Remote Sens. Environ. **145**, 25–37 (2014)
9. Pisek, J., Ryu, Y., Sprintsin, M., He, L., Oliphant, A.J., Korhonen, L., Kuusk, J., Kuusk, A., Bergstrom, R., Verrelst, J., Alikas, K.: Retrieving vegetation clumping index from Multi-angle Imaging Spectro Radiometer (MISR) data at 275 m resolution. Remote Sens. Environ. **138**, 126–133 (2013)
10. Mousivand, A., Menenti, M., Gorte, B., Verhoef, W.: Multi-temporal, multi-sensor retrieval of terrestrial vegetation properties from spectral-directional radiometric data. Remote Sens. Environ. **158**, 311–330 (2015)
11. De Moura, Y.M., et al.: Seasonality and drought effects of Amazonian forests observed from multi-angle satellite data. Remote Sens. Environ. **171**, 278–290 (2015)
12. Hall, F.G., Hilker, T., Coops, N.C.: Data assimilation of photosynthetic light-use efficiency using multi-angular satellite data: I. Model formulation. Remote Sens. Environ. **121**, 301–308 (2012)
13. Yang, G., Huang, W., Wang, J., Xing, Z.: Inversion of forest leaf area index calculated from multi-source and multi-angle remote sensing data. Chin. Bull. Bot. **45**(5), 566–578 (2010). (In Chinese with English abstract)
14. Li, X., Zhang, Y., Luo, J., Jin, X., Xu, Y., Yang, W.: Quantification winter wheat LAI with HJ-1CCD image features over multiple growing seasons. Int. J. Appl. Earth Obs. Geoinf. **44**, 104–112 (2016)
15. He, L., Song, X., Feng, W., Guo, B.-B., Zhang, Y.-S., Wang, Y.-H., Wang, C.-Y., Guo, T.-C.: Improved remote sensing of leaf nitrogen concentration in winter wheat using multi-angular hyperspectral data. Remote Sens. Environ. **174**, 122–133 (2016)
16. Xiao, L., Feng, M., Yang, W., Ding, G.: Estimation of water content in winter wheat (Triticum aestivum l.) and soil based on remote sensing data-vegetation index. Commun. Soil Sci. Plant Anal. **46**(14), 1827–1839 (2015)
17. Chen, X., Meng, J., Du, X., Zhang, F., Zhang, M., Wu, B.: The monitoring of the winter wheat leaf area index based HJ-1 CCD dada. Remote Sens. Land Resour. **25**(2), 55–57 (2010)
18. Wang, L., Zhou, X., Zhu, X., Guo, W.: Inverting wheat leaf area index based on HJ-CCD remote sensing data and random forest algorithm. Trans. Chin. Soc. Agric. Eng. (Trans. CSAE) **32**(3), 149–154 (2016). (In Chinese with English abstract)

19. Wang, Y., Qi, Y., Chen, Y., Xie, F.: Prediction of soil organic matter based on multi-resolution remote sensing data and random forest algorithm. Acta Pedol. Sin. **53**(2), 342–353 (2016). (In Chinese with English abstract)

20. Reulke, R., Säuberlich, T.: Image quality of optical remote sensing data. In: Proceeding of SPIE 9250, Electro-Optical Remote Sensing, Photonic Technologies, and Applications VIII; and Military Applications in Hyperspectral Imaging and High Spatial Resolution Sensing II, p. 92500L, 13 October 2014. doi:10.1117/12.2068538

21. Wu, H., Li, Z.-L.: Scale issues in remote sensing: a review on analysis. Process. Model. Sens. **9**(3), 1768–1793 (2009)

Assessment of Wavelet Base Based on Analytic Hierarchy Process in Remote Sensing Image De-noising

Yongmei Zhai[1], Shenglong Chen[2(✉)], Fuzhen Wang[3], and Qi Zhao[4]

[1] Shanghai Institute of Disaster Prevention and Relief,
Tongji University, Shanghai 20092, China
zymww@tongji.edu.cn
[2] College of Civil Engineering, Tongji University, Shanghai 200092, China
zensenlon@yeah.net
[3] East China Electric Power Design Institute, Shanghai 200063 China
wfzcqu@126.com
[4] Design Institute of Xi'an University of Architecture and Technology,
Xi'an 710055, China
cvzhao@126.com

Abstract. In the image de-noising based on wavelet transform, the selection of wavelet base remarkably influences the result. In the paper, a new quantitative method using analytic hierarchy process (AHP) to assess wavelet base was introduced. Seven indexes were selected to assess the result of image de-noising; a relative importance judgement-matrix was built and weights of each indexes were quantitatively analyzed based on AHP; the basic frame to assess wavelet bases was built. Living examples show that the method proposed is subjective, convenient and practical which can select the optimal wavelet base intuitively and assess the selection of wavelet bases effectively. The method conquers the shortcomings of traditional methods that are determined depending on experiences and lack of subjective quantitative assessment, which is of practical significance.

Keywords: Analytic hierarchy process (AHP) · Image de-noising · Wavelet base · Decomposition level · Comprehensive evaluation

1 Introduction

With the development of remote sensing technology, the information extraction based on High Resolution Remote Sensing Image (HRSS) was widely applied in building seismic hazard prediction which generally de-noises images firstly and then obtains attribute information extracted from marginal information of buildings. The essence of remote sensing image de-noising is the process to restrain the useless components in

Foundation Support: The National Natural Science Foundation of China, No. 51178351.

H. Yuan et al. (Eds.): GRMSE 2016, Part I, CCIS 698, pp. 310–320, 2017.
DOI: 10.1007/978-981-10-3966-9_35

signal, strengthen useful components and rebuild original signal. Traditional de-noising methods commonly pass noise-interfered images through a filter to remove the frequency components of noise. This kind of method might be effective but has its limitation to pulse signal, white noise or non-stationary process signal [1].

Applying wavelet analysis to de-noise can achieve good results and retain major details and background information [2]. Johnstone and Donoho proposed the shrinking method of wavelet coefficient in 1994 and derived the VisuShrink Threshold Formula [3]. Kivanc, John, Xu and they proposed different de-noising method as well [4, 5]. De-noising by thresholding has become the most applied and researched method. The method suggests that the wavelet coefficient corresponding to signal contains important information with high amplitude but small amount. On the other hand, the wavelet coefficient of noise is of uniform distribution with low amplitude but big amount [6]. Based on this theory, wavelet coefficient can be traded off to reconstruct signals.

The distribution of wavelet coefficient is a decisive factor of noise removal results in de-noising by thresholding. Moreover, the selection of wavelet basis will influence the distribution of wavelet coefficient in some extent. Therefore, the effect of de-noising is affected directly by the quality of wavelet basis selection [7]. Zhao quantitatively analyzes the effect of wavelet basis selection to de-noising which is of significant influence [8]. In a wide range of wavelet basis function such as the Harr wavelet, DB wavelet and Mayer wavelet, the mathematical features of different wavelet basis vary widely [9]. Most of existing researches only proposed the general principle of wavelet basis selection [7, 10]. Generally, the higher the zero moment of wavelet basis function is, the smaller the size of compact support is which is of benefit to de-noising. If there are more textures of images, noisy image with high noise level will be de-noised by biorthogonal wavelet base generally, but in practical application the specific function should be selected depending on experiences. How to determine the specific wavelet basis function is still a problem, there is no one opinion and method at present. Its essence is a decision problem that which wavelet basis function to select will acquire the best effecting of de-noising. Based upon above, a wavelet basis selection method based on AHP is proposed to solve the problem in this paper.

2   Fundamental of AHP and Quantitative Analyzing Method

2.1   Fundamental of AHP

Analytic Hierarchy Process (AHP) was a kind of qualitative and quantitative analysis of multiplicative objective decision proposed by American operations researcher Saaty in 1970s. The main idea is that based on the problem's nature and general goal firstly decompose problems into varied factors according to level classification. Then the weight (degrees of importance) of each factors within the same level L can be get by judging between each factors. The weight of next level L+1 relies on not only itself but also the weight factor of previous level. Therefore, generally we should calculate the combination weight level by level until the last one. At final the relative importance between plans can be reflected by the weight of internal factors [11].

2.2   Quantitative Analyzing Method

Establishment of Hierarchical Structure Model. Firstly, break up the problem according to evaluation targets and divide the components into groups by attribute to form varied levels.

Establishment of Importance Judgment Matrix. Supposing that there are n factors $A_1, A_2, \ldots, A_n$. In order to assess their importance, factors can be compared one by one. For example, a_{ij} means the relative importance after comparing A_i and A_j. According the scale of the Table 1, if A_i is of equal importance as A_j, $a_{ij} = 1$; if A_i is a bit more important than A_j, $a_{ij} = 3$; otherwise, if A_j is a bit more important than A_i, $a_{ij} = 1/3$; if the importance of A_i to A_j is between equally importance and a bit more important, a_{ij} equals 2. In turn, all importance scale can be obtained. Obviously, $a_{ii} = 1$ and $a_{ij} = 1/a_{ii}$.

Table 1. Judgement of the importance scale [9]

Scale	Meaning
1	A_i is as important as A_j
3	A_i is a bit more important than A_j
5	A_i is more important than A_j
7	A_i is far more important than A_j
9	A_i is absolutely more important than A_j
2, 4, 6, 8	Median between two adjacent scale

Based on the method above, an n by n judgment matrix $A = (a_{ij})$ can be obtained after comparing all factors.

Weight Calculation. For a matrix **A**, firstly determine its largest eigenvalue $\lambda_{\mathbf{max}}$, then determine the corresponding standardized feature vector **W**.

$$A \cdot W = \lambda_{max} \cdot W. \tag{1}$$

The components of $W: w_1, w_2, \ldots, w_n$ is the corresponding importance of n factors, which are called weight or weight coefficient.

Consistency Check. In the establishment of judging matrix, the consistency of judgment is not asked for, because of the complexity of objective things and the diversity of cognition. But to avoid perverse results and logical error, general consistency of judging matrix is needed. Meanwhile if judgment deviate from the consistency too much, it will cause problems when making decision upon the computation of sequencing weight vector. So consistency should be checked when $\lambda_{\max}$ is get. The detailed procedure is as followed.

1. Calculate the coincidence index C.I.

$$\text{C.I.} = \frac{\lambda_{max} - n}{n - 1}. \tag{2}$$

2. Calculate the average random consistency index R.I.
   Repeat calculating the eigenvalues of random judging matrix, then take arithmetic average as the average random consistency index.
3. Calculate the consistency ratio C.R.

$$\text{C.R.} = \frac{\text{C.I.}}{\text{R.I.}}. \tag{3}$$

It is generally believed that if C.R. ≤ 0.1, the consistency of judging matrix is acceptable.

3 Quality Assessment Method of Image De-noising and Enhancement

Research on image de-noising and enhancement will inevitably involves the quality problem. The assessment of image quality is of importance in image processing so that only with reliable assessment method of quality, treatment effect, solution techniques and system performance are likely to be evaluated correctly.

At present there are two methods to assess the image quality witch are subjective means and objective means. Subject means evaluates image quality according to the feelings of judges. The advantages of the method are that it is a human-oriented visual system getting an all-sided evaluation. But the results are affected greatly by some uncontrollable factors such as respective preference and self-quality of judges. In addition, it is hard to measure the visual psychological factors correctly by certain physical quantities, leading to high discreteness. Objective means is based on the establishment of objective assessment model that can evaluate image quality automatically. The models usually evaluate image quality by extracting one or some measurement indexes of image quality [12] and can be described by mathematical expression. So objective means is of convenient calculation and clear results and it can make quantitative analysis, applicable to machine processing. To make quantitative and objective comparative analysis, objective methods are applied in this paper.

In order not to adopt single index to take a one-sided evaluation of image quality, through analysis and comparison, 7 index followed are selected to comprehensively evaluate image quality in the paper.

Peak signal noise ratio (PSNR):

$$\text{PSNR} = 10\lg\left\{max\left[f^2(i,j)\bigg/\left(\frac{1}{MN}\right)\sum_{i=0}^{M-1}\sum_{j=0}^{N-1}[f(i,j)-g(i,j)]^2\right]\right\}. \qquad (4)$$

PSNR expresses the degree of noise elimination, and is the most important assessment index of de-noising. Higher values mean better effect of de-noising.

Mean square error (MSE):

$$\text{MSE} = \frac{1}{MN}\sum_{i=0}^{M-1}\sum_{j=0}^{N-1}[f(i,j)-g(i,j)]^2. \qquad (5)$$

MSE reflects the difference of grayscale among images, and is an important assessment index of de-noising. Lower values mean grater approximating level of processed images to original images.

Mean brightness:

$$m = \sum_{i=0}^{L-1} z_i p(z_i). \qquad (6)$$

readme Mean brightness expresses the degree of overall brightness and directly influences vision feeling. Higher values mean that images are more convenient to read.

Average contrast:

$$\mu_2 = \sum_{i=0}^{L-1} (z_i - m)^2 p(z_i). \qquad (7)$$

Average contrast expresses the richness of detail information and influences the clear degree of images. Higher values mean that images are easier to distinguish.

Third moment:

$$\mu_3 = \sum_{i=0}^{L-1} (z_i - m)^3 p(z_i). \qquad (8)$$

Third moment is used to judge the distribution characteristics of histogram to confirm whether histogram is symmetric or left (right) deflected. Lower values mean more even of image grayscale, which means better contrast.

Consistency:

$$U = \sum_{i=0}^{L-1} p^2(Z_i). \qquad (9)$$

Consistency expresses the smoothness of images. Higher values mean vaguer images.

Entropy:

$$e = -\sum_{i=0}^{L-1} p(z_i)log_2 p(z_i). \tag{10}$$

Entropy expresses the information content of images. Lower values mean less change of image grayscale which means better readability.

Above the formula, $f(i, l)$ represents original images; $g(i, l)$ represents images after de-noise processing; z represents the random variable of grayscale and $p(z_i)$ represents the corresponding histogram of z.

4   Example of Wavelet Basis Function Assessment Based on AHP

4.1   Sample Selection

In the paper high resolution remote sensing images on the Quick Bird are taken as experiment samples. The 3 images selected are as Fig. 1 shown.

a) Sample 1   b) Sample 2   c) Sample 3

Fig. 1. Experimental samples

4.2   Wavelet Basis Function

Generally speaking, in order to make the wavelet coefficient of images sparse after wavelet transform, when selecting the wavelet used to analyze, orthogonality, zero moment and dimension of support need to be considered. Considering that the texture of the images selected in the paper are informative, the biorthogonal wavelet: Daubechies Wavelet, Coiflet Wavelet and Symlet Wavelet are selected to be processed.

4.3   Weight of Image Assessment Index

According to the relative importance of each image quality assessment indexes, the relative importance judgment matrix is built as the Table 2 shown. The largest eigenvalue of matrix is calculated $\lambda_{max} = 7.563$. The feature vector is solved:

$$W = (w_1, w_2, \cdots, w_3)^T = (0.121, 0.141, 0.131, 0.124, 0.156, 0.162, 0.164)^T.$$

The weights of each assessment indictors are as the Table 3 shown.

Table 2. Importance judgement matrix of image quality indexes

	Mean brightness	Average contrast	Third moment	Consistency	Entropy	MSE	PSNR
Mean brightness	1	1/3	1/3	1	1/7	1/9	1/9
Average contrast	3	1	3	3	1/5	1/7	1/7
Third moment	3	1/3	1	1	1/5	1/7	1/7
Consistency	1	1/3	1	1	1/5	1/7	1/7
Entropy	7	5	5	5	1	1/3	1/3
MSE	9	7	7	7	3	1	1
PSNR	9	7	3	7	3	1	1

Table 3. Weight of image quality assessment indictors

	Mean brightness	Average contrast	Third moment	Consistency	Entropy	MSE	PSNR
Weight	0.121	0.141	0.131	0.124	0.156	0.162	0.164

When calculating consistency indictor, C.I. $= \frac{\lambda_{max}-n}{n-1} = 0.0938$. Gong Musen and them solve the 7 order average random consistency index R.I. $= 1.36$ after repeating calculating for 1000 times [13]. Calculate the average random consistency ratio C.R. $= \frac{C.I.}{R.I.} = 0.068 < 1$, so the consistency of judgment matrix is acceptable.

4.4   Calculation of De-noising Indexes

Adopt hard threshold function, select global-unified threshold and calculate the de-noising indexes of images by MATLAB, as the Table 4 shown.

Table 4. De-noising indexes of sample one

Wavelet	Mean brightness	Average contrast	Third moment	Consistency	Entropy	MSE	PSNR
DB1	171.7246	89.4029	−5.2083	0.1974	5.3771	78.2620	29.1953
DB2	171.7245	89.1326	−5.0815	0.1971	5.3776	78.1895	29.1993
DB3	171.6548	89.2936	−5.1348	0.1972	5.3779	78.2206	29.1976
DB4	171.9834	88.9061	−5.0594	0.1971	5.3765	78.1802	29.1998
DB5	171.7359	89.1682	−5.1026	0.1971	5.3779	78.2034	29.1985
DB6	171.5845	89.0871	−5.0160	0.1969	5.3779	78.1570	29.2011
DB7	172.0605	88.8048	−5.0362	0.1971	5.3758	78.1669	29.2060
DB8	171.9664	89.0564	−5.1236	0.1973	5.3763	78.2155	29.1979
DB9	172.0123	88.9338	−5.0808	0.1972	5.3761	78.1910	29.1992
DB10	171.9342	88.9890	−5.0818	0.1972	5.3762	78.1950	29.1990
SYM1	171.7246	89.4029	−5.2083	0.1974	5.3771	78.2620	29.1953
SYM2	171.7245	89.1326	−5.0815	0.1971	5.3776	78.1895	29.1993
SYM3	171.6548	89.2936	−5.1348	0.1972	5.3779	78.2206	29.1976
SYM4	171.5660	89.4123	−5.1620	0.1972	5.3784	78.2379	29.1966
SYM5	171.7808	89.1420	−5.1037	0.1972	5.3769	78.2044	29.1985
SYM6	171.7885	89.0838	−5.0795	0.1970	5.3778	78.1934	29.1991
SYM7	171.8024	89.0753	−5.0802	0.1971	5.3774	78.1922	29.1992
SYM8	172.3131	88.7359	−5.0833	0.1975	5.3730	78.1950	29.1990
COIF1	172.0613	88.9538	−5.1055	0.1973	5.3754	78.2054	29.1984
COIF2	171.6321	89.2100	−5.0881	0.1971	5.3780	78.1951	29.1990
COIF3	171.8612	89.0116	−5.0685	0.1971	5.3769	78.1868	29.1995
COIF4	172.2282	88.7641	−5.0697	0.1974	5.3742	78.1883	29.1994
COIF5	172.0126	88.9697	−5.0977	0.1973	5.3759	78.2015	29.1987

4.5 Choiceness of Wavelet Basis Based on Comprehensive Scoring Method

Comprehensive scoring method is a kind of data analysis method that sets several objectives for different plans and makes comprehensive evaluation by grading. It not only has the advantage of comprehensive qualitative analysis, but also takes multi-indexes as metrics, defining comprehensive quantitative assessment. So comprehensive scoring method is widely applied in decision-making theory and techno-economic appraisal. The basic expression is as followed:

$$P = \sum_{i=1}^{n} w_i P_i. \tag{11}$$

Above the formula, P represents total points; w_i represents the weight of each index, P_i represents the points of each item index. The extremum of each column at the m line P_m (whether it is maximum or minimum is based on analysis) is taken as the full score of the assessment index (100). The scores of other values are as followed:

$$P_j = \frac{P_i}{P_m} \times 100 (i \neq m). \tag{12}$$

Calculate the score of each wavelet basis according to the Formula 12. In the paper homologous wavelets and non-homologous wavelets are compared respectively. The final results are as the Table 5 shown.

Table 5. Comparison on comprehensive score of each wavelets

Wavelet	Sample 1	Sample 2	Sample 3	Mean value	Variance	Standard deviation	Range
DB1	99.3767	99.4288	99.3988	99.4014	0.0014	0.0262	0.0521
DB2	99.2291	99.7362	99.4996	99.4883	0.1287	0.2537	0.5071
DB3	99.4005	99.6077	99.6161	99.5414	0.0298	0.1221	0.2156
DB4	99.4928	99.7806	99.8358	99.7031	0.0678	0.1842	0.3430
DB5	99.5446	99.6848	99.6865	99.6386	0.0133	0.0815	0.1419
DB6	99.6069	99.9078	99.7320	99.7489	0.0457	0.1512	0.3010
DB7	99.9094	99.8382	99.6995	99.8157	0.0228	0.1068	0.2100
DB8	99.7745	99.6197	99.5061	99.6334	0.0363	0.1347	0.2684
DB9	99.5578	99.7245	99.7070	99.6631	0.0168	0.0916	0.1667
DB10	99.7401	99.7239	99.8627	99.7756	0.0115	0.0759	0.1387
SYM1	99.0429	98.1326	99.3843	98.8533	0.8372	0.6470	1.2516
SYM2	99.0952	98.4102	99.4851	98.9968	0.5922	0.5442	1.0749
SYM3	99.0673	98.2944	99.6017	98.9878	0.8639	0.6572	1.3072
SYM4	99.1318	98.2402	99.7731	99.0484	1.1854	0.7699	1.5329
SYM5	99.2291	98.3568	99.8291	99.1383	1.0962	0.7403	1.4723
SYM6	99.1920	98.4164	99.6860	99.0981	0.8191	0.6400	1.2696
SYM7	99.1879	98.4096	99.7859	99.1278	0.9525	0.6901	1.3763
SYM8	99.2652	98.3715	99.5931	99.0766	0.7995	0.6323	1.2216
SYM1	99.0429	98.1326	99.3843	98.8533	0.8372	0.6470	1.2516
SYM2	99.0952	98.4102	99.4851	98.9968	0.5922	0.5442	1.0749
COIF1	99.1356	98.3772	99.6564	99.0564	0.8276	0.6433	1.2792
COIF2	99.0396	98.4351	99.5182	98.9976	0.5892	0.5428	1.0831
COIF3	99.1436	98.4702	99.6229	99.0789	0.6705	0.5790	1.1526
COIF4	99.2490	98.4426	99.6160	99.1025	0.7207	0.6003	1.1734
COIF5	99.2301	98.3937	99.5957	99.0732	0.7594	0.6162	1.2020
COIF1	99.1356	98.3772	99.6564	99.0564	0.8276	0.6433	1.2792

Through analyzing the results above, we reached the following conclusions:

1. The effect of de-noising varies with wavelet basis. Even if homologous wavelets, the effect of de-noising varied with support length and zero moment.
2. Generally higher order of zero moment means better effect of de-noising. But it doesn't mean that the higher the better. In the results above, the wavelet with the

highest score isn't the wavelet with the highest order, because higher order means larger support length which is bad for de-noising.

3. Among the common wavelet basis, the variance and range of DB wavelet is the smallest, which means stable computation and small discreteness.

4. In the comparison of homologous wavelet and non-homologous wavelet, the de-noising effect of DB7 is the best. So it can be concluded that it is the optimal wavelet basis among research samples.

4.6 Analysis of De-noising Effect Through Remoting Sense Images

The before-and-after De-noised images (of Sample 1) are as Fig. 2 shown. Apply the optimal DB7 wavelet basis get through the analysis above, based on the Neyman-Pearson threshold; apply the border detection based on wavelet theory; select 10 building to calculate the extracted roof area and compare with measured data. The result is as the Table 6 shown.

a) Original Image   b) De-noised Image   c) Border Detection

Fig. 2. Before-and-after de-noised images

Table 6. Comparison on calculated value and measured value

No. of buildings	Calculate area	Actual area	Error
1	268.45	252.21	−6.44%
2	857.53	840.84	−1.98%
3	671.25	655.58	−2.39%
4	647.31	636.82	−1.65%
5	504.87	484.22	−4.26%
6	417.87	398.9	−4.76%
7	339.54	312.05	−8.81%
8	432.16	402.61	−7.34%
9	942.31	881.58	−6.89%
10	847.32	805.29	−5.22%

Through analyzing the computation above, we can know that the error mean is 4.97%, the error variance is 0.005 and the error range is 7.16%, so that the extraction effect on floorage of DB7 wavelet get by AHP is good. The mean errors are below 5%, so that we can get a more accurate result which meets the project's requirement. That the method has the smallest variance and range suggests that the computational stability is good and is able to adopt different applications. It displays as: because images are de-noised effectively, the impact of noise to extraction result is reduced greatly and scattered point, lines and small holes decrease obviously; the detectable rate of false border is significantly lowered; 'excessive kill' doesn't appear during de-noising and edge-preserving works well; images are clear and visual effect is fine.

5 Conclusion

According to objective quality assessment indictors of image de-noising and enhancement, based on AHP theory, the optimal wavelet basis assessment system of wavelet de-noising with AHP is established through introducing comprehensive score method; a new method to quantitative evaluate the selection of wavelet basis is proposed and 3 samples are selected to make experiment. Experimental results show that the selection method of optimal wavelet basis based on AHP is effective where optimal wavelet basis can be selected intuitively and using the wavelet basis above, the accuracy of later calculations can meet the requirement of practical construction, providing scientific and objective reference for practical operation, which is of certain directing significance.

References

1. Ni, L.: Wavelet Transform and Image Processing, pp. 5–14. University of Science and Technology of China Press, Hefei (2010)
2. Gong, X.: Digital Image De-noising Algorithm Research Based on Wavelet with Irregular Neighborhood. Tianjin University, Tianjin (2010)
3. Donoho, D.L., Johnstone, I.M.: Ideal spatial adaption by wavelet shrinkage. Biometrika 3, 425–455 (1994)
4. Pan, Q., Zhang, P., Dai, G.Z.: Two denoising methods by wavelet transform. IEEE Trans. Signal Process. 12, 3401–3406 (1999)
5. Xu, Y.S., Weaver, J.B., Healy, J.M.: Wavelet transform domain filters: a spatially selective noise filtration technique. IEEE Trans. Image Process. 6, 743–758 (1994)
6. Donoho, D.L.: De-noising by soft-thresholdings. IEEE Trans. Inf. Theory 41, 617–627 (1995)
7. Cai, D., Yi, X.: The selection of wavelet basis in image de-noising. J. Math. 25(2), 185–190 (2005)
8. Zhao, Q.: Research on the Application of Wavelet Analysis in the Building Extraction in RS Imaginary. Tongji University, Shanghai (2012)
9. Liang, X., He, J.: Wavelet Analysis. National Defense Industry Press, Beijing (2004)
10. Li, X., Zhu, S.: Survey of wavelet domain image de-noising. China J. Image Graph. 11(9), 1201–1209 (2006)
11. Xu, S.: The Principle of AHP-Practical Decision Method, pp. 6–13. Tianjin University Press, Tianjin (1988)
12. Zhu, Z.: Research on Image Enhancement. National University of Defense Technology, Changsha (2009)
13. Li, S.: The Principle and Technology of Decision Support System, pp. 311–316. Beijing Institute of Technology Press, Beijing (1996)

Estimation of Fishing Vessel Numbers Close to the Terminator in the Pacific Northwest Using OLS/DMSP Data

Tianfei Cheng[1], Weifeng Zhou[1(✉)], Hongyun Xu[1,2], and Wei Fan[1]

[1] Key Laboratory of Fishery Resources Remote Sensing and Information Technology, East China Sea Fisheries Research Institute, Chinese Academy of Fishery Sciences, Shanghai 200090, China
zhou_wf@hotmail.com
[2] College of Marine Sciences, Shanghai Ocean University, Shanghai 201306, China

Abstract. Squid (*Ommastrephes bartramii*) is a kind of economic fishery resources. Artificial light at night attracts and aggregates fish and squid because it mimics light produced by bioluminescent marine animals. The fish men use lamps to aggregate squid for squid jigging. So the squid fishing vessels can be detected by the sensor of Operational Linescan System (OLS) on Defense Meteorological Satellite Program (DMSP). This article mainly used OLS shimmer data from DMSP-F18 satellite covering the Pacific northwest ocean and estimated the quantity of squid jigging fishing vessels close to the terminator to expound the method that people can utilize the night-time satellite remote sensing data to estimate the quantity of light-trapping fishing vessels.

Keywords: OLS/DMSP · Squid · The Pacific northwest · Fishing vessels monitoring

1 Introduction

Nearly half of the aquatic products human require in the world mainly come from ocean fishing. Taking effective monitoring, control and supervision on fishing vessels and fishing activities is conductive to regulating the operation of fishing vessels and protecting fishery resources and ecological environment. Squid (*Ommastrephes bartramii*) is the ocean important cephalopod of economic value, mainly in temperate and subtropical waters of the three oceans, currently concentrated in commercial exploitation of waters in the Pacific Northwest. The fish men used lamps to lure Squid aggregation for squid jigging. Major fishing countries and regions are the Chinese mainland, China's Taiwan Province and Japan.

In general, satellite sensors mainly obtain the solar radiation signal of surface reflection. While the night-time satellite sensors use photomultiplier and utilize the strong ability of photoelectric magnification, so these night-time satellite sensors can detect radiation lower than visible channel as usual. Thus night-time satellite sensors can work and collect radiation signal produced by night lights, firelight and so on.

© Springer Nature Singapore Pte Ltd. 2017
H. Yuan et al. (Eds.): GRMSE 2016, Part I, CCIS 698, pp. 321–327, 2017.
DOI: 10.1007/978-981-10-3966-9_36

Light-trapping fishing way is a kind of important fishing way of Chinese pelagic fishery, mainly used in cephalopod fishing. For instance, China carried out catching of squid in the Pacific northwest Ocean, and light falling-net fishing has developed rapidly in recent years in the area of the South China Sea. Light-trapping fishing is a fishing method which makes use of cephalopods' phototaxis using fish lamp light, cooperates with fishing tools such as automatic line machine to trap. Light-trapping fishing way is generally used in the evening. The working way is static fishhook fishing or falling-net fishing and the fishing vessels don't produce wake current. So, visible spectral remote sensing can't monitor the fishing vessels and wake current in the evening. Therefore, it is important to find a new way that people can conduct remote sensing monitoring effectively and estimate the number of regional fishing vessels in view of static fishing vessels which is using light to trapping and fishing.

2 Material

This article mainly used OLS shimmer data of from DMSP-F18 satellite covering the Pacific northwest Ocean and estimated the quantity of squid jigging fishing vessels to expound the method that people can utilize the night-time satellite remote sensing data to estimate the quantity of light-trapping fishing vessels.

The United Stated Air Force (USAF) Defense Meteorological Satellite Program (DMSP) controls a series of satellites which equip the Operational Linescan System (OLS) that has a unique capability to monitor stable nighttime lights (NTL) emitted from the Earth's surface, including human settlement, gas flares and fishing vessels. By analyzing a time series of the NTL images, it is possible to provide a consistent measure for featuring trends of disorderly targets from regional to global scales. Therefore, the NTL dataset has been widely used in previous studies such as urbanization dynamics [1, 2], population density [3] as well as fishing vessels monitoring [4–7]. The DMSP-OLS has a ground swath of about 3000 km. It has two broad spectral bands, one covering the visible-near infrared band (0.5–0.9 µm) and the other is in the thermal infrared band (approximately 10 µm). The visible band images are composed of grid-based digital numbers (DN) values ranging from 0 (no radiance) to 63 (saturated radiance) with spatial resolution of approximately 1 km.

3 Method

Figure 1 shows the working flow of the whole work. Figure 2 is the raw image of OLS/DMSP on 09/01/2011 used in this paper.

3.1 Radiometric Calibration

For the visible light band, formula (1) can convert digital gray value, DN, into the observed radiant value, Radiance. The word of gain means the gain parameter of data in formula (1).

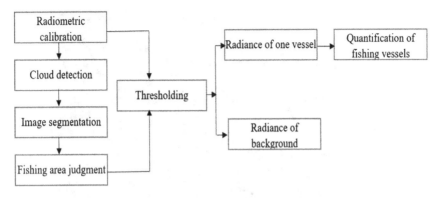

Fig. 1. The working flow of the whole work

$$Radiance = DN \times \frac{5.3}{63} \times 10^{-(6+\frac{gain}{20})} \tag{1}$$

3.2 Cloud Detection

Cloud is visible aggregate which is composed by icy particles or water droplets suspending in the atmosphere. If the area is covered by cloud, it is difficult for the satellite imaging equipment to obtain the information of underlying surface of the cloud sector, and then it would affect the extraction, interpretation and classification of image. Therefore, it is an important step for processing of remote sensing images that people remove the influence of the cloud and mist through cloud detection to extract the surface features information more accurately.

Because the OLS data selected in this example only has 11um of the thermal infrared bands and the surface feature of underlying surface is large bodies of water whose type is unitary, people can use ISCCP (International Satellite Cloud Climatology Project) method to carry out cloud detection.

3.3 Image Segmentation and Fishing Area Judgment

The radiance value of OLS image's background is not homogeneous. The contrast is poor and the boundary isn't clear due to the influence of solar flare, atmospheric transmission, noise and some environment factors. Thus we should carry out image segmentation and adopt local threshold method or adaptive threshold method. A group of thresholds which are related to pixel position are used to divide all parts of the image and distinguish whether these parts are fishing area or not respectively (Fig. 3).

Fig. 2. The raw image of OLS/DMSP on 09/01/2011

3.4   The Estimation on the Quantity of Fishing Vessels

Owing to squid jigging fishing vessels on the date of September 1st 2011 covering the Pacific northwest Ocean most are refitted from 8154-type trawlers with the light-trapping fishing technology, so the deployed light of each vessel is roughly similar. Thus it is reasonable to use formula (2) to estimate the quantity of squid jigging fishing vessels:

$$n = \frac{R}{r} \tag{2}$$

The n means the quantity of squid jigging fishing vessels, the R means total radiance value without background noise in fishing area, the r represents total radiance value without background noise for single fishing vessel.

There is the information of total radiance value for single fishing vessel in 5–8 area, the statistical results are shown in Table 1.

According to maximum interclass variance and scatter distribution diagram, the block whose maximum interclass variance is higher or equal to 11.5681 is solar flare area, like block 1–1, 1–2, 1–3, 2–1, 2–2, 3–1, 3–2. If one block's maximum interclass variance is less than 11.5681 and the variance is higher than 0.3617, the block is fishing area, like block 3–3, 4–4, 4–5, 4–6, 5–1, 5–4, 5–6, 5–8. The other blocks are not fishing areas.

$$r = 3.737815 \times 7 + 4.669826 \times 8 + 5.918207 \times 8 = 110.868969 \tag{3}$$

The threshold by OTSU of block 5–8 is assumed for T_{5-8}, the threshold by OTSU of other blocks is assumed for T, and then radiance value for single fishing vessel of other blocks is:

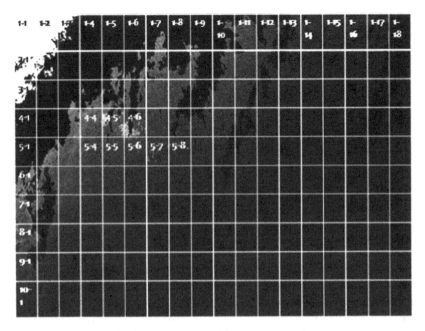

Fig. 3. The strategy of image segmentation

Table 1. Information of total radiance value for single fishing vessel

Radiance value	Quantities of pixel
3.737815	7
4.669826	8
5.918207	8

$$r = (3.737815 - (T_{5-8} - T)) \times 7 + (4.669826 - (T_{5-8} - T)) \times 8 + (5.918207 \\ - (T_{5-8} - T)) \times 8 \tag{4}$$

4 Result

The quantity of fishing vessels for each fishing area is calculated as shown in Table 2.

The number of fishing vessels for each fishing area is calculated is up to 134. Through the visual interpretation, the average estimation precision of fishing vessel numbers was 92.8%.

Table 2. The result of the **quantity** of fishing vessels for each fishing area

Blocks	Quantities of fishing vessels	Statistics via visual interpretation	Precision
3–3	0	0	100%
4–4	2	2	100%
4–5	44	38	86.4%
4–6	56	34	60.1%
5–1	0	0	100%
5–4	2	2	100%
5–6	21	20	95.2%
5–8	9	9	100%

5 Conclusion

It is not difficult to find that the method can conduct a large-area and synchronous remote sensing observation towards light-trapping fishing vessels and can estimate the quantity of fishing vessels quantificationally on a certain sea area on the basis of the characteristics of night-time working of light-trapping fishing way. The method also can provide third-party monitoring data of fishing vessels. Thus it not only can compensate for the lack of the data of Chinese current vessels position monitoring database to a certain extent, but it also is a supplement to the missing data about the quantity of other counties' light-trapping fishing vessels. This method can provide valuable reference for grasping accurately inherent law of fishery resources and environmental change and fishery forecast.

The method could be used to estimate the numbers of squid jigging vessels near the day/night terminator where the pixel is saturate. It could not only make up limitation of the existing database, but also make up the missing data of the other countries' squid jigging vessels operating in other seas, providing the reference for fish products market price trends and the fishing ground forecasting accurate.

Acknowledgments. This research is funded by National Natural Science Foundation of China (No. 31602206), Natural Science Foundation of Shanghai (16ZR1444700), The Key Technologies R&D Program of China (No. 2013BAD13B06), and Science and Technology Innovation Action Plan of Shanghai (15DZ1202201). Thanks are due to Kimberly Baugh of NOAA/National Geophysical Data Center for assistance with the data pre-processing.

References

1. Ma, T., Zhou, C., Pei, T., Haynie, S., Fan, J.: Quantitative estimation of urbanization dynamics using time series of DMSP/OLS nighttime light data: a comparative case study from china's cities. Remote Sens. Environ. **124**, 99–107 (2012)

2. Zhou, Y., Smith, S.J., Elvidge, C.D., Zhao, K., Thomson, A., Imhoff, M.: A cluster-based method to map urban area from DMSP/OLS nightlights. Remote Sens. Environ. **147**(18), 173–185 (2014)
3. Elvidge, C.D., Baugh, K.E., Dietz, J.B., Bland, T., Sutton, P.C., Kroehl, H.W.: Radiance calibration of DMSP-OLS low-light imaging data of human settlements. Remote Sens. Environ. **68**(1), 77–88 (1999)
4. Waluda, C.M., Griffiths, H.J., Rodhouse, P.G.: Remotely sensed spatial dynamics of the illex argentinus fishery, southwest Atlantic. Fish. Res. **91**(2–3), 196–202 (2008)
5. Cho, K., Ito, R., Shimoda, H., Sakata, T.: Technical note and cover Fishing fleet lights and sea surface temperature distribution observed by DMSP/OLS sensor. Int. J. Remote Sens. **20**(1), 3–9 (1999)
6. Waluda, C.M., Yamashiro, C., Elvidge, C.D., Hobson, V.R., Rodhouse, P.G.: Quantifying light-fishing for Dosidicus gigas in the eastern Pacific using satellite remote sensing. Remote Sens. Environ. **91**(2), 129–133 (2004)
7. Cho, K., Ito, R., Shimoda, H., Sakata, T.: Technical note and cover fishing fleet lights and sea surface temperature distribution observed by DMSP/OLS sensor. Int. J. Remote Sens. **20**, 3–9 (1999)

Similarities and Differences of Oceanic Primary Productivity Product Estimated by Three Models Based on MODIS for the Open South China Sea

Hongyun Xu[1,2], Weifeng Zhou[1(✉)], Anzhou Li[1,2], and Shijian Ji[1]

[1] Key Laboratory of Fishery Resources Remote Sensing and Information Technology, East China Sea Fisheries Research Institute, Chinese Academy of Fishery Sciences, Shanghai 200090, China
zhou_wf@hotmail.com
[2] College of Marine Sciences, Shanghai Ocean University, Shanghai 201306, China

Abstract. It's critical to carry out comparison among remotely sensed data products and choose appropriate oceanic primary productivity product to apply to evaluation of fishery resources, ocean carbon cycle and global change research. This paper mainly gives an intercomparison of three ocean net primary productivity (NPP) products for the open South China Sea (SCS) estimated by Standard VGPM, Eppley-VGPM and CbPM models from MODIS. Data preprocessing and intercomparison work including image reading, cropping, invalid values removing and calculating average were solved by MATLAB. There are significant differences among the annual ocean NPP averages estimated by the three models within one year. The differences of seasonal variation among the three products are different, the values estimated by Standard-VGPM are highest in winter and lowest in spring. The ocean NPP values estimated by the three models are all highest in 23°N. The NPP values are high in shallow offshore area and then decreasing with the increasing distance away from the coast, and the maximum and the minimum values of the three models are different. According to previous researches, NPP products estimated by Standard-VGPM compared with other two models are more suitable as the application data in the open SCS. In general, it's necessary to consider and improve the precision problem of the ocean NPP estimated by the algorithms based on inversion data from remote sensing.

Keywords: Intercomparison of products · Oceanic primary productivity · The open South China Sea

1 Introduction

The South China Sea (SCS) is rich in fishery resources as the largest epicontinental sea in China [1, 2], and there are still sovereignty disputes of some islands in the SCS. Therefore, it is one important way for maintaining sovereignty and ocean rights of the SCS by strengthening net primary productivity (NPP) research in the open SCS [3, 4].

© Springer Nature Singapore Pte Ltd. 2017
H. Yuan et al. (Eds.): GRMSE 2016, Part I, CCIS 698, pp. 328–336, 2017.
DOI: 10.1007/978-981-10-3966-9_37

And estimating oceanic primary productivity from satellite data becomes more and more significant for areas with large spatio-temporal scale [5]. While there are some unknown problems that if there are differences or how about the goodness of fit among products generated by different sensors, inversion algorithms or estimation models [6], which must be considered before choosing which kinds of NPP products as the open SCS's research materials. Because of existing some differences between inversion algorithms of different sensors, it's difficult to make comparative studies among NPP results estimated by different models and data sources. Thus it is necessary to do an intercomparison work among NPP products estimated by different kinds of models based on same remote sensing inversion data in the open SCS for understanding comprehensively differences of these products as well as the causes of these differences and choosing the appropriate product as research data.

Although the estimation works of NPP in different sea areas were done by multiple kinds of models from remote sensing, and homologous field validation studies were accomplished [7–11]. For instance, ocean primary production estimated by twelve algorithms from surface chlorophyll, temperature and irradiance was compared with daily production derived by ^{14}C uptake measurements in the second Primary Productivity Algorithm Round Robin (PPARR) activity, and the opinion that the performance of different models were independent of the algorithms' complexity was presented [7]. Then Scholars had done a comparison of global estimates of marine primary production from ocean color, which showed that current models were challenged by some special factors [8]. But these field validation data are limited to the coverage in space and time [6], which can't provide a criterion of magnitude for the open SCS's NPP. These previous studies showed that the differences among estimation of NPP existed in their studying area, and there is no complete comparative research of NPP products in the open SCS.

For this reason, this paper selects NPP products estimated by standard Vertically Generalized Production Model (Standard VGPM), an "Eppley" version of the VGPM (Eppley-VGPM) and Carbon-based Production Model (CbPM) based on MODIS inversion data as the research materials. The Standard VGPM just is the original VGPM, which is a common chlorophyll-based algorithm used to estimate regional and global ocean NPP [12]. Eppley-VGPM employs basic model structure and parameterization of VGPM. The only difference is that the Standard VGPM uses polynomial relationship, which is replaced with the exponential relationship described by Morel and Eppley for Eppley-VGPM to describe temperature-dependent photosynthetic efficiencies [12–14]. The CbPM is a carbon-based model, which estimates NPP by retrieving the satellite chlorophyll and the particulate backscattering coefficients to estimate phytoplankton carbon biomass for ocean NPP estimation [15, 16].

This paper aims at doing an intercomparison for the existing ocean NPP products estimated by the three models from MODIS data, and doing some analysis of their differences and the cause of the differences in the open SCS in order to improve the ocean NPP estimation algorithms and provide a reference for the data filtering and application.

2 Data and Methods

2.1 Data

The global ocean NPP data products estimated based on MODIS are from the website of the Oregon State University College of Science. The time span of data is from 2011 to 2014, and the spatial resolution is 1/6°. Here we employed 8 days and monthly average ocean NPP data.

2.2 Methods

Data Preprocessing. The global NPP data with HDF format were preprocessed including reading, image cropping, invalid values removing and calculating average in MATLAB, which were cropped in accordance with the scope of the studying area (105°–120°E, 5–25°N). The following text is the program code to preprocess the NPP product for one model:

```
Program of data preprocessing
for year=2011:2014;
for count=1:46;
    filename=[the path of the stored data\NPP product
files name. hdf'];
    npp(:,:,count) = hdfread(filename,'npp');
end
npp(npp==-9999)=nan;
npp_mean(:,:,year-2010)=nanmean(npp,3);

SCS_npp(:,:,year-2010)=npp_mean(390:510,1710:1800,year-20
10);
    end
```

Data Intercomparison. We used MATLAB to calculate averages per 8 day within one year for the three models products. The work of counting monthly averages of NPP for 4 years also were accomplished in MATLAB. The annual variation, the average differences in latitude and seasonal differences of the ocean NPP in the open SCS were analyzed by diagram in Excel [6]. Distributes comparison for annual high NPP and low NPP in open SCS were presented by ArcGIS.

3 Main Results

3.1 Annual Variation

The annual Variation differences of ocean NPP data products estimated by Standard VGPM, Eppley-VGPM and CbPM are very large for 4 years' average (Fig. 1).

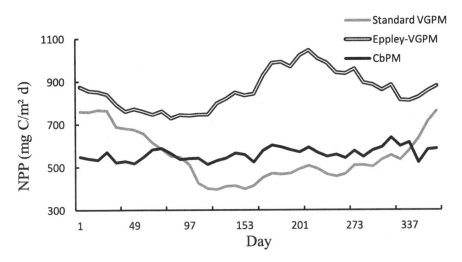

Fig. 1. The annual variation values of NPP average were compared in this figure for standard VGPM, Eppley-VGPM and CbPM models from 2011 to 2014.

The annual variation fluctuations range of ocean NPP products estimated by Standard VGPM and Eppley-VGPM are very large, while the estimations of CbPM within one year all distribute in 500–600 mg C/m^2 d. And the annual variation values of NPP estimated by Eppley-VGPM in the open SCS are very large compared with other two models results, but the trend of annual variation estimated by Eppley-VGPM in one year is almost keeping with the fluctuation of the annual variation estimated by CbPM. The annual averages of ocean NPP for the three models are 540.93 mg C/m^2 d, 865.57 mg C/m^2 d and 560.80 mg C/m^2 d respectively in the open SCS.

3.2 Seasonal Variation

The seasonal distributions calculated by the monthly NPP averages variations are different (Fig. 2). The NPP average values estimated by Standard VGPM are highest in winter (from December to February), reduce gradually in the spring (from March to May), start to rise in the summer (from June to August), drop in September, and then rise gradually during the month of 10–11. The seasonal distribution of the ocean NPP can be found from the fluctuation curves of the monthly NPP averages estimated by Eppley-VGPM in open SCS, the values are highest in summer and lowest in spring. While the NPP products values which are estimated by CbPM can't reflect significant changes at four seasons within one year. The monthly NPP average values have great deal of inconsistency estimated by the three models. The seasonal values of the ocean NPP estimated by Eppley-VGPM are also higher than other two models' estimations in the open SCS.

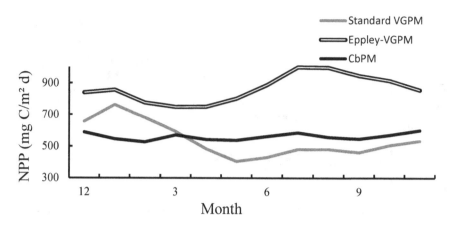

Fig. 2. This shows a figure consisting of different seasonal NPP average for standard VGPM, Eppley-VGPM and CbPM models from 2011 to 2014.

3.3 Distribution in Latitude

It can be seen in Fig. 3 that the trends of the ocean NPP estimated by three models are similar changing with latitude, especially the Standard VGPM and Eppley-VGPM. The largest difference of results estimated by CbPM compared with the other two models is that the values appear declining trend in the 25°N except differences in values. The NPP values are near peak both in 23°N for three models, because there are Leizhou Peninsula and the north continental shelf sea area in this latitude. The estimated ocean NPP values by Eppley-VGPM in latitude are higher than those by other two models.

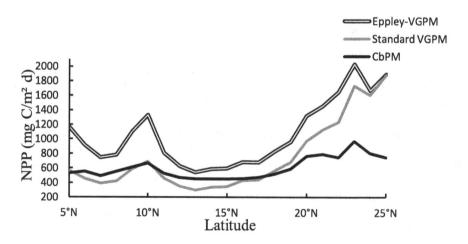

Fig. 3. The figure shows distribution of NPP for standard VGPM, Eppley-VGPM and CbPM models in different latitudes.

3.4   Spatial Distribution for High NPP and Low NPP

The estimations of the ocean NPP all can reflect the rule that the NPP is high in shallow offshore area and then decreasing with the increasing distance away from the coast (Fig. 4). The high NPP values appear in Pearl River Estuary, the shallow offshore of Leizhou Peninsula to Qiongzhou Strait, Gulf of Beibu and the north continental shelf sea area, the low values areas are mainly distributed in the deep water areas that are in the south of the north continental shelf sea area in the SCS. The ocean NPP values estimated by Standard VGPM, Eppley-VGPM and CbPM are concentrated in the range of 200–500 mg C/m^2 d, 500–700 mg C/m^2 d and 400–500 mg C/m^2 d respectively. The maximum values are 3800.78 mg C/m^2 d, 6417.49 mg C/m^2 d and 1829.64 mg C/m^2 d, and the minimum values are 252.04 mg C/m^2 d, 488.41 mg C/m^2 d and 377.69 mg C/m^2 d for the three models respectively.

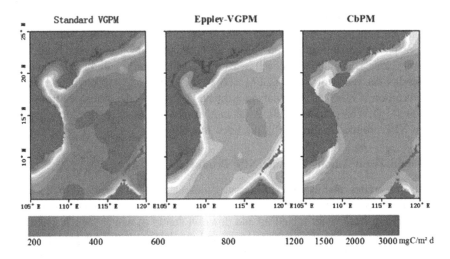

Fig. 4. The figure exported by ArcGIS is about spatial distribution of high NPP and low NPP for standard VGPM, Eppley-VGPM and CbPM models.

4   Discussion

4.1   Causes of Differences

The results above are obtained based on 4 years' average, but the differences are still very large. These differences show that the goodness of fit among products estimated by the three models in the open SCS is low. Complexity of the SCS's ocean environment and diversities of algorithms are two main reasons for these results [5, 17].

4.2   Complexity of the SCS's Ocean Environment

The SCS bounded by continent, gulf and some continental shelf, the largest marginal sea of China, is vast and rich in natural resources. Because of influence of the East

Asian Monsoon and the Kuroshio Current, the upper ocean circulation of the SCS follows the change of monsoons [17, 18]. There are some complicated physical phenomena in the SCS, such as mesoscale eddies, coastal upwelling, typhoon and Runoff [17, 19]. These cause seasonal and spatial distribution of NPP significantly in the open SCS. Strong regionalism and interaction between various factors increase the difficulty of estimating the open SCS's NPP from remotely sensed data, which is one reason for the difference of NPP products in the open SCS.

4.3 Diversity of Algorithms

The estimated ocean NPP values in coastal areas are very high, especially estimated by the Standard VGPM and Eppley-VGPM, the maximum estimated by the two models are more than 3000 mg C/m^2 d. The cause of the result may be related to that the two models all use chlorophyll concentration data as the input data to estimate NPP [5]. There are some considerations for NPP values in coastal areas, two reasons: (1) on the one hand, coastal upwelling and river runoff can carry rich nutrients, promote the growth of phytoplankton and improve coastal NPP, (2) but on the other hand, river runoff can carry silt and make costal sea turbid, which can result in uncertainty of chlorophyll data precision. The CbPM adopts the ratio of chlorophyll and phytoplankton carbon (Chl: C) as input data to estimate NPP, this is the difference from other two models. Diversity of algorithms is another reason, lead to the inconsistency between NPP products estimated by the three models.

4.4 Comparison Among Oceanic NPP Products

Previous researches mainly show different opinions in two aspects. Firstly, some research presented that the NPP in the coastal area is lower than the deep basin in winter [20], because inshore water mixed up and down violently in winter, which makes the mixing layer greater than the euphotic layer, and this is not helpful for phytoplankton photosynthesis [20]. While most researchers support the opinion that the NPP of coastal areas are higher than the deep basin within one year [17, 18], which is consistent with the result in this paper. Secondly, there are different understanding about seasonal variation of the SCS's NPP. The opinion that the NPP is highest in summer, and lowest in spring was presented in the research's paper [1]. While most researches showed that the values of NPP are highest in winter, and lowest in spring [17, 21, 22]. The NPP products estimated by Eppley-VGPM and CbPM can't reflect seasonal variation characteristics of the SCS. The seasonal changes of NPP products estimated by Standard-VGPM show consistence of the result of previous researches [17, 18]. What's more, the averages of the NPP are near the researchers' finding and in suit measurement [18, 23, 24], which is more suitable as the application data in the open SCS.

5 Conclusions

The differences among the estimated ocean NPP products based on inversion data from the same sensors are very large, the level of the ocean NPP estimation from remote sensing based on algorithms need to be improved in other word. The complexity of algorithms, parameterization process, parameters' precision, original input data and so on, all can be as the research directions in the further study. Thus there is a requirement for detailed comparison to analyze the causes of these differences and improve algorithms and parameterization setting.

Acknowledgements. This research was funded by the Key Technologies R&D Program of China (No. 2013BAD13B06), National Natural Science Foundation of China (No. 31602206), National Natural Science Foundation of China (31602206), Natural Science Foundation of Shanghai (16ZR1444700), and Science and technology innovation action plan of Shanghai (15DZ1202201).

References

1. Qiu, Y.S., Zeng, X.G., Chen, T., et al.: Fisheries Resources and Management in the South China Sea. China Ocean Press, Beijing (2008)
2. Chen, Z.Z., Qiu, Y.S.: Status and sustainable utilization of fishery resources of South China Sea. J. Hubei Agric. Coll. **22**, 507–510 (2002)
3. Che, B., Xiong, T.: Impact of south China sea dispute on fishery and countermeasures. Res. Agric. Mod. **30**, 414–418 (2009)
4. Zhang, P., Yang, L., Zhang, X.F., Tang, Y.G.: The present status and prospect on exploitation of tuna and squid fishery resources in South China Sea. South China Fish. Sci. **6**, 68–74 (2010)
5. Lee, Z., Marra, J., Perry, M.J., et al.: Estimating oceanic primary productivity from ocean color remote sensing: a strategic assessment. J. Mar. Syst. **149**, 50–59 (2015)
6. Liao, Y., Lü, D., He, Q.: Intercomparison of albedo product retrieved from MODIS, MISR and POLDER. Remote Sens. Technol. Appl. **29**, 1008–1019 (2014)
7. Campbell, J.W., Antoine, D., Armstrong, R., et al.: Comparison of algorithms for estimating ocean primary production from surface chlorophyll, temperature, and irradiance. Glob. Biogeochem. Cycles **16**(3), X-1–X-15 (2002)
8. Carr, M.E., Friedrichs, M.A.M., Schmeltz, M., et al.: A comparison of global estimates of marine primary production from ocean color. Deep Sea Res. Part II Topical Stud. Oceanogr. **53**(5–7), 741–770 (2006)
9. Saba, V.S., Friedrichs, M.A.M., Mary-Elena, C., et al.: Challenges of modeling depth-integrated marine primary productivity over multiple decades: a case study at BATS and HOT. Glob. Biogeochem. Cycles **24**(3), 811–829 (2010)
10. Saba, V.S., Friedrichs, M.A.M., Antoine, D., et al.: An evaluation of ocean color model estimates of marine primary productivity in coastal and pelagic regions across the globe. Biogeosci. Discuss. **7**(5), 489–503 (2011)
11. Lee, Y.J., Matrai, P.A., Friedrichs, M.A.M., et al.: An assessment of phytoplankton primary productivity in the Arctic ocean from satellite ocean color/in situ chlorophyll-a-based models. J. Geophys. Res. Oceans **120**(9), 6508–6541 (2015)
12. Behrenfeld, M.J., Falkowski, P.G.: Photosynthetic rates derived from satellite-based chlorophyll concentration. Limnol. Oceanogr. **42**(1), 1–20 (1997)
13. Eppley, R.W.: Temperature and phytoplankton growth in the sea. Fish. Bull. **70**, 1063–1085 (1972)
14. Morel, A.: Light and marine photosynthesis: a spectral model with geochemical and climatological implications. Prog. Oceanogr. **26**(3), 263–306 (1991)

15. Behrenfeld, M.J., Boss, E., Siegel, D.A., et al.: Carbon-based ocean productivity and phytoplankton physiology from space. Glob. Biogeochem. Cycles **19**(1), 177–202 (2005)
16. Westberry, T., Behrenfeld, M.J., Siegel, D.A., et al.: Carbon-based primary productivity modeling with vertically resolved photoacclimation. Glob. Biogeochem. Cycles **22**(2), GB2024 (2008). doi:10.1029/2007GB003078
17. Liu, H.X., Song, X.Y., Huang, H.H., Tan, Y.H., Huang, L.M.: A comparison study on primary production in typical low-latitude seas (South China Sea and Bay of Bengal). Acta Ecol. Sin. **32**(18), 5900–5906 (2012)
18. Ning, X., Chai, F., Xue, H., Cai, Y., Liu, C., Shi, J.: Physical-biological oceanographic coupling influencing phytoplankton and primary production in the South China Sea. J. Geophys. Res. **109**, C10005 (2004). doi:10.1029/2004JC002365
19. Huang, Q.Z., Wang, W.Z., Li, Y.S., Li, C.W., Mao, M.: General situations of the current and eddy in the south China sea. Adv. Earth Sci. **7**(5), 1–9 (1992)
20. Le, F.F., Ning, X.R., Liu, C.G., Hao, Q., Cai, Y.M.: Standing stock and production of phytoplankton in the north south China sea during winter of 2006. Acta Ecol. Sin. **28**(11), 5775–5784 (2008)
21. Hao, Q., Ning, X.R., Liu, C.G., Cai, Y.M., Le, F.F.: Satellite and in situ observations of primary production in the northern south China sea. Acta Oceanol. Sin. **29**(3), 58–68 (2007)
22. Gao, S.: Spatial and temporal distribution of ocean primary productivity and its relation with oceanic environments in the South China sea based on remote sensing. Chin. Acad. Meteorol. Sci., Beijing (2008)
23. Chen, Y.L.L., Chen, H.Y.: Seasonal dynamics of primary and new production in the northern south China sea: the significance of river discharge and nutrient advection. Deep Sea Res. Part I: Oceanogr. Res. Papers **53**(6), 971–986 (2006)
24. Song, X.Y., Liu, H.X., Huang, L.M., Tan, Y.H., Ke, Z.X., Zhou, L.B.: Distribution characteristics of basic biological production and its influencing factors in the northern south China sea in summer. Acta Ecol. Sin. **30**(23), 6409–6417 (2010)

Hydrological Feature Extraction of the Tarim Basin Based on DEM in ArcGIS Environment

Yaping Wei[1,2], Jinglong Fan[1], and Xinwen Xu[1(✉)]

[1] Xinjiang Institute of Ecology and Geography Chinese Academy of Sciences,
Urumqi 830011, China
sms@ms.xjb.ac.cn
[2] University of Chinese Academy of Sciences, Beijing 100049, China

Abstract. The main water of Tarim Basin is from snowmelt and precipitation in the mountainous area by surface runoff into the basin. The extraction of hydrological information and analysis of hydrological characteristics was conducted based on DEM, through studying the basic principle of watershed extraction, and using ArcGIS hydrology toolbox in this paper. The results show that the river network in the Tarim River Basin extracted by ArcGIS is basically the same, compared with original nine rivers flowing into the mainstream.

Keywords: DEM · ArcGIS · Watershed extraction · Tarim Basin

1 Introduction

The Tarim Basin is a fully enclosed large inland arid basin, surrounded by alpine zone and plateau, precipitation is mainly through the surface runoff into the basin [1]. Owing to the restriction of drought climate, the most important factor in the development of social economy and ecological environment for Basin is water resource. The Hydrology Model in ArcGIS was developed by Environmental Systems Research Institute, Inc. (ESRI) for extracting and analyzing of terrain and river network, which has been used for studies by many researchers. Integrating technology of Hec-GeoHMS and ArcView GIS was used to assess SRTM data by Abdelkader Mendas, then the result as a basic hydrological model was kept at last [2]. Cochrane [3] developed manual, hillslope, and flow path methods to apply Water Erosion Prediction Project (WEPP) using geographical information systems (GIS) to describe and evaluate three methods of integrating WEPP and GIS, and Wu [4] elaborated the hydrological simulation issues caused by the uncertainty of digital elevation models (DEM). There are few researchers who use GIS technology with DEM to extract hydrological feature than plenty of works for the water resources of Tarim Basin, this paper can be a supplement for the aspect that extracting hydrological feature with DEM by ArcGIS.

The Tarim Basin is the largest inland basin in China, which is located in the south area of Xinjiang and wide flat with West high East low topographic feature, tilted to the East, so that being similar to a diamond, terrain elevation difference there reaches 7833 m with desert widespread and oasis is distributed mainly over edge. The northern part of the basin is the Tianshan Mountains, and the south are and Altun Mountains,

© Springer Nature Singapore Pte Ltd. 2017
H. Yuan et al. (Eds.): GRMSE 2016, Part I, CCIS 698, pp. 337–341, 2017.
DOI: 10.1007/978-981-10-3966-9_38

while the west is the Pamir Plateau and East is Hexi Corridor, an area of about 56 million square kilometers. Basin belongs to the temperate continental climate, daily and annual temperature range is large with winter cold and summer heat accompanied by minimal annual precipitation, always is below 200 mm. Taklimakan Desert in the central part of the basin is mostly like basin in shape, covering an area of about 33 million square kilometers, annual precipitation here is less than 50 mm. Water resource is from snowmelt and precipitation in the mountainous area by surface runoff into the basin, short seasonal rivers with few tributaries and small flow, zero-flow of the river occurring in winter while flood in summer, runoff volume is 3.925×10^{10} m^3/a. River runoff is closely related to the area of glaciers and climate, therefore, the distribution of runoff is not uniform in the year. Summer runoff accounted for over 50% of the whole year, there are 14 rivers in which the annual runoff is more than 5×10^8 m^3/a, the total annual runoff accounts for 76.6% of the basin.

In the Tarim Basin, the river is bounded by the mountain pass, over the mountain pass to form a recharge area, with the river extending, the area of the catchment area increases and the amount of water increases. Below Pass is runoff dissipated area, in this interval rivers into the Gobi, alluvial plains and deserts, a large number of evaporation and infiltration along the river reduce the amount of water in a river, even to the extent that river finally disappeared in the desert [5].

2   Methods and Analysis

ArcGIS hydrology analysis is an important aspect of the digital terrain analysis, the main contents of surface hydrological analysis based on DEM is using hydrological analytical tools to extract the flow direction and sink flow accumulation, flow length, river network, network grading and basin analysis, etc.

This paper is based on Hydrology model of ArcGIS10.0 by DEM of Tarim River drainage basin and Chinese water system diagram to extract hydrological characteristics of basin, meanwhile in accordance with the data of previous studies, the hydrological characteristics of the Tarim River Basin were analyzed.

2.1   Data Conversion

According to the Xinjiang statistical yearbook data and the major sites as well as the previous studies of the Tarim Basin Water Resources in Xinjiang, statistical data from 1959 to 2009 and Xinjiang SRTM250 m resolution data download from. In this section, we reorganizing statistical data and transform coordinate projection from WGS1984 to BJ54.

2.2   DEM Without Sink

There would be sinking when using hydrology tools in ArcGIS according to grid DEM characteristics extraction theory to extract hydrologic features on the basis of Runoff Model [6], due to the existence of DEM error and particular terrain (such as Karst landform). This phenomenon often appears in the upper basin, when the valley width is less than the width of the unit, the lower valley low elevation of the unit [7].

Flow analysis was exploited to determine the flow direction of each unit, through the direction of water flow, the use of hydrological analysis in the sink to determine the sink. As can be seen from the Fig. 1, sink in the basin edge value is larger, mainly concentrated in the upstream river. Zonal Statistic and Zonal Fill used for calculating regional lowest elevation and outlet elevation of each sink, Raster Calculator used to calculate the depth of the sink. Finally, filling sink by comprising with the practical terrain to find the proper threshold, the results after filling the DEM as shown in Fig. 2.

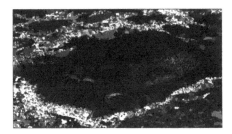 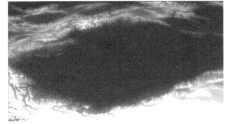

Fig. 1. Depth of sink   **Fig. 2.** DEM after filling sink

2.3 Drainage Network Extraction

In accordance with water from high flow to the natural law of the lowest. According to the regional topography in the direction of flow data calculates each point amount of water to gain the region flow accumulation which would be used for raster data vectorization by Stream to Feature (Fig. 3). Using Link Stream to segment the river and then grading of river network by Stream Order, different levels of river network represent different levels of accumulation, the higher the level, the greater the flow accumulation. In this paper, Strahler classification method of river network was used to grad the network.

2.4 Watershed Extraction

River basin is composed of watershed segmentation of catchment area. Through Basin command of Arc Hydro Tools in the Tools, determine the drainage Basin (Fig. 4) and the outlet (lowest point of catchment area), then combining with flow direction search out upstream all grids that through the outlet and find basin boundary, so that can determine watershed catchment area (Fig. 5).

It is easily can be found that when extracting river network and watershed, network extracted of Tarim Basin is consisting with the Nine River. But now, due to the human activities and climatic variation, there is the situation of four source streams.

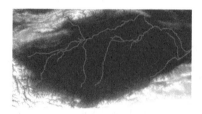

Fig. 3. River network

Fig. 4. Drainage Basin

Fig. 5. Catchment

Fig. 6. Water system of Basin

The basin area of "four sources and one trunk canal" is 2.503×10^5 km^2, with domestic area is 2.285×10^5 km^2 and foreign area is 2.18×10^5 km^2. The main conditions are shown in Table 1 (Fig. 6).

Table 1. "four sources and one trunk canal" in the Tarim Basin.

River	Length (km)	Drainage area ($\times 10^5$ km^2)		
		Basin	Mountain	Plain
Main Stream Basin of Tarim River	1321	1.76		1.76
Kaidu River-Kongque River Basin	560	4.96	3.30	1.66
Akesu River Basin	588	5.4(1.9)	3.8(1.9)	1.6
Yarkant River Basin	1165	7.98(0.28)	5.69(0.28)	2.29
Hotan River Basin	1127	4.93	3.80	1.13
Summation		25.03(2.18)	16.59(2.18)	8.44

Values in brackets represent the basin area in foreign countries

3   Conclusions

This paper obtains the three rivers basin area by ArcGIS hydrological analysis tool that Akesu River is 2.43×10^5 Km^2, Yarkant River is 7.84×10^5 Km^2 and Hotan River is 3.18×10^5 Km^2, which reflecting spatial distribution of water system in Tarim basin basically is consistent with statistical data (Table 1). The research findings can provide essential data for studies on eco-environmental evolution and offer an new method for spatial distribution of water system extraction.

The Hydrology tool in ArcGIS simulated flow direction by terrain, so the river network and water system extracted is under nature condition without taking into account human activities. In the catchment area, the Hydrology was adopted to describe water system under natural conditions without human activities, meanwhile, artificial river channels affect accuracy of the results. It is worth mentioning that how to select the appropriate threshold according to the terrain for determining the catchment area size and extraction of spatial correctness need further discussion.

References

1. Li, W.P., Hao, A.B., Liu, Z.Y.: A Study on the Potential Groundwater Resources at the Tarim Basin. Geological Publishing House, Beijing (2000). pp. 1–20
2. Mendas, A.: Erratum to: the contribution of the digital elevation models and geographic information systems in a watershed hydrologic research. Appl. Geomat. 2(1), 33–42 (2010)
3. Cochrane, T.A., Flanagan, D.C.: Assessing water erosion in small watersheds using WEPP with GIS and digital elevation models. J. Soil Water Conserv. 54(4), 678–685 (1999)
4. Wu, S., Li, J., Huang, G.H.: Characterization and evaluation of elevation data uncertainty in water resources modeling with GIS. Water Resour. Manag. 22(8), 959–972 (2007)
5. Youpeng, X.U.: The analysis of water resources exploitation and ecological environment protection in Tarim Basin. Arid Zone Res. 19(1), 7–11 (2002)
6. Liu, G., Zhao, R.: Lancang river basin hydrology information extracting and its application based on DEM. Geomat. World 5, 56–59 (2007)
7. Yan, Y.Q., Tao, X.: The hydrological analysis of Sanhuajiang based on ArcGIS9.2 and DEM. Yellow River 31, 56–57 (2009)

Extraction Method of Remote Sensing Alteration Anomaly Information Based on Principal Component Analysis

Nan Lin[1(✉)], Menghong Wu[2], and Weidong Li[1]

[1] College of Surveying and Prospecting Engineering,
Jilin Jianzhu University, Changchun, China
NANLIN@qq.com, WEIDONGLI@qq.com, 1039283970@qq.com
[2] College of Geoexploration Science and Technology,
Jilin University, Changchun, China
MENGHONGWU@qq.com

Abstract. Using ASTER image as data source, the data were preprocessed based on RS software, for different types of erosion and alteration minerals. Using the main principal component analysis (PCA) and band ratio together for alteration information extraction and comparing the extraction results with the known ore occurrences. The results show that: the method combining principal component analysis and band ratio together in extracting mineralized alteration has certain feasibility, the test of alteration information extraction in Qinghai Lalingzaohuo region confirms that the extract information with the known ore occurrences are in good agreement, which has certain reliability.

Keywords: Remote sensing · Principal component analysis · Alteration · Abnormal information

1 Introduction

Modern geological science development, depends on the advanced of detection means in a large extent, in many application fields of remote sensing technology, remote sensing is in a more rapid development and the application effect is one of the subjects which were more prominent, especially the application of geological structure interpretation, mineralization information extraction and lithology identification by remote sensing technology [1]. Among them, mineralization alteration information extraction can provide direct technical parameters for ore prospecting in the area, which is favorable for the geological prospecting in the area of difficult working conditions and large area. Using remote sensing data for geological mapping and prospecting work has been quite extensive in abroad, since the 1980s in domestic have also launched a experimental study on extraction of alteration information, and achieved fruitful results.

This paper use Qinghai Lalingzaohuo as the test area, using ASTER image as data source, according to the different types of altered mineral, choosing the main principal component analysis (PCA) and band ratio together for alteration information extraction, which can provide technical backstopping for metallogenic prognosis in Qinghai Lalingzaohuo region.

H. Yuan et al. (Eds.): GRMSE 2016, Part I, CCIS 698, pp. 342–349, 2017.
DOI: 10.1007/978-981-10-3966-9_39

2   Study Area

The study area locate on the west of Qinghai Province, the western part of East Kunlun, The south side of the Qaidam Basin. The administrative division belongs to Qinghai Golmud City URT Ren Xiang, geographical location: east longitude 93°00′93°30′, north latitude 36°20′36°40′, as Fig. 1 shows. Along the Suhaitu river, Lalingzaohuo, Lalinggaoli river valley, there has a sidewalk traffic, but most of the area in high mountains and deep valleys, eolian sand covered so much that the vehicles can not pass, we can only rely on a horse or foot travel.

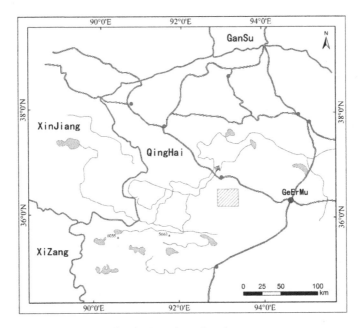

Fig. 1. Location of study area

3   Data Preprocessing

3.1   Atmospheric Correction

This paper chooses L1B type of Aster data, which can be only for atmospheric correction. Using the FLAASH atmospheric correction module of ENVI software for atmospheric correction to Aster satellite image data, the reflectance has no apparent absorption characteristics such as atmospheric, water vapor after atmospheric correction, so it can reflect the ground truth spectral features better.

3.2   Removing Interference Information

By using remote sensing image to extract alteration information often affected by many factors such as water, vegetation, shadow, cloud, snow. The interference information brightness values are special, distribution area is large, blocked some valuable information of the alteration.

By contrasting images in the study area we can find that: due to the higher elevation in the area, some areas have ice and snow all the year round. And the drainage system developed in the area, there are vegetative cover and half covered area, therefore, we remove the water, snow and vegetation on the Aster image.

By analyzing the spectrum characteristics of the interference information, using of band ratio, mask, cutting and other methods can remove the interference information from the image. Spectral characteristics of water found that water reflectance decreases with the increase of wavelength, corresponding to the ASTER data band set can Conduct ratio operation of the band 5 and the band 1, choose low value section of ratio image as mask to remove water interference. Snow and ice in the first 3 bands of Aster have a higher reflectivity, for this feature, you can mask the 3 bands of high value to remove snow and cloud interference. Corresponding to the ASTER data bands set, analysis the spectrum characteristics of vegetation, we can find the vegetation reflectance in Aster band 2 is a low value, and in band 3 showed high value [7], so by band ratio operation can ratio operation of the band 3 and band 2, using high value part of ratio image as mask to remove the interference of vegetation.

4   Method

4.1   Band Ratio Method

Band ratio method is a powerful tool for spectral information extraction, based on the principle of algebra, gets a new image data by ratio operation of the two band data, in the new image spectral information of ground objects are enhanced, and other features of the spectral information was inhibited. In the alteration information extraction, through the analysis of the spectrum characteristics of the rock mineral, the selection of the corresponding characteristic spectrum to deal with the ratio, which can enhance the lithology, alteration of the effect of information extraction.

4.2   Principal Component Analysis

All bands of remote sensing image reflect the spectral information of ground objects in different extent and range, but there are some correlations between the various bands, which make the information provided by the image, have a certain redundancy. Principal component analysis based on the relationship between the variables, under the premise of the conservation of the total amount of information, through orthogonal linear transformation to remove the related information of multi band image, which can reduce the influence to image information extraction of the relation between bands. After principal component analysis, the main components have no relation, and there is no duplication or redundancy between the main components. The principal component analysis is widely used in remote sensing alteration anomaly information extraction.

In a detailed analysis of the zone based on different spectral features of minerals, pick the optimal band combination of aster data to principal component analysis, then extracted anomaly information from the feature principal component image, finally, choose a reasonable threshold to extract the erosion becomes mineral information classification. According to the erosion in the study area, focusing on the extraction of a representative alteration information such as iron oxide, mud and propylitization [4, 8, 9].

- **Extraction of iron oxide alteration information**

Iron in the oxidizing environment, mainly in the form of $Fe2+$ and $Fe3+$, the spectral curve at 0.49 m and 0.87 m has two absorption Valley, corresponds to the band 1 and band 3 of ASTER data; between 0.60 and 0.70 μm has a reflection peak, corresponding to the band 2 of ASTER data.

Therefore, choosing the band 1, 2, 3, 4 to principal component analysis (PCA), using the high value of band 2 and negative high values of band 1, 3 as Criterion in alteration of information feature vector, from Table 1 can be seen the 4 principal

Table 1. Principal component analysis feature vector table of band 1, 2, 3, 4 of Aster data

Principal component	Band 1	Band 2	Band 3	Band 4
PC1	0.3928	0.5052	0.5359	0.5506
PC2	0.3255	0.3878	0.2451	−0.8267
PC3	−0.5029	−0.3243	0.7928	−0.1151
PC4	−0.6976	0.6993	−0.1554	0.0072

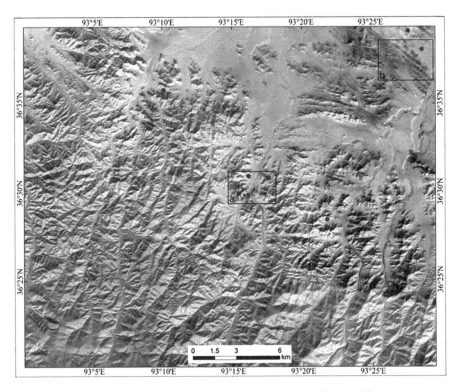

Fig. 2. Alteration information distribution map of iron oxide

components meet this criterion after the principal component analysis. So, in the PC4 image of the region of high brightness, which is a strong area of iron oxidation, and contrast alteration of the information distribution map (Fig. 2), it can be seen that the extracted iron oxide region is in good agreement with known mineralization points.

- **Extraction of alteration information of Argillic alteration**

The altered minerals of Argillic alteration are kaolinite, illite, smectite, sericite. Its related minerals are metal minerals contain Au, Ag, Cu, Pb, Zn [10], according to the results of the analysis of spectral characteristics of minerals, combining band 1, band4, band 6, band 7 o extract alteration information of Argillic alteration. Using the high value of band 7 and negative high values of band 6 as Criterion in alteration of information feature vector, from Table 2 can be seen the 4 principal components meet

Table 2. Principal component analysis feature vector table of band 1, 4, 6, 7 of Aster data

Characteristic value	Band 1	Band 4	Band 6	Band 7
PC1	−0.342	−0.564	−0.505	−0.556
PC2	0.937	−0.148	−0.204	−0.241
PC3	0.064	−0.804	0.4866	0.334
PC4	0.018	−0.111	−0.682	0.722

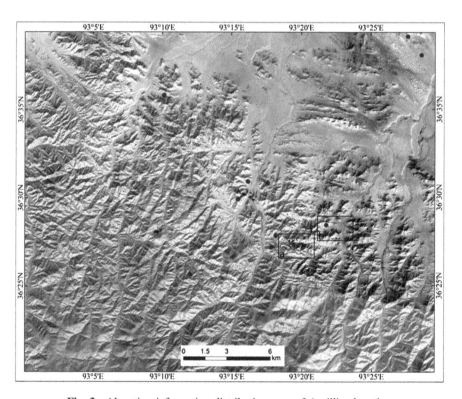

Fig. 3. Alteration information distribution map of Argillic alteration

this criterion after the principal component analysis. The high brightness areas of PC4 image are sericitization and kaolinite which was altered strongly, contrast Fig. 3, the extraction area of Argillic alteration have a closely relationship the known ore deposits.

- **Extraction of alteration information of Propylitization**

The altered minerals of Propylitization are chlorite, carbonate salt minerals, albite, epidote and so on. Its related minerals are metal minerals contain Au, Cu, Pb, Zn, Fe [11, 12]. According to the results of the analysis of spectral characteristics of minerals, combining six bands of ASTER data (Band 1, band 2, band 3, Band 5, Band 8, Band 9) to extract alteration information of Propylitization. Using the high value of band 9 and negative high values of band 8 as criterion in alteration of information feature vector, from the eigenvector matrix (Table 3) can be seen, the main contribution of PC5 comes from Band 8 (-0.63562) and Band 9 (0.66101), which meet the judgment criterion. Therefore, the PC5 image highlights chloritization and carbonation corrosion variable region strongly. Contrast alteration information distribution map (Fig. 4), we can see that propylitic alteration anomalies development is weak in this area, scattered in the eastern part of the study area.

Table 3. Principal component analysis feature vector table of band 1, 2, 3, 5, 8, 9 of Aster data

Characteristic value	Band 1	Band 2	Band 3	Band 5	Band 8	Band 9
PC1	−0.297	−0.346	−0.401	−0.442	−0.45	−0.481
PC2	0.451	0.488	0.435	−0.386	−0.295	−0.361
PC3	0.203	−0.036	−0.0692	0.795	−0.387	−0.411
PC4	−0.642	−0.049	0.666	0.105	−0.343	0.102
PC5	0.337	−0.169	−0.105	−0.071	−0.635	0.661
PC6	0.372	−0.78	0.435	−0.073	0.193	−0.143

Based on raster image computation makes the abnormal information disorderly distribution, often obscures the alteration remote sensing anomaly distribution rules, so we can handle the effective information to 3 * 3 median filter, filter processing can keep larger contiguous anomaly, eliminate extracted isolated anomaly; The study divide strong, medium and weak three kinds of abnormal information level using of $(X + k\sigma)$, X represents mean and K value range is 1.5–2.5, σ is the standard error. The various corrosion variable ratio of characteristic image of the mineral were summarized in this paper, alteration information distribution map of iron oxide (band 2 + band 3)/band 1; alteration information distribution map of Argillic alteration: (band 5 + band 7/band 6; alteration information distribution map of Propylitization: (band 7 + band 9/band 8). Finally, the alteration information superimposed on the RGB color composite image, the three kinds of minerals altered graded image have been obtained (Figs. 2, 3 and 4).

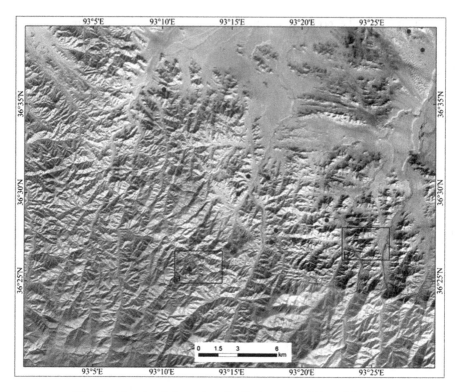

Fig. 4. Alteration information distribution map of Propylitization

5   Conclusions

The genesis of the ore deposit types are the skarn type and medium to low temperature liquid ore deposit, according to this characteristic, this study adopts the principal component analysis (PCA) for alteration information extraction, the geologic map and linear structure in the study area with alteration information we extracted are overlay, with the reference to related geological data. Analyzing the result of the extraction of alteration summarizes the mineralization alteration information distribution characteristics in the study area are as follows:

- The abnormal development of iron oxides is strong, and it is a strip along the Near East and North West, which is distributed in the Middle East of the study area, the ancient Proterozoic group and the late Devonian yak formation. Combined with the use of principal component analysis and ratio analysis and circle 2 iron oxide preferred region.
- Argillic alteration anomalies are mainly distributed in the study area southeast of the Jinshuikou group formation, nearly EW and NW linear structures set together better, and has a high agreement with the known ore occurrences in the area. So the corrosion becomes abnormal and hydrothermal mineralization is related closely.

- Propylitization anomaly in the region development is weak, scattered in the eastern part of the study area.

Acknowledgments. This work was financially supported by the Program of China Geological Survey (No. 1212010510613).

References

1. Lv, F., Xing, L., Fan, J.: Application study of remotely sensed alteration information extraction. Xinjiang Geol. **22**(4), 435–437 (2004)
2. Li, M., Xing, L., Pan, J., et al.: Research of combinatory analysis method in altered information extraction. Remote Sens. Technol. Appl. **26**(3), 303–308 (2011)
3. Jing, F., Chen, J.: The review of the alteration information extraction with remote sensing. Remote Sens. Inf. **2**, 62–65 (2005)
4. Yang, C., Jiang, Q., Liu, W., et al.: Hydrothermal alteration mapping using ASTER data in northern Dong Ujimqin, Inner Mongolia. J. Jilin Univ. (Earth Sci. Ed.) **39**(6), 1163–1167 (2009)
5. Zhang, Y., Zeng, C., Chen, W.: The methods for extraction of alteration anomalies from the ETM$^+$ data and their application: method selection and technological flow chart. Remote Sens. Land Resour. **56**(2), 44–49 (2003)
6. Zhang, Y., Yang, J.: The method of abstracting remote sending information of alterated rocks in the uncovered bedrocks area. Remote Sens. Land Resour. **36**(2), 46–53 (1998)
7. Wang, H., Zhang, T.: Discrimination of mineralization information based on apparent reflectivity image and its application. Acta Geosci. Sin. **26**(3), 283–289 (2005)
8. Zhang, Y., Zeng, Z., Chen, W.: The methods for extraction of alteration anomalies from ETM+(TM) data and their application: method selection and technological flow chart. Remote Sens. Land Resour. **15**(2), 44–49 (2003)
9. Zhang, J., Yang, Z., Hu, X., et al.: Remote sensing information extracting based on spectrum feature and metallogenic prediction. Miner. Resour. Geol. **18**(04), 346–349 (2004)
10. Yang, C., Jiang, Q., Zhu, Q., et al.: Altered minerals discriminating using ASTER data at low vegetable cover area. Geol. Explor. **45**(6), 761–766 (2009)
11. Yang, J., Jiang, Q., Zhao, J., et al.: Remoted-sensing mineralization information extracted based on ETM+ and ASTER data: a case study in Tarigenaobao, Inner Mongolia. J. Jilin Univ. (Earth Sci. Ed.) **38**(4), 153–158 (2008)
12. Geng, X., Yang, J., Zhang, Y., et al.: The application of ASTER remote sensing data for extraction of alteration anomalies information in shallow overburden area—a case study of the Baoguto porphyry copper deposit intrusion in western Junggar, Xinjiang. Geol. Rev. **54**(2), 184–191 (2008)

Geographical Situation Monitoring Applications Based on MiniSAR

Xuejing Shi[1,2(✉)], Gang Huang[2,3], Ming Qiao[4], and Bingnan Wang[4]

[1] Chinese Academy of Surveying and Mapping, Beijing 100830, China
XuejingShi@163.com, SHJ_shixuejing@163.com
[2] Beijing GEO-Vision Tech. Co., Ltd., Beijing 100070, China
GangHuang@163.com
[3] College of Resource Environment and Tourism, Capital Normal University, Beijing 100048, China
[4] Institute of Electronics, Chinese Academy of Sciences, Beijing 100190, China
MingQiao@163.com, BingnanWang@163.com

Abstract. For rapid development and wide application of UAV, high-resolution imaging radar suitable for UAV is more and more requirement. Miniature Synthetic Aperture Radar (MiniSAR) can be equipped with medium/low altitude UAV platform, which applies to the remote sensing tasks in hazardous and harsh conditions. Thus, it plays a pivotal role in geographical situation monitoring. The merit of high-performance MiniSAR is light weight, low-cost, low power consumption. Moreover, it has high resolution, interferometric and fully polarimetric imaging capabilities. Currently this paper does introduce the hardware components, system functionality and property of MiniSAR. The system is employed for the flight testing in Jishan County of Shanxi to achieve fully polarization and interference SAR imagery with 0.3 m resolution. Considering the fact that the fully polarization SAR imagery contains abundance of ground target's backscattering characteristics, it can be available for terrain classification, crop monitoring and other fields. For the SAR interferometry, DSM and DOM are obtained after phase unwrapping, block adjustment and other interference processing. Finally, the accuracy is verified using checkpoints. The result meet the requirements of topographic mapping at a scale of 1:5000. The applicability of MiniSAR system in the application of geographic condition monitoring is verified.

Keywords: MiniSAR · POLSAR · InSAR · DSM · DOM

1 Introduction

At present, geographical situation as the one of important status of national economy, represents the development of China's social economy and the progress of science and technology [1]. Due to the vastness of China, there are many regions need fast and dynamic geographic information service and response. Optical remote sensing is hardly meet the requirements of the rapid response in the difficult areas of special climatic conditions. Synthetic aperture radar (SAR) with the all-day and all-weather observation characteristics is a significant technical support to improve the capability of response

© Springer Nature Singapore Pte Ltd. 2017
H. Yuan et al. (Eds.): GRMSE 2016, Part I, CCIS 698, pp. 350–355, 2017.
DOI: 10.1007/978-981-10-3966-9_40

and promote the remote sensing and geographic information industry. The miniature SAR (MiniSAR) that developed by Chinese Academy of Sciences Institute of electronics is a high performance system of synthetic aperture radar for low altitude flight platforms. It is small, light and low power consumption. The diversity of microwave remote sensing information are accessed flexibly and low-cost by MiniSAR that can be employed in unmanned and small light manned aircraft platform [2, 3]. This again is an important future development direction, with potential applications across many areas, including geographical situation monitoring.

2 MiniSAR System Introduction

MiniSAR system, which operates in the Ku-band, owns fully polarimetric mode and interference mapping function so that it can obtain abundant information of the ground objects and geographic data. As more SAR have been launched they have facilitated available application of polarization and interference technology in geographical situation survey, which represents one of the most challenging and active areas of development at the current time. Less than 2 kg of the radar host and the whole system is less than 6 kg, therefore it can be carried in all kinds of light aircraft and equipped with UAV that load platform above 6 kg. Naturally, the assignment data collection of airborne remote sensing radar can be executed easily by MiniSAR system with all-weather, multi-polarization, multi-function. A novel solution is provided in monitoring of difficult geographical conditions.

The hardware of MiniSAR system include radar host, IMU, GPS antenna, RF transceiver antenna, polarization switching switch, power supply batteries, cables and other related equipment. The main role of the radar host is to generate excitation signals, and accept the radar echo signal from the receiving antenna for fast gathering and high-speed storage; IMU mainly provides motion compensation for MiniSAR imaging; the action of GPS supplies GPS Data for PCS module; the function of RF transceiver antenna is to transmit and radiate the output of the microwave power of the transmitter to the ground, then, the echo signals of the ground targets are received through the feeder to the receiver, which is transmitted to the receiver via the feeder. MiniSAR system specifications are shown in Table 1.

Table 1. MiniSAR system specifications

Name	Interference pattern	Polarization mode
Maximum flying height	1500 m	1900 m
Center frequency/bandwidth	14.5 GHz/600 MHz	
Beam center angle	45°	45°
Polarization mode	Single polarization	Full polarization
Interference baseline	1 m	/
Maximum mapping bandwidth	1.2 km	1.4 km
Polarization isolation	Better than 25 dB	
Weight	≤5 kg	
Power consumption	≤65 W	

Here we present a brief discussion of the work flow of flight test by MiniSAR. (1) Route planning, flight routes are designed according to the mapping area and parameters of radar. (2) Data playback and POS processing, radar echo data are being played back to the local hard disks. (3) Quality inspection, checking the information is overwritten and fast imaging processing, and the data quality is evaluated. (4) Imaging processing, fine imaging of full resolution into SLC data. (5) Polarization synthesis, a pseudo color imagery is created by pauli synthesis after polarization and radiation calibration of full polarimetric data. (6) Interference processing, the DSM and DOM are generated after registration, filtering, phase unwrapping, joint scaling, geocoding and other steps.

3   MiniSAR Mapping

3.1   Survey Area

The flight test of MiniSAR was conducted in November, 2015. The survey area is located in Jishan County. The east-west direction of the zone is about 3.5 km and there is about 5.5 km from north to south. The terrain is dominated by hills and partly mountainous, the highest elevation of 470 m in the region and the minimum altitude of about 320 m. Here we set up six flights in this zone, and overlap between the strips at arte of 30%.

3.2   Full Polarization Mapping

Polarimetric SAR (Polarimetric SAR, POLSAR), as the frontier imaging radar in recent years, which can obtain richer target signatures information than single polarimetric SAR system and open up possibilities for POLSAR remote sensing applications. So it has unattainable advantages of the single polarimetric SAR in analyzing, extracting and inversion of the polarimetric scattering characteristics of ground objects [4].

Data acquisition of dual-channel mode is utilized in the miniature POLSAR system. It adopt the full-polarization antenna and the polarized switch, which is needed to switch the H/V polarization mode to obtain HH, HV, VH and VV four kinds of polarimetric signals. The test of full polarization mapping was carried out. The aircraft flight with 1500 m altitude, 45° side view angle and the viewing angle of 21°. A plurality of corner reflectors is arranged in the middle strip of the survey area. There are two sets of polarization scalers, each group contains dihedral reflector of 0°, dihedral reflector of 45°, trihedral corner reflector, meanwhile, other triangular reflectors are adopt to make polarimetric calibration. The image is processed to obtain exquisite polarization image with the accuracy of 0.3 m after data playback. As shown in Fig. 1.

In order to facilitate the visual interpretation of radar images, the pauli synthesis is available for usage of polarization image to construct a pseudo color image. As shown in Fig. 2, the farmland, roads, waters, residential areas can be distinguished clearly. Obviously, the sophisticated fully polarized images of MiniSAR would be offered the vital sources of data for urban monitoring, resource census, land use and land cover [5].

Fig. 1. HH polarized image

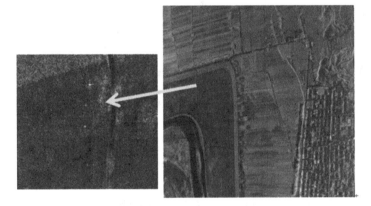

Fig. 2. Pauli synthesis of MiniSAR image (the purple point is trihedral reflector, red point is dihedral reflector of 0° and green point is dihedral reflector of 45°) (Color figure online)

3.3 Interference Mapping

The gray-scale images and terrestrial three-dimensional information can be get after imaging and interferometric processing of Synthetic Aperture Radar Interferometry (InSAR) data [6].

The work mode of interference MiniSAR selected dual-channel acquisition. In this test, the selected antennas is a pair of HH single-channel and a fully polarized antenna that is used as a single-polarized. The two antennas installed in the wing of the aircraft are employed to acquire the basic processing data that two SAR images in the same area, while the interference image is created by calculating the phase difference of the two SAR images [7]. Eventually, the terrain elevation data is generated through phase unwrapping, joint calibration and other related handling.

This interferometry test with aviation altitude 600 m, side view angle of 60°, the baseline length of 0.5 m. There are six corner reflectors are arranged in the pilot site, and four scalers are selected as control points for block adjustment. The block adjustment technology of InSAR can not only reduce the demand for control points, but also improve the accuracy of the coordinates of the connection point and the integrity of the regional network [6, 8]. It largely improves the precision of interferometry. Figure 3 shows the results of interferometry.

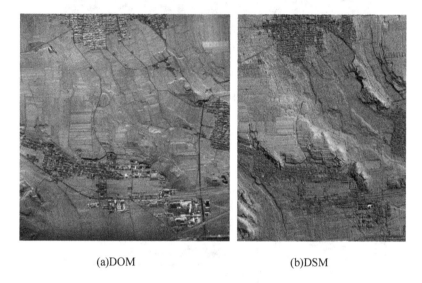

(a)DOM   (b)DSM

Fig. 3. The results of interferometry

To verify the accuracy of interferometry, in addition to the 4 control points, the other two points are used as check points in the six corner reflectors arranged in the survey area. While more check points of the required to do accuracy analysis are selected. In the figure, 17 feature points measured by GPS. As shown in Table 2, the results of the MiniSAR interferometric mapping meet the requirements of 1:5000 national standard accuracy, but it is closer to 1:2000 than those of the national standard accuracy requirements of hilly areas (Table 3). Selecting feature points artificially, what makes it difficult is the complexity of SAR imaging. The plane accuracy is not particularly high since the man-made factors. However, due to the choice of checkpoints are in flat terrain where the accuracy of elevation is relatively high.

Table 2. MiniSAR interference mapping accuracy

Point class	Error in plane (m)	Elevation error (m)
Control point	0.32	0.37
Corner reflector checkpoint	0.37	0.34
Spike checkpoint	1.44	0.44

Table 3. The national standards accuracy of hilly areas

Mapping scale	Point class	Error in plane (m)	Elevation error (m)
1:2000	Control point	0.6	0.26
	Checking point	1.0	0.4
1:5000	Control point	1.5	0.8
	Checking point	1.75	1.0

4 Conclusions

In this paper, the hardware composition and function index of MiniSAR system are described. The flight test of full polarization and interference mode is carried out. Interferometry achieves the positional precision of DSM and DOM at scale of 1:5000 in the case of sparse control points. It can realize important geographic information industry such as basic surveying and mapping, land-based monitoring, geographical situation survey and so on. Next we will further improve the accuracy of MiniSAR mapping and promote the continuous progress of MiniSAR to more miniaturization, more low power consumption, lower cost, higher resolution. It provides a broader space and prospects for the effectively application of MiniSAR in geographical situation monitoring.

Acknowledgments. This work was supported by the subject that Study on Air-ground integration of Holographic three-dimensional Data Acquisition and Demonstration Area Experiment, Chinese Academy of Surveying and Mapping, and National High Technology Research and Development Plan, under grant 2013AA092105.

References

1. Wang, J., Qiao, M., Wei, Y.: Geographical situation monitoring applications based on the UAV airborne SAR. Geomat. Spat. Inf. Technol. **39**(5), 61–64 (2016). HeiLongjiang
2. Wang, Y., Liu, C., Zhan, X.: An ultrafine multifunctional unmanned aerial vehicle SAR system. J. Electron. Inf. Technol. **35**(7), 1569–1574 (2013)
3. De Wit, J.J.M., Meta, A., Hoogeboom, P.: Modified range-Doppler processing for FM-CW synthetic aperture radar. IEEE Geosci. Remote Sens. Lett. **3**(1), 83–87 (2006). IEEE Press, New York
4. Lu, L., Zhang, J., Huang, G., Su, X.: Land cover classification and height extraction experiments using Chinese airborne X-band PolInSAR system in China. Int. J. Image Data Fusion **7**(3), 282–294 (2016). Taylor & Francis, Abingdon
5. Xie, L., Zhang, H., Liu, M., Wang, C.: Identification of changes in urban land cover type using fully polarimetric SAR data. Remote Sens. Lett. **7**(7), 691–700 (2016). Taylor & Francis, Abingdon
6. Yue, X., Han, C., Dou, C.: Mathematical model of airborne InSAR block adjustment. Geomat. Inf. Sci. Wuhan Univ. **40**(1), 59–63 (2015). The University of Wuhan, Wuhan
7. Pieraccini, M., Luzi, G., Atzeni, C.: Terrain mapping by ground-based interferometric radar. IEEE Trans. Geosci. Remote Sens. **39**(10), 2176–2181 (2001). IEEE Press, New York
8. Huang, G., Yang, S., Wang, N.: Block combined geocoding of airborne InSAR with sparse GCPs. Acta Geod. Cartogr. Sin. **42**(3), 397–403 (2013). Surveying and Mapping, Beijing

New Reduced-Reference Stereo Image Quality Assessment Model for 3D Visual Communication

Ying Wang[1], Kaihui Zheng[1], Mei Yu[1,2(✉)], Baozhen Du[1],
and Gangyi Jiang[1]

[1] Faculty of Information Science and Engineering,
Ningbo University, Ningbo 315211, China
yumei2@126.com
[2] National Key Lab of Software New Technology,
Nanjing University, Nanjing 210093, China

Abstract. Visual quality assessment of stereo image plays an important role in three dimensional visual communication. Considering the processing of binocular perception in viewing stereo image, we present a reduced-reference (RR) stereo image quality assessment (SIQA) model based on binocular perceptual characteristics. Firstly, stereo images are divided into binocular fusion portion and binocular rivalry portion with internal generative mechanism. Then, cyclopean view is generated according to the binocular fusion portion and binocular rivalry portion with binocular perception, and the Gaussian scale mixture RR features from cyclopean view and the binocular rivalry portion for SIQA. Finally, the quality indicators of cyclopean view and binocular rivalry portion are computed to obtain the final SIQA score. The proposed model is tested on the LIVE 3D IQA database. Experimental results show that compared with the state-of-the-art methods, the proposed model has high correlation with subjective perception and can evaluate human stereo visual properties effectively.

1 Introduction

With the rapid development of related products and applications in three dimensional (3D) visual communication, accurately measuring stereo image quality plays a very important role in 3D media applications [1]. In the objective image quality assessment (IQA), reduced-reference (RR) method has been an research hotspot for its advantages that reduces the amount of data for transmission than full-reference (FR) method and ensures better evaluation accuracy than no-reference (NR) method [2, 3].

In recent years, researches of image communication and applications had made great progress [4–7]. In stereo image quality assessment (SIQA), Bensalma et al. [8] considered the difference of binocular energy and proposed an FR-SIQA model. Maalouf et al. [9] proposed a cyclopean view based SIQA method. However, when obtaining the cyclopean views for the stereo images, the detail information of textures and edges may be lost. Chen et al. [10] proposed a SIQA model with cyclopean view fusion, which only considered binocular perception effect from the cyclopean view but

© Springer Nature Singapore Pte Ltd. 2017
H. Yuan et al. (Eds.): GRMSE 2016, Part I, CCIS 698, pp. 356–364, 2017.
DOI: 10.1007/978-981-10-3966-9_41

ignored other factors. Jin et al. [11] proposed a FR 3D image quality model with three quality components to measure image quality, that is, cyclopean view, binocular rivalry, and the scene geometry. Hewage et al. [12] proposed an RR quality metric model for 3D depth map transmission, by extracting the edges and contours information of the depth map to measure the 3D image quality. However, binocular perception is insufficiently considered in the above models.

Based on human binocular perception, a reduced-reference stereo image quality assessment (RR-SIOA) model is proposed in this paper. Considering internal generative mechanism (IGM) of visual signal [13], stereo image is divided into predicted portion and disorderly portion called as binocular fusion portion and binocular rivalry portion. The predicted portion is primary visual information of stereo images including outline and structure, which is fused into a cyclopean view in the processing of binocular perception. The binocular rivalry portion retains the texture details of stereo image. Then, the Gaussian scale mixture (GSM) [14] RR features are extracted from the cyclopean view and the binocular rivalry portion to measure the stereo images quality. Compared with the existing models, the proposed model has better performance for SIQA.

2 Proposed RR-SIQA Method

Some studies of the binocular perception have showed that visual information can be handled by two visual pathways at the same time, that is, cyclopean view and binocular view [11].

Based on binocular perception, a new RR-SIQA model is proposed as shown in Fig. 1. Firstly, the original and distorted stereo images are decomposed into the predicted portions and the disorderly portions, which correspond to binocular fusion portion and binocular rivalry portion in this paper, by using IGM model at the sender and the receiver, respectively. Secondly, the RR features are extracted from the predicted portion and the disorderly portion by using GSM model. Finally, the binocular fusion quality and binocular rivalry quality are obtained by comparing the RR features and pooled as the final quality score of stereo image.

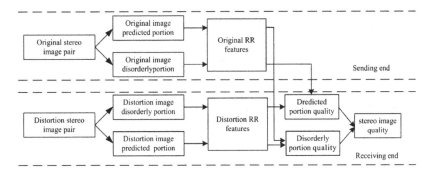

Fig. 1. The framework of the proposed RR-SIOA model

2.1   Visual Signal Decomposition with IGM

Studies of IGM have shown that human brain actively predicts the visual sensation and avoids the residual uncertainty. Based on IGM method, Wu et al. [15] proposed a model to decompose image into predicted portion and disorderly portion. The predicted portion is primary visual region such as edge and structure information, and it is the portion to fuse into a cyclopean view. The disorderly portion is texture and detail information of image that cannot fuse easily. It is the portion to present the binocular view. The predicted portion and disorderly portion of images, decomposed by IGM model, are shown in Fig. 2. Two stereo images are selected from the LIVE 3D IQA database (phase I). The selected original stereo images are shown as the Figs. 2(a) and (d). Figures 2(b), (e) and (c), (f) are the corresponding predicted portion and disorderly portion, that is, binocular fusion portion and binocular rivalry portion. We view the predicted portion images and disorderly portion images about different types of distortion in the LIVE 3D IQA database (phase I) to find the perceptive characteristics in these portions. The process of subjective test is strict accordance with the ITU.P910 [16]. Through the subjective test, it is found that there are different perceptive characteristics of different types of distortion at the binocular fusion portion and binocular rivalry portion. Figures 2(b) and (e) contain primary visual information of the image, that is, the main body of the edges and structures in image. The distortions on this portion will cause edge blur and structure distortion, but will rarely cause stereo mismatch and uncomfortable of binocular perception. So at this portion, the human perceptive characteristic presents as binocular fusion. Related study [10] showed that the difference stimuli received by the two eyes from each other causes match failure and results in binocular rivalry to affect stereo perception. Figures 2(c) and (f) contain texture and detail information and disturbance from distorted image, which present as inconsistent at the retinal location. Distortions on this portion mainly cause uncomfortable perception of stereo image impression.

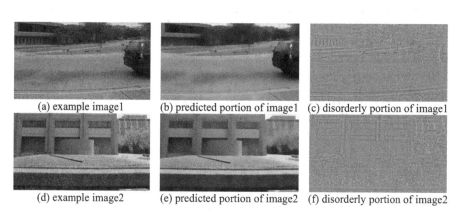

(a) example image1   (b) predicted portion of image1   (c) disorderly portion of image1

(d) example image2   (e) predicted portion of image2   (f) disorderly portion of image2

Fig. 2. The predicted portions and disorderly portions of images

2.2 Cyclopean View Generation

As a binocular fusion scheme, cyclopean view model is used to depict binocular perception of human vision usually. The usual cyclopean view models [9–11] can be used to match the left and right views directly to generate the cyclopean view. But these models may cause the loss of texture detail information. Here, we propose a cyclopean view model which obtains the cyclopean view by matching the predicted portion of the left and right views. The cyclopean view C is defined by

$$C = \omega_l E_l + \omega_r E_r \tag{1}$$

where E_l and E_r are the matching blocks at the predicted portion of stereo images, and ω_l and ω_r are weights for the matched left and right image blocks, respectively. For a symmetrically distorted stereo image, both of ω_l and ω_r are set to 0.5. In order to match the left and right views accurately, a block with the size of $a \times a$ is used to match the center matching pixels.

2.3 RR Feature Extraction and Stereo Image Quality Estimation

As an effective transform consistent with the multi-channel characteristics of human visual system, discrete wavelet transform (DWT) has been widely used in image processing. Statistics for DWT sub-band coefficients may be conducive to extract the RR indicators. Through analyzing DWT high-frequency sub-band coefficients, it is found that it is similar to the distribution shape. Here, the GSM model is used to normalize the coefficients in the DWT high-frequency sub-bands. Then, the divisive normalization transform is used to make the marginal distribution of the DWT coefficients to conform to Gaussian distribution. After normalization transform, the distribution is fitted with a Gaussian model as follows

$$P(x) = \frac{1}{\sqrt{2\pi\sigma^2}} \exp(-\frac{(x - \mu)^2}{2\sigma^2}) \tag{2}$$

where σ is the standard derivation and μ is the mean of distribution function. The Gaussian model is used to describe DWT sub-bands' features with σ and μ.

Here, σ and μ of the cyclopean view about predicted portion and disorderly portion is obtained as the RR indicators. Let $D_\mu(\mu_{org}^n, \mu_{dis}^n)$ denote the quality indicators μ of about the n-th DWT sub-band, which can be defined by

$$D_\mu(\mu_{org}^n, \mu_{dis}^n) = \left| \frac{\mu_{org}^n - \mu_{dis}^n}{\mu_{org}^n + \mu_{dis}^n} \right| \tag{3}$$

where μ_{org}^n or μ_{dis}^n denotes μ value of the n-th DWT sub-band in the original image or distorted image, respectively. Similarly, let $D_\sigma(\sigma_{org}^n, \sigma_{dis}^n)$ denote the quality indicators of σ about the n-th DWT sub-band, which can be defined by

$$D_\sigma(\sigma^n_{org}, \sigma^n_{dis}) = \left| \frac{\sigma^n_{org} - \sigma^n_{dis}}{\sigma^n_{org} + \sigma^n_{dis}} \right| \qquad (4)$$

where σ^n_{org} or σ^n_{dis} denote's σ of the n-*th* DWT sub-band in the original image or distortion image, respectively.

Then, we compare the difference about σ and μ indicators of the original and distortion images. Let q^n_{cyc} denote the cyclopean view quality indicators of binocular fusion portion, and it is computed by

$$q^n_{cyc} = \alpha \cdot D_\mu(\mu^n_{org}, \mu^n_{dis}) + \beta \cdot D_\sigma(\mu^n_{org}, \mu^n_{dis}) \qquad (5)$$

where α and β are the weights of quality indicators, $\alpha + \beta = 1$. The values of α and β are obtained by training in Ningbo University's stereo image database [17] which includes 12 original stereo images and 312 distorted stereo images with five types of distortions: JPEG, JPEG2000 (JP2K), Gaussian blur (Gblur), white noise (WN) and H.264 distortions. Through testing values of α and β for the most effective correlation coefficient between the stereo image quality and difference mean opinion scores (DMOS), α and β are set as 0.1 and 0.9, respectively.

The contrast sensitive function (CSF) is used for combining DWT sub-bands, so the cyclopean view quality Q_{cyc} can be gained through the combination of the n-*th* sub-band indicator, $q^n_{dis,l}$. The indicators q^n_{cyc} can efficiently reflect the change of different DMOS. In Fig. 3, the horizontal axis is the DMOS of distorted image and the vertical axis denotes the indicator q^n_{cyc}. It is found that the larger the image distortion (the larger the DMOS) is, the greater the indicator is in each sub-band.

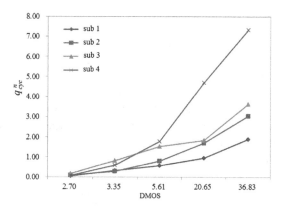

Fig. 3. The relationship between DMOS and sub-bands' quality indicators

In the disorderly portion, the quality indicators of disorderly portion in left and right views are denoted as $q^n_{dis,l}$ and $q^n_{dis,r}$, respectively, which can be obtained in the same way as q^n_{cyc}. The disorderly portion quality of the left and right image can be calculated by CSF. We denote them as $Q_{dis,l}$ and $Q_{dis,r}$. Let Q_{dis} be disorderly portion quality of stereo image, which can be calculated as follows

$$Q_{dis} = wQ_{dis,l} + (1 - w)Q_{dis,r} \tag{6}$$

where w is the weight between $Q_{dis,l}$ and $Q_{dis,r}$. For symmetrically distorted stereo image, w is set as 0.5.

Finally, the cyclopean view quality Q_{cyc} and disorderly portion quality Q_{dis} are combined as the final stereo image quality score Q.

$$Q = f(Q_{cyc}, Q_{dis}) \tag{7}$$

Taking into account that it may be difficult to combine Q_{cyc} and Q_{dis} for coinciding exactly with human visual system, the support vector regression method is used to establish $f()$ as a weighting function.

3 Experimental Results and Analyses

In the experiments, the proposed method was tested on the LIVE 3D IQA database (phase I) [18] to verify the consistency with subjective assessment. The symmetrically denoted database includes 20 original stereo images and 365 distorted stereo pairs with five types of distortion: the JPEG compression, JP2K compression distortion, additive white Gaussian noise (WN), Gaussian blur (Gblur) and a fast-fading (FF) modeled with the Rayleigh fading channel distortion. There are three criteria used for performance comparison of objective methods, that is, Pearson linear correlation coefficient (LCC), Spearman rank correlation coefficient (SROCC) and root mean squared error (RMSE). Both the values of LCC and SROCC range from 0 to 1. The better the performance of the method is, the closer SROCC and LCC values are to 1 and RMSE value is to 0.

To validate effectiveness of the proposed RR-SIQA method, it has been compared with several existing models, including Benoit's, Hewage's, You's, Gorley's, Shen's, Yang's, Zhu's, Akhter's methods which have been tested on LIVE 3D IQA database (phase I) [18]. The other two state-of-the-art Bensalma's SIQA method of cyclopean view [8] and Wang's RR-SIQA method of GSM have also been compared with the proposed RR-SIQA method. In these models, Hewage's and Wang's are the RR methods, and the others are FR methods. The related comparison results of these models are listed in Tables 1, 2 and 3.

As shown in Tables 1, 2 and 3, the proposed model can achieve the top three performances among the models for all distortion types. Especially the case of the distortion 'ALL', the proposed model achieves the best performance among all of the above models.

Compared with Bensalma's and Wang's RR-SIQA methods of, the proposed model has a better performance on 3 criteria, LCC, SROCC and RMSE, except the 'Blur' distortion. From Tables 1 and 2, it can be found that Wang's model has the better performance on the criteria LCC and SROCC for 'Blur' distortion. The reason may be as follow: the 'Blur' distortion main causes structure distortion in image, and almost all

Table 1. Performance of SIQA methods: LCC

Algorithm	JP2K	JPEG	WN	Blur	FF	ALL
Benoit	**0.9398**	**0.6405**	**0.9253**	**0.9488**	**0.7472**	**0.9025**
Hewage	0.9043	0.5305	0.8955	0.7984	0.6698	0.8303
You	0.8778	0.4874	**0.9412**	0.9198	0.7300	0.8814
Gorley	0.4853	0.3124	0.7961	0.8527	0.3648	0.4511
Shen	0.5039	0.3899	0.8988	0.6846	0.4830	0.5743
Yang	0.2012	0.2738	0.8701	0.6261	0.2824	0.3909
Zhu	0.8073	0.3790	0.5178	0.7770	0.5038	0.6263
Akhter	0.9059	**0.7294**	0.9047	0.6177	0.6603	0.4270
Bensalma	0.8389	0.3803	0.9147	0.9369	0.7339	0.8874
Wang	**0.9162**	0.5697	0.9133	**0.9574**	**0.7833**	**0.8921**
Proposed	**0.9364**	**0.6632**	**0.9390**	**0.9470**	**0.8285**	**0.9375**

Table 2. Performance of SIQA methods: SROCC

Algorithm	JP2K	JPEG	WN	Blur	FF	ALL
Benoit	**0.9103**	**0.6028**	**0.9292**	**0.9308**	**0.6989**	**0.8992**
Hewage	0.8558	0.5001	0.8963	0.6900	0.5447	0.8140
You	0.8598	0.4388	**0.9395**	0.8822	0.5883	0.8789
Gorley	0.4203	0.0152	0.7408	0.7498	0.3663	0.1419
Shen	0.2133	0.2440	0.8917	0.6586	0.2665	0.0679
Yang	0.1501	0.1328	0.8471	0.3266	0.1426	0.0785
Zhu	0.7708	0.2929	0.4651	0.7935	0.4752	0.6388
Akhter	0.8657	**0.6754**	0.9137	0.5549	0.6393	0.3827
Bensalma	0.8171	0.3283	0.9055	0.9157	0.6500	0.8747
Wang	**0.8832**	0.5420	0.9066	**0.9246**	**0.6548**	**0.8890**
Proposed	**0.9233**	**0.6631**	**0.9392**	**0.9225**	**0.7945**	**0.9366**

Table 3. Performance comparison of SIQA methods: RMSE

Algorithm	JP2K	JPEG	WN	Blur	FF	ALL
Benoit	**4.4266**	**5.0220**	**6.3076**	**4.5714**	**8.2578**	**7.0617**
Hewage	5.5300	5.5431	7.4056	8.7480	9.2263	9.1393
You	6.2066	5.7097	**5.6216**	5.6798	8.4923	7.7463
Gorley	11.3237	6.2119	10.1979	7.5622	11.5691	14.6350
Shen	12.2754	6.0216	7.2939	10.5547	10.8820	13.5473
Yang	12.6979	6.2894	8.2002	12.1291	11.9462	15.2481
Zhu	7.6813	6.0684	14.7201	9.1270	10.7362	12.7828
Akhter	5.4836	**4.4736**	7.0929	11.3872	9.3321	14.8274
Bensalma	7.0493	6.0477	6.7249	5.0595	8.4399	7.5585
Wang	**5.1890**	5.3741	6.7772	**4.1777**	**7.7245**	**7.4081**
Proposed	**4.2289**	**3.8838**	**5.4885**	**3.8540**	**6.3091**	**5.2694**

of this distortion is divided into predicted portion. When fusing the cyclopean view, the loss of image information is inevitable and it influences the results of quality assessment.

Compared with Bensalma's method, the proposed method achieves a better performance. It benefits from that stereo image is divided into predicted portion and disorderly portion. The RR features can be obtained with respect to different features of predicted portion and disorderly portion to keep enough image characteristics for quality assessment.

In addition, as a RR-SIQA model, the proposed model only generates small amount of information from the original stereo images. It reduces the amount of data for transmission and ensures the evaluation quality at the same time.

4 Conclusion

In this paper, a reduced-reference stereo image quality assessment model based on binocular visual perception of human visual system has been proposed. In this model, the internal generative mechanism theory is used to divide stereo images into predicted portion and disorderly portion. Then we extract the RR features from them. The stereo image quality is gained by comparing the RR features of the original image with that of the distortion image. Experimental results show that the proposed method is highly consistent with the subjective perception of binocular vision. Except the IGM, the best decomposition scheme needs to be found, and future research will explore to find more effective decomposition model for binocular vision. In the further research work, we can extend the proposed method into stereo video quality assessment by combining temporal characteristics.

Acknowledgments. This work was supported by Natural Science Foundation of China (61671258) and the Natural Science Foundation of Zhejiang Province, China (LY15F010005).

References

1. Shao, F., Li, K., Lin, W., Jiang, G., Yu, M., Dai, Q.: Full-reference quality assessment of stereo-scopic images by learning binocular receptive field properties. IEEE Trans. Image Process. **24**, 2971–2983 (2015)
2. Zhu, T., Karam, L.: A no-reference objective image quality metric based on perceptually weighted local noise. EURASIP J. Image Video Process. **2014**(5), 1–8 (2014)
3. Soundararajan, R.: Bovik, A.C: Video quality assessment by reduced reference spatio-temporal entropic differencing. IEEE Trans. Circuits Syst. Video Technol. **23**, 684–694 (2013)
4. Memon, M.H., Li, J.P., Memon, I., et al.: Efficient object identification and multiple regions of interest using CBIR based on relative locations and matching regions. In: International Computer Conference on Wavelet Active Media Technology and Information Processing, pp. 247–250, 16 June 2016

5. Memon, M.H., Khan, A., Li, J.P., et al.: Content based image retrieval based on geo-location driven image tagging on the social web. In: International Computer Conference on Wavelet Active Media Technology and Information Processing, pp. 280–283, 30 March 2014

6. Memon, I., Chen, L., Majid, A., et al.: Travel recommendation using geo-tagged photos in social media for tourist. Wirel. Pers. Commun. **80**(4), 1347–1362 (2015)

7. Shaikh, R.A., Li, J.P., Khan, A., et al.: Biomedical image processing and analysis using Markov random fields. In: The International Computer Conference on Wavelet Active Media Technology and Information Processing, ICCWAMTIP 2015, pp. 179–183, 16 June 2016

8. Bensalma, R., Larabi, M.C.: A perceptual metric for stereoscopic image quality assessment based on the binocular energy. Multidimension. Syst. Sig. Process. **24**, 281–316 (2013)

9. Maalouf, A., Larabi, M.C.: CYCLOP: a stereo color image quality assessment metric. In: IEEE International Conference on Acoustics, Speech and Signal Processing, pp. 1161–1164 (2011)

10. Chen, M., Su, C., Kwon, D., Cormack, L., Bovik, A.C.: Full-reference quality assessment of stereoscopic images by modeling binocular rivalry. In: Asilomar Conference on Signals, Systems and Computers (ASILOMAR), pp. 721–725 (2012)

11. Jin, L., Boev, A., Egiazarian, K., Gotchev, A.: Quantifying the importance of cyclopean view and binocular rivalry-related features for objective quality assessment of mobile 3D video. EURASIP J. Image Video Process. **2014**(6), 1–18 (2014)

12. Hewage, C.T.E.R., Martini, M.G.: Reduced-reference quality metric for 3D depth map transmission. In: 3DTV-Conference, Tampere, pp. 1–4, 7–9 June 2010

13. Knill, D.C., Pouget, A.: The Bayesian brain: the role of uncertainty in neural coding and computation. Trends Neurosci. **27**(12), 712–719 (2004)

14. Wang, X., Liu, Q., Wang, R., Chen, Z.: Natural image statistics based 3D reduced reference image quality assessment in contourlet domain. Neurocomputing **151**, 683–691 (2015)

15. Wu, J., Lin, W., Shi, G.: Perceptual quality metric with internal generative mechanism. IEEE Trans. Image Process. **22**, 43–54 (2013)

16. Telecommunication Standardization Sector of ITU, Subjective Video Quality Assessment Methods for Multimedia Applications, Recommendation ITU-T P. 910 (2008)

17. Zhou, J., Jiang, G., Mao, X., et al.: Subjective quality analyses of stereoscopic images in 3DTV system. In: IEEE Conference on Visual Communication and Image Processing, Taiwan, pp. 1–4, 6–9 November 2011

18. Moorthy, A., Su, C., Mittal, A., Bovik, A.C.: Subjective evaluation of stereoscopic image quality. Sig. Process. Image Commun. **28**, 870–883 (2013)

New Tone-Mapped Image Quality Assessment Method Based on Color Space

Hao Song, Gangyi Jiang, Hua Shao, and Mei Yu[✉]

Faculty of Information Science and Engineering,
Ningbo University, Ningbo 315211, China
yumei2@126.com

Abstract. High dynamic range image can provide wider dynamic range and more image details, it is needed a tone-mapping operator in order to be showed on an ordinary display, how to evaluate the tone-mapped image becomes an important problem to be solved. The distortion of tone-mapped image is different from the traditional image distortion, so, this paper proposes an objective quality evaluation algorithm of tone-mapped image based on color space which considers the difference between the reference and test images, the structural fidelity, the color distortion and the naturalness of the test image. Finally, the support vector machine is used as the pooling strategy to set up the quality assessment model. The experimental results show that the Pearson linear correlation coefficient of the proposed method is about 0.86, the Spearman rank correlation coefficient is about 0.84, which means that the proposed method is consistent with human visual perception.

Keywords: High dynamic range · Tone-mapped image · Quality assessment

1 Introduction

Human visual system (HVS) can accept luminance dynamic range of about 10^4, the luminance range of high dynamic range (HDR) image is about 10^{-4} cd/m^2–10^5 cd/m^2, it can present the true feeling of natural scene [1]. Display device is mainly based on 8 bit display depth currently, tone mapping operators are used to convert HDR image to low dynamic range (LDR) image for visualization, and the tone-mapped images should retain HDR image perceptive quality as much as possible [2, 3]. Tone-mapped image quality assessment (IQA) is an important issue to be solved.

Both typical full-reference (FR) objective LDR IQA approaches which include GMSD [4], FSIM [5], GSSIM [6], MS-SSIM [7] etc. and HDR IQA approaches which include HDR-VQM [8], HDR-VDP-2 [9] etc. assume the reference and test images to have the same dynamic range, and tone mapping mainly produce fidelity distortion and color distortion which are different from traditional images' distortion, thus the LDR IQA methods cannot be directly applied to evaluate tone-mapped images. It is necessary to consider new IQA methods for tone-mapped images. Yeganeh et al. proposed a tone-mapped quality index (TMQI) [10] which combines structure fidelity with a statistical naturalness. TMQI only considers the luminance component and ignores the chrominance component. Hossein et al. proposed a feature similarity index

© Springer Nature Singapore Pte Ltd. 2017
H. Yuan et al. (Eds.): GRMSE 2016, Part I, CCIS 698, pp. 365–376, 2017.
DOI: 10.1007/978-981-10-3966-9_42

for tone-mapped (FSITM) [11] with the locally weighted mean phase maps of HDR images and its corresponding tone-mapped images. Though it takes into account chrominance, three channels of RGB color space are non-orthogonal.

Considering dynamic range, chrominance and naturalness, this paper proposes an objective tone mapped image quality assessment (TM-IQA) method. First, HDR image's luminance is processed for the luminance difference between HDR images and tone-mapped images, and then the magnitude and phase information are extracted in the gradient domain to be gotten the structure similarity in which the HDR image's gradient amplitude are used as weights. Second, chrominance information is directly evaluated. Then, the naturalness of tone-mapped image is also taken into account. Finally, the features are fused to get TM-IQA models by the support vector machine. Experimental results show that the proposed method is better consistent with visual perception, compared with other methods.

2   The Proposed Method

Luminance is much sensitive to HVS than chrominance, thus most existing IQA methods just consider luminance. But chrominance played an important role in the tone mapping, so it cannot be ignored in TM-IQA. Figure 1(a) is a HDR image, Figs. 1(b) and (c) show two tone-mapped images produced by different TMOs. Obviously, the structures of Fig. 1(b) are similar to Fig. 1(c), while the color of Fig. 1(b) is more vivid than Fig. 1(c), and Fig. 1(c) is brighter than Fig. 1(b). Here, TMQI and FSITM are used to calculate the two tone-mapped images, Table 1 shows the results. The quality of Fig. 1(b) is lower than Fig. 1(c) calculated by the TMQI because of its focus on the luminance. FSITM also focuses on the chrominance, so the colorful Fig. 1(b) is better than light Fig. 1(c). It is easy to find that chrominance is also important in TM-IQA. And using the proposed method in which chrominance is considered in this paper, the scores are also shown that Fig. 1(b) is better than Fig. 1(c).

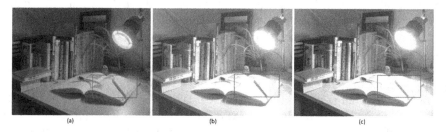

(a)   (b)   (c)

Fig. 1. HDR and its corresponding tone-mapped images generated with different TMOs: (a) HDR image. (b) The first tone-mapped image. (c) The second tone-mapped image. (Color figure online)

HSI color space is on the basis of the HVS, using hue, saturation and intensity to describe color. We convert HDR image and tone-mapped images into HSI color space, and extract their hue and saturation components, shown in Figs. 2 and 3. It is satisfied

Table 1. Tone-mapped images quality in Fig. 1 computed by different algorithms

Model	Figure 1(b)	Figure 1(c)
TMQI	0.7919	0.8103
FSITM	0.8040	0.7857
Proposed	0.7813	0.7745

that the dynamic range of two components of the HDR image and tone-mapped image are consistent, ranging from 0 to 1. Obviously, the structures of all images are similar. Figure 2(a) is clearer than Figs. 2(b) and (c), and Figs. 2(b) and (c) appear some distortion like the area in the red frame, and the areas in Fig. 2(c) is more than Fig. 2 (b). Figure 3(a) is brighter than Fig. 3(b) while Fig. 3(b) is brighter than Fig. 3(c), similarly, details in Fig. 3(a) is more than those in Fig. 3(b) while details in Fig. 3(b) is more than those in Fig. 3(c). These phenomena indicate that HDR image is more colorful than tone-mapped image, because of a certain extent distortion, and the more flat the color is, the more serious the distortion is. These phenomena indicate that the characteristics of the hue and saturation maps also represent HVS.

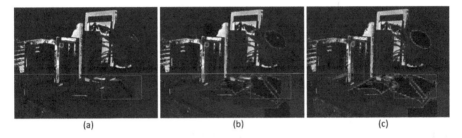

Fig. 2. The hue maps of the HDR and its corresponding tone-mapped images in Fig. 1: (a) The hue map of the HDR image. (b) The hue map of the first tone-mapped image. (c) The hue map of the second tone-mapped image. (Color figure online)

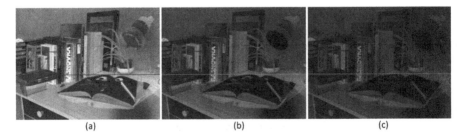

Fig. 3. The saturation maps of the HDR and its corresponding tone-mapped images in Fig. 1: (a) The saturation map of the HDR image. (b) The saturation map of the first tone-mapped image. (c) The saturation map of the second tone-mapped image. (Color figure online)

Generally, YUV space is used in IQA [12], but it is not applicable for comparison of the images with different dynamic ranges. In TM-IQA, if HSI color space is used, only the luminance needs to be transformed, and if the YUV space is used, the luminance and chrominance information both need complicated conversion in advance. There is no doubt that using YUV space will increase complexity. As a result, it is reasonable to evaluate the tone-mapped image quality in HSI color space.

In this paper, a new TM-IQA method is proposed by comparing similarity of the three components of the HDR and tone-mapped images respectively, and calculating a tone-mapped image's naturalness, then using SVM to fuse all features and predicting the tone-mapped image quality. The proposed method is shown in Fig. 4.

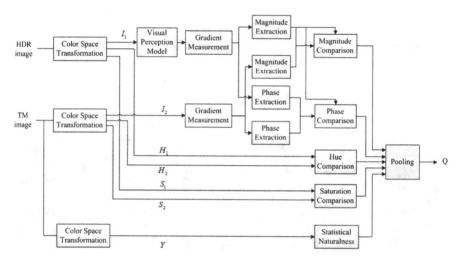

Fig. 4. The framework of the proposed algorithm. I1, H1, S1 and I2, H2, S2 represent the HDR and tone-mapped images' intensity, hue, saturation components respectively, Y represents the luminance of the tone-mapped image in Yxy color space.

2.1 Structural Fidelity

Figures 5(a), (b) and (c) are the intensity of the HDR and tone-mapped images corresponding to three images in Fig. 1. Obviously, the bright areas in Fig. 5(a) are particular bright, the dark areas are particular dark. Figure 6 is the three pixel values curves of the three lines in Fig. 5. The red curve represents HDR image's intensity, the other two curves represent two tone-mapped images' intensity. Obviously, compared with the tone-mapped images whose curves are relatively smooth, the curve of the HDR image has much wider dynamic range, and its fluctuation is extremely intense.

The response of the HVS to luminance has logarithmic property, and it is a monotonous nonlinear system [13]. Though the logarithmic transformation, HVS could represent the luminance range of 10^8. This nonlinear effect occurs before the mutual effects between the visual signals of different visual cells. Thus the logarithmic transformation can be used to transform the luminance of large dynamic range into

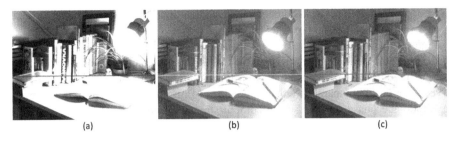

Fig. 5. The intensity maps of the HDR and its corresponding tone-mapped images in Fig. 1: (a) The intensity map of the HDR image. (b) The intensity map of the first tone-mapped image. (c) The saturation map of the second tone-mapped image. (Color figure online)

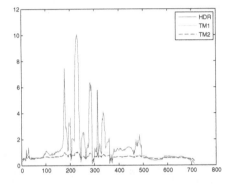

Fig. 6. Profile of the signals along the three lines in Fig. 5. The red, green, blue curves are for the lines in Figs. 4(a), (b) and (c), respectively. (Color figure online)

appropriate value to be perceived by the HVS. Though this transformation, the low input which has narrow range will be mapped to the output which has wide range, on the contrary, for the high input, too. Thus it can be used to extend the pixels in the dark areas and compress the pixels in the bright areas in the HDR images.

Image gradient contains edge information, so it is used as a characterization of the image's structure to measure image distortion. Image gradient is divided into gradient magnitude and gradient phase. Here, we use Sobel operators, h_x and h_y, in the horizontal and vertical directions, I represents an image. The gradient magnitudes in two directions, denoted by G_x, G_y are computed by $G_x = I \otimes h_x, G_y = I \otimes h_y$. The gradient magnitude at (i, j) is computed by $G(i,j) = \sqrt{G_x^2(i,j) + G_y^2(i,j)}$, and the gradient phase at (i, j) is computed by $\theta(i,j) = \arctan(G_y^2(i,j)/G_x^2(i,j))$.

Figure 7 shows the gradient phase maps of the HDR and tone-mapped images corresponding to three images in Fig. 5, the images in Fig. 8 are the corresponding gradient magnitude maps. Obviously, the details of the HDR image are more clear than the tone-mapped because of the distortion of intensity. As shown in the yellow frame in Figs. 7 (b) and (c), these smooth areas details are caused by serious distortion, and these areas usually too bright or too dark. The corresponding distorted areas in Figs. 8(b) and (c) also have not any details. These phenomenon indicate that the high and low luminance areas are easy to produce distortion in tone mapping.

Figures 9(a) and (b) are the three pixel values curves of three lines in Figs. 7 and 8, respectively. Obviously, three curves have the similar fluctuation, in other words, the gradient trend of the tone-mapped image is the similar to HDR image. Especially, these curves have more comparability than the curves of original intensity. So, the gradient magnitude and phase can be used to compute the similarity of intensity.

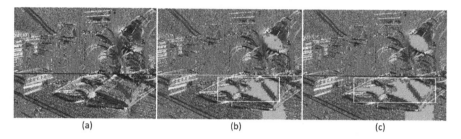

Fig. 7. The gradient phase maps of the intensity of the HDR and tone-mapped images in Fig. 5: (a) The gradient phase map of the image in Fig. 5(a). (b) The gradient phase map of the image in Fig. 5(b). (c) The gradient phase map of the image in Fig. 5(c). (Color figure online)

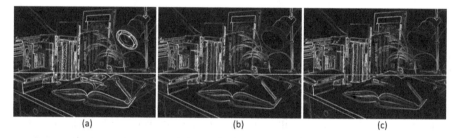

Fig. 8. The gradient magnitude maps of the intensity of the HDR and tone-mapped images in Fig. 5: (a) The gradient magnitude map of the image in Fig. 5(a). (b) The gradient magnitude map of the image in Fig. 5(b). (c) The gradient magnitude map of the image in Fig. 5(c). (Color figure online)

For the two-dimensional image signal, multi-resolution decomposition can be used to deal with the image from the coarse resolution to fine resolution, which is called band-pass property. This paper adopts a multi-resolution structure similarity model to evaluate the intensity component. The algorithm of gradient phase similarity frame is shown in Fig. 10. The original SSIM contains three components – luminance, contrast and structure. HDR and tone-mapped images' gradient phase maps are used as the two main inputs, and considering that the larger the gradient magnitude is, the more sensitive the HVS is, HDR image gradient magnitude map is used as the input of weights. And the inputs in next scale are obtained after the original inputs are low-pass filtered and downsampled.

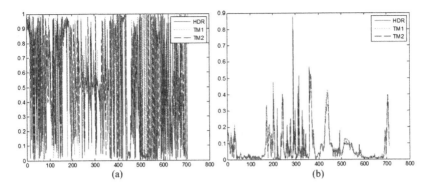

(a)   (b)

Fig. 9. Profile of the signals along the three lines in Figs. 7 and Fig. 8.: (a) The red, green, blue curves are for the lines in Fig. 7(a), (b), (c) respectively. (b) The red, green, blue curves are for the lines in Fig. 8(a), (b), (c) respectively. (Color figure online)

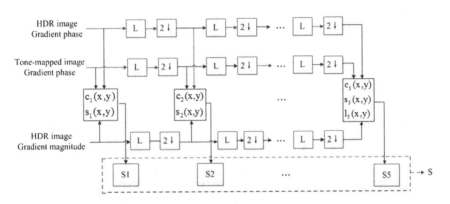

Fig. 10. The framework of the multi-scale similarity

In the first scale, computing the similarity of the contrast and structure between the two main inputs, noted by $c_1(x, y)$ and $s_1(x, y)$. Then, dividing $c_1(x, y) * s_1(x, y)$ into 7×7 blocks, and computing every block's mean, denoted as $cs_1(i)$. Next, the inputs of weights denoted by $g_1(x, y)$ is dealt with the same way, the obtained block mean should be divided by the sum of the total HDR image gradient magnitude, denoted as $g_1(i)$, which identifies the importance of blocks. Finally, the similarity in the first scale, denoted as S_1, is computed by

$$S_1 = \sum_{i=1}^{B} cs_1(i) * g_1(i) \tag{1}$$

where i is the i th block, and B is the number of blocks.

The similarities in the second, third, fourth scales are computed by the same ways as above, noted by S_1, S_2, S_3, S_4. In the fifth scale, the similarity of the two main inputs' luminance, noted by $l_5(x, y)$, is taken into account. Then, dividing $c_5(x, y) * s_5(x, y) * l_5(x, y)$ into blocks, and computing their means, noted by $cs_5(i)$. The similarity in the fifth scale, denoted by S_5, is computed by

$$S_5 = \sum_{i=1}^{B} csl_5(i) * g_5(i) \tag{2}$$

The gradient phase similarity, noted by QIP, is computed by pooling the five scales' similarities, and the weights of every scale are set as $\{\beta l\} = \{0.0448, 0.2856, 0.3001, 0.2363, 0.1333\}$ [7].

$$Q_{IP} = S = \prod_{l=1}^{5} S_l^{\beta_l} \tag{3}$$

Setting the two main puts as HDR and tone-mapped gradient magnitude maps, the gradient magnitude similarity, denoted as QIG, can be gotten in the same way.

2.2 Chrominance Similarity

Figures 11(a) and (b) are the three pixel values curves of the three lines in Figs. 2 and 3, respectively. Obviously, three curves in Fig. 11(a) are reflecting the same trend which indicates that the hue components of the HDR and tone-mapped images are similar, just a few areas produce obvious distortion. The curves in Fig. 11(b) have the similar fluctuation while their absolute values have certain disparity, the saturation of HDR images is generally higher than tone-mapped images which indicates that the saturation distortion is the general distortion in tone mapping. As a result, hue and saturation components can be used as a measurement in TM-IQA. GMSD is used to calculate the similarities of the hue and saturation between HDR and tone-mapped images, the two similarities are noted by QH and QS.

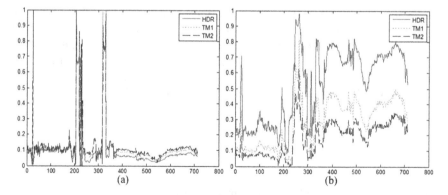

Fig. 11. Profile of the signals along the three lines in Figs. 2 and 3: (a) The red, green, blue curves are for the lines in Fig. 2(a), (b), (c) respectively. (b) The red, green, blue curves are for the lines in Fig. 3(a), 3(b), and 3(c), respectively. (Color figure online)

2.3 Statistical Naturalness

A high quality tone-mapped image should not only preserve the details of the HDR image as much as possible, but also look natural [14]. At present, the naturalness is used as a measurement by many NR IQA method [15]. Generally, brightness and contrast are the most important factors effected the naturalness.

This model is on the basis of the statistical information of luminance and contrast of 3000 high quality natural images. First, the natural image is transformed into Yxy color space, extracted the luminance components, and divided into 11×11 blocks. Next, computing the local mean and standard deviation of every block, noted by M_p and N_p, then the global mean and standard deviation of the image, noted by m and d, is pooled by averaging the total local values. As a result, the means and standard deviations of 3000 images can be well fitted by a Gaussian function and a Beta probability density function defined as follows:

$$P_m(m) = \frac{1}{\sqrt{2\pi}\sigma_m} \exp\left[-\frac{m - \mu_m}{2\sigma_m^2}\right], \; P_d(d) = \frac{(1 - d)^{\beta_d - 1} d^{\alpha_d - 1}}{B(\alpha_d, \beta_d)} \tag{4}$$

where $B(\cdot)$ is the Beta function. By a series of experiments, the parameters are set as $\mu_m = 115.94$, $\alpha_m = 27.99$, $\sigma_d = 4.4$, and $\beta_d = 10.1$.

Finally, the statistical naturalness measure is computed by $N = P_m P_d / K$, where K is a normalization factor changed with both P_m and P_d, defined as K = $\max\{P_m, P_d\}$ to make the statistical naturalness feature N normalized.

2.4 Pooling

So far, five features, including Q_{IP}, Q_{IG}, Q_H, Q_S and N, are extracted. Machine learning has been applied in IQA widely. Here, five features and the MOS values are the inputs, SVM are used as the pooling strategy to set up an TM-IQA model to predict the tone-mapped image quality.

3 Experimental Results

In order to verify the validity of the proposed algorithm, we compare our objective quality with subjective score on the TMID database [9]. The database contains 15 original HDR images and each HDR image has 8 tone-mapped images produce by 8 different TMOs. The subjective quality of the tone-mapped images are scored by 20 individuals, ranked from 1 (best quality) to 8 (worst quality). Finally, each image's MOS value is the mean of the 20 scores. Pearson linear correlation coefficient (PLCC), Spearman rank-order correlation coefficient (SROCC) and Kendall's rank-order correlation coefficient (KROCC) are used to compute the correlation performance. The higher the three scores are, the more accurate the algorithm is.

120 tone-mapped images are divided into two subsets, one is the training set which consists of 80% of the 120 images, the other is the testing set which consists of the rest

20%. To test the robustness of the proposed algorithm, the two subsets are selected randomly for 1000 times, then we will get 1000 values for each performance, and the final value is the median of 1000 values.

In Table 2, the first four models are traditional LDR IQA based on luminance. It is clear that the traditional LDR IQA models are not suitable for TM-IQA.

Table 2. Performance of the existing and proposed methods

Model	PLCC	SROCC	KROCC
MS-SSIM	0.4478	0.2927	0.1956
FSIM	0.3008	0.1874	0.1171
GSSIM	0.4886	0.3830	0.2577
GMSD	0.3204	0.1202	0.0723
TMQI [10]	0.7715	0.7407	0.5585
FSITM [11]	0.7496	0.7028	0.5160
Proposed$_0$	0.8086	0.7771	0.6062
Proposed$_I$	0.8226	0.8010	0.6377
Proposed$_H$	0.7488	0.6756	0.4783
Proposed$_S$	0.7443	0.6608	0.4643
Proposed$_C$	0.8689	0.8474	0.6764

TMQI combines the modified MS-SSIM and the statistical naturalness model which is used to evaluate the tone-mapped images' naturalness, the performance is much higher than MS-SSIM. TMQI ignores chrominance components, thus the performance has yet to be improved. FSITM is based on the FSIM, but it transforms the HDR images' values due to the factor of the dynamic range and also considers the chrominance component, the FSITM performs better than the FSIM. But the three channels of RGB color space are non-orthogonal, so there exists mutual influence among different channels so that the performance of the FSITM is not very good.

The last five methods are the proposed tone-mapped approaches in this paper. Proposed$_0$ is the approach which is only considering intensity component, and it is based on the GSSIM. First, considering the limit of dynamic range, it compresses the HDR images according to the human visual perception model, then extracts the gradient phase and magnitude features to compare similarity used the MS-SSIM respectively because of the consistent of the dynamic range in the gradient domain, next adds a statistical naturalness model, finally uses the SVM to predict quality. Obviously, the performance of the Proposed$_0$ is much higher than GSSIM, and also it performs better than the TMQI and FSITM. The Proposed$_0$ adopts the traditional SSIM model which is averaging the similarities of blocks to get the overall similarity in each scale. Generally, the areas of higher gradient magnitude are more sensitive to human, so the weights of these areas should be higher. The Proposed$_I$ improves the pooling strategy in the SSIM in each scale through using the HDR gradient magnitude as the weights' metric, it is more suitable for the HVS. It is easy to found that the Proposed$_I$ performs better than the **Proposed$_0$**. The Proposed$_H$ is the approach which is only considering the hue component, the **Proposed$_S$** is the approach which is only considering the saturation component, it can be found that it can't perform well if only consider chrominance because the HVS is much more sensitive to luminance than to chrominance. Finally, the **Proposed$_C$** combines luminance and chrominance, which is based on the **Proposed$_I$**, **Proposed$_H$** and **Proposed$_S$**. Obviously, it performs better than the single approaches, and the performance of the proposed method is better than the TMQI and the FSITM.

4 Conclusion

This paper proposed an objective TM-IQA approach based on color space. Since it considers the suitable color space in which has less algorithm complexity than other color spaces, the transformation of the luminance dynamic range, structural fidelity, modified pooling strategy, color distortion and naturalness, it performance better than TMQI and FSITM and fits well with HVS. However, the FR IQA on the situation that the reference and test images' dynamic are greatly different has two much limitations. So, reduced-reference or blind TM-IQA methods will be developed in the further works, and aesthetics of images should be considered seriously.

Acknowledgement. This work was supported by Natural Science Foundation of China (61671258) and the Natural Science Foundation of Zhejiang Province, China (LY15F010005).

References

1. Artusi, A., Mantiuk, R.K., Richter, T., et al.: JPEG XT: a compression standard for HDR and WCG images. IEEE Signal Process. Mag. **33**(2), 118–124 (2016)
2. Ma, K., Yeganeh, H., Zeng, K., et al.: High dynamic range image compression by optimizing tone-mapped image quality index. IEEE Trans. Image Process. **24**(10), 3086–3097 (2015)
3. Eilertsen, G., Mantiuk, R.K., Unger, J.: Real-time noise-aware tone mapping. ACM Trans. Graph. **34**(6), 198:1–198:15 (2015)
4. Xue, W., Zhang, L., Mou, X., et al.: Gradient magnitude similarity deviation: a highly efficient perceptual image quality index. IEEE Trans. Image Process. **23**(2), 684–695 (2014)
5. Zhang, L., Zhang, L., Mou, X., et al.: FSIM: a feature similarity index for image quality assessment. IEEE Trans. Image Process. **20**(8), 2378–2386 (2011)
6. Liu, A.M., Lin, W.S., Narwaria, M.: Image quality assessment based on gradient similarity. IEEE Trans. Image Process. **21**(4), 1500–1512 (2012)
7. Wang, Z., Simoncelli, E.P., Bovik, A.C.: Multi-scale structural similarity for image quality assessment. Asilomar Conf. Signals Syst. Comput. **2**, 1398–1402 (2003)
8. Narwaria, M., Silva, M.P.D., Callet, P.L.: HDR-VQM: an objective quality measure for high dynamic range video. Sig. Process. Image Commun. **35**, 46–60 (2015)
9. Mantiuk, R., Kim, K.J., Rempel, A.G., et al.: HDR-VDP-2: a calibrated visual metric for visibility and quality predictions in all luminance conditions. ACM Trans. Graph. **30**(4), 76–79 (2011)

10. Yeganeh, H., Wang, Z.: Objective quality assessment of tone-mapped images. IEEE Trans. Image Process. **22**(2), 657–667 (2013)
11. Nafchi, H.Z., Shahkolaei, A., Moghaddam, R.F., et al.: FSITM: a feature similarity index for tone-mapped images. IEEE Signal Process. Lett. **22**(8), 1026–1029 (2015)
12. Xie, J.: Principles and Applications of Vision Bionics. Science Press, Beijing (2013)
13. Cadik, M., Slavik, P.: The naturalness of reproduced high dynamic range images. Int. Conf. Inf. Vis. **24**(11), 920–925 (2005)
14. Appina, B., Khan, S., Channappayya, S.: No-reference stereoscopic image quality assessment using natural scene statistics. Sig. Process. Image Commun. **43**, 1–14 (2016)
15. Besrour, A., Abdelkefi, F., Siala, M., et al.: Luminance and contrast ideal balancing based tone mapping algorithm. In: Proceedings of SPIE, vol. 9598 (2015)

A Modified NCSR Algorithm for Image Denoising

Diwei Li$^{(\boxtimes)}$, Yunjie Zhang, and Xin Liu

Department of Mathematics, Dalian Maritime University, Dalian 116026, China
18841158062@sina.cn

Abstract. In this paper, a modified nonlocally centralized sparse representation method is introduced, which is suitable for removing both the non-sparse noise and sparse noise such as salt and pepper noise, periodic noise, and mixed noise in particular. In the proposed method the conventional median filtering is embedded in nonlocally centralized sparse representation. The main advantage is that it can attain better performance for various common noise, and significantly superior for mixed noise. The effective and efficient of the proposed method is demonstrated experimentally.

Keywords: Image denoising · Sparse noise · Nonlocally centralized sparse representation

1 Introduction

In recent years, sparse representation has attracted considerable attention in areas of image and vision processing. Sparse representation model can effectively describe the intrinsic structure of the image because many images can be represented using only a few non-zero coefficients in a suitable basis or dictionary [1–3].

The purpose of image denoising is to reconstruct the original image by removing unwanted noise from a corrupted image. In previous decades, numerous contributions addressed denoising problem from many and diverse points of view. Generally, image denoising approaches can be categorized as spatial domain, transform domain, and sparse representation (simultaneous dictionary learning) according to the image representation. Specifically, sparse representation based methods have become a trend for image denoising [4, 5].

The general idea of sparse representation-based denoising approaches is to construct or learn an appropriate over-complete dictionary that can accurately fit the local structures of images, such that the estimated image can be expressed as a linear combination of only few atoms chosen out from this dictionary. Representative methods are the K-clustering with singular value decomposition (K-SVD), learned simultaneous sparse coding (LSSC), clustering-based sparse representation (CSR), and nonlocally centralized sparse representation (NCSR) [5].

Nonlocally centralized sparse representation (NCSR) model was proposed by Dong et al. in [6, 7] for image restoration tasks, including denoising, deblurring and

© Springer Nature Singapore Pte Ltd. 2017
H. Yuan et al. (Eds.): GRMSE 2016, Part I, CCIS 698, pp. 377–386, 2017.
DOI: 10.1007/978-981-10-3966-9_43

super-resolution. The main contribution of the NCSR method lies in integrating local sparsity and the nonlocal sparsity constraints into a variational framework for optimization. It exploits the image nonlocal self-similarity to obtain good estimates of the sparse coding coefficients of the original image, and then centralize the sparse coding coefficients of the observed image to those estimates.

Sparse representation-based denoising algorithms have been proven to have a strong ability to denoise additive Gaussian white noise [8]. However, it performs suboptimally for the salt and pepper noise, periodic noise, and mixed noise. In this paper, we propose a modified image denoising approach which embedding conventional median filtering in the NCSR. The proposed method is considering to remove both the non-sparse noise and the sparse noise. Empirical experiments on noisy images illustrate that the modified algorithm is effective and efficient.

The rest of the paper is organized as follows. The sparse representation and NCSR algorithm are reviewed in Sect. 2. Section 3 introduces the proposed modified approach. Section 4 describes the conducted experiments, verifying the efficiency of the algorithm on noisy images. Finally, conclusions are drawn in Sect. 5.

2 Sparse Representation and NCSR Model

Assuming that $x \in \mathbf{R}^N$ denotes an image, the sparse representation model can be formulated as $x \approx \Phi\alpha$, where $\Phi \in \mathbf{R}^{N \times M}$ ($M > N$) is an over-complete dictionary, and most entries of the coding vector α are zero or close to zero. Motivated by the sparsity, the model aims to solve the following l_0-minimization problem

$$\alpha_x = \arg\min_{\alpha} ||\alpha||_0, \text{ s.t. } ||x - \Phi\alpha||_2 \leq \varepsilon \tag{1}$$

According to the theory of sparse optimization [3], if the solution is sparse enough, the l_0-minimization problem is equivalent to the l_1-minimization problem, which can be generally formulated in the following Lagrangian form

$$\alpha_x = \arg\min_{\alpha} \{ ||x - \Phi\alpha||_2^2 + \lambda||\alpha||_1 \} \tag{2}$$

where constant λ denotes the regularization parameter.

Let y be the noisy version of x, it can be generally formulated by $y = x + n$, where n indicates the independent additive noise [4, 5]. In the scenario of denoising based on sparse representation, to recover x from y, first y is sparsely coded with respect to Φ by solving the l_1-minimization problem

$$\alpha_y = \arg\min_{\alpha} \{ ||y - \Phi\alpha||_2^2 + \lambda||\alpha||_1 \} \tag{3}$$

and then x is reconstructed by $\hat{x} = \Phi\alpha_y$.

Furthermore, if $x_i = R_i x$ is used to represent an image patch of size $\sqrt{n} \times \sqrt{n}$ extracted at pixel i, where R_i is the matrix extracting patch x_i from x at pixel i, then each patch can be sparsely represented as $x_i \approx \Phi \alpha_{x,i}$ by solving an l_1-minimization problem

$$\alpha_{x,i} = \arg \min_{\alpha_i} \{ ||x_i - \Phi \alpha_i||_2^2 + \lambda ||\alpha_i||_1 \} \tag{4}$$

The image x is reconstructed by averaging each reconstructed patch of x_i, i.e.

$$x \approx \Phi \circ \alpha_x = \left(\sum_{i=1}^{N} R_i^T R_i \right)^{-1} \sum_{i=1}^{N} \left(R_i^T \Phi \alpha_{x,i} \right) \tag{5}$$

where α_x denotes the concatenation of all $\alpha_{x,i}$. Thus, the patch-based sparse denoising model can be represented as $\hat{x} = \Phi \circ \alpha_y$, where

$$\alpha_y = \arg \min_{\alpha} \{ ||y - \Phi \circ \alpha||_2^2 + \lambda ||\alpha||_1 \} \tag{6}$$

and y is the noisy version of x. The dictionary Φ can be learned by using algorithms such as KSVD. Obviously, this model can better adapt to local image structures [4].

Based on the fact that natural images often contain repetitive structures, i.e., the rich amount of nonlocal redundancies, the NCSR approach searched the nonlocal similar patches to the given patch i in order to obtain nonlocal estimate of unknown sparse code [6, 7]. Thus, replacing the local sparsity term $||\alpha||_1$ in Eq. (6) with the nonlocally centralized sparsity term $\sum_i ||\alpha_i - \beta_i||_1$, the nonlocally centralized sparse representation (NCSR) model can be built by solving

$$\alpha_y = \arg \min_{\alpha} \{ ||x_i - \Phi \circ \alpha||_2^2 + \lambda \sum_i ||\alpha_i - \beta_i||_1 \} \tag{7}$$

where the parameter λ that balances the fidelity term and the nonlocally centralized sparsity term should be adaptively determined for better performance, and β_i is an estimation of the unknown sparse coefficients α_i, which can be obtained by using the nonlocal redundancy of natural images.

3 Modified NCSR Method

Sparse representation-based denoising algorithms have been proven to have a strong ability to denoise additive Gaussian white noise [8]. However, it performs suboptimally for the salt and pepper, periodic and mixed noise. For example, we add Gaussian white noise, salt and pepper noise, mixture of Gaussian and salt & pepper, and periodic noise to the *House* image, *Peppers* image, *Cameraman* image respectively. The SNRs of noisy image and denoised image by NCSR are listed in Tables 1, 2 and 3.

Table 1. The House image performance of NCSR algorithm for various noise types

Noise type	Gaussian	Salt & Pepper	Mixed	Periodic
Original PSNR	22.54	22.52	22.61	22.60
NCSR PSNR	30.22	25.42	27.89	22.61
Raised ratio	1.34	1.13	1.23	1.00

Table 2. The Peppers image performance of NCSR algorithm for various noise types

Noise type	Gaussian	Salt & Pepper	Mixed	Periodic
Original PSNR	28.14	28.10	28.15	28.12
NCSR PSNR	30.07	28.32	31.95	28.91
Raised ratio	1.07	1.01	1.13	1.03

Table 3. The Cameraman image performance of NCSR algorithm for various noise types

Noise type	Gaussian	Salt & Pepper	Mixed	Periodic
Original PSNR	28.25	28.12	28.18	28.22
NCSR PSNR	31.79	27.80	30.86	27.66
Raised ratio	1.13	0.99	1.10	0.98

The above occurrence is mainly caused by both the forming mechanism of noise and the applicability of algorithm. Gaussian noise is non-sparse noise, which can be removed well by NCSR algorithm. However, salt and pepper noise and periodic noise are sparse in the identity and the Fourier basis, respectively. Since both the images and the noises are sparse over some dictionaries [8], NCSR algorithm has a serious drawback which is difficult to overcome for the sparse noises, and may not be appropriate in some cases.

The median filtering is a nonlinear image smoothing and enhancement technique, which considers each pixel in the image in turn and replaces it with the median of neighboring pixels. The median filter removes both the noise and the fine detail since it can't tell the difference between the two. It implies that filtered noise by median filter is likely to appear non-sparsity, and suitable to use NCSR algorithm to filter.

Above these motivate us to consider a two stage method to improve the NCSR algorithm. The proposed method first uses adaptive median filter to reduce sparsity of noise, and then uses NCSR to denoise the resulting data of the first stage from the remaining noise.

The pseudo-code of the modified NCSR algorithm with respect to implementation is illustrated in Table 4.

Table 4. The modified NCSR algorithm

Input: Noisy image y.

Output: Denoised image x.

Stage 1

Given maximum allowed size S_{max} of the neighborhood

For each pixel y_{ij} in the image y

Level A:

$A_1 = y_{med} - y_{min}$, $A_2 = y_{med} - y_{max}$

if $A_1 > 0$ AND $A_2 < 0$, go to level B

else increase the window size

if window size $< S_{max}$, repeat level A

else output y_{ij}

Level B:

$B_1 = y_{ij} - y_{min}$, $B_2 = y_{ij} - y_{max}$

if $B_1 > 0$ AND $B_2 < 0$, output y_{ij}

else output y_{med}

Stage 2

Initialization:

(a) Set the initial estimate as $x^{(0)}$

(b) Set initial regularization parameter λ and δ

Iterative Denoising:

Outer loop: iterateon $l = 1, 2, ..., L$

Update the dictionaries $\{\Phi_k\}$ via K-means and PCA

Inner loop: iterateon $j = 1, 2, ..., J$

(a) $x^{(j+1/2)} = x^{(j)} + \delta(y - x^{(j)})$, where δ is the pre-determined constant

(b) Compute

$$v^{(j)} = [\Phi_{k_1}^T R_1 x^{(j+1/2)}, ..., \Phi_{k_N}^T R_N x^{(j+1/2)}]$$

where Φ_{k_i} is the dictionary assigned to patch $x_i = R_i x^{(j+1/2)}$

(c) Compute $\alpha_i^{(j+1)}$ using the shrinkage operator

$$\alpha_i^{(l+1)} = S_\tau(v_{i,j}^{(l)} - \beta_i(j)) + \beta_i(j)$$

(d) If $mod(j, J_0) = 0$ update the parameters $\lambda_{i,j}$ and $\{\beta_i\}$ using

$$\lambda_{i,j} = \frac{2\sqrt{2}\sigma_n^2}{\sigma_{i,j}}, \quad \beta_i = \sum_{q \in \Omega_i} w_{i,q}\alpha_{i,q}, \quad w_{i,q} = \frac{1}{W}\exp(-\|x_i - x_{i,q}\|_2^2/h)$$

(e) Image estimate update: $x^{(j+1)} = \Phi \circ \alpha_y^{(j+1)}$ using

$$x \approx \Phi \circ \alpha_x = \left(\sum_{i=1}^N R_i^T R_i\right)^{-1} \sum_{i=1}^N (R_i^T \Phi \alpha_{x,i})$$

End of Inner loop.

End of Outer loop.

---

4   Experimental Results

To evaluate the performance of the proposed algorithm, we conduct a set of comparisons between NCSR and modified NCSR (MNCSR) algorithm for image denoising. The basic parameter setting of modified NCSR algorithm is as follows: the maximum allowed size of the neighborhood is $S_{max} = 5$ in Stage1; the patch size is 7×7, $K = 70$, $\delta = 0.02$, $L = 3$, and $J = 3$ in Stage2, which is same with the NCSR algorithm to make a fair comparison. The *House* image (Fig. 1), the *Peppers* image (Fig. 2) and the *Cameraman* image (Fig. 3) are used for the comparison study. The PSNR results of the two methods are reported in Tables 5, 6, 7, 8, 9, 10, 11, 12, 13, 14, 15 and 16.

Fig. 1. House image

Fig. 2. Peppers image

Fig. 3. Cameraman image

Table 5. With different Gaussian noise levels from 10 to 100 for House image

Noise levels	10	30	50	70	100
Original PSNR	28.10	18.65	14.56	12.19	10.13
NCSR PSNR	35.35	21.36	15.31	12.57	10.35
MNCSR PSNR	34.06	27.41	19.77	16.33	13.80

Table 6. With different Gaussian noise levels from 10 to 100 for Peppers image

Noise levels	10	30	50	70	100
Original PSNR	28.25	19.10	14.87	12.43	10.26
NCSR PSNR	31.79	21.72	15.69	12.84	10.49
MNCSR PSNR	29.18	25.59	19.50	16.16	13.57

Table 7. With different Gaussian noise levels from 10 to 100 for Cameraman image

Noise levels	10	30	50	70	100
Original PSNR	28.14	18.76	14.68	12.30	10.20
NCSR PSNR	33.06	21.21	15.40	12.68	10.42
MNCSR PSNR	30.29	25.85	19.42	16.21	13.71

Table 8. With different Salt & Pepper noise levels from 0.05 to 0.8 for House image

Noise levels	0.05	0.1	0.3	0.5	0.8
Original PSNR	18.48	15.38	10.67	8.50	6.41
NCSR PSNR	19.61	16.31	10.94	8.66	6.50
MNCSR PSNR	34.83	34.69	33.69	31.38	23.44

Table 9. With different Salt & Pepper noise levels from 0.05 to 0.8 for Peppers image

Noise levels	0.05	0.1	0.3	0.5	0.8
Original PSNR	18.42	15.36	10.60	8.33	6.85
NCSR PSNR	19.43	16.17	10.85	8.48	6.96
MNCSR PSNR	25.23	25.16	24.84	23.89	22.47

Table 10. With different Salt & Pepper noise levels from 0.05 to 0.8 for Cameraman image

Noise levels	0.05	0.1	0.3	0.5	0.8
Original PSNR	18.14	14.97	10.26	8.09	6.60
NCSR PSNR	19.02	15.69	10.50	8.23	6.70
MNCSR PSNR	29.28	28.92	27.08	24.44	21.84

Table 11. With different periodic noise levels from 10 to 90 for House image

Noise levels	10	30	50	70	90
Original PSNR	28.12	18.57	14.11	11.23	9.05
NCSR PSNR	28.70	19.14	14.16	11.30	9.21
MNCSR PSNR	33.03	20.82	14.17	11.61	12.99

Table 12. With different periodic noise levels from 10 to 90 for Peppers image

Noise levels	10	30	50	70	90
Original PSNR	28.12	18.57	14.11	11.23	9.05
NCSR PSNR	28.92	19.10	14.14	11.26	9.19
MNCSR PSNR	31.30	20.76	14.14	11.58	12.87

Table 13. With different periodic noise levels from 10 to 90 for Cameraman image

Noise levels	10	30	50	70	90
Original PSNR	28.12	18.57	14.11	11.23	9.05
NCSR PSNR	27.80	18.93	14.10	11.27	9.24
MNCSR PSNR	28.72	20.50	14.08	11.60	12.88

Table 14. Different mixed noise levels for House image

			Gaussian noise levels from 10 to 100				
			10	30	50	70	100
Salt & Pepper noise levels from 0.05 to 0.8	0.05	Original PSNR	18.09	15.72	13.25	11.45	9.73
		NCSR PSNR	19.41	16.87	13.79	11.76	9.94
		MNCSR PSNR	34.05	26.32	19.05	15.90	13.59
	0.1	Original PSNR	15.18	13.86	12.20	10.81	9.36
		NCSR PSNR	16.11	14.52	12.59	11.07	9.55
		MNCSR PSNR	34.01	25.51	18.56	15.58	13.35
	0.3	Original PSNR	10.66	10.24	9.60	8.95	8.14
		NCSR PSNR	10.93	10.47	9.80	9.13	8.29
		MNCSR PSNR	33.00	23.56	17.39	14.79	12.86
	0.5	Original PSNR	8.45	8.27	7.97	7.66	7.22
		NCSR PSNR	8.61	8.42	8.11	7.79	7.33
		MNCSR PSNR	30.35	22.25	16.70	14.31	12.50
	0.8	Original PSNR	6.42	6.40	6.32	6.23	6.10
		NCSR PSNR	6.51	6.50	6.41	6.32	6.18
		MNCSR PSNR	22.45	18.92	15.12	13.08	11.54

Table 15. Different mixed noise levels for Peppers image

			Gaussian noise levels from 10 to 100				
			10	30	50	70	100
Salt & Pepper noise levels from 0.05 to 0.8	0.05	Original PSNR	17.99	15.61	13.26	11.52	8.31
		NCSR PSNR	19.21	16.60	13.77	11.82	8.46
		MNCSR PSNR	29.71	24.94	18.74	15.79	13.19
	0.1	Original PSNR	15.00	13.84	12.22	10.84	9.40
		NCSR PSNR	15.79	14.46	12.60	11.10	9.59
		MNCSR PSNR	29.28	24.17	18.23	15.43	13.29
	0.3	Original PSNR	10.45	10.09	9.50	8.86	8.08
		NCSR PSNR	10.69	10.31	9.70	9.03	8.23
		MNCSR PSNR	26.66	22.13	17.40	14.62	12.74
	0.5	Original PSNR	8.30	8.13	7.85	7.53	7.10
		NCSR PSNR	8.45	8.28	7.98	7.66	7.21
		MNCSR PSNR	25.16	20.75	16.35	14.13	12.36
	0.8	Original PSNR	6.28	6.24	6.17	6.07	5.94
		NCSR PSNR	6.37	6.33	6.26	6.16	6.03
		MNCSR PSNR	20.16	17.47	14.58	12.86	11.31

Table 16. Different mixed noise levels for Cameraman image

			Gaussian noise levels from 10 to 100				
			10	30	50	70	100
Salt & Pepper noise levels from 0.05 to 0.8	0.05	Original PSNR	17.82	15.72	12.15	11.45	8.22
		NCSR PSNR	18.95	16.59	12.54	11.76	8.54
		MNCSR PSNR	29.11	24.80	18.34	15.90	13.27
	0.1	Original PSNR	14.79	13.74	12.20	10.81	9.13
		NCSR PSNR	15.53	14.36	12.59	11.07	9.44
		MNCSR PSNR	28.58	24.21	18.56	15.58	13.48
	0.3	Original PSNR	10.21	9.88	9.33	8.95	8.12
		NCSR PSNR	10.45	10.09	9.51	9.13	8.45
		MNCSR PSNR	26.77	22.33	17.26	14.79	12.46
	0.5	Original PSNR	8.07	7.92	7.66	7.66	7.16
		NCSR PSNR	8.21	8.06	7.79	7.79	7.25
		MNCSR PSNR	24.27	20.72	16.47	14.31	12.37
	0.8	Original PSNR	6.01	6.01	5.91	6.23	5.74
		NCSR PSNR	6.10	6.10	5.99	6.32	6.13
		MNCSR PSNR	19.28	17.15	14.51	13.08	11.21

These tables summarize the denoising results on *House* image, *Peppers* image and *Cameraman* image with gaussian noise, salt and pepper noise, periodic noise, and mixture of Gaussian and salt & pepper (Gaussian after salt and pepper), respectively. According to the data listed in tables, the proposed algorithm is more effective than the original algorithm, and the proposed algorithm is significantly superior for mixed noise. Compared with the NCSR algorithm, the PSNR of the modified NCSR (MNCSR) have increased by average 1.203, 2.507, 1.121 and 2.052 times, respectively (Fig. 4).

(a)   (b)   (c)   (d)

Fig. 4. Denoising results of different methods on House image. (a) Original image. (b) Image corrupted by Gaussian noise (e.g. noise level is 30). Denoising results of image in (c) by NCSR and (d) by WNCSR.

5 Conclusion

Based on nonlocally centralized sparse representation (NCSR) method, this paper presents a modified NCSR (MNCSR) algorithm for image denoising, which is suitable for removing both the non-sparse noise (such as Gaussian noise) and the sparse noise (such as salt and pepper noise, periodic noise and mixed noise). The main motivation behind this is that sparse representation-based denoising algorithms, including NCSR, are difficult to completely describe the sparse structure of the unknown original image, because both the images and the noises are sparse over some dictionaries. Experimental results on image denoising illustrate that the proposed approach can achieve highly competitive performance to the NCSR, and outperform much for mixed noise.

Acknowledgements. This work has been supported by the Fundamental Research Funds for the Central Universities 3132016220.

References

1. Bruckstein, A.M., Donoho, D.L., Elad, M.: From sparse solutions of systems of equations to sparse modeling of signals and images. SIAM Rev. **51**(1), 34–81 (2009)
2. Eldar, Y.C., Kutyniok, G.: Compressed Sensing: Theory and Applications. Cambridge University Press, New York (2012)
3. Elad, M.: Sparse and Redundant Representations-From Theory to Applications in Signal and Image Processing. Springer Science+Business Media, New York (2010)
4. Elad, M., Aharon, M.: Image denoising via sparse and redundant representations over learned dictionaries. IEEE Trans. Image Process. **15**(12), 3736–3745 (2006)
5. Shao, L., Yan, R., Li, X., Liu, Y.: From heuristic optimization to dictionary learning: a review and comprehensive comparison of image denoising algorithms. IEEE Trans. Cybern. **44**(7), 1001–1013 (2014)
6. Dong, W., Zhang, L., Shi, G.: Centralized sparse representation for image restoration. In: Proceedings of the 2011 IEEE International Conference on Computer Vision (ICCV 2011), pp. 1259–1266 (2011)
7. Dong, W., Zhang, L., Shi, G., Li, X.: Nonlocally centralized sparse representation for image restoration. IEEE Trans. Image Process. **22**(4), 1620–1630 (2013)
8. Yu, N., Qiu, T., Ren, F.: Denoising for multiple image copies through joint sparse representation. J. Math. Imaging Vis. **45**(1), 46–54 (2013)

Aviator Hand Tracking Based on Depth Images

Xiaolong Wang$^{(\boxtimes)}$ and Shan Fu

School of Electronic Information and Electrical Engineering,
Shanghai Jiao Tong University, Shanghai, China
1210264601@qq.com, sfu@sjtu.edu.cn

Abstract. Detecting and tracking aviator's hand in the cockpit is a fundamental task on analyzing and identifying the behavior of aviators. Due to the complicated conditions in the cockpit - such as the lighting varies, the space of Cockpit is narrow, the operation of aviator is sophisticated - tracking the hand in Aircraft Cockpit has more difficulties than tracking the hand in human-machine interaction. We propose a hand tracking method to track the aviator's hand based on depth images. In our experiment, most of the common flight operations are tested. The average error of hand position tracking is 6.4 mm and the ratio of losing tracking is only 1.4%, which indicate that the proposed algorithm has the ability to tracking the aviator's hand in the aircraft cockpit accurately.

Keywords: Aircraft cockpit · Hand tracking · Depth image · Kalman filter · Region growing

1 Introduction

The ratio of aviation accidents caused by aircraft itself has dramatically reduced [1]. The safety and reliability of civil aviation aircraft have been greatly improved with the development of science and technology. The aviation industry has formed a consensus: the mechanical reliability of modern aircraft is far greater than reliability of human operator. However, in recent years, many aircraft accidents are related to human factors [2], and the aviator operation is an important component of human factors. In order to identify and analyze the aviator operation, tracking the aviator's hand accurately in aircraft cockpit becomes a fundamental and pivotal task.

Previous works about hand tracking were often based on skin color information [3, 4]. But, color information is very vulnerable in condition that the light varies constantly in the cockpit and skin color varies across different human races of aviators. It is difficult to learn a general skin model since the skin color has large variations caused by different illumination conditions and human races. According to our previous experiences, tracking aviator's hand in the cockpit based on skin color can't get a satisfactory effect.

There are some achievement on hand tracking based on depth images have published [5–7]. Manders computed hand probability map for a Camshift tracker from a joint probability function based on depth image [8]. The depth information can improve the hand localization obviously. Nanda and Fujimura extracted edges in the depth images and created distance-transformed (DT) images, followed by fitting an oval-

© Springer Nature Singapore Pte Ltd. 2017
H. Yuan et al. (Eds.): GRMSE 2016, Part I, CCIS 698, pp. 387–395, 2017.
DOI: 10.1007/978-981-10-3966-9_44

shaped contour on the DT images for hand tracking [9]. As the hand may have large variation of shapes in the cockpit resulting in too many contours of different shapes that need to be taken into account, their algorithm is not effective for a robust hand tracking in the cockpit. Chen employed a region growing technique with the aid of geodesic distance to track the hand center in 3D space [10]. The author also proposed a hand click detection method to initialize the hand tracking. However, the proposed algorithm needs the camera be mounted right in front of the person, which is only available in human-machine interaction, but not suitable for hand tracking in the cockpit.

We proposed a novel hand tracking algorithm in the flight cockpit based on depth images. In our system, we are able to track the location of the hand center in the aircraft cockpit effectively. The Intel RealSense Camera R200 was employed in our research. It contains a HD camera, an infrared camera and an infrared laser projector. The dimensions of the camera is 130 mm × 20 mm × 7 mm, which is small enough for us to mount in the cockpit. The working range of the camera is from 0.7 to 4 m. The camera can provide 628 × 468 resolution depth image at 60 FPS and 1920 × 1080 resolution color image at 30 FPS, which is sufficient for hand tracking in the cockpit in our system. With the Intel RealSense Camera, we are able to obtain the hand region of the aviator as well as the location of the hand center. Our experiments results demonstrate that the average error of our method is 6.4 mm and ratio of losing tracking is only 1.4%, which indicates the proposed method has the ability to tracking the aviator's hand in the aircraft cockpit accurately.

In the remainder of this paper, Sect. 2 presents the proposed hand tracking algorithm. Experiments are conducted to evaluate the proposed algorithm in Sect. 3. And conclusions are in Sect. 4.

2   The Proposed Solution

In this section, we describe the proposed hand tracking algorithm. First, we reduce the noise with the spatial filter and the region growing technique. Then, we predict the hand center with Kalman filter and acquire the hand center position. Last, a method is proposed to initialize the hand position in the last part of this section.

For each frame of depth images, each pixel indicates the distance from this point to the plane of depth sensor in real space. Since the intrinsic parameters of the depth sensor are known, each pixel in the depth image can be transformed into a 3D point $p = (x, y, z)$. The origin of coordinate system is the center of depth sensor.

2.1   Noise Reduction

The depth image from the Intel RealSense Camera, has various sources of noise such as reflectance and mismatched patterns, which will have a great impact on the effect of the hand tracking algorithm. Therefore, noise reduction should be performed before hand detection. We use a spatial filtering and region growing technique for noise reduction, the result is shown in Fig. 1.

The spatial filtering: A 5 × 5 aperture median filter is used for spatial filtering. The median filter provides excellent salt and pepper noise reduction with considerably less

 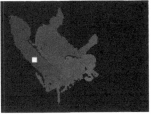

(a) Before noise filtering   (b) after noise filtering

Fig. 1. Effect of the noise reduction (Color figure online)

blurring. As the noise pattern of the motion image is very similar to salt and pepper noise, the median filter is very effective.

The region growing filtering: In our method, we use the region growing to remove the noise and leave the area which we need in the flowing algorithm. The region starts from some certain pixels, then takes in the neighboring pixels gradually according to a definition of connectivity. The connectivity between two points p and q are defined as this:

$$connected(p,q) = \begin{cases} 1 & if\ d(p,q) < \lambda \\ 0 & otherwise \end{cases} \qquad (1)$$

Where $d(p,q)$ denotes the Euclidean distance between p and q. λ is a setting threshold specifying the distance between each two connected points. We set $\lambda = 15$ mm in our experiments.

Then, we need to choose a region as the certain pixels. In our experiment, we choose a square area of aviator's seat shown in Fig. 1. Once the camera is fixed, the spatial coordinates of the pixels in the red square are unchanged. We choose a square area but not a single pixel, because a single pixel is unreliable. For example, when the spatial coordinates of single pixel is much smaller than its real value, the region growing will fail certainly.

2.2 Hand Center Tracking Algorithm

(1) *Kalman filter for predicting*: A In our method, we use the Kalman filter to predict the current position of hand center $P_{predict}$. The Kalman filter is used for object tracking in many applications and it has advantages for hand tracking. The main procedure of Kalman filter is to estimate the state, then refine the state from the error.

We need to measure the position and velocity of the hand center in three dimensions. So, the State vector of Kalman filter is defined as follows:

$$x_t = (x(t), y(t), z(t), v_x(t), v_y(t), v_z(t)) \qquad (2)$$

$x(t), y(t), z(t)$ represents the 3D coordinate position of the hand center, $v_x(t), v_y(t), v_z(t)$ represents the velocity of the hand center in the current frame t. And we define the observation vector of the Kalman filter as follows:

$$y_t = (x(t), y(t), z(t)) \tag{3}$$

According to the definition above, we can provide the Kalman system equations as follows:

$$x_t = Ax_{t-1} + Bw_t \tag{4}$$

$$y_t = Cx_t + v_t \tag{5}$$

Where A is the state transition matrix, B is the driver matrix, C is the observation matrix. w_t is the systematic error of the state vector x_t, which is correspondent to Gaussian distribution. v_t is observation error, which denotes the error between the measure position of the hand center and the real position of the hand center.

In our method, we assume that when the time interval between two consecutive frames is very small, the motion of each hand center between two consecutive frames can be approximated by a uniform linear motion. A, B and C can be defined as follows:

$$A = \begin{bmatrix} 1 & 0 & 0 & \Delta t & 0 & 0 \\ 0 & 1 & 0 & 0 & \Delta t & 0 \\ 0 & 0 & 1 & 0 & 0 & \Delta t \\ 0 & 0 & 0 & 1 & 0 & 0 \\ 0 & 0 & 0 & 0 & 1 & 0 \\ 0 & 0 & 0 & 0 & 0 & 1 \end{bmatrix}$$

$$B = \begin{bmatrix} 0 & 0 & 0 & 1 & 0 & 0 \\ 0 & 0 & 0 & 0 & 1 & 0 \\ 0 & 0 & 0 & 0 & 0 & 1 \end{bmatrix}^T \qquad C = \begin{bmatrix} 1 & 0 & 0 & 0 & 0 & 0 \\ 0 & 1 & 0 & 0 & 0 & 0 \\ 0 & 0 & 1 & 0 & 0 & 0 \end{bmatrix}$$

Then, we define the Kalman filter's time update equation and state update equation in the following manner.

Time update equation:

$$x_t^* = Ax_{t-1} + Bw_{t-1} \tag{6}$$

$$P_t^* = AP_{t-1}B^T + Q \tag{7}$$

State update equation:

$$K_t = P_t^* C_t^T (C_t P_t^* C_t^T + R)^{-1} \tag{8}$$

$$x_t = x_t^* + K_t(y_t - C_t x_t^*) \tag{9}$$

$$P_t = (I - K_t C)P_t^* \tag{10}$$

Where, x_t^* and x_t denote the priori estimates and the posteriori estimates respectively. P_t^* and P_t denote the priori error covariance and the posteriori estimates covariance respectively. The 6×6 diagonal matrix Q is a covariance matrix of w_t, the 3×3 diagonal matrix R is a covariance matrix of v_t. In our experiments, we set the value of the diagonal elements of Q as 10^{-4}, and the value of the diagonal elements of R as the Intel RealSense Camera's sampling rate.

(2) *Acquire the hand center:* When we got the $P_{predict}$ predicted by Kalman filter, we use the region growing to segment the hand region R_{hand}. The connectivity is shown as Eq. (1). And the connectivity will stop when the distance from the current point to the seed point is longer than 250 mm. After we get the hand region, the boundary points can be determined at where the growing stops.

Then we calculate the hand center position as follows:

$$\left(x_{centorid}, y_{centorid}, z_{centorid}\right) = \left(\frac{\sum P.x}{N_{R_{hand}}}, \frac{\sum P.y}{N_{R_{hand}}}, \frac{\sum P.z}{N_{R_{hand}}}\right)$$

Where $(p.x, p.y, p.z)$ denotes the coordinate of points in R_{hand}, $N_{R_{hand}}$ is the number of pixels in R_{hand}. Figure 2 illustrates the effect of the proposed method.

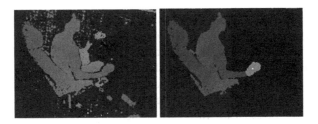

Fig. 2. Effect of the proposed solution

2.3 The Initial Hand Center Position

In order to activate the hand tracking algorithm, we need an approach to provide the initial hand center position. In our method, We take the center of steering wheel right handle as the initial hand center position $P_{initial}$, shown in Fig. 3(a). Aviator holding the handle is the activation of tracking algorithm. First, we measure the real depth value of $P_{initial}$. We denote the real depth value as α. When aviator doesn't hold the handle, the depth measure by the Intel RealSense Camera will be zero. When aviator is holding the handle, the depth measure by the Intel RealSense Camera will be close to α. We denote the measured depth value as β. When $abs(\alpha - \beta) <= 10$ mm, we deem that aviator is holding the handle, and hand tracking algorithm has been activated. Figure 3(b) and (c) show an illustration of $P_{initial}$ in the color image and depth image.

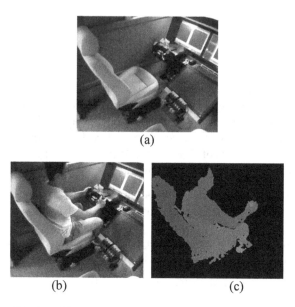

(a)

(b)   (c)

Fig. 3. Initial hand center position (Color figure online)

3   Experiments and Results

3.1   Experimental Environment

The whole environment contains a cockpit-simulator and a screen that can simulate the scene outside the cockpit. They are illustrated in Fig. 4. The layout of devices in cockpit-simulator is similar to the layout in a real cockpit. The simulator has the ability to simulate various aircraft types, various flights routes and various flight stages. We mounted the RealSense camera in the right corners of the simulator. The distance from the aviator to the camera is about 2 m. We can deem that aviator' hand and all devices are within the work range of the RealSense camera.

Fig. 4. Cockpit-simulator

3.2   Experimental Setup and Results

In our experiment, the cockpit-simulator was set as Table 1. We set the RealSense camera as Table 2. In the experiment, experimenter' operations include the common flight operations such as controlling the steering wheel, operating the joystick, operating the roof console and pressing buttons on the instrument panel. After we got the depth image and color image sequences of the scenario, we used the proposed method to track the experimenter's hand, and record the 3D position of the hand center and the outline of hand in each frame.

The effect of our algorithm is shown in Fig. 5. The whole time of the experiment is 6200 s, and the number of images is 18600. The average error of our method, that is,

Table 1. Setting of cockpit-simulator

Aircraft type	Rout	Wind speed	Weather
CRJ-200 7900	Capital International Airport to Shanghai Hongqiao International Airport	150 m/s	Sunny

Table 2. Setting of camera

Camera type	Resolution of color image	Sampling frequency of color image	Resolution of depth image	Sampling frequency of depth image
RealSense R200	640 × 480	30 FPS	628 × 468	30 FPS

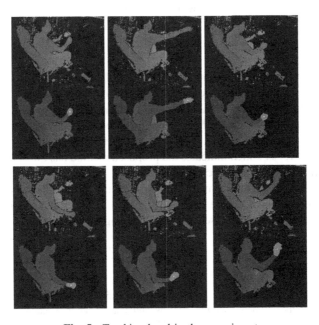

Fig. 5. Tracking hand in the experiment

the average distance from the ground truth to the tracked hand positions, is 6.8 mm, as shown in Table 3. Figure 6 shows an example of the error plot for the sequence. Note that the tracking errors is greater than 1 cm in several frames. We define that a frame whose tracking error is greater than 1 cm is the frame losing of the tracking. The rate of losing tracking is only 1.4%. We found that the tracking usually losses at the moments when the hand overlap with legs, or when the noise around the hand is so strong that region growing technique fails.

The results of our experiments demonstrate that, the proposed algorithm is effective and able to tracking the hand in the cockpit.

Table 3. Result of tracking

Time	Frame	Average error	Rate of losing tracking
6200 s	18600	6.4 mm	1.4%

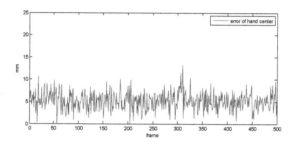

Fig. 6. Error plot

4 Conclusions

We have presented an algorithm for hand tracking in the aircraft cockpit based on depth images. With the region growing technique, we were able to obtain the body region and hand region of the aviator as well as the location of the hand center. We also proposed an approach to provide the initial hand center position and activate the hand tracking algorithm. Our experiments results show that the average error is 6.4 mm and rate of losing tracking is 1.4%, which indicates the proposed method has the ability to tracking the aviator's hand in the aircraft cockpit accurately.

Our further study is to overcome the difficult that the hand overlap with legs, and to cut down the influence of device noise on tracking. Another topic is to identify the various operations of aviators based on hand tracking information.

References

1. Wiegmann, D.A., Shappell, S.A.: Human error analysis of commercial aviation accidents: application of the Human Factors Analysis and Classification System (HFACS). Aviat. Space Environ. Med. **72**, 1006–1016 (2001)

2. Billings, C.E., Reynard, W.D.: Human factors in aircraft incidents: results of a 7-year study. Aviat. Space Environ. Med. **55**, 960–965 (1984)

3. Shan, C., Wei, Y., Tan, T.: Real time hand tracking by combining particle filtering and mean shift. In: Proceedings of the IEEE International Conference Automatic Face and Gesture Recognition (ICAFGR 2004), pp. 669–674. IEEE Press, May 2004. doi:10.1109/AFGR.2004.1301611

4. Wren, C.R., Azarbayejani, A., Darrell, T.: Pfinder: real-time tracking of the human body. IEEE Trans. Pattern Anal. Mach. Intell. **19**, 780–785 (1997)

5. Qian, C., Sun, X., Wei, Y.: Real time and robust hand tracking from depth. In: Proceedings of the IEEE Conference Computer Vision and Pattern Recognition (CVPR 2014), pp. 1106–1113. IEEE Press, June 2014. doi:10.1109/CVPR.2014.145

6. Frati, V., Prattichizzo, D.: Using Kinect for hand tracking and rendering in wearable haptics. In: IEEE World Haptics Conference (WHC 2011), pp. 317–321. IEEE Press, June 2011. doi:10.1109/WHC.2011.5945505

7. Park, S., Yu, S., Kim, J.: 3D hand tracking using Kalman filter in depth space. EURASIP J. Adv. Sig. Process. **2012**, 1–18 (2012). doi:10.1186/1687-6180-2012-36

8. Manders, C., Farbiz, F., Chong, J.H.: Robust hand tracking using a skin tone and depth joint probability model. In: IEEE International Automatic Face and Gesture Recognition (AFGR 2008), pp. 1–6. IEEE Press, September 2008. doi:10.1109/AFGR.2008.4813459

9. Nanda, H., Fujimura, K.: Visual tracking using depth data. In: IEEE Computer Society Conference Computer Vision and Pattern Recognition Workshop (CVPR 2004), p. 37. IEEE Press, June 2004. doi:10.1109/CVPR.2004.476

10. Chen, C.P., Chen, Y.T., Lee, P.H., Tsai, Y.P., Lei, S.: Real-time hand tracking on depth images. In: Proceedings of the IEEE Visual Communications and Image Processing (VCIP 2011), pp. 1–4. IEEE Press, November 2011. doi:10.1109/VCIP.2011.611598

Reachability Problem in Temporal Graphs

Kaiyang Liu$^{(\boxtimes)}$ and Xincan Fan

Department of Computer Engineering, Shenzhen Polytechnic, Shenzhen, China
{liukaiyang,horsefxc}@szpt.edu.cn

Abstract. Reachability over a massive graph is a fundamental graph problem with numerous applications. However, the concept of a classic reachability algorithm is insufficient for a temporal graph, as it does not consider the temporal information, which is vital for determining the reachability between two nodes. In this paper, we propose an efficient algorithm for answering the reachability problem in a temporal graph. Our approach fully explores the time constraints among edges in a temporal graph, and utilizes an index to speed up the computation. Furthermore, online graph update is supported. Various experimental results demonstrate that our algorithm outperforms competitors greatly.

Keywords: Reachability · Temporal graph · Algorithm

1 Introduction

With the exploding development in location-based services, social networks, and other graph-dependent system, there have been more and more research interests in graph data problem. Among various graph research fields, reachability problem is the fundamental problem, which answers the problem of whether there exists a path between two nodes in a graph. However, the classic reachability only considers static graphs, whose edges do not have a time constraint. In real life, many graphs are temporal graphs with time constraints, in which the connections between two nodes only occur at specific time periods. For example, A has visited a viewpoint at time t; A has sent a message to B at time t. Figure 1(a) shows a temporal graph that records the social inter-connections (i.e., phone call/email/message) between people. For simplicity, we assume the edge recording phone calls and use integers to represent the timestamps.

In Fig. 1(a), a directed edge with timestamp 1 between node A and B shows that A has made a phone call to B at timestamp 1. Another example is the two edges between E and F, which shows that E has called F twice at 2 and 7 respectively. A reachability query may ask whether there exists any social inter-connection between A and B during a specified timestamp period [2, 6] (which we call it a temporal reachability query), whose answer is false in Fig. 1(a), since there does not exist a valid temporal path between A and B during period [2, 6]. (The reason will be explained latterly).

Classical reachability efforts will convert a temporal graph into a static graph by discarding all temporal information, as shown in Fig. 1(b). Note that there is only one edge between E and F in Fig. 1(b), compared to two edges in Fig. 1(a). For the

© Springer Nature Singapore Pte Ltd. 2017
H. Yuan et al. (Eds.): GRMSE 2016, Part I, CCIS 698, pp. 396–402, 2017.
DOI: 10.1007/978-981-10-3966-9_45

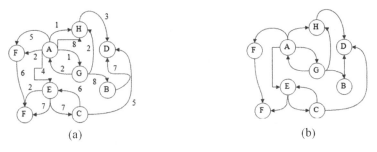

Fig. 1. A temporal graph and the corresponding static graph

previous temporal reachability question, the answer is true for the graph in Fig. 1(b), which is a false answer. Therefore, condensing a temporal graph into a static graph by discarding the vital temporal information may present erroneous information and lead to incorrect understanding of relationships between nodes.

The reachability problem may be solved by applying the techniques to solve path problems in a temporal graph, which has been studied in [8]. However, [8] considered a temporal graph as a graph stream, and proposed a linear algorithm to answer various path problems, i.e., earliest-arrival paths, latest-departure paths, and fastest paths. Under the worst cases, the algorithms in [8] might need to scan the whole graph stream, since they did not utilize any speedup measure. Another approach in [8] is to transform a temporal graph into a normal static graph by preserving the original temporal information, which still suffers from the costly BFS/DFS operation. Furthermore, as we may notice, in a real life temporal graph the order of a node may be very high, as one may make numerous calls or sent a bunch of emails to others at different time period. For such temporal graphs, the graph transformation technique employed in [8] proves to be very inefficient, or even inapplicable, as the sizes of the result graphs may grow up by many order of magnitudes.

By observing the underlying time constraints inherent in the temporal graph, we employ a B^+-tree to speed up the traversal and quickly determine the reachability between two nodes.

The rest of this paper is organized as follows. Section 2 defines notions and notations of a temporal graph and temporal paths. Section 3 defines the reachability problem in a temporal graph, and presents our approach. Section 4 shows the experimental results comparing our approach and the competitors.

2   Definition of Temporal Graphs and Paths

We define a temporal graph as followed. Let $G = (V, E)$ be a temporal graph, where G is the set of nodes, and E is the set of edges. An edge in a temporal graph G is a quadruple (u, v, t, λ), where $u, v \in V$, t is the starting time, and $\lambda \geq \geq 0$ is the traversal time to go from u to v at time t. If $\lambda = 0$, we abbreviate the notation of an edge to (u, v, t). To simplify the following discussions, we use $t(e)/\lambda(e)$ to represent the starting/traversal time of e, respectively. Furthermore, we focus on directed graphs. However, extending our approach to handle undirected graphs can be straightforward.

In a temporal graph, the number of edges between two nodes u and v can be numerous, as in the case of phone call, transportation, and social interactions. Similar to [8], we use $\Pi(u, v)$ to denote all edges between u and v, and denote the number of edges between u and v as $\pi(u, v) = |\Pi(u, v)|$.

Given the definition of a temporal graph, the definition of a temporal path is as follows: a temporal path P in a temporal graph G is a sequence of edges $P = \{v_1, v_2, ..., v_{k+1}\}$, where $(v_i, v_{i+1}, t_i, \lambda_i) \in E$, and $t_i + \lambda_i \leq t_{i+1}$ for $1 \leq i \leq k$. Furthermore, we define the starting/ending time of P as $start(P) = t(e_1)/end(P) = t_k + \lambda_k$, respectively. Therefore, the duration of P is defined as $dura(P) = end(P) - start(P)$. Taking the temporal graph in Fig. 1(a) as an example, $\{A, G, B\}$ is a valid temporal path, since $(A, G, 1, 0)$ and $(G, B, 8, 0)$ are two edges in the temporal graph. However, (G, B, D) is not, since there does not exist an edge e between B and D such that $t(e) \geq 8$.

3 Reachability in a Temporal Graph and Our Approach

In this section, we define the reachability problem in a temporal graph, and propose our approach to efficiently solve it. Given a temporal graph G and a time constraint $TC = [t_s, t_e]$, a temporal reachability query $TR = (G, u, v, TC)$ tries to find out whether the two nodes u and v are reachable by satisfying the time constraints in TC. If the answer for a temporal reachability query $TR = (G, u, v, TC)$ is true, then there exists a temporal path $P = \{u, ..., v\}$ such that $start(P) \geq t_s$ and $end(P) \leq t_e$. Otherwise, the answer for TR is false.

Taking the temporal graph in Fig. 1(a) as an example, by assuming $TP = [5, 8]$, then F can be reached from A through edges $(A, I, 5, 0)$ and $(I, F, 6, 0)$. However, neither of B, D, nor G is reachable from A during the specified TP. Although paths exist from A to all of B, D, and G in the corresponding static graph in Fig. 1(b), they do not satisfy the time constraint specified by the given TP.

From the above example, it is clear that the reachability problem in a temporal graph is completely different from that in a static graph. To solve a temporal reachability query, not only do we need to find out whether there exists a path between u and v, but we have to determine that the found path satisfies the given time constraint also.

To solve the temporal reachability problem, a straightforward solution is to apply the BFS/DFS traversal over the temporal graph, whose cost is forbiddingly high for a graph of massive size. Although we can directly adopt the approach in [8] to solve the temporal reachability problem, there are a few limitations:

1. The time cost of the approach in [8] is still linear time to the size of the graph (specifically, the number of edges in a graph). Compared with the BFS/DFS traversal one, the improvement is limited.
2. The approach in [8] aimed at various path problems in a temporal graph, not directly dealing with the temporal reachability problem. Therefore, it is limited in employing some speedup measures.

To answer the temporal reachability query between u and v, we need to find out whether there exists a valid temporal path between u and v. One important observation is that as long as a temporal path is found for u and v, then the reachability query is solved. However, to ascertain v is not reachable from u for a given TC, one may need to check all possible temporal paths between u and v, until no such path is found. This straightforward way is naïve and of poor performance. By considering the strictly ascending order of the starting time in a temporal path, we can determine the reachability between two nodes as early as possible, as illustrated by the following example.

Example 1: *Still taking the two nodes G and D in the temporal graph in Fig. 1(a) as an example, we can easily decide that D is reachable from G for the period [1, 4]. But for the period [8, 9], the answer is false. To verify the correctness of this answer, we firstly check three outgoing edges of G, of which only (G, B, 8) satisfies the time constraint of [8, 9]. If we further check the outgoing edges of B, we can safely ascertain that D is not reachable from G for the period [8, 9], since no edge e satisfies $8 \leq t(e) + \lambda(e) \leq 9$.*

Although we still need BFS/DFS traversal in a temporal graph to solve the reachability problem, we can utilize the time constraints to terminate the traversal as early as possible, so as to greatly improve the performance. To evaluate a TR for a given $TC = [t_s, t_e]$, we only need to check those edges with $t(e) + \lambda(e) \in [t_s, t_e]$. If $\lambda(e) = 0$, then the above constraint equals to $t(e) \in [t_s, t_e]$.

Therefore, for each node u, we propose to use a B^+-tree to index the $t(e)$ of all outgoing edges of u. Figure 2 presents a 2-order B^+-tree for all outgoing edges of node A in Fig. 1(a). Since the starting time of edges in a temporal path is of strictly ascending order, we add links between the leaf nodes to facilitate the bulk retrieval of edges.

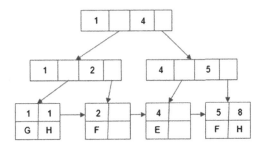

Fig. 2. A B^+-tree for node A in Fig. 1(a)

Given a temporal reachability query $TR = (G, u, v, TC)$, we perform a BFS traversal along the temporal graph. During each iteration, we utilize the B^+-tree to speed up the retrieval of next qualified edges. The detailed algorithm is as follows:

Algorithm 1. Temporal Reachability Query Algorithm

Input: A temporal graph G = (V, E),a temporal reachability query TR = (G, u, v, TC) with TC = [t_s, t_e]
Output: true or false
1   Initialize a stack S, and S.Push(e), where e is an outgoing edges of n with t(e) $\in$ [t_s, t_e]
2   While S is not empty
3   e= {n, n′, t} ←S.pop()
4   If n′ = v
5   Return true
6   End for
7   Look up the B+-tree of n′, and let E′ = {(n′, q, t′)$\in$ E, t′ $\in$ [t_s, t_e], t′ $\geq$ t}
8   if Q= $\emptyset$
9   continue
10   else
11   S.push(e), for $\forall$e = (n′, q, t) $\in$ E′,e $\notin$ S, q is unvisited
12   End for
13   End while
14   Return false

Algorithm 1 is based on a BFS traversal, as it can fully utilize the links between leaf nodes in the corresponding B$^+$-tree to quickly retrieve all qualified edges, thus gains a performance lead over a DFS based algorithm. In the algorithm, we start with the source node u, and insert all qualified edges into the stack S (Line 1). For each loop, we get an edge e from the top of stack (Line 3), and retrieve those outgoing edges of n' with $t' \in [t_s, t_e]$ and $t' \geq t$ (Line 7). $t' \geq t$ enforces the strictly ascending order of a temporal path. Line 11 excludes those edges that have been visited or are currently in the stack, so as to avoid the circular loop in the temporal graph. We repeat these procedures, until we encounter the ending node v (return *true* in Line 5) or the stack is empty (return *false* in Line 13).

Given a $TR = (G, A, B, TC)$ where $TC = [1, 4]$, firstly we search in the B$^+$-tree in Fig. 2, retrieve and insert edges $(A, G, 1)$, $(A, H, 1)$, $(A, I, 2)$, and $(A, E, 4)$. Next, we pop $(A, G, 1)$ from the stack, and insert $(G, H, 2)$ into the stack. The other two edges $(G, A, 2)$ and $(G, B, 8)$ are discarded, as A is already visited or the starting time $8 \notin [1, 4]$. The next edge to be processed is $(A, H, 1)$, and we insert $(H, D, 3)$. When processing $(A, I, 2)$, no outgoing edge of F will be pushed into the stack. We repeat the procedures, until the stack is empty and the algorithm returns *false*. But for anther temporal query $TR = (G, A, B, TC)$ where $TC = [1, 8]$, the algorithm will correctly find a temporal path (A, G, B), with edges $(A, G, 1)$ and $(G, B, 8)$.

4   Experimental Results

In this section, we compare our approach with competitors, namely, the one-scan algorithm and the transformed graph approach in [8]. We ran all the experiments under Windows 7 on a machine with 3.0 GHz CPU and 4 GB RAM.

For the testing temporal graph, we use 2 real temporal datasets in our experiments: *dblp* (DBLP coauthor network) and *enron* (email network), which are from Koblenz

Large Network Collection (http://konect.uni-koblenz.de/). The following Table 1 gives some statistics of the datasets. Compared with the *dblp* dataset, the *enron* dataset has a relative high order of edges between nodes, as the value of its π is higher than that of *dblp*'s by up to two orders of magnitudes.

Table 1. Real life dataset statistics (K = 10^3)

Dataset	$|V|$	$|E|$	π
Dblp	1103 K	11957 K	38
Enron	87 K	1135 K	1024

We create a synthetic query generator to randomly generate temporal reachability queries. Each testing set contains 1000 queries, and its average performance of all queries is measured. As the interval of *TC* can affect the overall performance significantly, we evaluate the performance of all algorithms for various intervals of $TC = [t_s, t_e]$, i.e., $I = t_e - t_s$. Four *I*s are used, $I_4 = 2I_3 = 2I_2 = 2I_1$, meaning that the interval of I_2 is as double as that of I_1 and so on. Figure 3 presents the experimental results (measured in seconds).

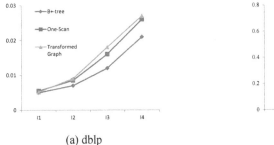 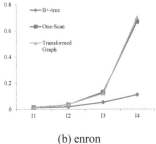

(a) dblp   (b) enron

Fig. 3. Performance comparison

From the above Fig. 3, clearly our approach (labeled B$^+$-tree in Fig. 3) outperforms the other two competitors in all experiments by utilizing a B$^+$-tree to speed up the traversal of edges. With the increasing of the interval of *TC*, the query time of all algorithms is increasing as they need to traverse more and more edges. However, the performance of our approach scales much better than others, especially for the *enron* dataset. The reason is that the orders of nodes in *enron* is much high than that of *dblp*. Therefore, our approach benefits more greatly from the utilization of B$^+$-tree to reduce the time cost for retrieving the set of qualified edges of one node to $O(\log \pi)$, whilst that of the other two is linear to π. The higher the π, the greater the performance improvement achieved by our approach.

5 Conclusion

Although the reachability is a fundamental problem for the graph dataset, few efforts has been conducted at the reachability problem for a temporal graph. Due to the time constraints enforced by the temporal information, algorithms for a static graph are either inefficient or inapplicable. In this paper, by observing some important characteristic of the temporal reachability query, we propose to use a B$^+$-tree to index the starting time of each edge, so as to improve the performance of the temporal reachability query. The experimental results confirm that our approach outperforms competitors significantly, and scales much better with the increasing either of the order of nodes or of the interval of the query time period.

References

1. Cheng, J., Huang, S., Wu, H., Fu, A.W.-C.: TF-label: a topological-folding labeling scheme for reachability querying in a large graph. In: Proceedings of the SIGMOD Conference, pp. 192–204 (2013)
2. Cheng, J., Ke, Y., Chu, S., Cheng, C.: Efficient processing of distance queries in large graphs: a vertex cover approach. In: Proceedings of the SIGMOD Conference, pp. 457–468 (2012)
3. Cheng, J., Shang, Z., Cheng, H., Wang, H., Yu, J.X.: Efficient processing of k-hop reachability queries. VLDB J. **23**(2), 227–252 (2014)
4. Kempe, D., Kleinberg, J.M., Kumar, A.: Connectivity and inference problems for temporal networks. J. Comput. Syst. Sci. **64**(4), 820–842 (2002)
5. Tang, J., Musolesi, M., Mascolo, C., Latora, V.: Temporal distance metrics for social network analysis. In: Proceedings of the ACM Workshop on Online Social Networks, pp. 31–36 (2009)
6. Xiang, L., Yuan, Q., Zhao, S., Chen, L., Zhang, X., Yang, Q., Sun, J.: Temporal recommendation on graph via long- and short-term preference fusion. In: Proceedings of the KDD, pp. 723–732 (2010)
7. Xuan, B.-M.B., Ferreira, A., Jarry, A.: Computing shortest, fastest, and foremost journeys in dynamics networks. Int. J. Found. Comput. Sci. **14**(2), 267–285 (2003)
8. Wu, H., Cheng, J., Huang, S., Ke, Y., Lu, Y., Xu, Y.: Path problems in temporal graphs. In: Proceedings of the VLDB Conference, pp. 721–732 (2014)

Research on Rapid Extraction Method of Building Boundary Based on LIDAR Point Cloud Data

Minshui Wang[✉], Guodong Yang, Xuqing Zhang, and Liji Lu

School of Earth Science and Technology,
Jilin University, Changchun, Jilin, China
1543383519@qq.com, 271950315@qq.com,
2631867384@qq.com, zxq@jlu.edu.cn

Abstract. LIDAR integrates digital camera, global positioning system (GPS), inertial navigation system, laser ranging and other advanced technology, using GPS to control the attitude of flying platform, obtaining the distance information between receiver and landmark via laser, capturing the image information by digital camera. At last, the distance is expressed as the 3D coordinates of the target landmark, and the image information is expressed as the surface characteristics on the target landmark. In this paper, firstly, the 3D coordinates are speckle noise filtered to realize point cloud de-noising and classification; then the point cloud of classification is plain rasterized to achieve the transformation from plane coordinates to image coordinates, and the image coordinates is processed to be corresponding elevation of grid unit by incremental insertion and median filter, at the same, the elevation is filled to the corresponding grid to generate DSM depth image by Gray-Scale transformation. Finally, the SOBEL detection operator is quickly and efficiently used to extract the building boundary characteristics which are applying in digital city construction, three dimensional entity model construction, urban planning and construction.

Keywords: Point cloud denoising · Incremental insertion · DSM depth image · Edge extraction

1 Preface

LIDAR integrates digital camera, global positioning system (GPS), inertial navigation system, laser ranging and other advanced technology, using GPS to control the attitude of flying platform, obtaining the distance information between receiver and landmark via laser and the distance is quickly and efficiently expressed to be 3D coordination information of the measured object by coordination transformation. The massive data obtained by this method are the basics in feature extraction of landmark. The building is an important symbol of city, its boundary information are important applied in the urban planning and construction, disaster prevention, three-dimensional digital simulation and many other fields. Therefore, how to quickly and accurately extract the building boundary information from the massive LIDAR point cloud data [5] has become a hot research topic at present.

© Springer Nature Singapore Pte Ltd. 2017
H. Yuan et al. (Eds.): GRMSE 2016, Part I, CCIS 698, pp. 403–413, 2017.
DOI: 10.1007/978-981-10-3966-9_46

Guo et al., University of Tokyo, Japan [1] combine the high resolution satellite images and airborne laser scanning data to extract buildings. You et al. [2] interpolate and encrypt the 3D point cloud data to generate DSM depth image and extract the building roof boundary by image data segmentation, boundary extraction and normalization and many other steps. Zeng et al. [9] build TIN model using the discrete point cloud data obtained from LIDAR, filter the boundary line of the whole triangulation network by using the spatial geometric relationship of the TIN model point, and obtain the initial boundary point; regard the point that change significantly on the boundary slope as turning point; further, expand the turning point in the range of point cloud to detect all the expansion points in the boundary; at last, use the expansion point to construct the building roof boundary. In the point cloud de-noising, Xiao et al. [10] put forward the multi-temporal polarization SAR data speckle noise filtering algorithm based on scattering model. The method uses the multi-temporal polarization SAR data to realize the speckle noise suppression on the basis of keeping the main scattering characteristic of the pixel.

Based on the de-noising and classification of LIDAR point clouds, this paper rasterizes the classified building point cloud to construct DSM depth image [4], then uses edge detection operator to extract the building boundary and obtains the building's total external contour. The process of building boundary extraction is shown in Fig. 1:

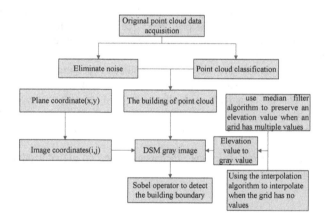

Fig. 1. Flow chart of building boundary extraction

2 Point Cloud Denoising and Classification

2.1 Principle and Method of Point Cloud Denoising

By using speckle noise filtering of multi-temporal point cloud, the speckle noise can be effectively removed. De-noising principle: choose the pixel that is same in the main scattering of the central pixel in the window, and averagely suppress the speckle noise through the covariance. The realized process: obtain the polar covariance matrix from the polar SAR image scattering matrix; decompose the polar covariance matrix in the way of Freeman-Durden to get 3 scattering components including surface scattering, volume scattering and double scattering; classify the pixel by using unsupervised classification; choose 9×9 moving window to do the speckle noise filter processing in the basic of classification, choose the pixel that is same in the main scattering of the central pixel in the window, and averagely suppress the speckle noise through the covariance. In order to keep up the point cloud features, remain the central pixel which has the max total power and is surface scattering or double scattering in the moving window. This can effectively remove the noise, but also maintain the edge texture information and the characteristics of the target point.

2.2 Classification Principle of Point Cloud Data

The principle of classification of the building: firstly, the landmark is divided into separate and small objects. The smaller division values are, the more careful in the subdivision. Then disengage the trees by normalized difference vegetation index and

Fig. 2. Original point cloud

Fig. 3. Building point cloud data

distinguish the roads and the buildings by using rectangular degree tool. Lastly, combine the nearby elements into an object and disengage the building point cloud (Figs. 2 and 3).

3   DSM Grayscale Image Generation

3.1   The Principle of Generating DSM Depth Image

DSM (Digital Surface Model) depth image is a kind of raster data, which has the advantage of two-dimensional plane reflecting three-dimensional features. LIDAR acquires discrete vector point. In order to generate the DSM, the horizontal coordinate (X, Y) of discrete vector point needs to be transformed to image coordinates (I, J), and realize the grid transformation of the vector point cloud. In raster data, each pixel (I, J) has the corresponding elevation value. In the actual operation, there will be a pixel corresponding a number of LIDAR points, on this occasion, use the median filtering algorithm to calculate elevation value which is regarded as the pixel value; When a pixel does not correspond to any LIDAR point, use the incremental inserting algorithm to insert the pixel value in order not to introduce a new elevation. In the end, the elevation value is filled to the corresponding grid to generate DSM depth image and grid transformation Digital Surface Model (DSM).

3.2   The Step of Generating DSM

In order to achieve the grid transformation of the vector point quickly, first of all, convert the plane coordinate of the point cloud that converts the X direction of the plane coordinate system into I direction and the Y direction of the plane coordinate system into J direction. Conversion formula as formula (1.1)

$$i = int((X - Xmin)/S)$$
$$j = int((Y - Ymin)/S)$$

S is grid resolution in the formulate above and $S = 1/\sqrt{n}$, n is the original point cloud density of the LIDAR. This can insure that 1 m in the X direction respondent to $1/step = \sqrt{n}$ in the I direction on the grid. (It is the same to the Y direction.) For the target area, every square metre correspondent with the grid of $\sqrt{n} * \sqrt{n} = n$. And the target areas exactly have n vector points per square metre. It is that n points corresponding to n grid. When a pixel corresponding to multiple point, calculate the pixel corresponds to the elevation values by using the median filtering method. Implementation Methods: Median filtering calculations usually use the mid-value among the height of every point as the height of the grid. Assuming that the window contains n points, order the elevation values from small to large, and then get the sequence of the median as the mid-value. And consider this value as the output value of the median filter (Figs. 4, 5 and 6).

When a pixel corresponding to no point, in order not to introduce new elevation values, we adopt point-by-point interpolation [11] to find out the elevation values that

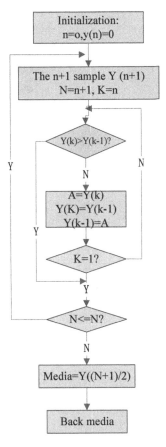

Fig. 4. Flow chart of median filter

correspond to the pixels. Implementation Methods: Take the adopt which need interpolation as the center of the field, choose the nearby eight points in order to determine the scope of the neighborhood, then calculate the height of the current point through the interpolation arithmetic with the point that falls within the neighborhood in this range. Point by point interpolation model can be expressed as (Figs. 7 and 8):

In order to generate gray image, firstly, read the maximum of x, y coordinates of building point cloud to get the data range; then, determine the side length of the plane grid in the range of point cloud data to create an empty two-dimensional matrix and fill the plane vector point into the related grid. In the end, read the points' elevation and the elevation maximum of building point cloud, i.e. Zmax and Zmin. According to gray conversion formula, the gray value related to elevation value. Conversion formula:

$$G(I, J) = (Z(I, J) - Zmin)/(Zmax - Zmin) * 255$$

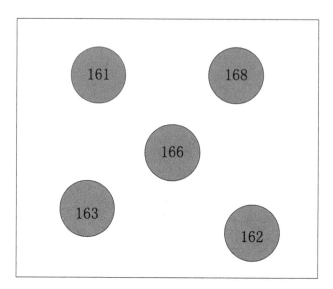

Fig. 5. The model of median filtering

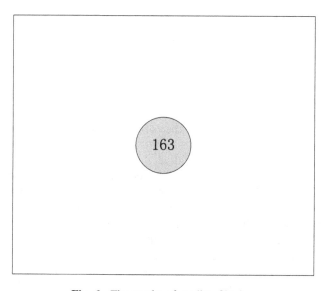

Fig. 6. The results of median filtering

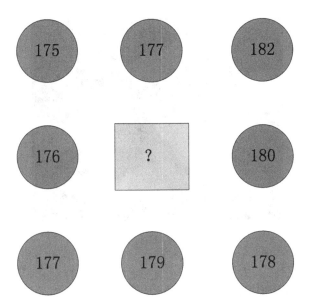

Fig. 7. The model of point interpolation

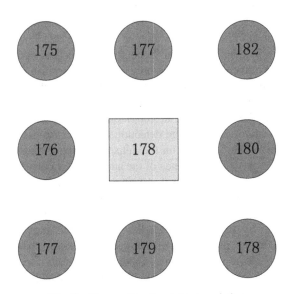

Fig. 8. The results of point interpolation

G(I, J) is the gray value of the point (I, J), Z(I, J) is the elevation value in DSM. The higher the elevation is, the greater the gray value is, the results on the image that will be brighter (Fig. 9).

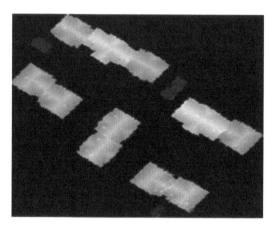

Fig. 9. Gray image display

4 Building Boundary Detection

4.1 Edge Detection Operator [6, 8]

Boundary is the most prominent position in the change of the gray level in an image, it is also the important feature that image segmentation depends on, the foundation and key information resources of the shape and texture characteristics. The boundary features of the image are also a key position of the image matching. There are many classical algorithms in edge detection. Sobel edge detection operator is a typical algorithm of first derivative which can not only eliminate the noise but can also smooth it. Roberts's edge detection operators finds the edge using the principle of partial difference, and detect the edge position through the gradient change of the two adjacent pixels in the diagonal direction. So the detection effect in the vertical direction is better than the oblique detection. Prewitt edge detection operator also use first differential to carry on the edge detection, it carrys on the detection using the extreme value that is acquired among the gray-value difference of the adjacent point that is located on the up and down or so of the pixel point. Thus it can smooth the noise to some extent. Laplace edge detection operator is a kind of second differential operator, and it is suitable for the situation that only cares about the edge position and don't consider the surrounding gray level difference by using the properties of isotropy. Laplace edge detection operator does well in the procession of the noiseless images because it responses strongly to the independent pixel response. Candy edge detection operator is a kind of edge detection operator that calculates the gradient magnitude and direction through the finite difference. First of all, it does the convolution operation for the images to reduce the noises by carrying out the Gaussian filter, then inhibit the maximum of the gradient's size, and keep the biggest point during the gradient change, finally adopt the double threshold algorithm for testing and the edge of the connection.

4.2   Sobel Operator Theory and Testing Results [3]

Sobel operator is adopted for edge detection through the practical comparisons on a variety of edge detections [5, 7]. It's strength is: Detect the edge points and suppress the noise effectively at the same time, the detection of the rear edge width is two pixelsat least. The sobel is obtained through the convolution of two convolution kernel $g1(x, y)$ and $g2(x, y)$ that is carried out on the original image $f(x, y)$. It's mathematical expression is:

$$S(x, y) = \max \left[\sum_{m=1}^{M} \sum_{n=1}^{N} f(m, n) \, g1 \, (i - m, j - n), \sum_{m=1}^{M} \sum_{n=1}^{N} f(m, n) \, g2 \, (i - m, j - n) \right].$$

$$(1)$$

In fact, the algorithms of the Sobel edge operator´are carrying on the weighted average firstly, and then carry on the differential operation.

We can use difference instead of the first order partial derivative operator, it's computing method is:

$$\Delta x \, f(x, y) = [f(x - 1, y + 1) + 2f(x, y + 1) + f(x + 1.y + 1)]$$
$$- [f(x - 1, y - 1) + 2f(x, y - 1) + f(x + 1, y - 1)].$$

$$(2)$$

$$\Delta y \, f(x, y) = [f(x - 1, y - 1) + 2f(x - 1, y) + f(x - 1, y + 1)] - [f(x + 1, y - 1)$$
$$+ 2f(x + 1, y) + f(x + 1, y + 1)].$$

$$(3)$$

The calculation template of the sobel operator's vertical direction and horizontal direction is as shown in figure, the former can detect the image edge of the horizontal direction, and the later can detect the image edge of the vertical direction, in the course

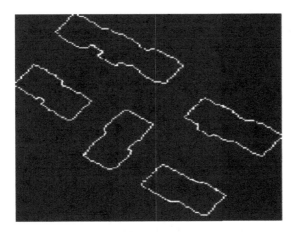

Fig. 10. The effect of detection using Sobel operator

of practical application, each pixel of the image is done with the two convolution kernels for convolution operation, and get the maximize as a result to put out, and the calculation result is an image that reflects the edge gradient (Fig. 10 and Tables 1 and 2).

Table 1. Horizontal direction

-2	-1	-2
0	0	0
2	1	2

Table 2. Vertical direction

-2	0	2
-3	0	1
-1	0	2

5 Conclusions

As an important symbol of a city, buildings play a vital role in the process of urban digitization. The extraction of the building boundary can help people to grasp the building area grandly, provides the basis for urban planning and the construction of 3D models. Our paper generates DSM depth image based on the de-noising of point cloud classification the rasterization of plane coordinates and the conversion of height gray, finally extracts the complete building boundary. Through the analysis of this thesis, we can see:

a. LIDAR technology can process the acquisition of the building's point cloud efficiently and quickly.
b. It is a good way to retain boundary information using the above methods for building boundary extraction.
c. The selection of multi-resolution mesh influences the effect of the boundary extraction. At the same time, it will produce serrated burrs in the process of extracting boundary that need to be further optimization for practical application.

References

1. Guo, T., Yoshifumi Y.: Combining high resolution satellite imagery and airborne laser scanning data for generating Bareland and DEM in urban areas. In: International Workshop on Visualization and Animation of Landscape, Kunming, China, 26–28 February 2002. XXX IV Part: 5/W3
2. You, H., Su, L., Li, S.: Automatic extraction of DSM data from airborne 3D imaging instrument. J. Wuhan Univ. **27**(4), 408–413 (2002)

3. Cui, J., Sui, L., Xu, H., Zhao, D.: LiDAR data building extraction based on edge detection algorithm. J. Surv. Mapp. Sci. Technol. **02**, 98–100 (2008)
4. Wang, B.: The Extraction of LiDAR Point Cloud Data Based on the Edges of the Building. Chang'an University, Xi'an (2014)
5. Wei, Z., Yang, B., Li, Q.: Fast extraction of the boundary of the building in the vehicle borne laser scanning point cloud. J. Remote Sens. **02**, 286–296 (2012)
6. Meng, F., Li, H., Wu, K.: Building feature line LIDAR point cloud data extraction. Sci. Surv. Mapp. (15405), 97–99+108 (2008)
7. Zhao, Y.: Research on Urban Building Extraction Based on Airborne LIDAR Data. Jilin University, Changchun (2011)
8. Yang, P.: Research on Building Extraction and Modeling Based on Airborne LiDAR Point Cloud Data. Liaoning Technical University, Fuxin (2013)
9. Zeng, Q., Mao, J., Li, X., Liu, X.: Roof boundary extraction of building LiDAR point cloud. J. Wuhan Univ. (Inf. Sci. Ed.) **3404**, 383–386 (2009)
10. Xiao, K., Wang, Zuo, T., Song, C., Yang, R., Han, X., Wang, Z.: The model of multiple phase polarization SAR speckle noise filtering algorithm. Surv. Mapp. Sci. **2**, 99–103+124 (2014)
11. Lan, Y., Wang, M., Liu, S., Dai, X.: Point by point interpolation method to establish the DEM study. Sci. Surv. Mapp. **1**, 214–216 (2009)

Absorption Band Spectrum Features Extraction for Minerals Recognition Based on Local Spectral Continuum Removal

Wei Zhou[✉], Qichao Liu, and Zhikang Xiang

School of Computer Science and Technology, Nanjing University of Science and Technology, Nanjing, Jiangsu, China
1309446202@qq.com

Abstract. Hyperspectral mineral identification methods have a wide application in field. In this paper, we propose a new pipeline of mineral identification by using absorption band spectrum features. In the pre-processing, a local spectral continuum removal algorithm is used to normalise the corresponding spectral data of the absorption band of different minerals. After that, the polynomial fitting is applied to remove the spectral outliers. Then, by using the preprocessing spectral data, the absorption band spectrum features are extracted to establish a rules based interring system for minerals recognition. Experimental results on hyperspectral images demonstrate that the proposed method has good performance in minerals recognition.

Keywords: Hyperspectral image · Rules based interring · Local spectral continuum removal · Spectrum features

1 Introduction

Hyperspectral imaging remote sensing technique is known as spectroscopy. It is one of the major technological breakthroughs of the 20th century in the aspect of earth observation. Hyperspectral remote sensing can get hundreds of pieces of a very narrow segment of the electromagnetic remote sensing images [1]. By using the spectrum features of hyperspectral images, we can accurately distinguish between kinds of minerals. Traditional mineral recognition methods need a priori spectral information to match the spectrum [2]. However, they do not make full use of the spectrum features. In this paper, we propose a new pipeline of mineral identification by making full use of the spectrum features. In the pre-processing, we use a local spectral continuum removal algorithm to normalise the corresponding spectral data of the absorption band of different minerals. After that, the polynomial fitting is applied to remove the spectral outliers. Then, by using the preprocessing spectral data, the absorption band spectrum features are extracted to establish a rules based interring system for minerals recognition. Finally, we use rules based inferring to infer minerals. As shown in the figure below. In Fig. 1 we use a local spectral continuum removal algorithm and polynomial fitting to enhance spectrum features in the preprocessing. Then we extract absorption band spectrum features and establish a rules based. Lastly we infer minerals by using rules based inferring.

© Springer Nature Singapore Pte Ltd. 2017
H. Yuan et al. (Eds.): GRMSE 2016, Part I, CCIS 698, pp. 414–422, 2017.
DOI: 10.1007/978-981-10-3966-9_47

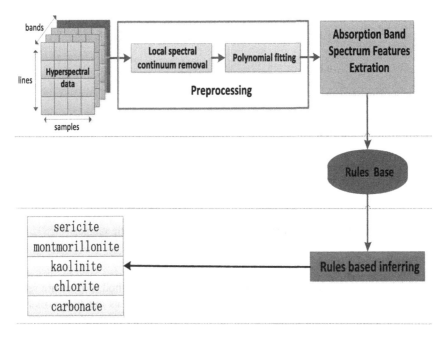

Fig. 1. Design flow chart

2 Preprocessing of Spectral Data

2.1 Local Spectral Continuum Removal

In spectral analysis, diagnostic absorption features can indicate the existence of specific materials. Absorption parameters such as absorption center, absorption width, and absorption depth can be used in identification and quantitative analysis of minerals. Continuum removal (CR) is commonly used to extract absorption features. However, for a band range containing more than one absorption contribution factors, the feature extracted by CR could be a result of comprehensive effect of different factors. In this paper, a new spectral feature extraction method named local spectral continuum removal is proposed [4]. Given the reference spectral background, local spectral continuum removal can eliminate the influence of unwanted contribution factor, and extract the absorption feature of target contribution factor. Using local spectral continuum removal, the basic absorption feature parameters including the absorption center, absorption width, and absorption depth are extracted. The results are compared with those obtained from the CR. It is shown that local spectral continuum removal can effectively extract pure absorption features of target material, while more accurate absorption parameters can also be achieved. In Fig. 2, the red line represents spectral data of wavelength within 900 nm to 1000 nm. The green line represents data of envelope line. The blue line represents data of local spectral continuum removal.

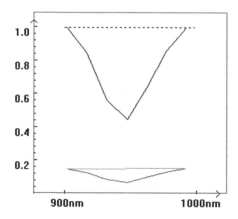

Fig. 2. Local spectral continuum removal (Color figure online)

2.2 Polynomial Fitting

Polynomial fitting [5] is approximate function of numerical method. It does not require the same approximated function at each node. It only requires as much as possible to reflect the underlying trend given data points. Therefore, polynomial fitting is used widely in application. Algorithm works as follows: Assuming that a given data points (xi, yi) (i = 0, 1, 2, ..., m), n represents time of fitting, Where $n \leq m$. To solve $p_n(x) = \sum_{k=0}^{n} a_k x^k$, made

$$\min\{\sum_{i=0}^{m} [p_n(x_i) - y_i]^2\} \tag{1}$$

We can easily get a_j and calculate $p_n(x)$. In Fig. 3, blue line shows data of local spectral continuum removal. Red line shows data of second time fitting. In Fig. 4, blue

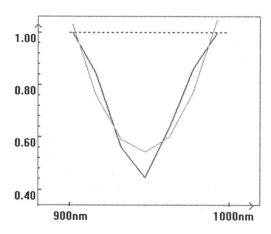

Fig. 3. Second time polynomial fitting (Color figure online)

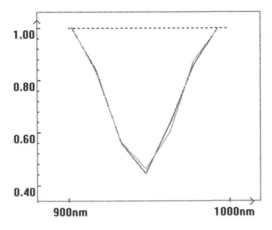

Fig. 4. Fourth time polynomial fitting (Color figure online)

line shows data of of local spectral continuum removal. Red line shows data of fourth time fitting.

3 Absorption Band Spectrum Features Extraction and Recognition

3.1 Absorption Band Spectrum Features Extraction

The absorption band spectrum features can use a series of waveform parameters to express. Spectrum features of different minerals have different degrees. Absorption band spectrum features of main parameters are the absorption position, full width at half-maximum and absorption depth [6]. The absorption depth has a quantitative relationship with mineral abundance. So the quantitative analysis of the spectrum features has become one of the main method of hyperspectral data mineral identification [7]. In the paper we use a local spectral continuum removal algorithm and polynomial fitting to enhance spectrum features in the preprocessing. After that, we calculate spectral features. In Fig. 5, the blue line shows data of local spectral continuum removal. The red line shows data of fourth times fitting. The method of calculating spectral features is described below.

(1) Absorption position: we call the position of minimum reflectivity as absorption position. The value of w represents absorption position in Fig. 5.
(2) Absorption depth: we call the distance of between 1 and minimum reflectivity as absorption depth. The value of d represents absorption depth in Fig. 5.
(3) Full width at half-maximum: we call width of half absorption depth as full width at half-maximum. The value of wh represents full width at half-maximum in Fig. 5.

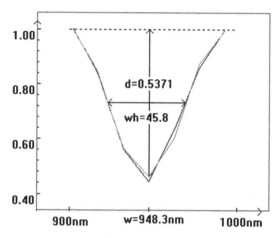

Fig. 5. Calculate spectral features after fourth time polynomial fitting (Color figure online)

3.2   Rules Based Minerals Inferring

Expert system is a hot spot in the research of the artificial intelligence. System simulates the way of expert's thinking. It can reason and judge the problems need to be solved. From the point of view of its structure composition, the deposit can be simplified to a rules base and a inferring engine. Basic structure is shown below (Figure 6).

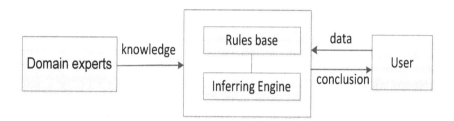

Fig. 6. Basic structure of expert system

Rules base and inferring engine are the key of expert system. Knowledge of the rules base learn from experience of the experts. Inferring engine is used to infer minerals. At present, Rule-based Reasoning (RBR) is widely used in expert system. It is mainly based on the knowledge of domain experts and turns knowledge into the form of a rule. All rules have the same structure, namely 'IF...THEN...'. It is easy for us to understand rules. Therefore, minerals recognition method is based on RBR. Listed below are parameters. These parameters will be used in rules.

(1) Al-OH band: wavelength ranges from 2180 nm to 2225 nm. d2.20 represents the depth of the band. Yd2.20 represents depth threshold of band. The value of Yd2.20 is generally 0.03.

(2) Mg-OH band: wavelength ranges from 2300 nm to 2365 nm. d2.34 represents the depth of the band. Yd2.34 represents depth threshold of band. The value of Yd2.34 is generally 0.03.

(3) 2165 nm band: wavelength ranges from 2152 nm to 2185 nm. d2.16 represents the depth of the band. Yd2.16 represents depth threshold of band. The value of Yd2.16 is generally 0.03.

(4) R1 = d2.20/d2.34, with the depth ratio of two absorption bands. YR1 represents the minimum pure sericite. The value of YR1 is generally 1.245.

(5) R2 = d2.16/d2.20, with the depth ratio of two absorption bands. YR2 represents the minimum pure Kaolinite. The value of YR2 is generally 0.08.

The different letters represent different minerals and represent the content of mineral (band depth as the measures). S represents sericite. M represents montmorillonite. K represents kaolinite. L represents chlorite. T represents carbonate.

Listed below are rules.

Rule 1: IF d2.20 < Yd2.20 THEN S = M = K = 0

Rule 2: IF d2.20 ≥ Yd2.20 d2.34 < Yd2.34 THEN S = L = T = 0

Rule 3: IF d2.20 ≥ Yd2.20 d2.34 < Yd2.34 d2.16 ≥ Yd2.16 THEN K = d2.16 M = d2.20-(YR2-R2) * d2.16

Rule 4: IF d2.20 ≥ Yd2.20 d2.34 < Yd2.34 d2.16 < Yd2.16 THEN K = 0 M = d2.20

Rule 5: IF d2.20 ≥ Yd2.20 d2.34 ≥ Yd2.34 d2.16 < Yd2.16 THEN K = 0

Rule 6: IF d2.20 ≥ Yd2.20 d2.34 ≥ Yd2.34 d2.16 ≥ Yd2.16 THEN K = d2.16

Rule 7: IF d2.20 ≥ Yd2.20 d2.34 ≥ Yd2.34 d2.16 ≥ Yd2.16 R2 < YR2 THEN S = d2.20-(YR2-R2) * d2.16

Rule 8: IF d2.20 ≥ Yd2.20 d2.34 ≥ Yd2.34 d2.16 ≥ Yd2.16 R2 ≥ YR2 THEN M = S = 0

Rule 9: IF d2.20 ≥ Yd2.20 d2.34 ≥ Yd2.34 d2.16 < Yd2.16 R1 < YR1 THEN M = d2.20-(YR1-R1) * d2.34

Rule 10: IF d2.20 ≥ Yd2.20 d2.34 ≥ Yd2.34 d2.16 < Yd2.16 R1 ≥ YR1 THEN M = 0 S = d2.20

After rules ares built and then hyperspectral data is imported by the user. Inferring engine derives minerals and the result is feedback to the user.

4 Experiments and Results

The algorithm uses MFC draw in VS2010 platform. Compared with traditional identification of minerals, we can prove the accuracy and feasibility of the algorithm. In this paper, we use NASA's AVIRIS imaging hyperspectral remote sensing image data (the southern Nevada desert Cuprite data). The image has been atmospheric correction and has been converted to a spectral reflectance image. The original data has 224 bands. We remove 1–3, 105–115, 150–170, 115–224 several noise segments and remain the other bands. We clip image 450 × 450 pixels in size from the original image to be tested.

Hyperspectral image is shown in Fig. 7. We use sericite mineral to map. The threshold value is 0.094. In Fig. 8 it shows result by using the traditional method to map. In Fig. 9 it shows result by using algorithm proposed to map.

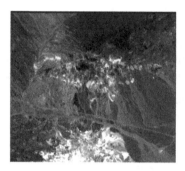

Fig. 7. Original image

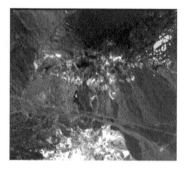

Fig. 8. Traditional methods

Fig. 9. This algorithm identify sericite

Comparison of results between traditional identification methods and the algorithm identification method shows in Table 1. We use different minerals to map in different regions. Accuracy of identification has been improved obviously after using the algorithm identification method on the whole. We think that the improvement of the identification accuracy is due to local spectral continuum removal and polynomial fitting we use. These operations can remove the spectral outliers and enhance spectrum features. We use the more accurate spectrum features to identify minerals. Through Table 1, the algorithm identification method is correct and has a higher accuracy of identification.

Table 1. Minerals identification results comparison table of two methods.

Minerals	Traditional identification methods	The algorithm identification method
Sericite	71.3%	73.0%
Montmorillonite	67.6%	72.3%
Kaolinite	77.3%	79.8%
Chlorite	59.3%	64.5%
Carbonate	61.1%	64.2%

5 Conclusion

In order to highlight spectrum features this paper uses local spectral continuum removal and polynomial fitting. This method can reduce spectral outliers and make the calculation more accurate. After that, we calculate features and establish a rules base. Finally we infer minerals by the rules based inferring. But there is also a need to improve in this paper. For example, we can consider using B-spline fitting to fit. B-spline fitting can make result more accurate. This algorithm will be more widely used. There is much work to be done.

References

1. Tong, Q.X., Zhang, B., Zheng, L.F.: Hyperspectral Remote Sensing Principle, Technology and Application. Higher Education Press, Beijing (2006)
2. Zhang, B., Gao, L.: Hyperspectral Image Classification and Target Detection. Science Press, China (2011)
3. Gan, F., Wang, R., Ma, A., et al.: The development and tendency of both basis and techniques of discrimination for minerals and rocks using spectral remote sensing data. Remote Sens. Technol. Appl. **17**(3), 140–147 (2002). (in Chinese)
4. Rodger, A., Laukamp, C., Haest, M., et al.: A simple quadratic method of absorption feature wavelength estimation in continuum removed spectra. Remote Sens. Environ. **118**(4), 273–283 (2012)

5. Xu, N., Hu, Y.X., Lei, B., et al.: Based on improved spectral characteristics of mineral information extraction of hyperspectral data fitting method. Spectrosc. Spectral Anal. **31**(6), 1639–1643 (2011)
6. Wen, X.P., Hu, G.D., Yang, X.F.: Based on spectral feature extraction of hyperspectral remote sensing image vegetation coverage. J. Geogr. Geogr. Inf. Sci. **24**(1), 27–30 (2008)
7. Liu, T.L.: Based on the Mineral Spectral Feature Analysis and Extraction of Hyperspectral Remote Sensing. China University of Geosciences, Beijing (2009)

Analysis of Seasonal Variation of Surface Shortwave Broadband Albedo on Tibetan Plateau from MODIS Data

Zihan Zhang, Shengcheng Cui[(⊠)], and Xuebin Li

Key Laboratory of Atmospheric Composition and Optical Radiation,
Anhui Institute of Optics and Fine Mechanics,
Chinese Academy of Sciences, Hefei 230031, China
{zzh,csc,xbli}@aiofm.ac.cn

Abstract. Surface albedo is one of the main factors in climate modeling, and is widely used in the energy balance analysis for the coupled atmosphere and surface system. While the surface albedo of the Tibetan Plateau, located in the southwest of China, is of great significance to the study of China and the global climate. This paper aims to analysis seasonal changes of surface albedo in Tibetan Plateau by using MODIS (MODrate-resolution Imaging Spectroradiometer) global albedo product and land cover type dataset, and to obtain statistically the spatial and seasonal variations over this region.

Keywords: Surface albedo · Tibetan plateau · MODIS · IGBP

1 Introduction

The amount of solar radiation reflected by the earth surface mainly depends on the surface albedo. Energy transferred from surface to atmosphere generally comes from longwave radiation from surface, which is heated by solar radiation [1]. Thus, affecting radiative forcing of climate system, surface albedo is one of the most discussed topic in climate modeling and climate research. For example, the change of surface albedo can lead to the adjustment of energy budget of the surface-atmosphere system, and consequently influence general atmospheric circulations [2]. It is defined as the ratio of the total solar radiation reflected by the earth's surface in all directions and the total radiation flux on the surface of the object. Due to the complexity and diversity of surface and their characteristics, the surface albedo is usually calculated as a regional mean in previous studies [3].

The importance role of Tibetan Plateau played in regional climate is embodied by its high sensitivity to climate change. Tibetan Plateau covers more than 2,000,000 km^2 with an average of 4.0–4.5 km altitude. Also, it has complex orography and large diurnal variations in different meteorological elements. The systematic research conducted by Chinese researchers from 1972 shows that climate in China is strongly influenced by dynamic and thermal effect of Tibetan Plateau [4]. In recent years, there is an increase in local surface albedo in Tibetan Plateau arising from glacial degradation and desertification caused by the trend toward dry and warm climate [5]. The increase

© Springer Nature Singapore Pte Ltd. 2017
H. Yuan et al. (Eds.): GRMSE 2016, Part I, CCIS 698, pp. 423–428, 2017.
DOI: 10.1007/978-981-10-3966-9_48

of surface albedo on Tibetan Plateau can change the heat source, thus affecting the East Asian summer atmospheric circulation. In 1994, Xiaodong investigated the relation between surface albedo and monsoon circulation. He found that climate effect becomes significant after 0.05 unit of surface albedo rise. It is expected that in the time scale of years or decades, the surface albedo changes on the Tibetan plateau would be greater, thus result in greater climatic effects [5]. Hence, it is necessary to further study the surface albedo tendency on Tibetan Plateau, which is meaningful to the improvement of accuracy of climate models.

2 Data and Method

Two datasets are used in this study: MODIS/Terra + Aqua 30 arc sec global gap-filled snow-free BRDF (Bidirectional Reflectance Distribution Function) parameter product and MODIS land cover types dataset.

SZN-extended MODIS/Terra + Aqua 30 arc sec Global Gap-Filled Snow-Free Bidirectional Reflectance Distribution Function (BRDF) parameter product is generated from V005 30 arc second Climate Modeling Grid product MCD43D by University of Massachusetts Boston. The original dataset is acquired by MODIS from both the Terra and Aqua polar orbiting satellite platforms. MODIS has 490 detectors and 36 bands, covered from visible band to thermal infrared band. Data from both satellite platforms result in the global dataset of spatially complete maps, which use the first seven MODIS bands, three additional broad bands, a visible band and a near-infrared band. Usually, data retrieved from visible band (0.4–0.7 μm) and near-infrared band (0.7–3.0 μm) are applied as inputs for atmospheric general circulation models, while data from total albedo (0.25–0.5 μm) is used as a tool for surface energy budget research.

The corresponding quality assurance (QA) map is produced on the same 30 arc second resolution grid. Based on the QA map, high quality data obtained by MODIS in MCD43D products can be used for time fitting operation, while the lower quality data is used to fill the region of long-term data loss. Data gap is filled using space fitting method. With the latitude of 1° as the moving space, surface albedo is filled by the same type of albedo. The albedo is corrected by BRDF for view and illumination angle effects.

MODIS Land Cover Types dataset provides the geographic distributions. Following the International Geosphere-Biosphere Program (IGBP) classification system, 17 classes of land cover showed in the table [7] (Table 1):

Here we utilize the shortwave broadband albedo along the quality assurance map from 2003 to 2014 with a time resolution of 8 day, and a spatial resolution of 0.0083°. To investigate the trend of regional albedo change and feedback of climate, shortwave albedo (0.3–0.5 μm) is used [1], especially black sky albedo (BSA, directional hemispherical reflectance at local solar noon [8]) and white sky albedo (WSA, bihemispherical reflectance) [6].

The albedo distribution on Tibet Plateau can be analyzed by land cover distribution. Utilizing MODIS Land Cover Types dataset, Fig. 1 shows the various orographic

Table 1. IGBP land cover legend

IGBP Land cover legend	
Value	Description
0	Water bodies
1	Evergreen needleleaf forest
2	Evergreen broadleaf forest
3	Deciduous needleleaf forest
4	Deciduous broadleaf forest
5	Mixed forest
6	Closed shrublands
7	Open shrublands
8	Woody Savannas
9	Savannas
10	Grasslands
11	Permanent wetlands
12	Croplands
13	Urban and built-Up
14	Cropland/natural vegetation mosaic
15	Permanent snow and ice
16	Barren or sparsely vegetated
17	Unclassified

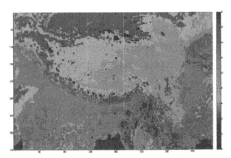

Fig. 1. IGBP land cover types in Tibet

surface on Tibetan Plateau. In northwest of the Tibetan Plateau, there are a large area of open shrubland, while grasslands mainly dominate in southeast.

Figure 2 shows BSA in the Tibetan Plateau in four seasons, from which we can see that the zonal distribution of surface albedo in Tibet area is quite obvious. The surface albedo in Tibet area was mainly in the South higher than in the North. The region of open shrublands and grass lands has similar albedo, ranging from 0.1 to 0.2. In contrast, the plateau glacier and frozen soil region has a higher shortwave albedo which can reach about 0.8.

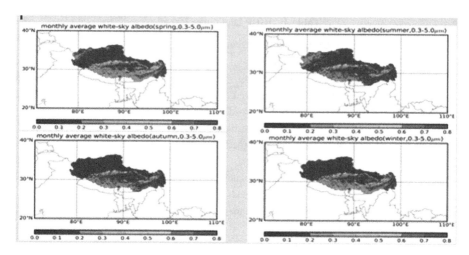

Fig. 2. Seasonal average distribution of black-sky albedo in the four seasons in Tibet area (upper left: spring (March, April and May), upper right: summer (June, July and August), lower left: autumn (September, October and November), lower right: winter (December, January and February))

Also, the surface shortwave albedo reflects pronounced seasonal variations [9]. Consistent with previous studies [9], the albedo peaks in the winter and reaches lowest in the summer. Especially in the grasslands area, surface albedo varies significantly at different the seasons. Due to the seasonal snow melts in the eastern part of the Tibetan Plateau [10], the surface albedo differences between spring and summer is largest around this area with value about 0.3–0.4.

Since the white-sky albedo (WSA) is the further integral of BSA with respect to the incidence angles, compared with BSA, it is closer to the general sense of the albedo [1]. It can be seen from Fig. 3 that the year average of surface albedo in Tibet is 0.4. At the

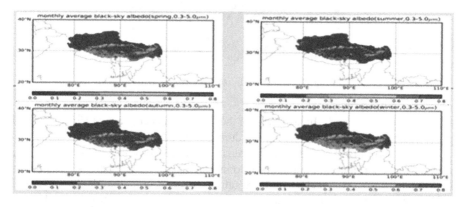

Fig. 3. Seasonal average distribution of white-sky albedo in the four seasons in Tibet area (upper left: spring (March, April and May), upper right: summer (June, July and August), lower left: autumn (September, October and November), lower right: winter (December, January and February))

same time, generally with increasing latitude, showing all the year round in the North South high-low distribution, the Tibetan topography and overall from the northwest to the southeast tilt related. However, in terms of the seasonal variation, the largest average surface albedo is shown in winter, with an extreme of more than 0.8 in the eastern part of Tibet. Following the seasons change, the minimum surface albedo appears in summer. From spring to autumn, the surface albedo in the northwest, central and southeast regions shows significant reduce, while the southern region remains roughly constant.

3 Conclusion

(1) Surface albedo is mainly determined by the physical properties of land surface, such as the color, the humidity, the level of roughness of land surface etc. The surface albedo distribution in the eastern Tibet is higher than that in the west, the specific albedo on the open shrublands area is lower than average, and the grassland area is higher. The cold climate, long snow period, dry surface and sparse vegetation lead to the higher surface albedo than average. Besides, albedo changes significantly with latitude and altitude.

(2) As for time scales, in the eastern Tibet, the surface albedo varies significantly with the four seasons, while is relatively stable around the western region. In summer, the surface albedo decreases due to the snow melting and vegetation growth. In winter, snow cover is an important influence factor on albedo increase.

(3) Due to the snow, the local average albedo increases sharply in the winter. Entering the spring, with the gradual increase of the temperature, the soil moisture increased with snow began to melt, which resulted in the decline albedo. The lowest albedo occurred in summer when plateau went into the rainy season. The incident solar radiation is reflected and heat exchange between the earth and atmosphere is reduced [10]. As a cold source, Tibetan Plateau is bound to affect other areas by general atmospheric circulation.

The investigation of the seasonal variability of the surface albedo in Tibet has great significance for research of climate model and monsoon precipitation. It will be of great interest to deeply study the inter-annual variation of surface albedo in the Tibetan Plateau.

Acknowledgments. The research was supported by the National Natural Science Foundation of China (No. 41305019). Many thanks to Prof. Crystal Schaaf from University of Massachusetts Boston, who provided the MODIS albedo dataset.

References

1. Wang, Y., Zhu, B.: Trend of surface albedo changes in China in last decade. Meteorol. Sci. Technol. **39**(2), 147–155 (2011)
2. Wang, G., Han, L.: Progress in the research of surface albedo. Plateau Mount. Meteorol. Res. **30**(2), 79–83 (2010)

 3. Sheng, P.: Atmospheric Physics. Peking University Press, Beijing (2003)
 4. Zhu, Q.: Principles and Methods of Synoptic Meteorology. China Meteorological Press, Beijing (2000)
 5. Liu, X.: Numerical experiments of influences of surface albedo variation in Qinghai-Xizang plateau on East-Asia summer monsoon. Plateau Meteorol. **13**(4), 468–472 (1994)
 6. Roesch, A., Schaaf, C., Gao, F.: Use of moderate-resolution imaging spectro radiometer bidirectional reflectance distribution function products to enhance simulated surface albedos. J. Geophys. Res. Atmos. **109**(109), 933–946 (2004)
 7. Loveland, T.R., Belward, A.S.: The IGBP-DIS global 1 km land cover data set, DISCover: first results. Int. J. Remote Sens. **18**(15), 3289–3295 (1997)
 8. Salomon, J.G., Schaaf, C.B., Strahler, A.H.: Validation of the MODIS bidirectional reflectance distribution function and albedo retrievals using combined observations from the aqua and terra platforms. IEEE Trans. Geosci. Remote Sens. **44**(6), 1555–1565 (2006)
 9. Cai, F., Zhu, Q., He, H.: Estimation and spatio-temporal distribution of monthly mean surface albedo in China. Resour. Sci. **27**(1), 116–120 (2005)
10. Xu, X., Tian, G.: Dynamic distribution and albedo change of snow in China. J. Remote Sens. **4**(3), 178–182 (2000)

A Novel Multiple Watermarking Algorithm Based on Correlation Detection for Vector Geographic Data

Yingying Wang[1], Chengsong Yang[2(✉)], Changqing Zhu[1], Na Ren[1], and Peng Chen[3]

[1] Key Laboratory of Virtual Geographic Environment, Ministry of Education, State Key Laboratory Cultivation Base of Geographical Environment Evolution, Jiangsu Center for Collaborative Innovation in Geographical Information Resource Development and Application, Nanjing Normal University, NJNU, Nanjing, China
{wyychs, rennal026}@163.com, 649397417@qq.com
[2] Institute of Field Engineering, PLA University of Science and Technology, PLAUST, Nanjing, China
36946046@qq.com
[3] PLA 91206 Troops, Qingdao, China
45563508@qq.com

Abstract. The multiple watermarking algorithm is present for vector geographic data based on the characteristics of vector data. In the embedding progress multiple watermarks were embedded additively in the cover data following additivity rule. In the detection progress the additive watermarks were extracted and then the contents of watermarks were detected on the basis of correlation detection together with discriminant analysis. In the experiments the robustness and adaptability were analyzed. The results show that the proposed multiple watermarking algorithm is robust against common attacks and suit for the vector geographic data with small amount of vertexes.

Keywords: Vector geographic · Digital watermark · Multiple watermarking · Correlation detection · Robustness

1 Introduction

Vector geographic is the important digital achievement for nations and society, and it plays a very important role in national economical construction. Owning to the development of information technology, the coping, transforming and sharing of vector geographic become more and more easy. How to share vector geographic data, to protect the owner copyright and to trace infringement act has become an important issue.

C. Zhu—Co-corresponding author.

© Springer Nature Singapore Pte Ltd. 2017
H. Yuan et al. (Eds.): GRMSE 2016, Part I, CCIS 698, pp. 429–436, 2017.
DOI: 10.1007/978-981-10-3966-9_49

Digital watermarking technology is an effective way to solve the above problems. Since this century many researchers have devoted in digital watermarking for vector geographic data and produced many results which have been used in industrial applications [1–4]. Most of the results are on single watermarking, the multiple watermarking is scarce [5–7]. In practical implementations, the need for multiple copyright protections and the whole process tracing is urgent. The multiple watermarking gives a better hand than single watermarking to the problems.

The multiple watermarking less focuses on the vector geographic data at present. Li Qiang et al. exploited an algorithm that the additional information was generation when the watermark was embedded in cover data [8]. Cao Jianghua proposed a multiple watermarking combining block dividing, domain dividing, frequency dividing with zero watermarking [9]. Cui Hanchuan present the multiple watermarking based on bidirectional mapping, segmented watermarks, quadtree dividing and so on [10].

The existing multiple watermarking for vector geographic data concentrates on the way of dividing domains or blocks, which the watermarks are embedded in the different domains or blocks of the cover data. It plays good performance in several watermarks embedding and detecting. But it is restricted when the cover data is little. For the argument above, the novel multiple watermarking is designed based on the overlaying and correlation detection. The algorithm and the experimental results in detail are described in the following.

2 Multiple Watermarking Scheme for Vector Geographic Data

2.1 Watermak Generation Scheme

Watermark can be classified into two categories according to the actual meaning of the watermark: meaningful watermark and meaningless watermark. The meaningful watermark has meaningful definite meaning, for instance, characters, images, voice, video and so on. The meaningless watermark has no definite meaning, for instance, chaos sequence, pseudorandom sequence and so on. In practical implementations, the length of the meaningless watermark is short and the statistical characteristic is better, that is good for correlation detection. Considered with the features of vector geographic data, watermark capacity, the data quantity, and the need for correlation detection, etc., we chose the meaningless watermark.

Let W_j be the being embedded arbitrary watermark, $W_j = \{w_j[i], 0 \leq i < N\}$, $0 \leq j < M$, where $w_j[i]$ denotes the single bit of watermark, $w_j[i] \in \{-1, 1\}$, and $P(w_j[i] = -1) = \frac{1}{2}$, $P(w_j[i] = 1) = \frac{1}{2}$, N denotes the length of the watermark, M denotes the number of the watermark being embedding in the same cover data, i denotes the index of the $w_j[i]$, j denotes the jth watermark being embedding in the cover data. In consideration of the data quantity of vector geographic data and the practical applications, N is 200 and the max of M is 6 in the paper.

2.2   Watermark Embedding Scheme

To suit the multiple watermarking for little vector geographic data listed above, the multiple watermarks are embedded in the cover data based on the watermarks overlaying.

Let P be the data points set which are chosen from the cover data, $P = \{p[k], 0 \leq k < L\}$, where L denotes the number of the points, $p[k]$ denotes the kth point in the set. The x and y coordinates of $p[k]$ are $p[k].x$ and $p[k].y$. The principles of choosing the points for the robustness of the algorithm are followed. One is the number of vertex coordinates, L, is integer times bigger than the length of the watermarks, N, the other one is that the choosing progress is random and the chosen vertex coordinates are different. The principles are ready against deletion attack and clipping attack.

The embedding procedure can be represented by

$$P \oplus W_j = \{p[k].x + w_j[k\%N]\} \tag{1}$$

Where $\oplus$ denotes the rules of watermarks embedding, $\%$ denotes the modular arithmetic, which the mapping relationships between i and the coordinates are established. The multiple watermarks can be embedded in the cover data by the (1) when the vertex coordinate set is fixed.

The original being embedding watermarks coordinates set, P, must be saved for the watermarks extraction and detection on reason for the additive embedding rule.

2.3   Watermark Extraction and Detection Scheme

The steps of watermarks detection are twofold: one is watermarks extraction; the other is to decide the content of the extracted watermarks by use of correlation detection, called watermark detection.

Watermark Extraction Scheme. Because the embedding rule is additive, the watermark extraction is the inverse operation of the embedding process. Let P' be the vertex coordinate set corresponding to the original points set, P, $P' = \{p'[k], 0 \leq k < L\}$. Where $L = c * N$, c is a positive integer, which means that the number of the vertexes in the set is c times bigger than the length of the watermark. The specific extraction steps are stated as follows.

- Let $k = 0$, and $W' = \{w'[i] = 0, 0 \leq i < N\}$.
- $w'[k\%N] = w'[k\%N] + (p'[k] \cdot x - p[k] \cdot x)$.
- Let $k = k + 1$. If $k < L$, it skips to the second step. If $k > L$, it skips to the next step.
- Let $W' = W'/c$, i.e. $w'[i] = w'[i]/c$. The extraction is done.

If the original vertex coordinates corresponding to the bits of watermarks couldn't be extracted because of the attacks, the random bits of watermarks, -1 and 1, are instead of $(p'[k] \cdot x - p[k] \cdot x)$ in the second step. If the watermarked vertex coordinates were not attacked, the extracting result, $W' = \sum_{j=0}^{j=M-1} W_j = \{\sum_{j=0}^{j=M-1} w_j[i], 0 \leq i < N\}$, is the summarization of extracting watermarks bits.

Watermark Detection Scheme. As stated in above section, the summarization of extracting watermark bits is expressed by

$$W' = \sum_{j=0}^{j=M-1} W_j = \{ \sum_{j=0}^{j=M-1} w_j[i], 0 \leq i < N \} \qquad (2)$$

The content of the watermarks is difficult to be decided from (2). The correlation detection offers help to detect the specific watermarks from the additive watermarks. The number of the embedding watermarks can be judged based on the statistical property of W'. Let the number of being embedded in cover data is M. The specific content of watermarks is confirmed according to the (3).

$$cor = \frac{W' * W}{M * N} = \frac{(\sum_{j=0}^{j=M-1} W_j) * W}{M * N} \qquad (3)$$

W means the original watermark in (3). According to whether the extracted watermarks include W or not, the statistical distributions after correlation detections are different. The details are represented as followed.

The extracted watermarks include W. Let $W = W_j$. The correlation coefficient, cor, between W' and W is caculated by

$$cor = \frac{W_k * W + (\sum_{j=0}^{j=k-1} W_j + \sum_{j=k+1}^{j=M} W_j) * W}{M * N}$$

$$= \frac{W_k * W}{M * N} + \frac{(\sum_{j=0}^{j=k-1} W_j + \sum_{j=k+1}^{j=M} W_j) * W}{M * N} \qquad (4)$$

$$= \frac{1}{M} + \frac{(\sum_{j=0}^{j=k-1} W_j + \sum_{j=k+1}^{j=M} W_j) * W}{M * N}$$

The statistical distribution after correlation detections is accordance with normal distribution as indicated in (5) according to the (4) and the statistical property of the watermarks.

$$cor \sim N(\frac{1}{M}, \frac{M-1}{M^2 N}) \qquad (5)$$

The extracted watermarks don't include W. The statistical distribution is accordance with normal distribution as indicated in

$$cor \sim N(0, \frac{1}{M * N}) \tag{6}$$

The detection threshold value, T0, can be calculated by (7) by use of discriminant analysis, (5) and (6).

$$T_0 = \frac{\frac{1}{M} * \sqrt{\frac{1}{M*N}}}{\sqrt{\frac{1}{M*N}} + \sqrt{\frac{M-1}{M^2N}}} \tag{7}$$

Consequently, the result whether the extracted watermarks include W or not can be estimated by

$$\begin{cases} if \ cor \geq T_0 & Including \ W \\ if \ cor < T_0 & Non - including \ W \end{cases} \tag{8}$$

3   Experimental Validation

The above algorithm has been evaluated by exploiting in VC++ 6.0 for testing the robustness and efficiency. The test focused on the data clipping attacks and dynamic multiple watermarking.

Figure 1 shows the original two vector maps with the scale 1:1000000 and the unit meter, where Fig. 1(a) is polylines including 27848 coordinate points, Fig. (b) is points including 804 coordinate points.

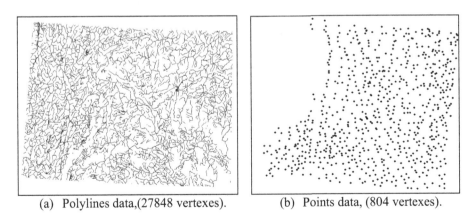

| (a) Polylines data,(27848 vertexes). | (b) Points data, (804 vertexes). |

Fig. 1. Original maps (a) Polylines data, (27848 vertexes). (b) Points data, (804 vertexes).

3.1 Robustness for Attacks

The watermarks were embedded in the cover data as showed in the Fig. (a) of Fig. 1, and then the robustness of the watermarking algorithm was evaluated for different attacks. The steps of the experiment were followed. First, the two different watermarks, Mark1 and Mark2, were embedded in cover data as showed in Fig.(a). Then several kinds of attacks attacked the watermarked data. Finally, the data in the second step was detected to evaluate the robustness. In this experiment, the length of the watermarks, N, is 400, L proposed in C of the part II is 1600. Table 1 shows whether watermarks were detected or not after attacks and the percent of every listed attack, where $\sqrt{}$ means there is Mark1 or Mark2 in the data and $\times$ means there is not being detected watermark in the data.

Table 1. Rubustness experiments.

Attacks	The percent and detection results					
Data compression	Compression Percent (%)	91.2	64.3	49.8	36.2	21.5
	Mark 1		✓	✓	✓	×
	Mark 2	✓	✓	✓	✓	×
Random adding coordinates	Adding Percent (%)	25	43	56	74	96
	Mark 1	✓	✓	✓	✓	✓
	Mark 2	✓	✓	✓	✓	✓
Random deleting coordinates	Deleting Percent (%)	28	46	57	71	88
	Mark 1	✓	✓	✓	✓	×
	Mark 2	✓	✓	✓	✓	×
Random deleting feature	Deleting Percent (%)	82.7	69.4	43.9	26.8	14.3
	Mark 1	✓	✓	✓	✓	×
	Mark 2	✓	✓	✓	✓	×
Data clipping	Clipping Percent (%)	84.9	62.3	44.5	35.4	22.6
	Mark 1	✓	✓	✓	✓	×
	Mark 2	✓	✓	✓	✓	×

The watermarked data was compressed in the way of Douglas Peucker algorithm and every compression percent is showed in the above table. Random adding vertex coordinates is that the random vertex coordinates were added in the watermarked data and every adding percent is shown in Table 1. Random deleting vertex coordinates is that the random vertex coordinates were deleted from the watermarked data and every deleting percent is shown in Table 1. Random deleting feature is that the random features were deleted from the watermarked data and every deleting percent is shown in Table 1. Data clipping is that random parts of the watermarked data were clipped and every clipping percent is shown in Table 1.

Table 1 shows that the proposed method has good robustness against common data compression, random vertex coordinates adding, random vertex coordinates deleting, random features deleting and data clipping. On reason that the watermarks were embedded in some coordinates, the algorithm is marginally robust to the strong attacks of which the above percent are bigger.

3.2   Watermark Capacity for the Data with Small Amount of Vertexes

As proposed above, the development of multiple watermarking is restricted because of the watermark capacity. The common method of multiple watermarking is embedding watermarks in the many blocks into which the cover data was divided. It is not suitable for the data with small amount of vertexes. The algorithm in the paper solve this problem based on the additivity multiple watermarking. The experimental steps are followed.

Several watermarks were embedded in the cover data as shown in Fig.(b). The number of the watermarks is random and is bigger than or equal to 2, but is smaller than or equal to 6. Then the watermarked data was detected. The experimental results of 10 experiments comparing the extracted watermarks with the original watermarks are present in Table 2.

Table 2. Experimental results.

No.	The number of original watermarks	The number of extraction watermarks	Detect result
1	3	3	✓
2	5	5	✓
3	5	5	✓
4	3	3	✓
5	6	6	✓
6	4	4	✓
7	2	2	✓
8	2	2	✓
9	4	4	✓
10	3	3	✓

Table 2 shows that all watermarks can be detected in watermarked data. The multiple watermarking is good for the vector geographic data with small amount of vertexes.

4   Conclusion and Prospect

In this paper, we have presented a multiple watermarking for vector geographic data based on overlaying and correlation detection, then analyzed its robustness through the experiments. The experiment results denote that the algorithm has good robustness for the attacks of less strength: data compression, adding coordinates, deleting coordinates, deleting features and data clipping. Moreover, the algorithm is suit for the multiple watermarking of the vector geographic data with small amount of vertexes, which is better than the algorithm based on dividing.

The embedding rule is additivity in the paper, the proposed scheme requires the original data coordinates. The algorithm is non-oblivious. Future work will include exploiting the oblivious algorithm on the basis of the previous work.

Acknowledgment. The work was supported by the National Natural Science Foundation of China (grant No. 41401518), and the Natural Science Foundation of Jiangsu Province (grant No. BK20140066), the special scientific research of public service industry on state surveying and mapping (grant No. 201512019), a project funded by the Priority Academic Program Development of Jiangsu Higher Education Institutions.

References

1. Voigt, M., Busch, C.: Watermarking 2D vector data for geographical information systems. In: Proceedings of SPIE-The International Society for Optical Engineering, January 2002, pp. 621–628 (2002). doi:10.1117/12.465322

2. Chengsong, Y., Changqing, Z., Yingying, W.: Research on self-detection watermarking algorithm and its applications for vector geo-spatial data. Geomat. Inf. Sci. Wuhan Univ. **36**, 1402–1405 (2011). doi:10.13203/j.whugis2011.12.012. (杨成松,朱长青,王莹莹, "矢量地理数据自检测水印算法及其应用研究." 武汉大学学报信息科学版, vol. 36, Dec. 2011, pp. 1402–1405)

3. Sukhwan, L., Xiaojiao, H., Kiryong, K.: Vector watermarking method for digital map protection using arc length distribution. IEICE Trans. Inf. Syst. **E97-D**, 34–42 (2011). doi:10.1587/transinf.E97.D.34

4. Abubahia, A., Cocea, M.: A clustering approach for protecting GIS vector data. In: Zdravkovic, J., Kirikova, M., Johannesson, P. (eds.) CAiSE 2015. LNCS, vol. 9097, pp. 133–147. Springer, Heidelberg (2015). doi:10.1007/978-3-319-19069-3_9

5. Sleit, A., Abusharkha, S., Etooma, R., Khero, Y.: An enhanced semi-blind DWT-SVD-based watermarking technique for digital images. Imag. Sci. J. **60**, 29–38 (2012). doi:10.1179/1743131X11Y.0000000010

6. Gaurav, B., Qm Jonathan, W.: A new robust and efficient multiple watermarking scheme. Multimed. Tools Appl. **74**, 8421–8444 (2015). doi:10.1007/s11042-013-1681-8

7. Lizhi, X., Zhengquan, X., Yanyan, X.: A multiple watermarking scheme based on orthogonal decomposition. Multimed. Tools Appl. **75**, 5375–5377 (2016). doi:10.1007/s11042-015-2504-x

8. Qiang, L., Lianquan, M., Hongzhi, H., Yongqiang, Y.: A solution research on multiple watermark embedding. Sci. Surv. Mapp. **3**, 119–120 (2011). doi:10.16251/j.cnki.1009-2307.2011.02.022. (李强,闵连权,何宏志,杨永强, "一种多重水印嵌入的解决方案研究," 测绘科学, vol. 36, Mar. 2011, pp. 119–120)

9. Jianghua, C.: Research on the multiple watermarking for GIS vector data (2011). (unpublished [曹江华, "GIS矢量数据多重水印研究,"2011])

10. Hanchuan, C.: Research on sharing security of vector geography data (2013). (unpublished [崔翰川, "面向共享的矢量地理数据安全关键技术研究,"2013])

A Spatial SQL Based on SparkSQL

Qingyun Meng, Xiujun Ma, Wei Lu$^{(\boxtimes)}$, and Zerong Yao

Department of Machine Intelligence, School of EECS,
Peking University, Beijing, China
alex_meng_el@126.com, maxj@cis.pku.edu.cn,
{luwei16, zearom32}@pku.edu.cn

Abstract. The volume of spatial data increased tremendously, and growing attention has been paid to the research of distributed system for spatial data analysis. Spark, an in-memory distributed system which performs much better than Hadoop in speed and many other aspects, lacks spatial SQL query extensions. In this paper, we study the technology framework of Spark SQL, and implement the spatial query extension system tightly combined with the native Spark system. The extensions in the system include spatial types, spatial operators, spatial query optimizations and spatial data source formats. The spatial extension system on Spark retains the scalability and can be further extended with more query optimizations and data source formats. In this paper, the spatial data type system and spatial operator system follow OGC standards. In addition, the extension method is also a general method of query extensions on Spark SQL in other fields.

Keywords: Spark SQL · Spatial operators · Distributed system extensions

1 Introduction

With the emergence of massive scale spatial data, it brings to us a challenge of understanding and utilizing such data. Spatial data is very important and widely exist, and its volume tends to increase tremendously. Building a platform to make sense of spatial data is in urgent demand. Such platforms will be beneficial for applications that may have a huge impact on society.

Apache Spark [2] is an in-memory cluster computing engine, and is the most active open source project for big data processing. Spark offers a new abstraction called resilient distributed datasets [7] (RDDs) that enables efficient data reuse in a broad range of applications. Each RDD is a collection of Java or Python objects partitioned across a cluster. RDD is built using parallelized transformations (filter, join or groupBy) that could be traced back to recover RDD data. Spark has more efficient fault tolerance, faster computation speed and more convenient to use compared to existing models (MapReduce). However, Spark does not provide support for spatial data and its operations. And the most fundamental spatial operation is the spatial SQL.

This paper will focus on the spatial SQL extension on Spark. As Spark SQL is a Spark module for structured data processing, our spatial SQL extension is based on Spark SQL. This paper introduces SpatialSparkSQL an in-memory cluster computing

© Springer Nature Singapore Pte Ltd. 2017
H. Yuan et al. (Eds.): GRMSE 2016, Part I, CCIS 698, pp. 437–443, 2017.
DOI: 10.1007/978-981-10-3966-9_50

system for processing large-scale spatial data. The key contributions of this paper are as follows: (1) Design and implement spatial type system following OGC standards, and integrate them into the extension system. (2) Implement spatial SQL operators following OGC standards and integrate them into the system in two ways: user-defined functions (UDF) and domain specific language(DSL). (3) Implement the spatial query optimization and spatial data source format extension through mechanisms in Spark SQL.

2 Background and Related Work

Spatial database systems are the spatial extensions to some traditional database systems, such as Oracle Spatial and SQLServer Spatial. These systems extend mainly on the spatial data types, spatial SQL query, spatial index, and more.

As the development of distributed data processing system, Hadoop has become the most successful and influential open source distributed systems. Initially Hadoop does not support SQL, bringing inconvenience to the study and use of Hadoop. NewSQL databases which support SQL make an impact on Hadoop, and then a significant amount of the work has been devoted to SQL-on-Hadoop systems that make Hadoop support a series of SQL functions. Hadoop-GIS [6] is a scalable and high performance spatial data warehousing system for running large scale spatial queries on Hadoop. It utilizes global partition indexing and customizable on demand local spatial indexing to achieve efficient query processing. SpatialHadoop [5], a comprehensive extension to Hadoop, has native support for spatial data by modifying the underlying code of Hadoop. GeoSpark [15] is an in-memory cluster computing framework for processing large-scale spatial data based on Spark. System users can leverage the newly defined Spatial Resilient Distributed Datasets (SRDDs) to effectively develop spatial data processing programs in Spark. However, GeoSpark does not offer relational interfaces to spatial data. Thus We build the SpatialSparkSQL system based on the 1.4 version of Spark SQL. SpatialSparkSQL supports relational data processing in Spark and efficiently executes spatial query processing.

3 Spatial SQL Design on SparkSQL

3.1 Spark SQL Overview

Spark SQL [11] builds on the earlier SQL-on-Spark effort called Shark [3]. Due to some serious defects, Shark has been abandoned and is replaced by Spark SQL. Spark SQL provides a DataFrame API that can perform relational operations on both external data sources and Spark's built-in distributed collections. This API is similar to the widely used data frame concept in R [16], but evaluates operations lazily so that it can perform relational optimizations.

Spark SQL [17] introduces a novel extensible optimizer called Catalyst. Catalyst's general tree transformation framework is applied in four phases of query planning in Spark SQL: (1) analyzing a logical plan to resolve references, (2) logical plan optimization, (3) physical planning, and (4) code generation to compile parts of the query to Java bytecode.

There are several ways to interact with Spark SQL including SQL and the Dataset API. When computing a result the same execution engine is used, independent of which API/language one is using to express the computation.

3.2 Spatial SQL System Architecture

As depicted in Fig. 1, we provide support for spatial data types in every layer of Spark SQL. This paper mainly focuses on the spatial type and operators support on the DataFrame API layer and above.

Through Spark SQL user-defined types and user-defined functions, we design and implement a set of spatial data types and spatial operators. System users can read the spatial data directly from the JSON file and then stored it in DataFrame. System users can also use the domain specific language or SQL language to operate on the spatial data. In the end, our spatial operators passed the test samples designed according to the OGC standards [12].

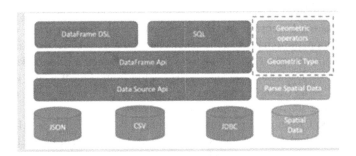

Fig. 1. Architecture of spatial extension on Spark SQL.

3.3 Spatial Data Types Extension

In Spark SQL, the built-in data types are not sufficient for spatial data. User-defined types (UDTs) allow advanced analytics in Spark SQL. We use this interface to extend spatial data types in Spark.

In Spark SQL, the DataFrame structure stores data in a columnar, compressed format for in-memory caching. Built-in types and user-defined types have a common parent class DataType. Since spatial data has different sizes, spatial data is suitable for columnar storage. Therefore it is appropriate to integrate spatial data types into Spark SQL through UDT interface.

The design of spatial data types is based on OGC standards [12]. It consists of 8 primary types. The public parent class Geometry represents all types of spatial types, and the class Geometry is integrated into the Spark SQL system through the user-defined type interface.

3.4   Spatial Operators Extension

The design of spatial operators is based on the definition of operators in OGC standards [12]. We decide on a total of 42 major spatial operators to implement. To ensure the correctness of the operators implementation, we apply Java Topology Suite [13] as the basis of algorithm implementation.

In Spark SQL, users can execute query by two means: SQL statements and DSL. And we extend spatial operators in both ways.

Spatial Operators Extension on SQL. In the first stage of query planning, Spark SQL builds an "unresolved logical plan" tree and uses Catalyst rules and a Catalog object to resolve references. When resolving functions, it first looks up in the user-defined functions. If unfound, it searches in SQL built-in functions. So through the registration mechanism of user-defined function, spatial operators can be integrated reliably. When naming a user-defined function, system users should pay attention not to override the built-in functions.

Therefore, we adopt the registration mechanism of user-defined functions to integrate spatial operators into Spark SQL. The naming of methods follows the OGC standards [12]. We apply the ST_ prefix to avoid overriding built-in functions and user-defined functions in other fields.

Spatial Operators Extension on DSL. In Spark SQL, system users can employ method invocation in domain specific language (DSL) to achieve the same effect in SQL. For example, "student.where("id" === 1).select ("name")" has the same effect as SQL statement on DataFrame student: "select name from student where id = 1". The operator "===" is an example of built-in functions, and the spatial operators after the integration should act similarly to the "===" operator.

Under the hood, relational operators are calculated between columns. All parameters involved in the calculation are nodes of the Catalyst logical plan tree. Either data or operator can be a node in the logical plan tree. Hence, the integration of DSL mode basically requires encapsulating spatial operators as built-in functions in class Column.

Similar to the spatial extension on SQL, spatial operators are encapsulated as methods in class Column. These operators will be a node in the Catalyst logical plan tree and integrate into the DSL query process.

3.5   Spatial Data Source Format Extension

After the integration of spatial data types and spatial operators, the extension system can meet the basic needs of spatial query. But there are still some supplements to enrich the spatial query processing.

Spark SQL has native support for JSON and Parquet, and plugins provided by Spark official can support XML and CSV data formats. At present, data formats supported by Spark SQL are not sufficient for the spatial extension system, so it is necessary to extend on spatial data source format.

In this paper, we take the GeoJSON [14] format as an example of the format extension. Special fields in GeoJSON cannot be resolved by the built-in format parsing method of JSON, thus it is necessary to implement the class GeoJSONRelation extending class BaseRelation to load GeoJSON format automatically. By inheritance we implement different methods to read and write GeoJSON format.

3.6 Spatial Query Optimization

In Catalyst workflow, the logical optimization phase applies standard rule-based optimizations to the logical plan. Since details about spatial data in the nodes are not accessible at the logical optimization stage, we can only optimize spatial queries in the physical optimization phase.

In the physical planning phase, Spark SQL takes a logical plan and generates one or more physical plans, then selects a plan using a cost model. Consequently, we add a PhysicalPlanner tailored to spatial computation as an example: once we find a join operation, then

- Each Row as a partition in RDD.
- Always selects the smaller number of data byte as the broadcast direction to avoid shuffle.

4 System Test and Experiments

4.1 Operators Correctness Verification

We design the test data and SQL statements based on OGC standards. In DSL test, we design the program in domain specific language according to the logic of SQL statements. We prepare 11 tables on spatial data and 46 test SQL statements. Table 1 exhibits an example of verifying the correctness of operators "ST_Dimension".

Table 1. Operators correctness verification

No	SQL	DSL	Expected result	Experiment result
1	select ST_Dimension (shore) as dimension from lakes where name = 'Blue Lake'	lakes.where ($"name" === "Blue Lake").select(($"shore" ST_Dimension) as "dimension")	2	2

4.2 Performance Test

Since our spatial extension on Spark SQL conforms to rules by the API and the source code of Spark, we hold that the spatial extension system should maintain the high performance of Spark. Thus we adopt the same way of performance test by Spark and compare Hadoop with Spark. The experiment steps are as follows:

- The program reads input data from HDFS.
- The program computes typical spatial queries.
- The program outputs results to HDFS.

In addition, to show the influence on the variance of computing scale, the computing scale of the test data increases progressively. The experiment result is shown in Tables 2 and 3 below:

Table 2. Binary operator: contains

Computing scale	Hadoop	Spark
1000	46	33
10000	55	45.8
100000	60	50.4
1000000	81.8	68.6
10000000	251.8	204.2

Table 3. Unary operator: convexhull.

Computing scale	Hadoop	Spark
100	40.8	37.8
1000	51.6	45.2
10000	54.8	55.2
100000	62.8	61.6
1000000	103	83.2
10000000	359	278.2

The reason why Spark does not display great computing advantage is that the calculation process accounts for a small portion of time. Spark has a great advantage in memory computing optimization and data scheduling over Hadoop. However, a large portion of time is spent on distributed task deployment and read/write the file. But with the increase in the computing scale, the portion of the calculation process grows and the performance advantage of Spark is growing prominent. In conclusion, our spatial extension system maintains the high performance of Spark system.

5 Conclusion and Future Work

This paper displays a detailed analysis of extension points on spatial query in Spark SQL. We research the extension mechanism of user-defined types, user-defined functions, domain specific language and data source format in Spark SQL. Then we implement spatial type system, spatial operators and data source format extension and spatial query optimization. Our work lay the grounds for a more comprehensive spatial extension on Spark SQL in the future. The ways to improve include extending on more spatial data formats and spatial query optimization rules.

Furthermore, compared to Hadoop, Spark is still in the initial stage of development. Program interface between different versions of Spark frequently changes. Therefore, the extension system of this paper cannot be applied to all Spark1.X versions and we can only adjust to a particular version of Spark. This problem is expected to be resolved in the unified interface provided by the latest release in the Spark2.0 version.

References

1. Dean, J., Ghemawat, S.: MapReduce: simplified data processing on large clusters. OSDI 2004: 1 (2004, to appear)
2. Zaharia, M., Chowdhury, M., Franklin, M.J., et al.: Spark: cluster computing with working sets. In: USENIX Conference on Hot Topics in Cloud Computing. USENIX Association, pp. 1765–1773 (2010)
3. Xin, R.S., Rosen, J., Zaharia, M., Franklin, M.J., Shenker, S., Stoica, I.: Shark: SQL and rich analytics at scale. In: SIGMOD Conference, pp. 13–24 (2013)
4. Apache spark: lightning-fast cluster computing. http://spark.apache.org
5. Eldawy, A., Mokbel, M.: A demonstration of SpatialHadoop: an efficient MapReduce framework for spatial data. Proc. VLDB Endow. 6(12), 1230–1233 (2013)
6. Aji, A., Wang, F., Vo, H., Lee, R., Liu, Q., Zhang, X., Saltz, J.: Hadoop-GIS: a high performance spatial data warehousing system over MapReduce. Proc. VLDB Endow. 6(11), 1009–1020 (2013)
7. Zaharia, M., Chowdhury, M., Das, T., Dave, A., Ma, J., McCauley, M., Franklin, M., Shenker, S., Stoica, I.: Resilient distributed datasets: a fault-tolerant abstraction for in-memory cluster computing. In: NSDI 2012 (2012)
8. Beckmann, N., Kriegel, H.-P., Schneider, R., Seeger, B.: The R*-tree: an efficient and robust access method for points and rectangles. In: Proceedings of the 1990 ACM SIGMOD International Conference on Management of Data (SIGMOD 1990), pp. 322–331. ACM, New York (1990)
9. Neis, P., Zipf, A.: Analyzing the contributor activity of a volunteered geographic information project—the case of OpenStreetMap. ISPRS Int. J. Geo-Inf. 1(2), 146–165 (2012)
10. Fox, A., Eichelberger, C., Hughes, J., Lyon, S.: Spatio-temporal indexing in non-relational distributed databases. In: Proceedings of the IEEE International Conference on Big Data, October 2013
11. Armbrust, M., Xin, R.S., Lian, C., et al.: Spark SQL: relational data processing in spark. In: Proceedings of the 2015 ACM SIGMOD International Conference on Management of Data, ACM (2015)
12. OpenGIS® Implementation Standard for Geographic information - Simple feature access - Part 2: SQL option. 2010 Open Geospatial Consortium, Inc
13. Java Topology Suite. http://www.vividsolutions.com/jts/main.htm
14. The GeoJSON Format Specification. http://geojson.org/geojson-spec.html
15. Yu, J., Wu, J., Sarwat, M.: Geospark: a cluster computing framework for processing large-scale spatial data. In: SIGSPATIAL International Conference (2015)
16. R project for statistical computing. http://www.r-project.org
17. Armbrust, M., Xin, R.S., Lian, C.: Spark SQL: relational data processing in spark. In: SIGMOD Conference (2015)

Ecological and Environmental Data Processing and Management

A Comparison of Four Global Land Cover Maps on a Provincial Scale Based on China's 30 m GlobeLand30

Xiaohui Ye[1], Jinling Zhao[1,2(✉)], Linsheng Huang[1,2],
Dongyan Zhang[1,2], and Qi Hong[1,2]

[1] Anhui Engineering Laboratory of Agro-Ecological Big Data,
Anhui University, Hefei 230601, China
aling0123@163.com
[2] Key Laboratory of Intelligent Computing and Signal Processing,
Ministry of Education, Anhui University, Hefei 230039, China

Abstract. To monitor the entire earth system, several global land cover mapping products have been produced based on different remote sensing imagery (e.g., AVHRR, MODIS, SPOT, HJ-1), including UMD Land Cover, Global Land Cover 2000, GlobCover 2009 and GlobeLand30. However, the application potential of those products has not been fully explored at a provincial scale. The primary objective of this study is to compare and investigate the potential of these products in Anhui Province, China. The area and spatial consistency were used to evaluate the four datasets based on China's GlobeLand 30 due to 30 m spatial resolution and the total accuracy of 83.51%. The Pearson's correlation coefficient (R) and the percentage disagreement (PD) were used to evaluate the area consistency. The spatial similarity coefficient (O) was here adopted in order to verify the accuracy of spatial positions. A total of eight cover classes including cropland, woodland, grassland, shrubland, wetland, water bodies, artificial surfaces and bareland were reclassified and mapped to perform the intercomparison. The analysis results show that the PD of GlobCover 2009 is respectively 15.36% and −8.43% for "cropland" and "woodland", while they are −20.20% and 2.10% for UMD. The "woodland" has better agreement percentage in comparison with "cropland". The O of UMD is the worst of 69.25%, indicating that the spatial consistency is weak for "cropland". Conversely, the O of GlobCover 2009 reaches up to 83.50%.

Keywords: Remote sensing · Global land cover product · Anhui Province · GlobeLand30 · GlobCover

1 Introduction

Land cover plays a major role in the climate and biogeochemistry of the Earth system, which is a considerable variable that impacts on and links many parts of the human and physical environments [1]. Nevertheless, it has always been a difficult problem to derive such a parameter, especially on national, continental and even global scale, since the emergence of remote sensing. Remote sensing technology has ability to represent

© Springer Nature Singapore Pte Ltd. 2017
H. Yuan et al. (Eds.): GRMSE 2016, Part I, CCIS 698, pp. 447–455, 2017.
DOI: 10.1007/978-981-10-3966-9_51

and identify land cover categories by means of pixel-based or object-oriented classification process. With the availability of multispectral remotely sensed data in digital form and the developments in digital processing, remote sensing supplies a new prospective for land-cover/land-use analysis [2]. Land cover is an intrinsic element of most remote sensing analysis, but not the ultimate goal of remote sensing studies [3]. It is often a useful aid for further analysis and many applications, including modelling climate change extent and impacts, conserving biodiversity and managing natural resources. Remote sensing has facilitated the presentation and identification of large-scale land cover classification.

The development of satellite sensors and application requirements in various earth observations have been required to produce derived products instead of original remote sensing imagery. Consequently, several land cover classification products have been developed by different agencies and organizations. Hansen et al. (2000) reported the production of a 1 km spatial resolution land cover classification using data for 1992–1993 from the Advanced Very High Resolution Radiometer (AVHRR) [4]. Loveland et al. (2000) developed of a global land cover characteristics database and IGBP DISCover from 1 km AVHRR data [5]. A new global land cover database for the year 2000 (GLC2000) has been produced by an international partnership of 30 research groups coordinated by the European Commission's Joint Research Centre [6]. Ran et al. (2010) compared and evaluated four land cover datasets over China, specifically, which include the IGBP DISCover, MODIS land cover map 2001, University of Maryland (UMD) global land cover map, and GLC 2000 [7].

To find out the differences among those products, some comparison studies have been performed. Hansen et al. (2000) made a preliminary comparison of the methodologies and results of the International Geosphere-Biosphere Programme Data and Information System (IGBP-DIS) DISCover and the UMD 1 km land cover maps [8]. McCallum et al. (2006) compared four satellite derived 1 km land cover datasets freely available from the internet [9]. Giri et al. (2005) compared the Global Land Cover 2000 (GLC 2000) and MODerate resolution Imaging Spectrometer (MODIS) global land cover data to evaluate the similarities and differences in methodologies and results [10]. We can find that the previous studies primarily investigated the land cover data products at a global scale. Consequently, some issues must be addressed when the products are used at an operational level. Taking Anhui Province, China as an example, this study aimed at investigating the application potential and comparing the differences of currently available global land cover products on a provincial scale.

2 Materials and Methods

2.1 Study Area

Anhui Province is located in southeastern China, lying in the lower reaches of both the Yangze River and the Huaihe River, with a longitude between 114°54′ E and 119° 37′ E and a latitude between 29°41′ N and 34°38′ N (Fig. 1). The total area of Anhui is 139,000 km^2 with a quite diverse topography, in which the plain, mountain, plateau, water account for 49.6%, 15.3%, 14.0%, 13.0%, 8.1%, respectively. The province is naturally divided by Huai River and Yangze River into three natural regions, from north to south: Huaibei Plain, Jianghuai Hill and Wannan Mountain. Consequently, the topography shows generally that it is higher in the west and south and lower in the east and north. It lies in the transitional climatic zone between warm temperature zone and subtropical zone, with a warm and moist climate as well as distinct four seasons. The average temperature is between 14 °C and 17 °C and the yearly rainfall is between 770 mm and 1700 mm.

2.2 Production of Four Global Land Cover Maps

- UMD Global Land Cover Classification was generated by the University of Maryland Department of Geography in 1998. The remote sensing imagery derived from the AVHRR satellites acquired between 1981 and 1994 to distinguish fourteen land cover classes. This product is available at three spatial scales: 1 degree, 8 km and 1 km pixel resolutions (Data source: http://www.landcover.org).
- GLC 2000 was coordinated and implemented by the Global Land Cover 2000 Project (GLC 2000) in collaboration with a network of partners around the world. The product makes use of the VEGA 2000 dataset: a dataset of 14 months of pre-processed daily global data acquired by the VEGETATION instrument on board the SPOT 4 satellite (Data source: http://bioval.jrc.ec.europa.eu/products/glc2000/products.php).
- ESA Global land cover map (ESA GlobCover) was generated using 19 months worth of data from Envisat's Medium Resolution Imaging Spectrometer (MERIS) instrument working in Full Resolution Mode to provide a spatial resolution of 300 m. The maps cover 2 periods: December 2004–June 2006 and January–December 2009. (Data source: http://due.esrin.esa.int/globcover/).
- GlobeLand30 was the 30 m GLC data product with 10 classes for years 2000 and 2010 within a four year period, which is derived from a GLC mapping project of China in 2010. The images utilized for GlobeLand30 classification are multispectral images with 30 m, including the TM5 and ETM+ of America Land Resources Satellite (Landsat) and the multispectral images of China Environmental Disaster Alleviation Satellite (HJ-1). (Data source: http://www.globallandcover.com/GLC30 Download/index.aspx).

2.3 Land Cover Classification Schemes

Land cover classification scheme is the first to be determined for any GLC product. In general, the scheme is always different for different GLC products due to the use of different remotely sensed imagery. In addition, the availability of training data is also the primary influence factor to derive the number of land cover types. For instance, for the UMD GLC product, a total of fourteen land cover types are classified and there is a data file for each land cover type classification layer (with numbers from 0 (Water) to 13 (Urban and Built)). To identify the pixels to be used for training of the 1 km

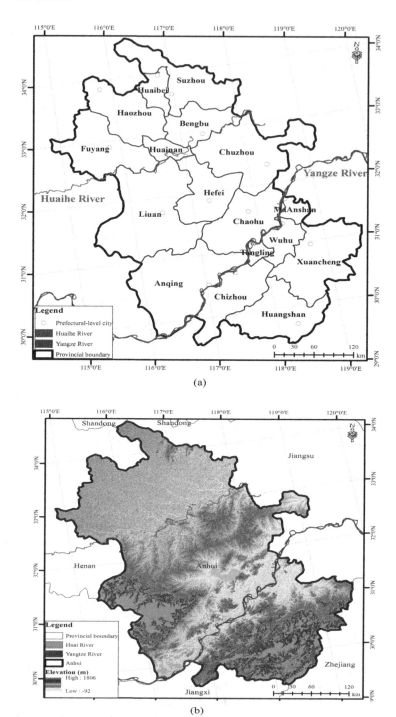

Fig. 1. Geographic location (a) and DEM (b) of the study area

AVHRR Pathfinder data, a total of over 200 high resolution scenes were collected including the Landsat Multispectral Scanner System (MSS), the Landsat Thematic Mapper (TM) and the Linear Imaging Self-Scanning Sensor (LISS) [4]. To compare the GLC products, it is highly important to reclassify the land cover categories prior to the evaluation of consistency [11]. A total of eight cover types were acquired in ENVI (The Environment for Visualizing Images) by harmonizing the various land cover classification schemes (Table 1). In our study, the C4.5 decision tree algorithm to achieve our goal based on expert knowledge [12].

Table 1. Harmonized land cover classification schemes for four GLC products.

No.	GLC product	Number of classes	Reclassified class name (Code)
1	UMD	14	Cropland (1)
2	GLC 2000	22	Woodland (2)
3	GlobCover 2009	22	Grassland (3)
4	GlobeLand30	10	Shrubland (4)
			Wetland (5)
			Water bodies (6)
			Artificial surfaces (7)
			Bareland (8)

2.4 Evaluation of Area and Spatial Consistency

Area consistency and spatial consistency are selected to evaluate the consistency among the four GLC products. The Pearson's correlation coefficient (R) and the percentage disagreement (PD) are used to evaluate the area consistency. Specifically, R represents the overall correlation degree (Eq. 1) and PD (Eq. 2) represents the correlation degree on each type. The pixel-by-pixel comparison method was here adopted in order to verify the accuracy of spatial positions (Eq. 3).

$$R = \frac{\sum\limits_{k=1}^{n} (x_k - \bar{x})(y_k - \bar{y})}{\sqrt{\sum\limits_{k=1}^{n} (x_k - \bar{x})^2 \sum\limits_{k=1}^{n} (y_k - \bar{y})^2}} \times 100\% \tag{1}$$

$$PD = \frac{x_k - y_k}{x_k + y_k} \times 100\% \tag{2}$$

where n is the classification number; x_k and y_k are, respectively, the total area of type k in the four GLC products and GlobeLand30; and $\bar{x}$ and $\bar{y}$ are, respectively, the average area of all land cover types in the four GLC products and GlobeLand30.

$$O = \left(\frac{A}{A + B + C}\right) \times 100\% \tag{3}$$

where O is the spatial similarity coefficient, and A, B and C are, respectively, the total number of pixels of three types of classification data: cropland/cropland, cropland/non-cropland, and non-cropland/cropland, respectively.

3   Results and Discussion

3.1   Comparison of Global Land Cover Mapping Databases

Global land cover maps represent important sources of baseline information to a wide variety of users. Generally, the global land cover maps were produced by a range of different agencies and/or for different periods of time. Different remotely sensed imagery (sensors), methodologies, and validation techniques were always employed [13]. It is highly necessary to identify and quantify uncertainty and spatial disagreement in the comparison of Global Land Cover for different applications [14]. In our study, the UMD, GLC 2000, GlobCover 2009 and GlobeLand30 were reviewed and summarized to provide a clear overview of these differences by category (Table 2).

Table 2. Comparison of the four global land cover maps.

	UMD	GLC 2000	GlobCover 2009	GlobeLand30
Sensor	AVHRR	SPOT-4 VGT	ENVISAT MERIS	Landsat, HJ-1
Collection date	April 1992–March 1993	November 1999–December 2000	January 2009–December 2009	January 2010–December 2010
Spatial resolution	1 km	1 km	300 m	**30 m**
Input data	41 metrics derived from NDVI and bands 1–5	4 spectral bands and NDVI	13 Spectral bands and NDVI composites	Spectral bands and other ancillary dataset
Classification method	Supervised decision tree	Unsupervised classification with ISODATA algorithm	Per-pixel supervised (urban and wetland) and unsupervised	Hierarchical classification strategy
Validation method	None	Statistical validation	Statistical validation	Statistical validation
Total accuracy	–	68.6%	67.5%	**83.51%**

3.2   Analysis of Area Consistency

Four land cover maps of Anhui Province are produced after harmonizing the land cover classification schemes (Fig. 2). Nevertheless, it is relatively difficult to compare the spatial consistency of all the land cover types among different GLC products, due to various land cover classification schemes. Some classes are always disappeared although the land cover categories have been reclassified in accordance with the specified classes of GlobeLand30. For instance, "wetland" is disappeared in the UMD map (Fig. 2a), "bareland" is disappeared in the GLC 2000 (Fig. 2b), and "shrubland" is disappeared in the UMD map (Fig. 2d). Consequently, the spatial consistency of "woodland" is just considered in our study. It can be found that the spatial distribution of woodland shows highly different in the four maps. The O of UMD is the worst of 69.25%, indicating that the spatial consistency is weak for "woodland". The O of GlobCover 2009 reaches up to 83.50%.

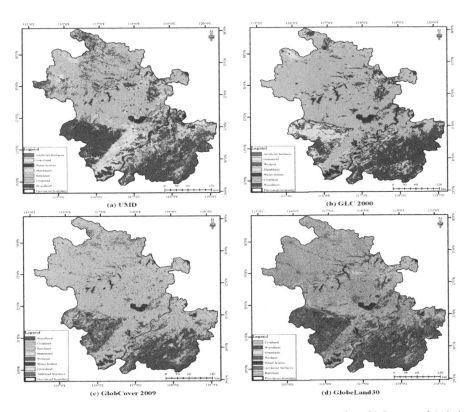

Fig. 2. Comparison of spatial consistency of land cover types among four GLC maps of Anhui Province. (a) UMD, (b) GLC 2000, (c) GlobCover 2009, and (d) GlobeLand30.

3.3   Comparison of Global Land Cover Mapping Databases

It is also an important indicator to analyze the area consistency of four GLC maps in addition to comparing the spatial consistency (Fig. 3). In this study, due to the temporal differences of four maps, two groups of GLC maps are respectively compared between GlobeLand30 and GlobCover 2009 (Fig. 3a) and between UMD and GLC 2000 (Fig. 3b). It can be found that "cropland" and "woodland" show similar performance for the two groups. Taking "woodland" as the example, the overall R value among the four maps is only 0.45, but it shows relatively good performance between UMD and GLC 2000 and between GlobCover 2009 and GlobeLand30. GLC 2000 is used as the reference to evaluate UMD and GlobeLand30 is used to assess GlobCover 2009. The results show that the *PD* is respectively 15.36% and −8.43% for "cropland" and "woodland" (Fig. 3a), while they are −20.20% and 2.10% (Fig. 3b). The "woodland" has better agreement percentage in comparison with "cropland".

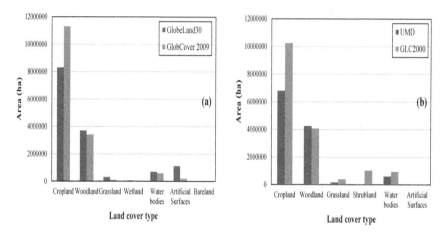

Fig. 3. Comparison of area values between GlobeLand30 and GlobCover 2009 (a) and between UMD and GLC 2000 (b).

4 Conclusion

The freely distributed global land cover maps have facilitated the identification and understanding of land surface processes and modelling. In addition, the spatial resolution of GLC products has gradually increased from 1 degree (e.g. UMD) to 30 m (e.g. GlobeLand30). It is of fundamental importance in studies of dynamic changes in monitoring Earth's natural resources. Consequently, quantitative assessments of the map quality and classification accuracy for existing land cover maps will help to improve accuracy in future land cover mapping. Nevertheless, the GLC maps were produced using different classification systems, which made the comparison difficult. They must be first aggregated by reclassifying them using a unified legend system. In our study, a total of eight classes were reclassified including cropland, woodland, grassland, shrubland, wetland, water bodies, artificial surfaces and bareland according to the reference dataset (GlobeLand30). The area and spatial consistency are usually different in different GLC maps for a certain land cover type. It is highly necessary to evaluate the suitability and accuracy of a certain type or all the types when carrying out thematic applications.

Acknowledgements. The work presented here was supported Scientific Research Training Programme of Anhui University (ZLTS2015178), Anhui Provincial Science & Technology Project (1604a0702016) and Anhui Provincial Science and Technology Major Project (16030701091). The data set is provided by Environmental and Ecological Science Data Center for West China, National Natural Science Foundation of China (http://westdc.westgis.ac.cn).

References

1. Foody, G.M.: Status of land cover classification accuracy assessment. Remote Sens. Environ. **80**, 185–201 (2002)
2. Bektas, F., Goksel, C.: Remote sensing and GIS integration for land cover analysis, a case study: Bozcaada Island. Water Sci. Technol. **51**, 239–244 (2005)
3. Aplin, P.: Remote sensing: land cover. Prog. Phys. Geogr. **28**, 283–293 (2004)
4. Hansen, M.C., Defries, R.S., Townshend, J.R.G., Sohlberg, R.: Global land cover classification at 1 km spatial resolution using a classification tree approach. Int. J. Remote Sens. **21**, 1331–1364 (2000)
5. Loveland, T.R., Reed, B.C., Brown, J.F., Ohlen, D.O., Zhu, Z., Yang, L., Merchant, J.W.: Development of a global land cover characteristics database and IGBP DISCover from 1 km AVHRR data. Int. J. Remote Sens. **21**, 1303–1330 (2000)
6. Bartholomé, E., Belward, A.S.: GLC2000: a new approach to global land cover mapping from earth observation data. Int. J. Remote Sens. **26**, 1959–1977 (2005)
7. Ran, Y.H., Li, X., Lu, L.: Evaluation of four remote sensing based land cover products over China. Int. J. Remote Sens. **31**, 391–401 (2010)
8. Hansen, M.C., Reed, B.: A comparison of the IGBP DISCover and university of Maryland 1 km global land cover products. Int. J. Remote Sens. **21**, 1365–1373 (2000)
9. McCallum, I., Obersteiner, M., Nilsson, S., Shvidenko, A.: A spatial comparison of four satellite derived 1 km global land cover datasets. Int. J. Appl. Earth Obs. **8**, 246–255 (2006)
10. Giri, C., Zhu, Z., Reed, B.: A comparative analysis of the global land cover 2000 and MODIS land cover data sets. Remote Sens. Environ. **94**, 123–132 (2005)
11. Liang, D., Zuo, Y., Huang, L.S., Zhao, J.L., Teng, L., Yang, F.: Evaluation of the consistency of MODIS land cover product (MCD12Q1) based on Chinese 30 m GlobeLand30 datasets: a case study in Anhui Province. China. ISPRS Int. J. Geo-Inf. **4**, 2519–2541 (2015)
12. Ahmad, A.: Decision tree ensembles based on kernel features. Appl. Intell. **41**, 855–869 (2014)
13. Congalton, R.G., Gu, J., Yadav, K., Thenkabail, P., Ozdogan, M.: Global land cover mapping: a review and uncertainty analysis. Remote Sens. **6**, 12070–12093 (2014)
14. Fritz, S., See, L.: Identifying and quantifying uncertainty and spatial disagreement in the comparison of global land cover for different applications. Glob. Change Biol. **14**, 1057–1075 (2008)

Research Progress on Coupling Relationship Between Carbon and Water of Ecosystem in Arid Area

Xiang Huang[✉]

State Key Laboratory of Desert and Oasis Ecology,
Xinjiang Institute of Ecology and Geography, CAS, Urumqi 830011, China
huangx@ms.xjb.ac.cn

Abstract. Carbon and water cycle of terrestrial ecosystem is the frontier scientific issues of global change and carbon cycle research, the process mechanism of carbon and water coupling analysis is the scientific basis of climate change mechanism, climate change prediction, and climate change adaptation strategy formulation. For the arid and semi arid region, which accounts for 30–50% of the total land area in the earth, the carbon source problem still has a lot of uncertainty, and the combination research of the two processes of carbon and water is very little. The analysis about research results from domestic and foreign related ecosystem carbon and water coupling relations not only can understand the carbon and water coupling effect in different scale and internal relations but also contribute to a profound understanding of the effect of temperature and precipitation patterns change on the water and carbon balance of ecological system in arid area under the context of global change, which will help to grasp the dynamic ecological system in arid area. This paper analyzes the coupling relationship between carbon and water, water use efficiency and carbon and water coupling model, and focus on carbon and water coupling relationship of ecological system in arid area. Meanwhile, this paper studies carbon and water fluxes in different spatial and temporal scales of desert riparian forest ecosystem, compares the water efficiency of desert riparian forest ecosystem system (WUE) in different scales, and explore carbon and water coupling relationship of desert riparian forest ecosystem and water use efficiency change under the background in order to study the coupling model based on process.

Keywords: Coupling relationship between carbon and water · Terrestrial ecosystem · Carbon and water fluxes

1 Introduction

Carbon and water cycle of terrestrial ecosystem is a complex interaction coupling relationship and interaction of the interface between the soil-plant-atmosphere continuum (SPAC) (Jasechko et al. 2013; Keenan et al. 2013), its process is the core of

Work is supported by Xinjiang Autonomous Region Outstanding Youth Science and technology talent project (2014711002) and National natural science foundation of China (41371128).

© Springer Nature Singapore Pte Ltd. 2017
H. Yuan et al. (Eds.): GRMSE 2016, Part I, CCIS 698, pp. 456–465, 2017.
DOI: 10.1007/978-981-10-3966-9_52

material energy cycle in terrestrial ecosystem, and is the link of geosphere-biosphere-atmosphere interactions (Yu and Sun 2006), and its change characteristics and environment control mechanism is the basin to reveal the variation of carbon and water fluxes, understanding the carbon and water cycle process of ecosystem in temporal and spatial scale, and clarifying the carbon sink potential development model of ecological system. Under the background of global change, the change of temperature and precipitation in time and space will affect the carbon, water process of ecosystem and the carbon water balance, and ultimately affect the carbon sink strength of terrestrial ecosystem. So it has become a hot issue for the academic circles to study the climate change and human activity response and adaptation to the key process of carbon water coupling relationship in terrestrial ecosystem (Chen et al. 2009).

At present, most research related to carbon and water cycle used to be separated into two processes, mainly focusing on soil respiration, vegetation photosynthesis and transpiration rate, environmental factor response separately (Wang et al. 2014), and the research which combine the carbon and water process is relatively small, which focus on studying the carbon water coupled relationship in temperate and tropical forest ecosystems (Baldocchi 1997; Michelot et al. 2011; Tong et al. 2014). The arid and semi arid regions accounting for 30–50% of total land area in the earth, especially for arid areas where still exist great uncertainty in the carbon sinks, has rarely reports about water and carbon flux and their coupling relationship (Asner et al. 2003; Liu et al. 2012a). Research has shown that the ecological system in arid area has a significant carbon absorption function (Jasoni et al. 2005; Wohlfahrt et al. 2008), but there are studies that reported desert ecosystems had limited carbon absorption capacity (Li et al. CO 2014). In addition, temperature, precipitation, grazing and other natural or man-made factors are likely to affect the carbon sink function of ecological system in arid region, and lead to carbon source/sink change in ecosystem (Bell et al. 2012; Gao et al. 2012; Liu et al. 2012b; Snyder et al. 2012). The comprehensive analysis of the current research results in domestic and foreign related carbon and water coupling relations in ecosystem not only can understand the carbon and water coupling effect of different scales and internal relations, but also help to construct the water carbon coupled model in arid region and the estimation methods of WUE coordination (Wang et al. 2014), and help to understand temperature and precipitation patterns change of ecological system in arid area which affect on carbon and water process and the relationship between water balance of carbon under the background of global change.

2 Study on the Coupling Relationship Between Carbon and Water

The research about global change and carbon cycle is including the impacts of climate warming, carbon dioxide concentration, precipitation pattern change and N deposition on the carbon sink of spatial and temporal changes in terrestrial ecosystem, which is a frontier science problem. The process mechanism of water and carbon cycle in terrestrial ecosystem is the scientific basis for analyzing the mechanism of climate change, predicting the climate change and making the adaptive strategy of climate change.

However, the current understanding of the carbon water balance and its response to climate change in terrestrial ecosystems is still limited.

The exchange between water, heat and carbon in terrestrial ecosystem and atmosphere is the key process of the material and energy cycle, which has been concerned by researchers all the time. Since twenty-first Century, especially with the increasing awareness of global warming, researchers pay more attention to climate change impact on the exchange process of carbon, water and its response to climate change (Li 2008). Vegetation influences the water, energy and carbon fluxes of the whole ecosystem through the physiological ecology, phenology and biological response in different environments (Singh et al. 2014). Carbon and water cycle is the key link of ecosystem function, which is the ecology process that is controlled by soil, atmosphere and vegetation of various biological and environmental factors together. The process regards the stomata as main coupling nodes, mainly for the CO_2 exchange process between vegetation and water vapor, which mean the coupling auspices of stomatal and the close contact between and transpiration (Cowan and Farquhar 1977). Photosynthesis is the original impetus of support of the whole ecosystem, which is the main way to carbon sequestration in terrestrial ecosystems, but also is the basis to understand the vegetation carbon sink formation mechanism, construction of carbon cycle model of terrestrial ecosystem. Evaporation effect is the process of plant and soil water loss associated with photosynthesis. Transpiration and photosynthesis are two closely related and mutually coupled processes, which lead the formation of physiological activities and yield of the crops. It is significant to study the relationship between water and carbon in different scales to improve water use efficiency (WUE) (Zhang et al. 2013).

Study on the relationship between carbon and water coupling must be based on the comprehensive understanding of the transpiration which is controlled by stomatal behavior and eco physiological processes related to carbon fixation, and focus on research the plant water use efficiency of physiological and ecological mechanism, and comprehensive reveal the water carbon cycle of physiological and ecological mechanism in leaf and patch scale of ecosystem, and investigate the coupling process of plant photosynthesis - stomatal behavior-transpiration. The lack of understanding of the various fluxes exchange process between the land surface and the atmosphere will not be conducive to the effective use of knowledge in the field of climate, hydrology, ecology and environmental chemistry to solve practical problems and decisions. How to combine these disciplines together to accumulate data and develop a comprehensive numerical simulation model for the study of land surface processes has become a new research in the international academic community (Wang et al. 2004). However, our understanding of the relationship of balance between water and carbon, and its response to climate change is very limited, the previous studies of carbon and water coupled mostly from the carbon cycle and water cycle angle independent, but in fact these two loops are coupled with each other. The relationship between carbon and water is an urgent need to be clear in the different ecosystem, and determine carbon and water Trade-offs in different ecosystem, and determine the optimal threshold of balance of water and carbon, and know the balance of water and carbon in spatial and temporal scales of ecological system. In order to make clear the impact of climate change future on water and carbon of ecosystem, carbon and water flux and its coupling relationship should be deepen to understand and quantify, and understand physiological and ecological processes of ecosystem in different spatial and temporal scales as well as the relationship among various scales (Hu et al. 2006).

3 Study on Water Use Efficiency (WUE)

The water use efficiency (WUE) is the ratio between water dissipation units from leaf transpiration of plant and the photosynthate amount assimilated by the same leaf, or the ratio between dry matter produced by photosynthesis (or vegetation net primary productivity, NPP) and water amount assumed by transpiration (ET)(Rosenberg 1983). Terrestrial plants absorb the CO_2 from the atmosphere by photosynthesis, and this process is accompanied by the loss of water from the leaf. The percentage of water loss and carbon capture - water use efficiency is a key feature of ecosystem function, which is the core of the global water, energy and carbon cycle. Plant WUE can be divided into leaf, individual, community or ecosystem scale according to the research scale. On the leaf and individual plant scales, the determination of plant WUE is usually acquired through gas exchange method and stable carbon isotope method. Gas exchange method is simple and quick, but the value is only WUE of instantaneous blade, and it will be affected by micro environment. The stable carbon isotope method is not limited by time and space, which can better reflect the water status of plants, so it is considered to be the best method for the determination of plant (Donovan et al. 2007). At the ecosystem scale, the eddy correlation method is usually used, but its only water and carbon fluxes at a point in the ecosystem, and in site level, the formation of atmospheric CO_2 concept of water use efficiency can be understood as the quality of unit water consumption to fixed vegetation productivity. Water use efficiency at the ecosystem level is an important physiological index to reflect the coupling relationship between water and carbon cycle (Tong et al. 2014). Changes in water use efficiency showed a change in the coupling relationship of the carbon and water cycle (Yu et al. 2004; 2008). At the ecosystem scale, WUE is defined as the ratio between gross primary productivity (GPP) and the evapotranspiration (ET) (Brümmer et al. 2012; Jassal et al. 2009), the water use efficiency can reflect the plant water use strategy in different living environment (Donovan and Ehleringer 1991) study on WUE of ecosystem level will be helpful to predict the impact of climate change on the carbon and water processes (Hu et al. 2006). In addition, the water use efficiency can be used to evaluate the impact of water resources on the carbon sink/source function of terrestrial ecosystems (Tong et al. 2014). In recent years, the study of water use efficiency is used to explain the interaction between water and carbon cycle in ecosystem, so as to predict the impact of global change on ecosystem function. Long-term research shows that carbon and water exchange of ecosystem, in the past twenty years, a substantial increase in the efficiency of utilization of the temperate and boreal forests of the northern hemisphere, which is explained the fertilization effect of CO2 by many systematic hypotheses (Keenan et al. 2013).

4 Study on Coupling Model of Carbon and Water

Many scholars used the eddy correlation technique to observe the ecosystem carbon and water fluxes (Baldocchi 2003), and the common carbon and water fluxes estimation model including statistical model (Shen et al. 2011), a comprehensive model of carbon and water coupled process (Leuning et al. 1995; Wang and Leuning 1998) and the model based on remote sensing. The advantages of the statistical model is simple and easy to get data, which has a wide range of guiding significance to simulate the carbon and water fluxes, but due to depend greatly on experience, it is difficult to spread to other areas, the physical meaning is not clear, do not have the ability to predict. The remote sensing model was late to start, its main sources of data from satellite remote sensing, remote sensing and remote sensing, its advantage is to value carbon and water fluxes of large area or even within the scope of global, but its disadvantage is to combine the local meteorological data and empirical parameters containing. The model which is based on the process of carbon and water flux coupling has a clear mechanism, which combines physiological characteristics, plant canopy structure, soil characteristics and climatic conditions together. The latter model could simulate different scales carbon and water fluxes including single scale (Leuning 1995; Sellers et al. 1992), canopy (Sellers et al. 1992) and ecosystem scale (Owen et al. 2007), and even applied to the regional scale and global climate models (Sellers et al. 1996).

At present, the research of water and carbon coupling model is mostly focused on the level of the leaf, but the research on total assimilation estimated by the transpiration of canopy still has many challenge (Yu et al. 2003). The nonlinear relation between photosynthesis and transpiration and solar radiation led to the big simulation error of leaf model hypothesis based on eco physiological leaf scale. When the existing water and carbon coupling model is applied to the ecological system, the parameters should mesh leaf scale into the canopy scale in order to each mesh size close to the leaf scale, which will simulate the radiation distribution and intensity of each grid. It is crucial to accurately describe the canopy structure which will determine the absorption of light leaf and canopy light distribution. Canopy radiative transfer is not only affected by the amount of leaf (LAI), but also affected by the distribution of leaves (Louarn et al. 2008). The discontinuous distribution of canopy structure expression of desert riparian forest canopy is more complex than the homogeneous canopy. Through comparative study of different stomatal conductance model, canopy conductance estimation method which is screened by the specific study area has a significant meaning to value accurately the water and carbon flux simulation (Zhang et al. 2008). Therefore, the comparative study of the simulated results with different stomatal conductance models can help to select and clarify the mechanism of the driving factors of physiological processes.

Accurate estimation of water and carbon fluxes in ecosystems is of great significance in evaluating the role of terrestrial ecosystems in the global carbon cycle. Mechanism model of ecosystem is an indispensable means and tool to estimate and predict the flux change of ecosystem water and carbon. The model of ecological system is based on the simulation of the water and carbon flux, Farquhar's photosynthesis equation and the stomatal conductance algorithm based on photosynthesis (Gu et al. 2006). Water and carbon flux coupling model integrates the wide application of photosynthetic Farquhar model (Farquhar and Von Caemmerer 1982; Farquhar et al. 1980) and Penman-Monteith transpiration model, and stomatal conductance model (Jarvis 1976) of the two model associated coupling of water and carbon fluxes simulation (Kim and Lieth 2003; Tuzet et al. 2003). Water vapor flux is more easily obtained in a certain degree, so if the water vapor flux can be determined, the CO_2 flux can be obtained through the coupling relationship between the two items. This process requires a more accurate correlation between two physiological processes (Moren et al. 2001), which is generally associated with stomatal conductance or water use efficiency. Canopy stomatal conductance is one of the most important parameters affecting the accuracy of water carbon flux simulation (Wullschleger et al. 2002).

5 Outlook

At present, the research work at home and abroad, due to the limitations of a variety of technical means, is still unable to integrate multi scale, multi way ecosystem carbon water interaction evaluation model. So building a carbon and water coupling model based on ecological process is an urgent need to simulate and evaluate the characteristics of the spatial and temporal characteristics of the ecosystem.

Under the background of climate change, the frequency and intensity of drought will strengthen in arid and semi-arid regions (IPCC 2007), the seasonal distribution of precipitation is the key factor in the area of carbon assimilation and water flux (Mielnick et al. 2005). In the study of global change, the impact of drought on the water and carbon cycle of the terrestrial water and carbon cycle has become one of the focal problems. To study the energy and material exchange between the atmosphere and the surface is of great significance for the water resources utilization and ecological environment protection in arid area (Bonan 2008; Wang et al. 2014). The water in arid area is the decisive control function of the biology, which causes the vegetation in the arid area to reside on the water, forming the desert riparian forest with unique regional characteristics. The Ecosystem Observation and experiment station are lack of research on water flux observation data, the coupling relationship between carbon there is a big problem, so clear the desert riparian forest ecosystem carbon and water coupling relationship and response to environmental factors, which has very important scientific significance for energy and ecological system to clarify the arid ecosystem productivity and ecological water consumption.

Desert riparian forest ecosystem is an important part of the ecological system in arid area, is a special type of mountain desert oasis in arid area of the composite system (and Xiangxiang roose1999 tour, Wang 2001), but also because of its unique landscape, in the process of carbon and water from other ecosystems, has special carbon and water cycle and transformation and balance laws, and because the matter and energy in the system, there are many uncoordinated, especially water heat exchange of extreme imbalance, environmental and biological factors are in the critical state of transformation, which makes the system extremely fragile environment, increase carbon sink uncertainty, so deep influence of desertification and oasis ecosystem stability two.

Therefore, using the method of field test and laboratory test and simulation model of combining the analysis of desert riparian forest evapotranspiration, carbon cycle and water use efficiency (WUE) of the temporal and spatial variation control factor, study the main influence mechanism of water and carbon cycle and the coupling between the carbon and water fluxes of desert riparian forest ecosystem analysis of temporal and spatial scales, and the influence mechanism of carbon water coupling relationship, discuss the future of climate change under the background of temperature and precipitation patterns change on carbon and water coupling relationship and may affect the use efficiency, the building suitable for the study area of carbon water coupled model, which reveals the special maintenance of desert riparian forest ecosystem stability the security mechanism, has important academic significance to the scientific questions in arid area of carbon and water cycle process related ecological evolution mechanism and regulation etc. Carbon and water fluxes, the interannual variation and process mechanism of desert riparian forest ecosystem is an important content to determine the relationship between carbon and water coupling ecosystem in arid area, spatial pattern and its change trend, find out the carbon intensity of desert riparian forest is a special ecological system, has important significance for the scientific understanding of climate change impacts on Ecological the system in arid area, this is the future of research on carbon and water coupling relationship of ecological system in arid area of focus and hot issues.

References

Asner, G.P., Archer, S., Hughes, R.F., Ansley, R.J., Wessman, C.A.: Net changes in regional woody vegetation cover and carbon storage in Texas drylands, 1937–1999. Glob. Change Biol. **9**, 316–335 (2003)

Baldocchi, D.: Measuring and modelling carbon dioxide and water vapour exchange over a temperate broad-leaved forest during the 1995 summer drought. Plant, Cell Environ. **20**, 1108–1122 (1997)

Baldocchi, D.D.: Assessing the eddy covariance technique for evaluating carbon dioxide exchange rates of ecosystems: past, present and future. Glob. Change Biol. **9**, 479–492 (2003)

Bell, T.W., Menzer, O., Troyo-Diéquez, E., Oechel, W.C.: Carbon dioxide exchange over multiple temporal scales in an arid shrub ecosystem near La Paz, Baja California Sur, Mexico. Glob. Change Biol. **18**, 2570–2582 (2012)

Bonan, G.B.: Forests and climate change: forcings, feedbacks, and the climate benefits of forests. Science **320**, 1444–1449 (2008)

Brümmer, C., Black, T.A., Jassal, R.S., Grant, N.J., Spittlehouse, D.L., Chen, B., Nesic, Z., Amiro, B.D., Arain, M.A., Barr, A.G., Bourque, C.P.A., Coursolle, C., Dunn, A.L., Flanagan, L.B., Humphreys, E.R., Lafleur, P.M., Margolis, H.A., McCaughey, J.H., Wofsy, S.C.: How climate and vegetation type influence evapotranspiration and water use efficiency in Canadian forest, peatland and grassland ecosystems. Agric. For. Meteorol. **153**, 14–30 (2012)

Cowan, I., Farquhar, G.: Stomatal function in relation to leaf metabolism and environment. Symposia Soc. Exp. Biol. **31**, 471–505 (1977)

Donovan, L.A., Dudley, S.A., Rosenthal, D.M., Ludwig, F.: Phenotypic selection on leaf water use efficiency and related ecophysiological traits for natural populations of desert sunflowers. Oecologia **152**, 13–25 (2007)

Donovan, L.A., Ehleringer, J.R.: Ecophysiological differences among juvenile and reproductive plants of several woody species. Oecologia **86**, 594–597 (1991)

Farquhar, G., Von Caemmerer, S.: Modelling of photosynthetic response to environmental conditions. In: Lange, O.L., Nobel, P.S., Osmond, C.B., Ziegler, H. (eds.) Physiological Plant Ecology II. Encyclopedia of Plant Physiology, vol. 12/B, pp. 549–587. Springer, Heidelberg (1982)

Farquhar, G., von Caemmerer, S., Berry, J.: A biochemical model of photosynthetic CO_2 assimilation in leaves of C3 species. Planta **149**, 78–90 (1980)

Gao, Y., Li, X., Liu, L., Jia, R., Yang, H., Li, G., Wei, Y.: Seasonal variation of carbon exchange from a revegetation area in a Chinese desert. Agric. For. Meteorol. **156**, 134–142 (2012)

IPCC: Intergovernmental panel on climate change. IPCC Secretariat Geneva (2007)

Jarvis, P.: The interpretation of the variations in leaf water potential and stomatal conductance found in canopies in the field. Philos. Trans. Roy. Soc. Lond. B Biol. Sci. **273**, 593–610 (1976)

Jasechko, S., Sharp, Z.D., Gibson, J.J., Birks, S.J., Yi, Y., Fawcett, P.J.: Terrestrial water fluxes dominated by transpiration. Nature **496**, 347–350 (2013)

Jasoni, R.L., Smith, S.D., Arnone, J.A.: Net ecosystem CO2 exchange in Mojave Desert shrublands during the eighth year of exposure to elevated CO2. Glob. Change Biol. **11**, 749–756 (2005)

Jassal, R.S., Black, T.A., Spittlehouse, D.L., Brümmer, C., Nesic, Z.: Evapotranspiration and water use efficiency in different-aged Pacific Northwest Douglas-fir stands. Agric. For. Meteorol. **149**, 1168–1178 (2009)

Keenan, T.F., Hollinger, D.Y., Bohrer, G., Dragoni, D., Munger, J.W., Schmid, H.P., Richardson, A.D.: Increase in forest water-use efficiency as atmospheric carbon dioxide concentrations rise. Nature **499**, 324–327 (2013)

Kim, S.H., Lieth, J.H.: A coupled model of photosynthesis, stomatal conductance and transpiration for a rose leaf (Rosa hybrida L.). Ann. Bot. **91**, 771–781 (2003)

Leuning, R.: A critical appraisal of a combined stomatal-photosynthesis model for C3 plants. Plant, Cell Environ. **18**, 339–355 (1995)

Leuning, R., Kelliher, F., De Pury, D.G., Schulze, E.D.: Leaf nitrogen, photosynthesis, conductance and transpiration: scaling from leaves to canopies. Plant, Cell Environ. **18**, 1183–1200 (1995)

Li, L.H., Chen, X., van der Tol, C., Luo, G.P., Su, Z.B.: Growing season net ecosystem CO2 exchange of two desert ecosystems with alkaline soils in Kazakhstan. Ecol. Evol. **4**, 14–26 (2014)

Liu, R., Li, Y., Wang, Q.-X.: Variations in water and CO2 fluxes over a saline desert in Western China. Hydrol. Process. **26**, 513–522 (2012a)

Liu, R., Pan, L.-P., Jenerette, G.D., Wang, Q.-X., Cieraad, E., Li, Y.: High efficiency in water use and carbon gain in a wet year for a desert halophyte community. Agric. For. Meteorol. **162–163**, 127–135 (2012b)

Louarn, G., Lecoeur, J., Lebon, E.: A three-dimensional statistical reconstruction model of grapevine (Vitis vinifera) simulating canopy structure variability within and between cultivar/training system pairs. Ann. Bot. **101**, 1167–1184 (2008)

Michelot, A., Eglin, T., Dufrene, E., Lelarge-Trouverie, C., Damesin, C.: Comparison of seasonal variations in water-use efficiency calculated from the carbon isotope composition of tree rings and flux data in a temperate forest. Plant, Cell Environ. **34**, 230–244 (2011)

Mielnick, P., Dugas, W.A., Mitchell, K., Havstad, K.: Long-term measurements of CO2 flux and evapotranspiration in a Chihuahuan desert grassland. J. Arid Environ. **60**, 423–436 (2005)

Moren, A.S., Lindroth, A., Grelle, A.: Water-use efficiency as a means of modelling net assimilation in boreal forests. Trees-Struct. Funct. **15**, 67–74 (2001)

Owen, K.E., Tenhunen, J., Reichstein, M., Wang, Q., Falge, E., Geyer, R., Xiao, X.M., Stoy, P., Ammann, C., Arain, A., Aubinet, M., Aurela, M., Bernhofer, C., Chojnicki, B.H., Granier, A., Gruenwald, T., Hadley, J., Heinesch, B., Hollinger, D., Knohl, A., Kutsch, W., Lohila, A., Meyers, T., Moors, E., Moureaux, C., Pilegaard, K., Saigusa, N., Verma, S., Vesala, T., Vogel, C.: Linking flux network measurements to continental scale simulations: ecosystem carbon dioxide exchange capacity under non-water-stressed conditions. Glob. Change Biol. **13**, 734–760 (2007)

Rosenberg, N.J.: Microclimate: The Biological Environment. Wiley, New York (1983)

Sellers, P., Berry, J., Collatz, G., Field, C., Hall, F.: Canopy reflectance, photosynthesis, and transpiration III. A reanalysis using improved leaf models and a new canopy integration scheme. Remote Sens. Environ. **42**, 187–216 (1992)

Sellers, P., Randall, D., Collatz, G., Berry, J., Field, C., Dazlich, D., Zhang, C., Collelo, G., Bounoua, L.: A revised land surface parameterization (SiB2) for atmospheric GCMs. part I: model formulation. J. Clim. **9**, 676–705 (1996)

Singh, N., Patel, N.R., Bhattacharya, B.K., Soni, P., Parida, B.R., Parihar, J.S.: Analyzing the dynamics and inter-linkages of carbon and water fluxes in subtropical pine (Pinus roxburghii) ecosystem. Agric. For. Meteorol. **197**, 206–218 (2014)

Snyder, K.A., Scott, R.L., McGwire, K.: Multiple year effects of a biological control agent (Diorhabda carinulata) on Tamarix (saltcedar) ecosystem exchanges of carbon dioxide and water. Agric. For. Meteorol. **164**, 161–169 (2012)

Tong, X.J., Zhang, J.S., Meng, P., Li, J., Zheng, N.: Ecosystem water use efficiency in a warm-temperate mixed plantation in the North China. J. Hydrol. **512**, 221–228 (2014)

Tuzet, A., Perrier, A., Leuning, R.: A coupled model of stomatal conductance, photosynthesis and transpiration. Plant, Cell Environ. **26**, 1097–1116 (2003)

Wang, Y.-P., Leuning, R.: A two-leaf model for canopy conductance, photosynthesis and partitioning of available energy I: model description and comparison with a multi-layered model. Agric. For. Meteorol. **91**, 89–111 (1998)

Wohlfahrt, G., Fenstermaker, L.F., Arnone, J.A.: Large annual net ecosystem CO2 uptake of a Mojave desert ecosystem. Glob. Change Biol. **14**, 1475–1487 (2008)

Wullschleger, S.D., Gunderson, C.A., Hanson, P.J., Wilson, K.B., Norby, R.J.: Sensitivity of stomatal and canopy conductance to elevated CO2 concentration - interacting variables and perspectives of scale. New Phytol. **153**, 485–496 (2002)

Yu, G.-R., Kobayashi, T., Zhuang, J., Wang, Q.-F., Qu, L.-Q.: A coupled model of photosynthesis-transpiration based on the stomatal behavior for maize (Zea mays L.) grown in the field. Plant Soil **249**, 401–415 (2003)

Yu, G.-R., Wang, Q.-F., Zhuang, J.: Modeling the water use efficiency of soybean and maize plants under environmental stresses: application of a synthetic model of photosynthesis-transpiration based on stomatal behavior. J. Plant Physiol. **161**, 303–318 (2004)

Yu, G.R., Song, X., Wang, Q.F., Liu, Y.F., Guan, D.X., Yan, J.H., Sun, X.M., Zhang, L.M., Wen, X.F.: Water-use efficiency of forest ecosystems in eastern China and its relations to climatic variables. New Phytol. **177**, 927–937 (2008)

Zhang, B., Kang, S., Li, F., Zhang, L.: Comparison of three evapotranspiration models to Bowen ratio-energy balance method for a vineyard in an arid desert region of northwest China. Agric. For. Meteorol. **148**, 1629–1640 (2008)

Chen, X., Ju, M., Chen, J., Ren, M.: Terrestrial ecosystem carbon water cycle interaction and its simulation. Ecol. J. **28**, 1630–1639 (2009)

Gu, F., Cao, M., Wen, X., Liu, Y., Lin, T.B.: Simulated water and carbon fluxes and subtropical coniferous and observational comparative study. China Sci. Ser. D, 224–233 (2006)

Hu, Z., Yu, G., Fan, J., Wen, X.: The drought research progress on terrestrial ecosystem carbon and water processes. Prog. Geogr. **25**, 12–20 (2006)

Li, Y.: Corn farmland water heat carbon flux dynamics and its environmental control mechanism research. Graduate University of Chinese Academy of Sciences (2008)

Shen, X., Zhang, M., Qi, X.: Estimation methods of regional forest carbon distribution based on regression and stochastic simulation. **47**, 1–8 (2011)

Wang, Q., Watanabe, M., Zhu, O., Yan, L., Li, Y., Zhao, X., Wang, K.: Different types of ecosystem water heat carbon flux monitoring and research. J. Geogr. **59**, 13–24 (2004)

Wang, Y.X.: The community characteristics of the oasis ecosystem: a case study of desert riparian forest ecosystem. Resour. Environ. Arid Area **15**, 22–24 (2001)

Wang, Y., Bai, J., Li, L., Jing, C., Chen, X., Luo, G., Ji, S.: Growth flux characteristics in arid central Asia 3 typical ecological system. J. Plant Ecol. (2014)

Yu, G., Sun, X.: The principle and method of the terrestrial ecosystem flux observation (2006)

Zhang, B., Liu, Y., Xu, D., Cai, J., Wei, Z.: Summer maize leaf and canopy scale water and carbon coupling simulation. Science Bulletin, 1121–1130 (2013)

Karst Rocky Desertification Dynamic Monitoring Analysis Based on Remote Sensing for a Typical Mountain Area in Southeast of Yunnan Province

Ling Yuan, Shu Gan[(✉)], Xiping Yuan, Ce Wang, and Da Yi

Kunming University of Science and Technology,
Kunming 650093, Yunnan, China
851649146@qq.com, 1193887560@qq.com

Abstract. Selecting a typical karst rocky mountain area as a study case located in southeast of Yunnan province, getting Landsat5 TM digital images detected from two different time point of 1990 and 2007 year, karst rocky desertification condition is monitored and change analysis is completed by use of remote sensing process and GIS spatial analysis technique integrated land use diagnoses. The results are shown that: the dynamic changes of the karst rocky desertification amount were not distinct from 1990 to 2007, but the amount of the different degree types of karst rocky desertification changes relative obvious, in general karst rocky desertification condition presents a severity trend. The study also demonstrates that the integrated technique application by remote sensing, GIS and land use diagnoses can get the past back data and landscape pattern with location and quantity. This is help to understand deeply the development characteristic of karst rocky desertification change, and be benefit to provide decision-making support for karst rocky desertification mountain land use and management.

Keywords: Karst rocky desertification · Remote sensing · Dynamic monitoring analysis · Transfer matrix · Southeast of Yunnan province

1 Introduction

Karst rocky desertification refers to the process of land degradation that soil was eroded seriously, bedrock was exposed widespread, bearing capability of land declined seriously, landscape similar to desert appears on the earth's surface under the disturbance of unreasonable social economic activities of human beings in the subtropical karst environment. At present, rocky desertification of karst mountain areas in Southwest China has become one of the fundamental problems of regional ecological environment in the implementation of western development strategy. The soil erosion and the rapid

Thanks to the support of National Fund Project (41561083, 41261092) and Natural Science Foundation of Yunnan Province (2015FA016).

© Springer Nature Singapore Pte Ltd. 2017
H. Yuan et al. (Eds.): GRMSE 2016, Part I, CCIS 698, pp. 466–476, 2017.
DOI: 10.1007/978-981-10-3966-9_53

desertification caused by excessive land use often makes the human environment system rapid collapse in karst area. Karst area in Southwest China is one of the most concentrated areas and the largest distribution areas of the poverty, which is quite difficult to get rid of poverty in China. As the main component of karst area in Southwest China, related research on rocky desertification in Southeast Yunnan has important scientific and practical significance in containing regional land degradation, maintaining the sustainable utilization of regional land resources and promoting regional sustainable development.

The rapid development of remote sensing technology provides a strong technical support for the effective detection of rocky desertification distribution and its degree, dynamic and so on. Remote sensing monitoring can provide an important basis for further revealing the nature and mechanism of rocky desertification and discussing the effective control of rocky desertification. Remote sensing technology has made a lot of achievements in the investigation of rocky desertification since the beginning of the new century.

However, it is not difficult to find that the application of remote sensing dynamic monitoring of rocky desertification in a particular area has obvious regional difference due to the complexity and diversity of rocky desertification environment. Therefore, this paper has chosen typical karst mountain area of Southeast Yunnan, which is a fragile karst environment with soil erosion, desertification and land degradation issues outstanding in GuangNan County, Yunnan Province, southwest corner of the local mountain as a case to deepen on Rocky Desertification of Remote Sensing Dynamic Monitoring and analysis.

2 Study Area and Data

2.1 General Situation of Study Area

The eastern Yunnan karst rocky desertification area is located in the southeast of Yunnan province. It adjacent to the northeast of GuiZhou and Southeast of Guangxi, the southern adjacent to Vietnam. Part of Southeast Yunnan of China belongs to the typical karst landform in southwest Karst, the typical karst landform in the high altitude areas is mainly represented in the Stone Forest and Stone bud, While the typical karst landform in low altitude region represented in peaks, peak cluster and Isolated peak. We selected 8 villages and towns in southwest Guan Nan County which is located in southeastern Yunnan angle and has a typical karst rocky desertification representative as the specific scope of research work, in order to deepen the research on the dynamic monitoring by remote sensing, Show in Fig. 1. It's an area of 93.9 km square, the geographical position varies at a latitude of 23°31′31.594″ North to 24°6′42.226″ North, and 104°31′26.273″ East to 105°16′9.083″ East. The average annual temperature in the region is 17.6 °C, which means the region belongs to a low-latitude plateau subtropical monsoon climate, with an average elevation varies 1000 m to 1400 m. The rainfall mainly concentrated in the May to October each year, and the average annual rainfall is 1400 mm, the Precipitation of the rainy season accounts for over 80% of annual rainfall. It is warm in winter and cool in summer and less seasonal, but dry, rain season is more obvious distinction from other seasons.

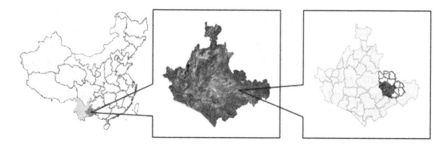

Fig. 1. Range diagram in the study area.

2.2 Research on Data and Data Pretreatment

The remote sensing data for the research is two temporal LandSat5TM data which is collected in August 17, 1990 and September 27, 2007. The Research area involves two-view image of the track in the region number 128/043 and 128/044 respectively. The remote sensing information acquisition of rocky desertification got a good comparability and reliability for the acquisition time of the two temporal image of remote sensing are in different time, the late summer and early autumn in years.

In the research working, in order to deepen the Southeast Yunnan Stone Desertification Remote Sensing Application Research of dynamic monitoring technology with the effectively help of multi-source heterogeneous data support, we has collected the research area about DEM, land use and relevance social economic data, organized a field trip and make a sample surveys and data analysis with the typical rocky desertification GPS positioning. For the accuracy of spatial positioning for monitoring results of research on safeguard, we controlled 1–5 10000 topographic map, select a typical feature control points, and carried out the geometric correction by using remote sensing image processing system, which makes the correct error in half a pixel, and complete the space coordinate of the image data and the study area of the basic geographic data matching processing. In addition, through the study area boundary mask processing, we got two temporal of the range remote sensing image data in the study area (Indicated in Fig. 2).

Remote sensing data in 1999   Remote sensing data in 2007

Fig. 2. Two phase of the remote sensing image in the study area.

3   Research Methods

3.1   Karst Rocky Surface Desertification Classification System to Be Determined

In view of complexity environment in karst rocky desertification research area, objectively and effectively to determine the classification system of rock desertification state is the foundation of carry out specific remote sensing monitoring for rocky desertification dynamic research. This research mainly refers to the classification system and other methods on some relevant documents of Karst rocky desertification in the domestic study; the paper has determined the classification system of desertification status for the extreme, severe, moderate, mild, non-rocky desertification and other five level types according to the difference of rocky desertification degree.

In specific research process, according to the technology characteristics of remote sensing, and combined with field surveys, we conclude that the bare surface condition or vegetation coverage is the most important index to characterize the degree surface of rocky desertification situation. The vegetation covers greater, the rocky desertification degree is lower in generally; on the contrary, vegetation cover small, rocky desertification degree is high. Therefore, from the remote sensing image processing to get index of vegetation coverage and combine with the bare surface feature band remote sensing monitoring spectral processing can better discrimination of rocky desertification degree; in addition, the rocky desertification degree often has closely related to the local anthropogenic land use condition or natural karst landform development situation. Furthermore, the relevant standards in the study of remote sensing integrated monitoring and diagnosis of the rocky desertification degree level type identification formulation and finishing as listed in Table 1.

Tab.1. Classification of rock desertification

Rocky desertification degree	Vegetation coverage	Associated with land use types of conditions	Karst landscape combined mode	Meaning analytical category
Extreme	0	Bare rock	Peaks - Feng Cong	Karst exposed large - scale, non-topsoil
Severe	10%	Other bare or weeds	Peaks - Feng Cong - Poe Valley	Karst exposed a large scale, with a thin topsoil
Moderate	20%	Sparse shrub land, grassland	Feng Cong - Peaks - Poe Valley	Karst exposed a certain size, with a thin topsoil
Mild	30%	Forest land, garden land and grassland	Poe Valley - depression - Feng Cong - Fenglin	Karst exposed a certain size, but has a thick topsoil
Non-rocky desertification	50%	Construction land, arable land, water	Depression - Feng Cong - Fenglin	Karst slightly exposed stone bud, but has a thick topsoil

3.2   Rocky Desertification Remote Sensing Dynamic Monitoring

Rocky desertification itself is the result of many factors; it is with natural factors lithology, elevation, slope, social factors in population density and degree of aggregation which are significantly correlated. Research identified uses the method which combined comprehensive land use investigation with remote sensing monitoring and applied it in rocky desertification dynamic monitoring study. The method is integrated use of remote sensing image processing, and based on various thematic maps, specially obtained fully understand for the study area of karst rocky desertification through combination of many land use fieldwork or thematic layers processing, then integration of remote sensing and thematic elements of space rock desertification auxiliary information to extract information on the spatial distribution of technical. This method has many advantages, such as based on remote sensing techniques to monitor the extent of surface coverage, judgment of potentially rocky desertification major categories and the control rang of spatial orientation distribution, the methods also comprehensive consider land-use survey analysis combination with thematic layers' auxiliary space, so it can better improve the accuracy of rocky desertification interpretation and identification.

The specific research used the main technical route and comprises steps on rocky desertification dynamic remote sensing monitoring as follows: (1) remote sensing image preprocessing; (2) analysis of multi band spectral characteristics of specific image data on the study area; (3) conduct remote sensing characteristic integrated analysis to typical categories of different rocky desertification degree on study area; (4) feature process about remote sensing vegetation indices and bare ground index and correlation analysis of karst rocky desertification categories evaluation; (5) integrated DEM terrain analysis and comprehensive land-use investigation and analysis of rocky desertification monitoring information extraction processing; (6) comparative and analysis of multi-temporal remote sensing monitoring results.

Overall, the dynamic monitoring of remote sensing is main used the "long direct classification method", respective to two remote sensing data conduct the classification information extraction of rocky desertification, and to verify the correct the reference current land use map and field survey results, thus obtained two layers of remote sensing data on the spatial distribution of surface rock desertification status. Carry out comparative analysis of the rocky desertification dynamic process on the basis of it which specific method is to use two monitoring results conducting comparative analysis of the spatial change, the level type of transfer matrix changes and related statistical indicators to achieve quantitative and orientation research.

4   Results Analysis

4.1   The Overall Characteristics Analysis of Dynamic Change

Rocky desertification classification of remote sensing monitoring method application in front of mentioned, after classification and interpretation, then further to check the reliability of the results through the typical area of the field, thereby modification possible emergence of false alarm or leaky judgment in the monitoring process, finally obtained two phase type of rocky desertification distribution layer in 1990 and 2007. Statistics result of rocky desertification remote sensing monitoring shown in Table 2 (Fig. 3).

Table 2. Two-point monitoring results statistics.

Degree	1990		2007		Change in number	
	Area	Proportion	Area	Proportion	Area	Proportion
Extreme	5902	2.15	13768	5.02	7865	2.87
Severe	60652	22.1	40239	14.66	−20414	−7.44
Moderate	42139	15.35	72604	26.45	30465	11.1
Mild	81383	29.65	67618	24.63	−13765	−5.01
Rocky desertification Subtotal	190077	69.25	194229	70.76	4152	1.51
Non- rocky desertification	84414	30.75	80262	29.24	−4152	−1.51
Total	274491	100	274491	100	0	0

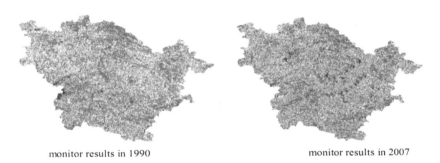

monitor results in 1990   monitor results in 2007

Fig. 3. Two phase spatial distribution of Rocky desertification classification.

Based on Table 2 to the proportion of different level rocky desertification amounts structure changes contrast which drawn in 1990 and 2007 in Fig. 4, comprehensive analysis can gain: (1) In 1990, the study area showed that non-rocky desertification have maximum of proportion on distribution structure, slight rocky land cover followed, a high proportion of the both structure more than 60%, while the moderate degree of rocky desertification is below 40%, the phenomenon indicated that the main surface overlay is non rocky desertification and slight rocky desertification at this time; (2) in 2007, although the study area surface landscape still showed that non-rocky desertification have largest structure number, but its proportion has dropped 1.51% which compared with 1990, the proportion of slightly rocky desertification amount have significantly decreased, while the number of moderate rocky desertification distribution is significantly increased, that rocky desertification status in this period

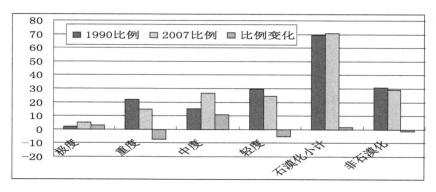

Fig. 4. Change status at all levels of rocky desertification comparison chart in 1990-2007.

appears to develop in the direction of the trend aggravated degree condition; (3) Overall, during 1990–2007, the proportion of non-rocky desertification is reducing, while rocky desertification is increasing, in particular, the number of extreme and moderate rocky desertification has increased while the severe and mild rocky desertification has decreased, it is indicating that the whole study area rocky desertification is more serious and the number of rocky desertification between different types appear more complex relationship. It can be seen since the 1990s, rocky desertification land changes performance for total area increased slightly in the study area, but the extent of rocky desertification has worse trend. (4) In terms of specifically identified five types of rocky desertification constituted, in the two tenses, karst mountain landscape matrix type advantages are not obvious in the study area.

4.2 Rocky Desertification Degree Category Transfer Change Analysis – Space Analysis of Rocky Desertification Category Transformation

Look at Tables 3 and 4. For accurate quantitative analysis the extent of rocky desertification landscape patterns in study area, get rocky desertification monitoring graph during the two periods of remote sensing to process spatial association, to calculate the number of changes between different periods of rocky desertification degree type

Table 3. Two-point monitoring results transfer matrix. (Hectare).

		2007 year					
		Extreme	Severe	Moderate	Mild	Non-desertification	Total
1990 year	Extreme	1064.7	7401.87	2147.31	2034.09	1119.6	13767.57
	Severe	915.03	20146.05	5411.7	12262.86	1503.27	40238.91
	Moderate	2415.96	20233.62	19648.71	24810.48	5495.67	72604.44
	Mild	962.73	9101.52	10679.67	27558.18	19315.71	67617.81
	Non-desertification	543.78	3769.38	4251.69	14717.34	56979.99	80262.18
	Total	5902.2	60652.44	42139.08	81382.95	84414.24	274490.9

Table 4. Two-point monitoring results transfer rate matrix type. (Hectare).

		2007 year					
		Extreme	Severe	Moderate	Mild	Non-desertification	Total
1990 year	Extreme	7.73	53.76	15.60	14.77	8.13	100.00
	Severe	2.27	50.07	13.45	30.48	3.74	100.00
	Moderate	3.33	27.87	27.06	34.17	7.57	100.00
	Mild	1.42	13.46	15.79	40.76	28.57	100.00
	Non-desertification	0.68	4.70	5.30	18.34	70.99	100.00
	Total	2.15	22.10	15.35	29.65	30.75	100.00

transfer matrix and the change rate of transfer matrix. Parsing table value connotation, the diagonal data presentation the number of rocky desertification situation not change between 1990 and 2007, but non-diagonal value is other rating categories development during 1990 to 2007.

Based on the result of transfer matrix and combined with the Tables 3 and 4. The changes of rocky desertification status were:

(1) 17 years from 1990 to 2007, extreme changes in the type of Rocky desertification increased to 13767.57 Hectares from 5902.20 Hectares, with 1064.70 Hectares that are not changed in same area and number. Which means that only 7.73% keeps the same, and 92.27% comes from other types. Therefore, the extreme changes of desertification during 17 years are so important, especially severe type transfer contribution is the largest proportion. Moderate, mild and non-diversion type of desertification following.

(2) 60652.44 Hectares of severe changes in the type of Rocky desertification in 1990 has declined to 40238.91 Hectares, with 20146.05 Hectares that are not changed in same area and number. Which means that only 50.07% keeps the same, and 49.93% comes from other types. The largest proportion of contribution is mild type, followed by moderate, non-diversion and extreme of desertification. Which changes of severe desertification is relatively stable compare to extreme changes during 17 years;

(3) 42139.08 Hectares of moderate changes in the type of Rocky desertification in 1990 increased to 72604.44 Hectares, with 19648.71 Hectares that are not changed in same area and number. Which means that only 27.06% keeps the same, and 72.94% comes from other types. The largest proportion of contribution is mild type, followed by mild, severe, non-diversion and extreme of desertification. Which changes of severe desertification is relatively stable compare to extreme changes during 17 years;

(4) 81382.95 Hectares of mild changes in the type of Rocky desertification in 1990 declined to 67617.81 Hectares, with 27558.18 Hectares that are not changed in same area and number. Which means that only 40.76% keeps the same, and 49.24% comes from other types. The largest proportion of contribution is non-diversion type, followed by moderate, severe, and extreme of desertification. Which changes of severe desertification is relatively stable compare to extreme changes during 17 years;

(5) 84414.24 Hectares of mild changes in the type of Rocky desertification in 1990 declined to 80262.18 Hectares, with 56979.99 Hectares that are not changed in same area and number. Which means that only 70.99% keeps the same, and 29.01% comes from other types. The largest proportion of contribution is non-diversion type, followed by moderate, severe, and extreme of desertification.

4.3   Discussion on the Spatial Changing Characteristics

Combination of the above discussion on corresponding number structure and spatial distribution of monitoring result analysis, can be drawn:

The spatial distribution pattern of landscape matrices in 1990 and 2007 year was consistent, which means the main type of non-Rocky desertification is mainly distributed in high vegetation coverage in mountain regions, as well as agricultural and construction of anthropogenic activity concentrated Karst depression - Valley areas, while the remaining landscape distribution of different rocky desertification degree level is inlaid into the matrix pattern. Also, two phrases of surface landscape monitoring by Remote Sensing rendered out specific pattern characteristics of Karst mountain, show that the history traces and spatial result in process of Rocky Desertification driven by natural and human driving force, that is distribution of high vegetation coverage in landscape surface and human frequent activities in depressions-valley. The extent of Rocky Desertification in Rocky Mountain tend to show spatial evolution of gradient change from lower stage into a higher one, namely evolution of mild- moderate- serve- extreme degree. In general, distribution of high available quality cultivated land is relatively stable, while the spatial dynamics of the transition edge of wood land and dry land is intense in land use type conversion. The above shows that once the fragile topsoil loss occurred in land use conversion, the land may deteriorate sharply into Rocky, and make it difficult to reserve and repair.

5   The Results and Discussion

Based on the analysis and research of dynamic remote sensing monitoring of rocky desertification in the typical case of Southeastern Yunnan, the results of the study are as follows:

Changes of the number of karst mountains desertification and non-rocky desertification in research area is relatively little between 1990 and 2007, but the change of the degree of karst rocky desertification is severe and complicated, and the degree of rocky desertification displays an aggravating trend. In terms of 5 types of rocky desertification degree determined by the research, the advantage of landscape type in karst mountain area is not obvious.

The work of this study further confirmed that land use and remote sensing image technology can well inversion analyze comprehensive information of rocky desertification in different historical periods using satellite remote sensing and GIS technology and combining with field survey and sampling. The method has the characteristics of reliable technology, convenient data acquisition and strong operability. The application of remote sensing dynamic monitoring technology is in favor of rapidly defined quantity change and spatial distribution pattern of different rocky desertification state types in the study area. This research is helpful to the effective control of the information change of karst rocky desertification and the law of related environmental factors, which can provide decision support for the development and management of karst mountain land. In addition, due to the influence of various factors, the results of dynamic remote sensing monitoring of rocky desertification in the study area exist the uncertainty in a certain scale. The construction of method system for monitoring and management of karst rocky desertification ecosystem is necessary, which uses remote sensing and GIS technology as platform, and field monitoring as a basis for verification and validation based on classical areas. Combining with the spatial information management technology, the model and the method, to analyze and understand karst mountain ecosystem status and trends, and ultimately improve the karst mountain ecosystem monitoring and research level requires further efforts in the direction of research and development.

Acknowledgments. Our sincere thanks to Nature Science Foundation of China (NSFC) (NO. 41561083, 41261092) and Natural Science Fund of Yunnan Province (NO. 2015FA016) for providing funding to carry put the research at Kunming University of Science and Technology, China. The authors would like to thank two anonymous reviewers for their constructive comments which were helpful to bring the manuscript into its current form.

References

1. Wang, S.: Concept deduction and its connotation of Karst rocky desertification. Carsologica Sinica **21**(2), 101–105 (2002)
2. Bulletin of the Chinese Academy of Sciences: Suggestions on advance comprehensive treatment of stony desertification in Karst mountainous area in southwest China. Adv. Earth Sci. **18**(4), 389–492 (2003)
3. Liu, Z., Liu, Y., Deng, C.: Karst systems and their influence upon the engineering in the western section of Hu-Rong expressway. J. Mt. Res. **22**(3), 357–362 (2004)
4. Wan, J., Cai, Y., Zhang, H., Rao, S.: Land use/land cover change and soil erosion impact of Karst area in Guanling county, Guizhou province. Scientia Geographica Sinica **24**(5), 0573-07 (2004)
5. Hu, Y., Liu, Y., Wu, P., Zou, X.: Rocky desertification in Guangxi Karst mountainous area: its tendency, formation causes and rehabilitation. Trans. Chin. Soc. Agric. Eng. **24**(8), 96–101 (2008)
6. Li, W., Yang, H., Liu, J.: Empirical research on classification and control modes of KarstRocky desertification zone in southeastern Chongqing based on remote sensing-a case study of Youyang in Chongqing. J. Shenyang Normal Univ. (Nat. Sci. Ed.) **24**(4), 490–494 (2006)
7. Wang, J., Yang, D., Yu, S., Yu, F., Yang, J.: Analysis on Karst rocky desertification in upper reaches of Pearl River based on remote sensing. Sci. Soil Water Conserv. **5**(3), 1–6 (2007)
8. Li, L., Tong, L., Li, X.: The remote sensing information extraction method based on vegetation coverage. Remote Sens. Land Resour. **84**(2), 59–62 (2010)

9. Wang, J., Li, S., Li, H., Luo, H., Wang, M.: Classifying indices and remote sensing image characters of rocky desertification lands: a case of Karst region in northern Guangdong province. J. Desert Res. **27**(5), 765–770 (2007)
10. Yang, C.: A discussion on the remote sensing analysis of Karst stone desertization in GuangXi. Remote Sens. Land Resour. **26**(2), 34–36 (2003)
11. Hu, B., Li, L., Jiang, S.: Spatial pattern analysis on Karst rocky desertification of GuangXi based on land scape spatial method. Earth Environ. **33**(Suppl.), 581–587 (2005)
12. Lan, A., Zhang, B., Xiong, K., An, Y.: Spatial pattern of the fragile Karst environment in southwest Guizhou province. Geogr. Res. **22**(6), 0733-09 (2003)

Guangxi Longtan Reservoir Earthquakes S-Wave Splitting

Lijuan Lu[1], Bin Zhou[1,2(✉)], Xiang Wen[1], Shuiping Shi[1],
Chunheng Yan[1], Sha Li[1], and Peilan Guo[1]

[1] Earthquake Administration of Guangxi Zhuang Autonomous Region,
Nanning 530022, China
yaya997@163.com
[2] Institute of Geophysics, Chinese Earthquake Administration,
Beijing 100081, China

Abstract. In order to understand and grasp the anisotropic characteristics of crustal media in 12 fixed seismic stations of Longtan Reservoir Digital Telemetry Seismic Network (LRDTSN) from impoundment in October 2006 to July 2013, on the basis of precise positioning of seismic events in Longtan Reservoir, using SAM comprehensive splitting analysis method, which contained correlation function calculation, time delay correction and polarization analysis and test, the effective dominant shear-wave polarization direction and delay time (shear-wave splitting parameters) in 9 fixed stations of LRDTSN were calculated. Analysis showed that under the influence and control of principal compressive stress and regional faults in the South China Block, fast shear-wave polarization directions in stations of LRDTSN had obvious local characteristics. The principal compressive stress directions within the scope of Longtan Reservoir included the dominant polarization directions of NW and NNE. Under the impact of loading and unloading water during impoundment of Longtan Reservoir, the spatial distribution of slow shear-wave normalized delay time was uneven, high in the northwest of dam area and slightly high in the periphery of dam area. Meanwhile, shear-wave splitting parameters in some stations had certain correspondence with the water level of reservoir.

Keywords: Longtan Reservoir area · Crustal shear-wave splitting · Fast wave polarization direction · Slow wave delay time

1 Introduction

Shear-wave splitting is also called shear-wave birefringence. When a horizontally polarized shear-wave propagated through effective solid with some form of elastic anisotropy, it was split into two approximately orthogonal polarized phases, with different velocities and different vibration directions [1]. The faster one is called fast

Fund Project: This work was supported by the science-technology plan of Guangxi (Project number: 1377002, 1598017-11, 14124004-4-8).

© Springer Nature Singapore Pte Ltd. 2017
H. Yuan et al. (Eds.): GRMSE 2016, Part I, CCIS 698, pp. 477–493, 2017.
DOI: 10.1007/978-981-10-3966-9_54

shear-wave, while the slower one is called slow shear-wave. There is a time lag between slow and fast shear-waves, known as (slow shear-wave) time delay. Ever since shear-wave splitting was discovered, domestic and foreign experts and scholars had achieved fruitful results in theories, analysis methods and applications, etc. of shear-wave splitting. In terms of the theories of shear-wave splitting, Crampin [2] argued that approximately vertical, directionally aligned and parallel microcracks pervaded the crustal rocks and proposed an EDA hypothesis. The so-called EDA referred to a structure of stress-oriented fluid-filled intergranular microcracks and pore space of small aspect ratio. With the deepening of research, temporal variations of shear-wave splitting and discreteness of observations, etc., cannot be fully interpreted by EDA theory. Zatsepin and Crampin [3] and Crampin and Zatsepin [4] put forward APE theory, which can describe the dynamic features of cracks and fluids in a critical state due to of stress, and explain dynamic changes in shear-wave splitting observations (over time). In terms of the analysis methods of shear-waves, there was synthetic seismogram, rotation and direct interpretation of seismogram, automatic analysis of particle motion projection, self-adaption analysis and maximum eigenvalue, etc. However, the most widely used was the polarization analysis method presented by Crampin [5]. Previously, Chinese scholars put forward a CDP shear-wave splitting analysis method (Gao and Zheng [6]), now known as SAM method [7], that is, to develop a systemic shear-wave splitting analysis method on the basis of correlation function calculations, including the calculation of correlation function, time delay correction and polarization analysis and test. In terms of the applications of shear-waves, according to APE theory, through a shear-wave splitting research on a small isolated earthquake swarm in Hainan, China, Gao et al. [8] concluded that stress accumulation can be identified through the reference quantity of shear-wave splitting time delay. Crampin et al. [9] made a successful earthquake prediction of an $M = 5$ earthquake in Iceland in 1998 through the shear-wave splitting by applying APE theory; Gao et al. [10], through a shear-wave splitting research on a small earthquake swarm in the east of Hainan Province in 1992, found that the time difference of arrival (TDOA) between fast and slow shear-waves changed with seismicity and stress accumulation and inferred that the principal compressive stress field in this area was NWW. From the existing literature, land shear-wave splitting research findings are abundant, while reservoir shear-wave splitting research is little, but the results are also remarkable. For example, Huang and Yang [11], Zou et al. [12] conducted shear-wave splitting researches on Xinfeng River Reservoir and Wenzhou Shanxi Reservoir, respectively and came to the conclusion that the shear-wave dominant polarization direction in this area was basically the same as the principal compressive stress direction. Shi and Zhao [13] got the relationship between fast shear-wave dominant polarization direction in this area and delay time and its relationship with variation in water level etc., through a shear-wave splitting research in Guangxi Longtan Reservoir area. But the biggest difference was different durations of seismic data and different seismic stations. In the present study, seismic data of a total of 12 fixed seismic stations of LRDTSN from 2006 to 2013 were adopted, while previously the seismic data of 15 temporarily encrypted mobile stations and 10 fixed stations from April, 2009 to April, 2010 were adopted, principally temporarily encrypted mobile stations. All these lay a foundation for obtaining the characteristics of shear-wave splitting 12 fixed stations of

LRDTSN, expanding the time span of study, understanding and grasping the anisotropic characteristics in Longtan Reservoir area comprehensively, as well as studying
its indication significance for earthquake prediction in future studies.

2 Geological and Tectonic Setting and Data

2.1 Geological and Tectonic Setting

Longtan Hydropower Station is located in the southeast of Yunnan-Guizhou Plateau.
The terrain is complex. Longtan Reservoir (24.7°N–25.5°N); 106.3°E–107.3°E is
located in the western margin of the South China Block. According to field geological
landform and seismic geological surveys, within the scope of study area, four groups of
faults, i.e., NW, NNW, NE and quasi-SN are developed (Fig. 1). Most of them belong
to fault structure formed with Indosinian folds and are limited to two wings or core of
the fold. The scale is small. The main structure near the dam area is Longfeng- Bala
Fault (F9), with a Tian'e box-like anticline, whose west wing is NNE-striking. This
fault was developed between Triassic and Permian Systems. The length inside is about
25 km. In the southwest of dam area, there is NNW- striking Wangmo- Luoxi Fault
(F7), which is composed of multiple secondary small faults and basement fractured
zones and about 28 km long inside, NW- striking Changli- Banan Fault (F8), which is
a dense cleavage belt and gentle fold deformation belt developed in Permian lime-
stones, and Dangming- Guahua Fault (F6). In the northwest of reservoir area, there is
NE- striking Luodian- Wangmo Fault (F1), which is about 26 km inside, Gaoxu-
Bamao Fault (F2), which is composed of fractured zones and up to 30–100 m wide and
Fengting- Xialao Fault (F3). In the due north of dam, main structures include Ma'er-
Lalang Fault (F4) and Daheng- Daliang Fault (F5), which have a Daliang anticline and
the east and west wings are quasi-SN-striking. 2 faults have opposite fracture ten-
dencies, but both of them have normal strike-slip motion properties. According to the
fault activities revealed in tectonic landforms and geologic sections and chronological
evidence obtained, F2, F3, F4, F6, F8 and F9 faults had different degrees of activities
during Early- Middle Pleistocene. Fractured zones develop. Fissures and karst fissure
springs are linearly distributed along fractured zones [14, 15].

2.2 Seismic Stations and Data

The construction of Longtan Reservoir Digital Telemetry Seismic Network (hereinafter
referred to as LRDTSN) began in 2004. At first, there were a total of 10 fixed stations,
including 8 stations in key monitoring areas in the reservoir head and 2 stations in the
middle of reservoir area. The average distance between stations near the dam was
20 km. The distance between stations outside the dam was 50 km. The M_L1.0 monitoring capacity was 50 km around the reservoir [16–18]. LRDTSN was completed in
November 2005 and officially put into use in March 2006. To strengthen the monitor of
seismic activities in the reservoir head, 2 additional stations were built in 2009, i.e.,
GAL and XIY stations. Therefore, LRDTSN consists of 12 stations, except that TIE
Station is equipped with a CMG-3ESPC broadband seismometer (the bandwidth is

0.03–40 Hz), others are equipped with a FSS-3B seismometer. The data collector is EDAS-24L. The sampling frequency is 100 Hz.

Longtan Hydropower Station is located in Tian'e County, Guangxi, in the upstream of the Hongshui River, the trunk of the Pearl River. It is a super-huge hydropower project in Xiluodu Power Station on the Jinsha River, which is under construction in mainland China now, second only to the Yangtze River Three Gorges Power Station. It is a landmark project in national western development and electricity transmission from west to east china. The capacity of reservoir is from 10 billion m^3 to more than 20 billion m^3. The construction of Longtan Hydropower Project began in July 2001. In October 2006, it began to impound water. Since impoundment, small and medium-sized earthquakes have often happened near the reservoir area. By July 2013, 4482 earthquakes with magnitudes greater than $M_L 0.0$ had been recorded, including 11 earthquakes with $M_L 3.0$–3.9 and 3 earthquakes greater than $M_L 4.0$. The biggest was an $M_L 4.8$ earthquake on the border between Tian'e County, Guangxi and Guizhou on

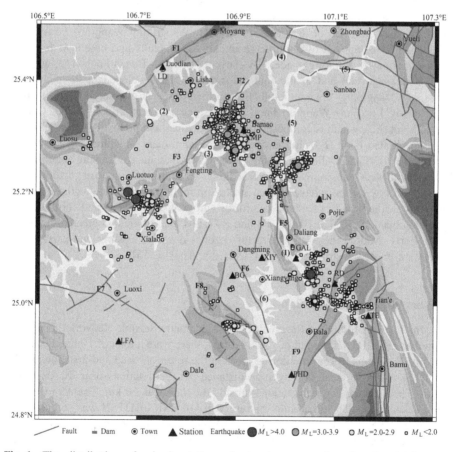

Fig. 1. The distribution of seismic stations, fractured zones and earthquakes in Longtan Reservoir area

September 18, 2010. The intensity in meizoseismal area was VI. It caused serious economic losses and social impact to the local area. In this paper, based on digital seismic waveform data of LRDTSN, strict data selection principles and precise positioning of earthquakes in the reservoir area, a study on the crustal shear-wave splitting was carried out.

Data selection principles: first of all, the selection of the scope of shear-wave window. When shear-wave propagating in homogeneous media makes incidence to a free surface, if the incident angle is greater than the critical angle, total reflection will happen. For Poisson media, the critical angle is about 35°, but due to the bending wave front and low-speed surface deposition, the shear-wave window can effectively increase to 40°–45°, which is precisely the limited scope of shear-wave window. With this limit scope, select data in which incident angle is less than or equal to 45°, and the focal depth is greater than the epicenter distance; secondly, the selection of waveform. To ensure the quality of seismic waveforms, seismic waveform data with clear original records, unlimited amplitude and high signal-to-noise ratio without deformation should be selected. According to the above two major data selection principles, 327 seismic waveforms are selected from 4482 earthquakes events in Longtan Reservoir area to calculate shear-wave splitting parameters. The distribution of seismic stations, fractured zones and earthquakes in Longtan Reservoir area is shown in Fig. 1.

3   Principle and Methodology

3.1   Calculation Method

Since the 1980s when shear-wave splitting in the earth's crust was observed, all kinds of analysis methods have been developed both at home and abroad, such as synthetic seismogram, rotation and direct interpretation of seismogram, automatic analysis of particle motion projection, self-adaption analysis and maximum eigenvalue, etc. However, the most widely used and most practical was the polarization analysis method presented by Crampin [5]. For calculation in this study, SAM comprehensive splitting analysis method was adopted, which contained correlation function calculation, time delay correction and polarization analysis and test. SAM method is featured with self-check, high objectivity and high accuracy.

Calculation was conducted, taking an $M_L2.3$ earthquake event (Fig. 2) at 21:35 on April 13, 2009 recorded by RD station for example. Figure 2A is a filtered waveform using butterworth filter, at a frequency range of 1.0–20 Hz and a response order of 4. By correlation function calculation of waveforms from the beginning of the shear-wave on the horizontal level, Fig. 2B obtains the initial value of fast shear-wave polarization direction and slow shear-wave time delay, which are corresponding to the maximum correlation coefficients. Figure 2C is the initial state of the shear-wave particle trajectory on the horizontal level. On this basis, according to the initial results calculated by the correlation function in Fig. 2B, the polarization angle and the time delay are eliminated, and then Fig. 2D is formed. If the calculation result of fast shear-wave polarization direction is not correct, and the fast and slow shear-waves are mixed together, separated fast and slow shear-waves will not be obtained. If the calculation

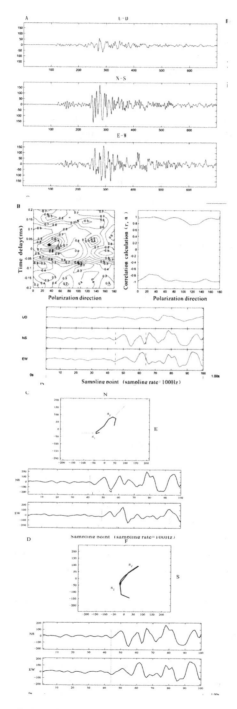

Fig. 2. The calculation of shear-wave splitting of an $M_L2.3$ seismic event at 21:35, April 13, 2009 in RD station

result of time delay is not correct, shear-wave polarization diagram excluding time delay effects will not be linear; and if so, it needs to be recalculated until motion trail is linear. The final results obtained are two parameters of the shear-wave anisotropy—the polarization direction and time delay. In this instance, the initial results of fast shear-wave polarization direction are accurate, but the calculation results of slow shear-wave delay time have the deviation of 0.01 s.

3.2   Reliability of the Results

By using SAM comprehensive splitting analysis method, the shear-wave splitting parameters of 9 seismic stations of LRDTSN were acquired. As the sampling frequency of seismic waves of LRDTSN was 100 Hz and the minimum resolution of delay time was 0.01 s, those calculation results with a delay time of less than 0.01 s were excluded. At the same time, when calculating both the precise positioning of earthquakes and incident angle, Southwest Guangxi structural belt model among South China crustal velocity structure models [19] was adopted, that is, the thickness of the first layer was 2 km, the thickness of the second layer was 10 km, the thickness of the third layer was 20 km, and the corresponding velocities were $3.7/km·s^{-1}$, $4.85/km·s^{-1}$ and $5.9/km·s^{-1}$ for the first layer, the second layer and the third layer. In order to enhance the reliability of calculation results, within the scope of less than or equal to 45° incident angle, only the splitting parameters results of shear-waves whose focal depth was greater than the epicenter distance were adopted. After the above processing of results, it was proved that shear-wave splitting parameters in 9 seismic stations of LRDTSN were reliable, and detailed parameter results can be seen in Table 1.

Table 1. Seismic stations and shear-wave splitting parameters

Station name	Numbers recorded	Average polarization direction (°)	Average slow shear-wave normalized delay time (ms/km)
RD	100	19.7 (NNE)	2.76 ± 1.43
		142.0 (NNW)	
GAL	76	33.87 (NNE)	2.67 ± 1.64
		126.4 (NW)	
TIE	52	57.96 (NE)	2.65 ± 1.47
		127.96 (NW)	
XMP	41	124.2	4.3 ± 2.23
LN	33	45.38 (NE)	2.35 ± 1.00
		137.65 (NW)	
LD	7	123.71	6.08 ± 2.56
PHD	2	60.5	0.87 ± 0.33
XIY	2	129.0	4.32 ± 0.51
JL	1	30.0	4.17

4 Shear-Wave Splitting Characteristics of Longtan Reservoir Earthquakes

In the present study, using SAM comprehensive splitting analysis method, it is concluded that fast dominant shear-wave polarization directions in 9 stations are different (Fig. 3). Fast dominant shear-wave polarization directions in 4 stations have 2 groups of obvious dominant directions, i.e., LN, GAL, RD and TE stations. LN station is characterized by NE and NW directions. TE station is characterized by NEE and NWW directions, while RD and GAL stations are characterized by NNE and NNW, NNE and NWW dominant directions, respectively. Among the remaining 5 stations, XMP, LD and XIY stations are characterized by NW. XMP station is slightly N-S oriented. The dominant polarization direction of PHD and JL stations is NE. The dominant polarization direction of TE, XMP and GAL stations is basically consistent with the findings of Shi and Zhao [13]. From the entire Longtan Reservoir area, fast shear-wave polarization directions in this area have two groups of dominant directions (Fig. 4), i.e.,

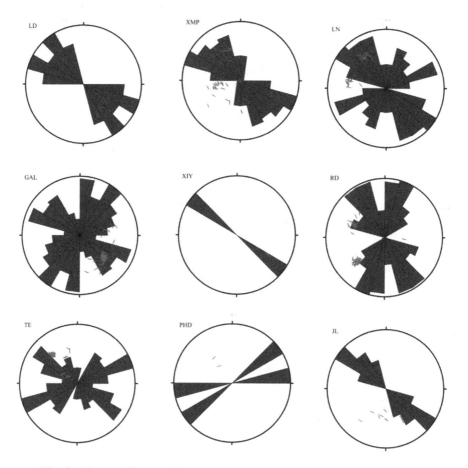

Fig. 3. The rose diagram of fast shear-wave polarization directions of LRDTSN

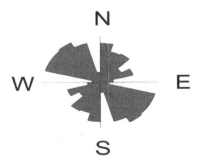

Fig. 4. The complex rose diagram of fast shear-wave polarization directions in Longtan Reservoir area

NW and NNE, but NW is slightly stronger than NNE. To sum, fast dominant shear-wave polarization directions in seismic stations of Longtan Reservoir area have obvious individual differences. The reservoir is controlled by 2 major stress directions, i.e., NW and NNE.

Among crustal shear-wave splitting parameters, delay time is one of the important research objects. Delay time refers to the time lag between fast and slow shear-waves. It is an important parameter that reflects the anisotropic characteristics of crustal cracks. For the purpose of quantification, delay time is normalized as the quantity of time delay per kilometer (unit: ms/km), known as normalized time delay. Through an extreme value analysis of shear-wave splitting correlation functions and time delay correction of seismic waveforms, shear-wave splitting parameters of Longtan Reservoir earthquakes were calculated in the present study. Among them, the distribution ranges of slow shear-wave normalized delay time in all stations were 0.9–6.13 ms/km (RD), 1.03–6.46 ms/k (GAL), 0–5.61 ms/km (TE), 0–9.69 ms/km (XMP), 0.81–4.42 ms/km (LN), 3.21–10.73 ms/km (LD), 0.54–1.19 ms/km (PHD) and 3.96–4.68 ms/km (XIY) respectively. The calculation results and standard deviations were (2.76 ± 1.43) ms/km (RD), (2.67 ± 1.64) ms/km (GAL), (2.65 ± 1.47) ms/km (TE), (4.3 ± 2.23) ms/km (XMP), (2.35 ± 1.00) ms/km (LN), (6.08 ± 2.56) ms/km (LD), (0.87 ± 0.33) ms/km (PHD), (4.32 ± 0.51) ms/km (XIY) respectively.

5 The Relationship Between Shear-Wave Splitting Parameters of Longtan Reservoir Earthquakes and Regional Settings

5.1 The Relationship Between Shear-Wave Splitting Parameters and Regional Tectonics in Longtan Reservoir Area

In Longtan Reservoir area, 4 groups of faults, i.e., NW, NNW, NE and quasi-SN, including 9 large faults, are developed in the southwest, northwest, due north and vicinity of the dam area. The southwest faults in the dam area are mainly NNW and NW-striking, with reversed strike-slip motion properties. The northwest faults are mainly NE-striking, with dextral strike-slip motion properties, in which fractured zones develop. The due north faults are mainly SN-striking, with normal strike-slip motion properties. The main structure near the dam area is Longfeng- Bala Fault, with a Tian'e box-like anticline and the west wing is NNE-striking. Some fractured zones develop. Fissures and karst fissure springs are linearly distributed along fractured zones. Fault structures in different directions during the Neotectonic Period also presented different degrees of differential activities [20].

The Relationship Between the Spatial Distribution of Fast Shear-Wave Polarization Directions and Regional Tectonics in Longtan Reservoir Area. The present study concludes shear-wave splitting parameters in 9 stations of LRDTSN, using SAM. From the spatial distribution map in Fig. 5, 9 stations are distributed in the south, northwest, due north and vicinity of the dam area. Most of them are corresponding to the distribution of major faults in Longtan Reservoir area. From the calculation results, in some stations, the fast dominant shear-wave polarization direction is consistent with fault strike. In some stations, the dominant direction is consistent with regional principal compressive stress direction. In some stations, it is consistent with both regional principal compressive stress direction and fault strike. On the whole, there are obvious regional differences. This may be associated with the feature that fast dominant shear-wave polarization direction mainly reflects the principal compressive stress of the earth's crust under seismic stations. Fault strike at a close distance to stations would control or influence the fast dominant shear-wave polarization direction [21]. This also suggested that local tectonics and fault distribution may complicate stress [22, 23]. According to a characteristic analysis of fast shear-wave polarization direction, a total

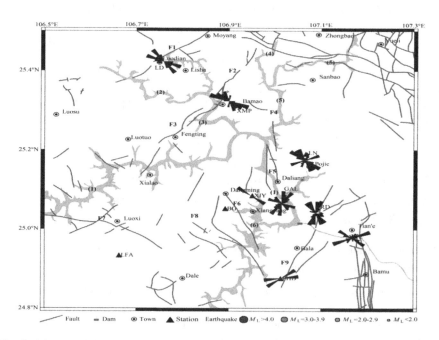

Fig. 5. The rose diagram spatial distribution of fast shear-wave splitting polarization direction in Longtan Reservoir area

of 4 stations, LN, GAL, RD and TE, have 2 groups of dominant directions. Among them, LN station is characterized by NE and NW directions. TE station is characterized by NEE and NWW directions. NEE is slightly stronger than NWW. RD and GAL stations are characterized by NNE and NNW, NNE and NWW, respectively. In GAL station, NNE is stronger than NWW. From geological locations and tectonics, LN, GAL, RD and TE stations are all close to the dam area and distributed in both sides of NNE Longfeng- Bala Fault. From a characteristic analysis of dominant shear-wave polarization directions in 4 stations, they are mainly controlled and influenced by NNE Longfeng- Bala Fault. The closer to this fault, the more evident the influence is. The shear-wave polarization directions in RD station, which is closest to the dam area, NNE and NNW, are well-matched in intensity, that is, they are consistent with both regional principal compressive stress direction and fault strike and the dominance is almost equal. This may be associated with the fact that it lies in the intersection between NNE Longfeng- Bala Fault (normal fault) with a Tian'e box-like anticline and quasi-SN-striking Tian'e-GuJin Fault (thrust fault).

The Relationship Between the Spatial Distribution of Slow Shear-Wave Normalized Delay Time and Regional Tectonics. Figure 6 shows an isoline spatial distribution map of slow shear-wave normalized delay time of LRDTSN. The spatial

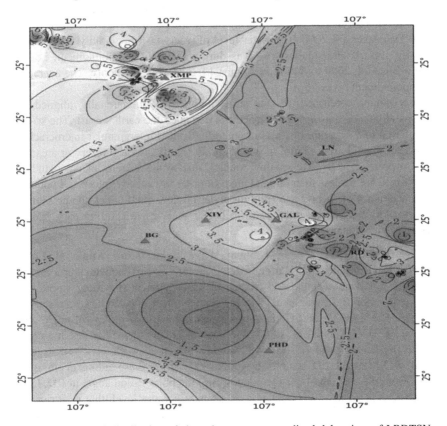

Fig. 6. The spatial distribution of slow shear-wave normalized delay time of LRDTSN

distribution of slow shear-wave normalized delay time of LRDTSN is uneven, high in the northwest of dam area. This may be associated with the fact that it passed through Gaoxu-Bamao Fault. Because their fractured and crushed zone disappeared, fissures and karst fissure springs are linearly distributed along the fractured zone, and they are vulnerable to the permeation of the reservoir water, rock fissure density decreases and the pore space expands. Slow-wave time delay is sensitive to variation of stress, and result in the increase of slow shear-wave delay time. The higher area of slow shear-wave normalized delay time is the area where moderate and small earthquakes are relatively active in Longtan Reservoir area. But the question whether the intensity of the earthquake activity in Longtan Reservoir area corresponds with the spatial distribution difference of slow shear-wave delay time needs not only deeper under-standing and grasping of regional tectonics, but also more support from the seismic data, to make the spatial distribution results of slow shear-wave delay time more comprehensive and accurate.

5.2 The Relationship Between Shear-Wave Splitting Parameters of Longtan Reservoir Earthquakes and Water Level

From October 2006 to November 2013, Longtan Reservoir had experienced seven rises of water level. The biggest rise was during the first impoundment, i.e., from October 1, 2006 to December 13, 2006. The water level of reservoir grew rapidly from 220 m to 320 m by 100 m. The smallest rise was during the fourth impoundment, i.e., from 338 m in June 24, 2009 to 360 m in August 14, 2009, by 22 m only. The average rise of 7 impoundments was 45 m.

Observations confirmed that a large number of directionally aligned EDA (extensive-dilatancy anisotropy) microcracks pervaded the earth's crust. The so-called EDA referred to a structure of stress-oriented fluid-filled intergranular microcracks and pore space of small aspect ratio. In this study, the study area is a reservoir area full of fluids. At present, in S-wave splitting studies, due to the scarcity of data, studies on reservoir area are far less than on land area. In order to further explore and understand the characteristics of shear-wave splitting under obvious fluid effect, the present study tries to explore the change relationship between shear-wave splitting parameters and water level in Longtan Reservoir area.

The Relationship Between Fast Shear-Wave Polarization Angle of Longtan Reservoir Earthquakes and Water Level. Figure 7 gives a sequential process of fast shear-wave polarization angle and change in water level in 6 stations (more than 2 events) in Longtan Reservoir area. From the change interval, it can be seen clearly from Fig. 7 that RD, LN, TE and GAL stations mainly have two groups of angle change intervals, i.e., 0°–80° and 100°–180°. The interval of 0°–80° mainly lies at a depth of 216–276 m. The interval of 100°–180°mainly lies at a depth of 296–356 m. The interval of XMP and LD stations is 100°–180°. The corresponding depth is 296–356 m. From the sequential change relationship, the polarization angle in RD, LN and XMP stations is positively correlated with change in water level, that is, it rise as the water level rises, and drops as the water level drops. It is, to a certain degree,

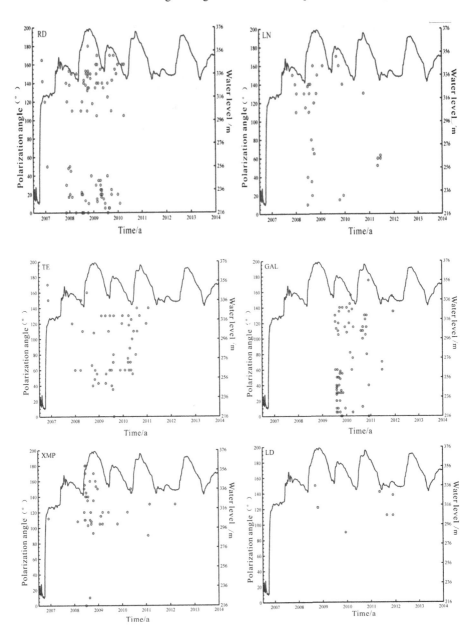

Fig. 7. The sequential process of fast shear-wave splitting polarization angle of Longtan Reservoir earthquakes and water level

synchronous with the change in water level. In conclusion, the fast shear-wave splitting polarization angles in some stations in Longtan Reservoir area has a certain relationship with water depth and change in water level, especially stations near developed fractured zones.

The Relationship Between Slow Shear-Wave Normalized Delay Time of Longtan Reservoir Earthquakes and Water Level. Figure 8 gives a sequential process of slow shear-wave normalized delay time and change in water level in 6 stations (more than 2 events) in Longtan Reservoir area. From the figure, the change in slow shear-wave normalized delay time in LN, XMP, LD and GAL stations is scattered. While the slow shear-wave normalized delay time of RD and TE stations, which are closest to the dam area, gradually increases with the water depth. Every time the water depth increases 20 or 30 m, normalized delay time will increase once as a whole. The increase of delay time is relatively fixed and doesn't change over time, suggesting a strong change relationship with water depth. Crampin et al. [24]'s study pointed out

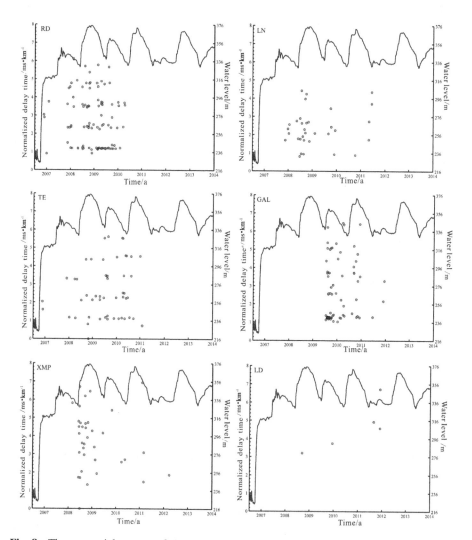

Fig. 8. The sequential process of slow shear-wave normalized delay time of Longtan Reservoir earthquakes and water level

that as for EDA media, the azimuth, density, aspect ratio and water content, etc. of cracks may affect the TDOA between fast and slow waves, but in different parts of the focal area, time delay has different responses to different changes in crack parameters. In some parts of the focal area, time delay mainly responds to crack density. While in other parts, it responds to crack aspect ratio. Therefore, changes in slow wave time delay observed in observation points in different positions of focal area may vary. Changes in crack density may lead to the increase or decrease of time delay as a whole. Perhaps for this reason, the slow shear-wave normalized delay time in RD and TE stations in Longtan Reservoir area increases with water depth successively.

6 Conclusion and Discussion

Using SAM comprehensive splitting analysis method and precise positioning data of seismic events of LRDTSN in nearly seven years from October 2006 to July 2013, preliminary results of a total of 314 valid shear-wave splitting in 9 stations of LRDTSN were acquired. Through analysis, the following conclusions are drawn:

1. The fast shear-wave polarization directions in Longtan Reservoir area have obvious local characteristics. Some of the fast dominant shear-wave polarization directions in 9 stations are consistent with faults near the stations, some are consistent with the regional principal compressive stress in the South China Block and some are consistent with both. There are obvious regional differences as a whole. All of the LN, GAL, RD and TE stations with 2 groups of dominant polarization directions are close to the dam area and mainly distributed in both sides of NNE Longfeng-Bala Fault. From LN station in the due north of the dam area, which is far from the dam, to TE station in the southeast of the dam area, the first and second dominant directions alternate, namely, the first dominant polarization direction gradually changes from NW to NE, suggesting that the *in situ* principal compressive stress directions of 4 stations are mainly controlled and influenced by NNE Longfeng-Bala Fault. The shear-wave polarization direction in RD station, which is closest to the dam area, i.e., 2 groups of dominant directions, NNE and NNW, are well-matched in intensity, suggesting that the principal compressive stress is subject to the distribution of faults in the intersection between NNE Longfeng- Bala Fault (normal fault) with a Tian'e box-like anticline and quasi-SN-striking Tian'e-GuJin Fault (thrust fault), as well as the regional principal compressive stress direction of South China Block.

2. Longtan Reservoir area as a whole is subject to the influence and control of principal compressive stress in the South China Block. From the complex rose diagram of shear-wave splitting in Longtan Reservoir area, it is known that the dominant polarization directions within the scope of Longtan Reservoir are NW and NNE. NW is slightly stronger than NNE. This is consistent with the directions of 4 groups of faults, NW, NNW, NE and quasi-SN4 developed within the scope in Longtan Reservoir area. The first dominant direction is NW. This may indicate that the principal compressive stress direction within the scope in Longtan Reservoir area is mainly principal compressive stress in the South China Block.

3. The spatial distribution of slow shear-wave normalized delay time in Longtan Reservoir area is uneven, high in the northwest of dam area. The isoline spatial distribution map of slow shear-wave normalized delay time of LRDTSN indicates that it is high near XMP station in the northwest of dam area and slightly high in the periphery of dam area. The spatial distribution is extremely uneven. Perhaps the reason is that rich fractured zones in these areas are vulnerable to the permeation of the reservoir water. The density of fracture decreases, pore space expands and stress changes, resulting in the increase of slow shear-wave delay time. The higher area of slow shear-wave normalized delay time is the area where moderate and small earthquakes are relatively active in Longtan Reservoir area. But this kind of correspondence still needs to be confirmed in subsequent studies through tracking and analysis.

4. The changes in shear-wave splitting parameters in some stations in Longtan Reservoir area have a relationship with water level. From the sequential process of fast shear-wave splitting polarization angle and water level, the polarization angles of RD, LN and XMP stations were positively correlated with change in water level, that is, it rises as the water level rises, and drops as the water level drops. It is, to a certain degree, synchronous with the change in water level, especially stations near developed fractured zones. This phenomenon suggests that the fast shear-wave splitting polarization angle of some stations in Longtan Reservoir area is sensitive to the change in water depth. From the sequential process of slow shear-wave normalized delay time and water level, the slow shear-wave normalized delay time of RD and TE stations, which are closest to the dam area, gradually increases with the water depth. Every time the water depth increases 20 or 30 m, normalized delay time will increase once as a whole, suggesting a strong change relationship with water depth. Based on the relationship between changes in shear-wave splitting parameters in some stations in Longtan Reservoir and water level, it is preliminarily analyzed that changes in the water level of Longtan Reservoir lead to obvious changes in media anisotropy in some areas in Longtan Reservoir area and result in change in stress.

References

1. Gao, Y.: Anisotropy and shear wave splitting in the crust of the earth. Earthquake parameter – the application of digital seismology in earthquake prediction. Monitoring and Forecasting Department of China Earthquake Administration, pp. 82–106 (2003)
2. Crampin, S.: Effective anisotropic elastic constants for wave propagation through cracked solids. Geophys. J. R. Astr. Soc. **76**, 135–145 (1984)
3. Zatsepin, S.V., Crampin, S.: Modelling the compliance of crustal rock-I: response of shear-wave splitting to differential stress. Geophys. J. Int. **129**, 477–494 (1997)
4. Crampin, S., Zatsepin, S.V.: Modelling the compliance of crustal rock-II: response to temporal changes before earthquakes. Geophys. J. Int. **129**, 495–506 (1997)
5. Crampin, S.: Seismic-wave propagation through a cracked solid: polarization as a possible dilatancy diagnostic. Geophys. J. R. Astr. Soc. **53**, 467–496 (1978)

6. Gao, Y., Zheng, S.H.: Study on shear wave splitting in Tangshan region(II): correlation function analysis. Earthq. Res. China **10**(Suppl.), 11–21 (1994)

7. Gao, Y., Crampin, S.: Temporal variation of shear-wave splitting in field and laboratory in China. In: Proceedings of the 10th International Workshop on Seismic Anisotropy, Tutzing (2002). J. Appl. Geophys. (2003, in press)

8. Gao, Y., Wang, P.D., Zheng, S.H., Wang, M., Chen, Y.T., Zhou, H.L.: Temporal changes in shear-wave splitting at an isolated swarm of small earthquakes in 1992 near Dongfang, Hainan Island, Southern China. Geophys. J. Int. **135**, 102–112 (1998)

9. Crampin, S., Volti, T., Stefansson, R.: A successfully stress-forecast earthquake. Geophys. J. Int. **138**, F1–F5 (1999)

10. Gao, Y., Zheng, S.H., Wang, P.D.: Study on the shear wave splitting of small earthquake swarm in the East of Hainan Province in 1992. Chin. J. Geophys. **39**(2), 221–232 (1996). (in Chinese)

11. Huang, T.L., Yang, M.L.: Preliminary study of S wave polarization in Xinfeng River. South. China J. Seismol. **21**(4), 22–26 (2001)

12. Zhou, Z.X., Li, J.L., Yu, T.H., et al.: Study on seismic S wave splitting in Wenzhou Shan Xi Reservoir. Acta Seismolog. Sinica **32**(4), 423–432 (2010)

13. Shi, H.X., Zhao, C.P.: The research of seismic shear wave splitting in Guangxi Longtan Reservoir. Seismol. Geol. **32**(4), 595–606 (2010)

14. Xiang, H.F., Zhou, Q.: Seismic parameter check of hydropower station project site on Guangxi Red River Longtan. Institute of Geology, China Earthquake Administration (2006)

15. Guo, P.L., Yao, H.: Analysis of seismic hazard on Longtan Reservoir. Plateau Earthq. **18**(4), 17–23 (2006)

16. Yao, H., Yang, S.W., Chen, B.: Broadband access system in the design of Longtan Hydropower Project digital telemetry seismic network channel optimization. Seismol. Geomagn. Obs. Res. **28**(6), 63–68 (2007)

17. Yao, H., Chen, X., Huang, S.S., et al.: The test of digital telemetry seismic network monitoring ability on Longtan Hydropower Project. Seismol. Geomagn. Obs. Res. **29**(4), 62–66 (2008)

18. Yao, H., Sun, X.J., Yang, C.Y.: The digital telemetry seismic network system of Longtan Hydropower Project. South. China J. Seismol. **28**(4), 53–62 (2008)

19. Zheng, Q.S., Zhu, J.S., Xuan, R.Q., et al.: Analysis of crustal velocity structure in Southern China area. Sediment. Geol. Tethyan Geol. **23**(4), 9–13 (2003)

20. Li, W.Q.: The relationship of new tectonic zoning with earthquakes in Guangxi. South. China J. Seismol. **9**(4), 22–26 (1989)

21. Gao, Y., Wu, J.: The principal stress field of the earth's crust is deduced by using shear wave anisotropy: example on metropolitan area. Sci. Bull. **53**(23), 2933–2939 (2008)

22. Gao, Y., Zheng, S.H., Zhou, H.L.: Fast shear wave polarization image and its change in Tangshan area. Chin. J. Geophys. **8**(3), 351–363 (1999)

23. Lai, Y.G., Liu, Q.Y., Chen, J.H., et al.: Shear wave splitting and crustal stress field in the capital circle region. Chin. J. Geophys. **49**(1), 189–196 (2006)

24. Crampin, S., Booth, D.C., Evans, R.: Changes in shear wave splitting at Anza near the time of the North Palm Springs earthquake. J. Geophys. Res. **95**(B7), 11197–11212 (1990)

Study on Inversion Forecasting Model for 2011 Tohoku Tsunami

Chao Ying[✉], Yong Liu, Xin Zhao, and Jinbin Mu

Zhejiang Institute of Hydraulics & Estuary, Hangzhou 310020, China
xinqing928@126.com

Abstract. Tsunami is a kind of wave with great destructive power, which has great impact on the environment. Numerical prediction is an important way to reduce the environmental disasters. The inversion-forecasting model is an important method for the prediction of tsunami. In this paper the Japan's east coast was divided into 18 unit sources to set up a tsunami database by using COMCOT numerical model. An inversion forecasting model was established by using least non-negative square method based on the database. The model was applied to 2011 Tohoku tsunami, the initial tsunami water level with 10 m increase and 3 m decrease calculated by the model were basically the same as previous research, the buoy level of prediction is in good agreement with measured data. Comparing with tsunami heights measured by tidal stations at coastal area of Zhejiang Province, the deviation of forecasted and measured value is large. But the prediction accuracy can be greatly improved by solving COMCOT nonlinear equations with source parameters inversed by the inversion forecasting model. This study has significance for reference of East China Sea tsunami forecasting mechanism.

Keywords: COMCOT · Tsunami · Inversion model

1 Introduction

Tsunami is a typical long wave in the ocean with powerful destructive, the wave travel fast and can cause coastal hazard. Once in motion, they can't be stopped, to prevent loss of life and damage we have to detect tsunami in advance and predict the strength. The primary goal of tsunami prediction is to produce efficient forecast of wave arrival time, height and inundation, reduce the computing time of the forecast model also very important.

The inversion model plays an important role on tsunami forecasting. The algorithm divides the source regions into several sub-faults, then generates a database of synthetic water level at tide gauge and some observe station near the source. Regression of recorded tsunami signals using the water level provides the slip distribution at the source and the expected waveforms. This methodology has been used as source input for tsunami forecasting models and has been shown to be feasible.

Titov [1] discussed the use of the inverse algorithm to determine the initial conditions for real-time simulation of transpacific tsunamis. Wei [2] describes a methodology to assess the severity of a tsunami in progress based on real-time water-level data

© Springer Nature Singapore Pte Ltd. 2017
H. Yuan et al. (Eds.): GRMSE 2016, Part I, CCIS 698, pp. 494–504, 2017.
DOI: 10.1007/978-981-10-3966-9_55

near the source, based on the earthquake source parameters from Johnson [3], Wei [2] predicted the tsunami waveforms away from the source, and forced on its potential threat to Hawaii. Yamazaki [4] using the algorithm to hindcasting two major tsunamis generated from the Japan–Kuril–Kamchatka source and the computed tsunami heights show good agreement with recorded water-level data. Li [5] applied the inversion model for tsunami forecasting, and hypothetical earthquake occurred at Manila trench, validated with the COMCOT model results. However, there have not been any studies that examined tsunami in East China Sea. We should take consideration of the performance of COMCOT and the inversion model applied in East China Sea. Also, we need pay attention to the compatibility of applying the inversion model to coastal area.

On March 11, 2011, a magnitude-9 earthquake shook northeastern Japan, unleashing a savage tsunami, causing a lot of casualties and loss. The tsunami also impacted east china sea, 6–8 h later of the earthquake, some stations along east china mainland recorded increase of water-level in 10–55 cm [6]. From examine the inversion model and COMCOT applied on 311 earthquakes, shows significance for reference of East China Sea tsunami forecasting mechanism.

2 The Inversion Method

The methods are organized as follow:

(1) Deploy of tsunameters

To keep the tsunami buoys safe and efficient, it should deploy near the source of earthquake in deep-ocean and located on flat terrain. The tsunami buoy consists of ocean-bottom pressure sensors and surface buoy, to get the data of bottom pressure and surface water level. After the earthquake, the buoy will work automatic and measure the real-time tsunami data every minute.

(2) Set up a database

We divided the source regions into several unit sources, get the deformation data of each unit source, and save to the database.

(3) Record of tsunami data and source inversion

After the earthquake, the buoy will record the tsunami data, after the measurement of the initial tsunami wave, we can use the tsunami database and the inversion algorithm to attain the tsunami wave height and arrive time in specific region, and reconstruct the detailed structure of the earthquake source.

3 Set up of Tsunami Database

3.1 Model Configuration

COMCOT (Cornell Multigrid Coupled Tsunami model) was developed in University Cornell; it is capable of studying the entire life-span of a tsunami including its generation, propagation run-up and inundation. The model adopts explicit staggered leap-frog finite difference schemes to solve Shallow Water Equations in both Spherical and Cartesian Coordinates system for numerical studies at different scales [7]. The model has been used to investigate several historical tsunami events, such as the 2004 Indonesia tsunami [8].

Our model has 3 nested multi-grid, the simulation domain (Fig. 1) include east china sea, Taiwan Straits, south china sea, japan and west pacific with a grid size of 2 min(ETOPO1), the depth of first layer were interpolated from ETOPO1, the depth of second and third layer were interpolated from in-situ measurements and Chart Datum. The detail of nested grid is shown in Table 1. The time step of the model was 2 s to satisfied CFL condition. The tsunami database was based on linear hypothesis, so the equations in the first layer were linear.

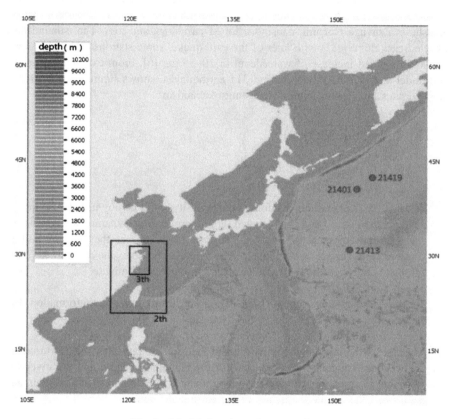

Fig. 1. Model domain and topography.

Table 1. Model grid set up and model configuration

Nest layer	Domain	Resolution	Configuration		
			Coordinates	Governing equations	Bottom friction
1	3°N–66.5°N, 105°E–175°E	2.0 min	Spherical	Linear	None
2	21°N–32°N, 118°E–128.5°E	0.5 min	Spherical	Nonlinear	0.0013
3	27°N–31.5°N, 120.5°E–122.5°E	6.0 s	Spherical	Nonlinear	0.0013

3.2 Initial Condition and Boundary Condition

The COMCOT model use the displacement of sea floor to compute the initial water level caused by the earthquake, with the Hypothesis of hydrostatic, the bed deformation and the water level change will happen at the same time. The initial condition of tsunami is determined via fault models. We used the Okada fault model, there are 9 parameters: epicenter (latitude, longitude), focal depth, length of fault plane, width of fault plane, dislocation, strike direction, dip angle and rake angle. Sketch shown in [13].

For the boundary condition, the radiation open boundary was adopted for boundaries in the water region, a reflective boundary (vertical wall) was assigned along the shoreline when use the linear shallow water equation; flip boundary was assigned when use the nonlinear shallow water equation.

3.3 Unit Sources in the Japan's East Coast

The PMEL of NOAA have built a propagation database, The Propagation Database is a collection of tsunami propagation model runs that have been pre-computed for tsunami source functions at selected locations along known and potential earthquake zones. The pre-defined source in the propagation database is referred to as a "unit source", Each unit source is equivalent to a deformation due to an earthquake with a fault length of 100 km, fault width of 50 km, and a slip value of 1 m, equivalent to a moment magnitude of 7.5 [10].

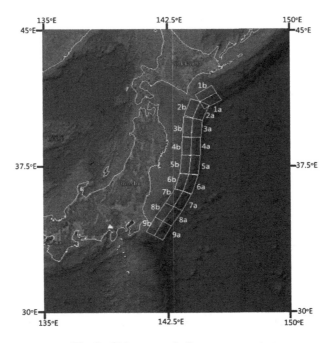

Fig. 2. Unit sources in Japan east coast

As we forced on 311 Japan Tsunami, we have chosen 18 "unit source" near the earthquake source to construct the tsunami database, location of unit source is shown in Fig. 2, parameters are shown in Table 2.

Table 2. Fault Parameters of unit sources

Unit sources	Epicenter (Lon)	Epicenter (Lat)	Strike angle	Dip angle	Focal depth
1a	144.304	42.163	242.00	25.00	23.73
1b	144.561	41.804	242.00	22.00	5.00
2a	143.286	41.334	202.00	21.00	21.28
2b	143.803	41.176	202.00	19.00	5.00
3a	142.980	40.349	185.00	21.00	21.28
3b	143.527	40.313	185.00	19.00	5.00
4a	142.884	39.454	185.00	21.00	21.28
4b	143.425	39.418	185.00	19.00	5.00
5a	142.762	38.584	188.00	21.00	21.28
5b	143.293	38.525	188.00	19.00	5.00
6a	142.532	37.783	198.00	21.00	21.28
6b	143.036	37.653	198.00	19.00	5.00
7a	142.132	37.027	208.00	21.00	21.28
7b	142.594	36.830	208.00	19.00	5.00
8a	141.597	36.264	211.00	21.00	21.28
8b	142.042	36.048	211.00	19.00	5.00
9a	141.055	35.433	205.00	21.00	21.28
9b	141.521	35.256	205.00	19.00	5.00

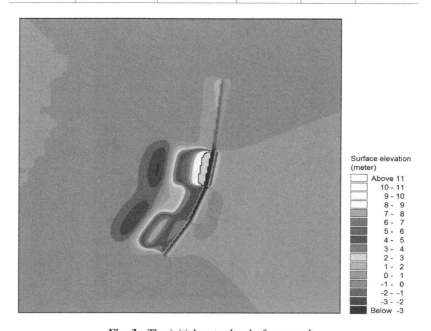

Surface elevation
(meter)
Above 11
10 - 11
9 - 10
8 - 9
7 - 8
6 - 7
5 - 6
4 - 5
3 - 4
2 - 3
1 - 2
0 - 1
-1 - 0
-2 - -1
-3 - -2
Below -3

Fig. 3. The initial water level of our results

4   The Validation of the Inverse Approach

4.1   The Method of the Inverse Approach

The inverse approach makes use of the linearity of the tsunami generation mechanism at the source and the propagation characteristics in the open ocean.

$$\eta(x, y, t) = \sum_i^{N_s} c_i G_i(x, y, t) \tag{1}$$

Where Ns is the number of unit source, $G_i(x, y, t)$ is the water level in position (x, y) at time t, $\eta(x, y, t)$ is the total water level, c_i is the weight coefficient of unit source, we can obtain from the inverse approach.

After the earthquake, the buoys will record the water level when the initial tsunami wave arrived, the dataset of buoy recorded are:

$$\{Z_k(x_0, y_0, t_k), k = 1, N_t\} \tag{2}$$

Where N_t is the number of recorded data, then we can get the weight coefficient c_i by using least non-negative square method, shown in (3). A is the pre-build tsunami database, c is the weight coefficient, and the b is the recorded data from buoys.

$$\min_c \|Ac - b\|_{2'} \tag{3}$$

$$A = \left\{ \begin{array}{ccc} G_1(x_0, y_0, t_1) & \cdots & G_{N_s}(x_0, y_0, t_1) \\ \cdots & \ddots & \cdots \\ G_1(x_0, y_0, t_{N_t}) & \cdots & G_{N_s}(x_0, y_0, t_{N_t}) \end{array} \right. \tag{4}$$

$$c = \left\{ \begin{array}{c} c_1 \\ \vdots \\ c_{N_s} \end{array} \right\} \quad b = \left\{ \begin{array}{c} Z_1(x_0, y_0, t_1) \\ \vdots \\ Z_{N_t}(x_0, y_0, t_{N_t}) \end{array} \right\} \tag{5}$$

It is noteworthy that a unique solution of c exists and is given by $N_t \geq N_s$, where N_t is the number of data recorded by buoys.

$z_i = F_i(x, y)$ is the deformation of the entire earthquake zone, when i is the number index of unit sources and $z = F(x, y) = \sum_{i=1}^{N_s} c_i F_i(x, y)$ is the deformation of total earthquake area.

4.2   The Validation of the Inverse Approach in Japan 311

After the earthquake of 311, tsunameter D21401 measured a water level pulse within 78 min of the earthquake. We use the unit source shown in Fig. 2 and Table 2 to construct our tsunami database, and compute inversion of the tsunami source, the weight coefficient of each unit source (Table 3) shown that 3b–7b are the main fault unit source. Figure 4 is the initial water level we obtained from Table 4 and OKADA model. Tang [11] used the deep-ocean pressure measurements and numerical models to inverse the earthquake source. Results showed that six tsunami source functions that give the best fit to the observations. The parameters of fault are shown in Table 4, accompany with the OKADA model as initial condition. Run-up and inundation are computed at the coastline of west coast of America, the method was validated by comparing the prediction of the amplitude time series with measurements at 30 tsunameter stations throughout the Pacific Ocean, by forecast/measurement comparisons of coastal impacts at 32 coastal sites, results provide reasonable regional forecasts. Wei [12] used the same fault setting to compute the tsunami run-up and inundation along the japan coast, the computed inundation penetration agrees well with survey data, giving a modeling accuracy of 85.5% for the inundation areas along 800 km of coastline between Ibaraki Prefecture (north of Kashima) and Aomori Prefecture (south of Rokkasho).

Table 3. Weight coefficient

Unit source name	3b	4a	4b	5a	5b	6a	6b	7a	7b
weight coefficient	3.9	0.0	3.9	24.9	32.1	9.1	15.7	23.3	11.8

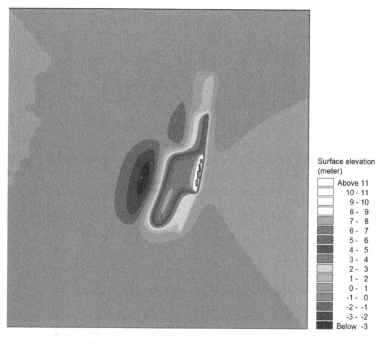

Fig. 4. The initial water level by Tang [11]

Table 4. Earthquake information inversed by Tang [11]

Epicenter	Focal depth (Km)	Strike angle (°)	Dip angle (°)	Rake angle (°)	Dislocation (m)
143.800°E, 40.290°N	13.14	185	19	90	4.66
143.698°E, 39.399°N	13.14	185	19	90	12.23
143.562°E, 38.498°N	13.14	188	19	90	21.27
143.029°E, 37.588°N	30.24	188	21	90	26.31
143.291°E, 37.588°N	13.14	198	19	90	4.98
142.784°E, 37.718°N	30.24	198	21	90	22.75

Using the earthquake source inversed by tang, the model initial water level is shown in Fig. 3, The comparison shows agreement between them, both of them shows run-up of 10 m and inundation of 3 m.

Further validation is needed. Our results are validated by comparing the prediction of the amplitude time series with D21401 D21409 and D21413 (Fig. 5). As shown in Fig. 5, The model accurately reproduced the first dominant wave. We can see some misfit between model results and observed data, from a forecasting perspective, the short-term forecasting of tsunamis described in this study is satisfied for the forecasting.

5   The Accuracy of the Inverse Approach in Coastal

Since our inverse model were based on the hypothesis that the tsunami wave is propagation linear in the ocean, linear shallow water equation is used when compute the tsunami database, it's hard to avoid the error when use (1) to predict the water levels along coast area. To overcome the error, we may apply the inverse algorithm reconstruct the information of earthquake source, then apply the nonlinear shallow water equation to simulate the coastal tsunami wave propagation.

We use methods mentioned above to simulate the tsunami wave of earthquake 311, A comparison of water level data in Kanmen, Shipu and Shenjiamen is made between two methods, they have the same model grid as in Sect. 3.2, location of tide-gauge station are shown in Fig. 6. Comparison results are shown in Table 5, both accurately reproduced the first dominant wave within the error of 20 min.

We conclude that errors in not includes nonlinear interaction between the tsunami waves and bathymetric features nearshore and ignore the run-up of tsunami wave, are responsible for the model underestimate of the first peak reproduced by the inverse algorithm.

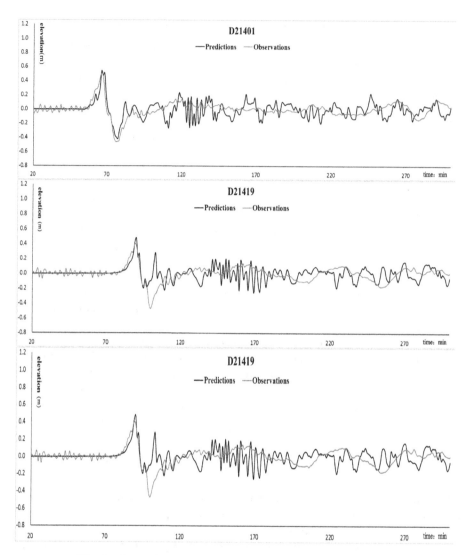

Fig. 5. Comparisons of recorded and com puted water level data

Use of the nonlinear model of COMCOT and the earthquake source reproduced by the inverse algorithm can increase the first peak of tsunami wave, the misfit of model results and observation are 20%, but the misfit in Shenjiamen are bigger than other stations, because the inaccuracies in low-resolution bathymetry data.

In conclusion, using the inverse algorithm can approximate get arrival time of tsunami wave. As propagation of tsunami wave in coastal area is a highly nonlinear process, the prediction accuracy can be greatly improved by solving COMCOT non-linear equations with source parameters inversed by the inversion forecasting model.

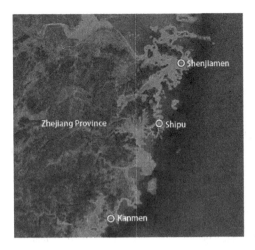

Fig. 6. Location of tide gauge along Zhejiang coast

Table 5. Comparisons of tide gauge data

	Observed data		Inverse approach computed		Inverse approach with COMCOT	
	First peak (cm)	Arrival time of first peak	First peak (cm)	Arrival time of first peak	First peak (cm)	Arrival time of first peak
Kanmen	30	21:24	14	21:36	25	21:30
Shipu	52	22:08	23	22:27	42	22:20
Shenjiamen	55	22:20	16	22:25	27	22:20

6   Conclusions

We have setting up a tsunami database in Japan's east coast by using COMCOT numerical model.

An inversion forecasting model was established by using least non-negative square method based on the database. The model was applied to 2011 Tohoku tsunami, results shown we can finish the forecasting in 2 min, the initial tsunami water level calculated by the model were basically the same as previous research. The prediction of buoys water level was in good agreement with measured data.

The arrive time of first peak of tsunami wave in coastal area computed by the inversion forecasting model shown good agreement, but the model underestimate the water level of first peak.

The prediction accuracy can be greatly improved by solving COMCOT nonlinear equations with source parameters inversed by the inversion forecasting model.

Acknowledgements. This work was supported by the Natural Science Foundation of Zhejiang Province (LY13E090001), and the Science and Technology Plan for Zhejiang Province (2015F50064).

References

1. Titov, V.V., Mofjeld, H.O., González, F.I., et al.: Offshore forecasting of Alaskan tsunamis in Hawaii. In: Hebenstreit, G.T. (ed.) Tsunami Research at the End of a Critical Decade, pp. 75–90. Springer, Heidelberg (2001)
2. Wei, Y., Cheung, K.F., Curtis, G.D., et al.: Inverse algorithm for tsunami forecasts. J. Waterw. Port Coast. Ocean Eng. **129**(2), 60–69 (2003)
3. Johnson, J.M.: Heterogeneous coupling along Alaska-Aleutians as inferred from tsunami, seismic, and geodetic inversions. Adv. Geophys. **39**, 1–116 (1998)
4. Yamazaki, Y., Wei, Y., Cheung, K.F., et al.: Forecast of tsunamis from the Japan–Kuril–Kamchatka source region. Nat. Hazards **38**(3), 411–435 (2006)
5. Li, L., Mao, X.: Tsunami forecasting based on inversion model in the south China sea (in Chinese). Chin. J. Hydrodyn. **27**(1), 62–67 (2012)
6. Wang, P.T., Yu, F.J., Zhao, L.D., et al.: Numerical analysis of tsunami propagating generated by the Japan Mw 9.0 earthquake on March 11 in 2011 and its impact on China coasts (in Chinese). Chin. J. Geophys. **55**(9), 3088–3096 (2012)
7. Philip, L.F.L., Woo, S.B., Cho, Y.S.: Computer Programs for Tsunami Propagation and Inundation (1998)
8. Wang, X., Liu, P.L.F.: An analysis of 2004 Sumatra earthquake fault plane mechanisms and Indian Ocean tsunami. J. Hydraul. Res. **44**(2), 147–154 (2006)
9. Pan, W., Wang, S.: Introduction and application of COMCOT model (in Chinese). Mar. Forecasts **26**(3), 45–52 (2009)
10. Gica, E., Spillane, M.C., Titov, V.V., et al.: Development of the Forecast Propagation Database for NOAA's Short-Term Inundation Forecast for Tsunamis (SIFT) (2008)
11. Tang, L., Titov, V.V., Bernard, E.N., et al.: Direct energy estimation of the 2011 Japan tsunami using deep-ocean pressure measurements. J. Geophys. Res. Oceans (1978–2012), **117**(C8) (2012)
12. Wei, Y., Chamberlin, C., Titov, V.V., et al.: Modeling of the 2011 Japan tsunami: lessons for near-field forecast. Pure Appl. Geophys. **170**(6–8), 1309–1331 (2013)
13. Wang, X.: User manual for COMCOT version 1.7. Cornel University, p. 65 (2009)

Remote Sensing Dynamic Monitoring and Driving Force Analysis of Grassland Desertification Around the Qinghai Lake Area

Yu'e Du[1,3], Baokang Liu[2(✉)], Fujiang Hou[3], and Zongli Wang[3]

[1] Natural Energy Research Institute, Gansu Academy of Sciences,
Lanzhou 730000, China
[2] Qinghai Institute of Meteorological Sciences, Xining 810001, China
liubk04@qq.com
[3] College of Pastoral Agriculture Science and Technology
of Lanzhou University, Lanzhou 730020, China

Abstract. According to the results of the remote sensing satellite data of HJ, TM and MSS in the recently 40 years. The area of grassland desertification was increasing trend around Qinghai lake area from 1975 to 2000, whose rate was 12 km²/a, the area change of grassland desertification was small during 2000 to 2008, which was stable. The area greatly reduced in 2008–2012, reaching to 56.90 km². This shows that the reduced area of grassland desertification is more obvious in the Qinghai Lake area in the past 10 years. Meanwhile the ecological environment tends to be improved. The main driving force of desertification area decreases is: The water level increased of the Qinghai lake is significant in nearly 10 years; The climate of Qinghai Lake basin shows warm and wet trend. It is particularly prominent after entering the 21st century; the runoff into of the increased lake is also obviously; Human activity is slowing around Lake area.

Keywords: Grassland desertification · Remote sensing dynamic · Driving force

1 Introduction

Grassland desertification refers to that different climate zones with surface environment (Sandy Lake area was characterized by silt, silty sand and silty loam) affected by wind erosion and water erosion, drought, pests and mice which were not economic factors such as grassland, long-term overgrazing, unreasonable reclamation, deforestation, excessive digging and cutting herbs, the natural grassland suffered varying degrees of damage, soil erosion, soil desertification became coarse, decreased the content of soil organic matter, nutrient loss, grassland productivity decreased, resulting in the original non desert grassland, the process of [1, 2] degradation in sandy desert like landscape activities as the main feature of grassland. Grassland desertification is a special type of grassland degradation [3].

Since 1960s, due to natural factors and human factors, the deteriorating ecological environment in Qinghai Lake basin, the water level dropped, grassland degradation, desertification, snow line rising, species diversity decreasing [4]. Lin Jiancai, according to the survey data of 1994 and 1999 and satellite image analysis, pointed out that the

© Springer Nature Singapore Pte Ltd. 2017
H. Yuan et al. (Eds.): GRMSE 2016, Part I, CCIS 698, pp. 505–514, 2017.
DOI: 10.1007/978-981-10-3966-9_56

grassland degradation, desertification land increased at an annual rate of 2.7% in the region around Qinghai Lake, the flow of sand increased faster, annual growth rate of 5%, nearly 8000 hm^2 of the water depth was less than 50 cm, probably in the last 5–8A into sandy land [5–8]. Yu Weiguo using aerial photos of 1956–1972 years, the TM data of 1986 and 2000 monitoring results showed that the average annual net increase of land desertification was of 18.07 km^2 from 1956 to 2000 in Qinghai Lake area, especially after 1980s expanded at a faster rate, the net increase was of 35.08 km^2 during 1986–2000 [9]. Gao Xiaohong County analyzed the change characteristics of the land use by using the TM data of 1986 and 2000 surrounding Qinghai Lake area, found that the land of cultivated, water area, urban and rural residential construction had increased, and woodland, grassland and unused land decreased, which was the deterioration of ecological environment from 1986 to 2000 in Qinghai Lake area [10].

However, the research results showed that the ecological environment of Qinghai Lake basin had continued to deteriorate since late 80s until 90s by using Multi-source Satellite of remote sensing within 40 years [11]. In the early 21st century the grassland desertification entered a relatively stable period around the Qinghai Lake area, at the same time, the grassland degradation area was relatively stable, the Qinghai Lake area had begun to rise slowly since 2005, especially in the beginning of 2010, the area of Qinghai Lake and the water level both had significant long rise trend, and the grassland productivity also increased, the grassland desertification and degradation area began to have a more obvious increase of dune moving distance, which was shortened obviously, grass height and coverage increased significantly. The fact that the warm wet climate of Qinghai Lake basin was obvious under the background of ecological environment in recent 10 years though there were still some fluctuations [12], but the overall trend showed a good.

2 Overview of the Study Area

Qinghai Lake is China's largest inland salt water lake, which is an important part of the Qinghai Tibet Plateau, with a sensitive area of global change and ecosystem typical vulnerable areas. In nearly 30 years, the desertification around Qinghai Lake was rapid expansion, zonary spread, forming a lake island and the southeastern edge of the larger lake, Lake East northeast of the Gan River, shore of Lake waves and sand zone and so on 4 gang.

Qinghai Lake area is in the range of 99°36′–100°16°E, 36°32′–37°15′N. Belongs to the arid area of Northwest China and the convergence zone of southwest, climate is cold temperature, no obvious summer, but cold and long winter [13]. The annual sunshine percentage is 68–69%, and the light resource is more abundant. The average annual precipitation is 300–400 mm; evaporation large annual evaporation is up to 1500 mm, subject to the climate and geographical location, the alpine meadow soil in the basin is the most widely distributed soil, to the northwest distribution of sandy soil, chestnut soil, chernozem, meadow soil, mountain meadow soil, mountain meadow soil and cold desert soil from the southeast, with a sandy more nutrient content is also low, serious soil erosion, Lake area types of vegetation in temperate steppe, desert shrub, alpine shrub, alpine meadow, vegetation diversity, distribution patterns of temperature vegetation and alpine vegetation coexistence [14]. The total land area is about

29690.58 km², the main land use types is arable land, woodland, grassland, swamps and some debris, bare land, sand, water, beaches, rock etc. (Fig. 1).

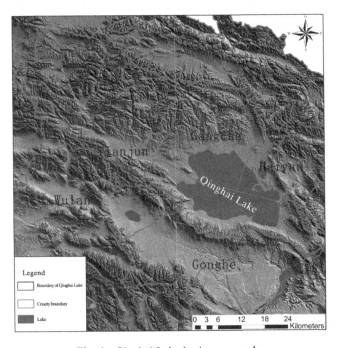

Fig. 1. Qinghai Lake basin topography

3   Three Data Sources and Methods

To analyze the spatial changes of grassland desertification of Qinghai lake area, this paper selected MSS data in 1975, 2000, 2008 and 2009 TM data and HJ_1A data (Table 2) in 2012 of the remote sensing data, after all the data were uniformly re-sampled, spatial resolution were 30 m. In addition to the MSS, the rest of remote sensing data selected 7–8 month period (Table 1).

Table 1. MS, TM and HJ data of the time

Serial number	Time phase	Types	Band
1	1975-10-20	MSS	1–7
2	2000-08-10	LandSat5	1–7
3	2008-07-07	LandSat5	1–7
4	2009-08-24	LandSat5	1–7
5	2012-08-03	HJ-1A	1–4

Using ENVI5.2 software to have a geometric correction for satellite remote sensing data, and using the nearest neighbor method to have Pixel re-sampling, after image correction, the image needs RMS test, controlling the error within the 1 pixels. And then we will establish a unified system of projection, projection transformation later again through data splicing, cutting and other processing, According to the field investigation and expert experience, this paper selected the typical area to establish the main land use type for the symbol of the interpretation, through the method of human-computer interaction to have the artificial interpretation of the sand, and get the annual grassland desertification area.

4 The Result Analysis

4.1 The Sand Area

Analysis remote sensing image interpretation results of 1975, 2000, 2009 and 2012, there were the flow, fixed and semi fixed beam nest like sand dunes around the Qinghai Lake East and North, meanwhile west Coast distribution, northeast, northwest edge with tall mountains. According to the satellite remote sensing monitoring, the desertification area increased, According to the satellite remote sensing monitoring, the desertification area of 1975–1980 showed an increasing trend, whose rate was of 10.85 km²/a, the desertification area variation of 2000–2008 showed a stable trend; the sandy land area decreased greatly from 2008 to 2012, being of 56.90 km², whose rate of 11.38 km²/a (Table 2).

Table 2. Qinghai lake basin within the scope of the sand changes (km²)

The year	2000	2008	2012	2015
Grassland sand area	461.90	452.50	371.93	366.78

4.2 The Spatial Variation of Desertification Grassland

From 1975 to 2000, the east coast of Qinghai Lake shoreline westward, a bird eye completely isolated Milton lake from Qinghai Lake, shoreline southward obviously [15]. In addition, a long sand beam in Northwest - southeast direction was the width

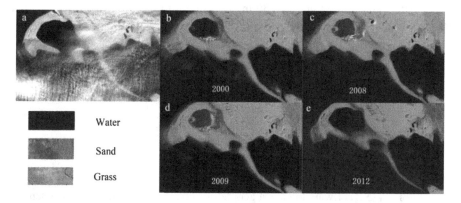

Fig. 2. The dynamics of grassland desertification area on the east shore of Qinghai Lake by 2000, 2008 and 2012

significantly between the Haiyan Bay lake and the Qinghai Lake locating in the lower right corner (southeast direction) in Fig. 2, which showed that the grassland desertified area increased along the Qinghai Lake in 2000 compared with 1975. Compared to 2000 and 2008, the grassland desertification area had no significant change, but the area of grassland desertification significantly increased in 2009 compared with 2000 and 2008. Compared with 2009, Milton Lake increased obviously in 2012, where the South River connected with Qinghai Lake, sand beam width narrowed between Qinghai Lake and Haiyan Bay Lake, the river water became wide between the Haiyan Bay lake and the Qinghai Lake, a small island was overwhelmed by the Hai Yan Wanhu into lake where Lake water had been rising, the desertification area decreased significantly comparing with in 2000, 2008 and 2009 (Fig. 3).

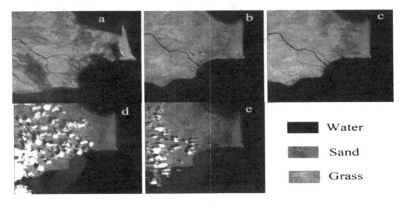

Fig. 3. The dynamic monitoring of the satellite remote sensing monitoring on the west shore of Qinghai Lake

Seeing from the five periods remote sensing images of west coast of Qinghai Lake, the grassland desertification area decreased significantly in 2000 compared to 1975, the shoreline northward and southward obviously, but the West grassland area in 2009 increased obviously compared to 2000 and 2008. In 2012 compared with 2009, grassland area increased significantly, the Qinghai Lake area continued to increase, the west shore line mainly northward, southward, grassland desertification area decreased significantly compared with 2009 (Fig. 4).

There was an increase of the grassland desertification area along north shore in Qinghai Hu in 1975 than in 2000, mainly due to the north shore of the Qinghai Lake shoreline southward, resulting in desertification area increase; in 2008 compared with 2000, there was no obvious change; Compared with 2008, the increase of grassland area is larger than that of 2009, and the area of grassland desertification decreased; In 2012 compared with 2009, the grassland area further increased, and the shoreline northward, the desertified land in north shore of Qinghai Lake had been transformed into grassland, indicates that the grassland ecological environment in north coast of Qinghai lake had improved [16] (Fig. 4).

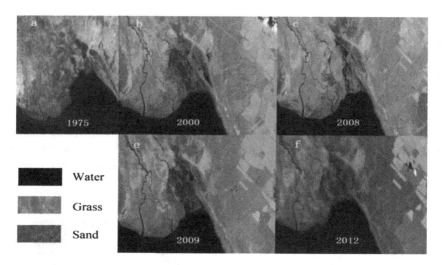

Fig. 4. The dynamic monitoring of grassland desertification in the northen shore of Qinghai Lake

5   Driving Force Factor Analysis

5.1   Water Level Change

In recent 50 years (1961–2010), the water level of Qinghai Lake continued to decline, whose rate is 0.65 m/10a. Among them, the water level fell by about 3.37 m, average being at a rate of 0.84 m/10a down from 1960s to the 21st Century, water level continued to rise again from 2004 to 2010 for six consecutive years, the water level reached 3193.59 m, the cumulative rise of 0.72 m (Figs. 5).

5.2   Climate Change

Affected by the global warming and increasingly strong influence of plateau monsoon, The climate change of Qinghai Lake basin showed a warm wet trend with temperature increase and precipitation increases from 1961 to 2010, this trend was particularly prominent in twenty-first Century [17]. Among them, the annual average temperature of 0.36 °C in every 10 °C, there was increased by 1.8 °C within 50 years; annual precipitation increase of 8.4 mm/10 years, summer precipitation increase was particularly significant (Fig. 6). From 2005 to 2010, the average precipitation in the Qinghai Lake basin was 407 mm, which increased by 18.6% from 1961 to 2004, compared with the climate change being by 12.3% from 1971 to 2000.

5.3   Into the Lake Runoff Increase

Inflow runoff was the main factor that affects the lake area and water level. Since the 21st Century, the basin temperature and precipitation increase resulted that snow melt water supply quantity and water supply quantity increased in the Buha River,

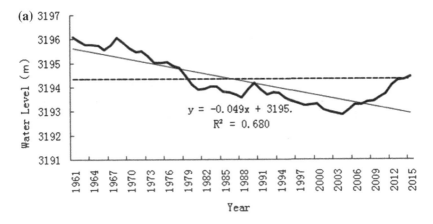

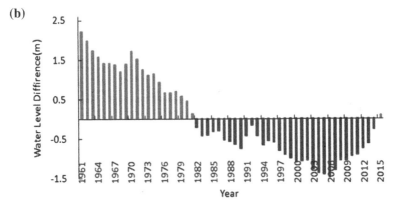

Fig. 5. The water level change trend (a) and the change of water level in Qinghai Lake from 1961 to 2015 (b)

meanwhile increased runoff, and runoff into the lake water level rose significantly greater than the increase rate, the increase of runoff into the lake water level rise had a significant "pulling effect", finally to the consecutive six years rise of the water level of Qinghai Lake in nearly 50 years (Fig. 7).

5.4   Human Activity Slowed Down

For a long time, the human activities of lake area had also caused some impact on the decline in the water level of Qinghai Lake, this is because the water consumption of the water reduced the supply of water. In mid 1980s, the irrigated land of lake area was about 20 thousand hectares, Grassland irrigation area of 4000 hectares, annual irrigation volume was 80 million cubic meters, 10 million cubic meters of water drinking, the water losses of years was about 450 million cubic meters, the human consumption

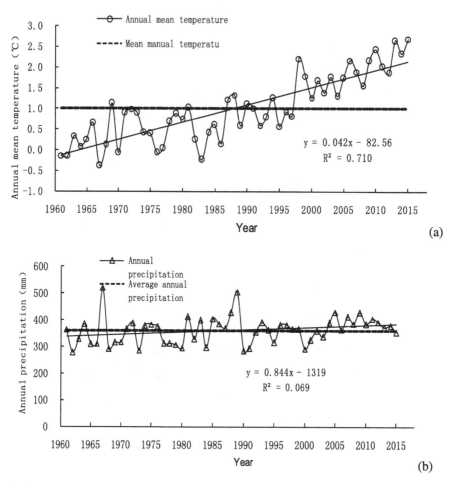

Fig. 6. The change of annual mean temperature (a) and annual mean precipitation (b) in the Qinghai Lake basin from 1961 to 2010

of water accounted for only 1/5; in 1990s, with a lot of land around the lake of the abandoned Qinghai Lake, the losses water of years was about 436 million cubic meters of. Human consumption of water accounted for the loss of water to 8.7%; In recent years with the returning farmland to forest (grass) implementation of ecological environmental protection project of the Qinghai Lake basin [18], which the ecological environment and the adverse effects on human activities further slew. Therefore, due to human activity slowing down the Qinghai Lake basin agriculture, water, which was conducive to the river runoff to the lake's supply, the rise of the water level had played a role in adding fuel to the flames.

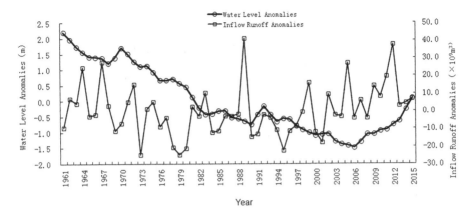

Fig. 7. The cumulative distance average variation of the water level and the runoff in the Qinghai Lake from 1960 to 2009

6 Conclusions

According to the Qinghai Lake area MSS, TM and HJ satellite remote sensing monitoring, the area grassland sandy around Qinghai Lake increased during 1975–1980. From 2000 to 2008, the grassland desertification area variation showed a stable reduced trend; However, during 2008–2012 the sandy land area decreased greatly, being of 56.90 km^2. This showed that the ecological environment in the Qinghai Lake area had tended to improve in the past 10 years, and the area of grassland desertification decreased significantly. The main driving force of desertification area is as follows: in recent 10 years, the water level increased significantly and the area of Qinghai Lake increased in continuous 8 since 2005, especially in the last 3 years, Qinghai Lake area and water level increase was very significant. Affected by the global warming and increasingly strong influence of plateau monsoon, the climate change around Qinghai Lake basin showed a warm wet trend of temperature increase and precipitation increases during 1961–2010, this trend was particularly prominent in twenty-first Century. Also, due to the Qinghai Lake basin temperature and precipitation increase, resulting to snow melt water supply quantity and water supply quantity increases in the Buha River, meanwhile increased runoff into the lake; In addition, over the past 10 years with the returning farmland to forest (grass) implementation of ecological environment protection engineering around the Qinghai Lake basin, the adverse effects of human activities on the ecological the environment was further reduced.

References

1. Yan, C., Wang, W., Feng, Y., et al.: Status and distribution characteristics of desert in Qinghai province in 1996. Arid Area Geogr. **23**(4), 104–108 (2004)
2. Qinghai Province Local Records Editorial Board. Qinghai Province Qinghai Lake Zhi. Qinghai People's Publishing House, Qinghai (1998)
3. Shen, Y., Li, P.: The Natural Geography of Qinghai Prince. Ocean Press, Beijing (1991). pp. 21–26
4. Wang, Y., Tang, Z.: Study on the present situation and comprehensive control of desertification in Qinghai Lake basin. Anhui Agric. Sci. **37**(5), 2267–2269 (2009)
5. Dong, Y., Liu, Y., Liu, Y.: Study on Several Issues of Desertification. Xi'an Map Publishing House, Xi'an (1995)
6. Yu, W., Chen, K.: Desertified areas around the Qinghai Lake of Saline Lake on dynamic. Remote Sens. **10**(4), 48–51 (2002)
7. Li, F.: Qinghai Lake lake Morphological changes and genetic analysis. Resour. Sci. **26**(1), 38–44 (2004)
8. Qinghai Lake river basin ecological environment comprehensive treatment project planning (2004)
9. Hou, G., Xu, C.: Use (RS) and (GIS) investigation on the distribution of desert around Qinghai Lake of Qinghai. Environment **15**(13), 105–107 (2005)
10. Xu, H., Li, X., Sun, Y., et al.: Analysis of climate change in the Qinghai Lake basin in past 47 years. J. Arid Meteorol. **25**(2), 52–53 (2007)
11. Li, L., Wang, Q., et al.: Climate change in Qinghai Lake area and its influence on water level. Meteorol. Sci. Technol. **33**(1), 58–62 (2005)
12. Li, F., Li, L.: Climate change and its environmental effects in the surround of Qinghai lake. Clim. Change **1**(4), 172–175 (2005)
13. Li, L., Zhu, W., Wang, Z., et al.: The impacting factor and its tendency forecasting of the variation of water level in Qinghai Lake in recent 42 years. J. Desert Res. **25**(5), 690–696 (2005)
14. Wang, Y., Zhou, X., Ni, S., et al.: Analysis of climate change in the Qinghai Lake in recent 40 years. Acta Meteorol. Sin. Nanjing **26**(2), 228–235 (2003)
15. Fu, Y., Li, F., Zhang, G., et al.: Analysis of natural grassland degradation and its environmental impact in Qinghai province. Glacial Frozen Soil **29**(4), 525–528 (2007)
16. Dong, G., The Gold Line, et al.: Desertification and Ways of Gonghe Basin in Qinghai. Science Press, Beijing (1993)
17. Chen, W., Yang, Z.: Qinghai agriculture and forestry science and technology. Effects Tourism Disturbance Biodiversity. Sandy Grassland. Qinghai Lake Science **22**(1), 1–4 (2012)
18. Zhang, Y., Chao, Z., Yang, Y., et al.: Effect evaluation of ecological environment protection in Qinghai Lake basin based on PSR model. Pratacultural Sci. **33**(5), 851–860 (2016)

Leaf Area Index Estimation of Winter Pepper Based on Canopy Spectral Data and Simulated Bands of Satellite

Dan Li, Hao Jiang, Shuisen Chen[(✉)], Chongyang Wang, Siyu Huang, and Wei Liu

Guangdong Open Laboratory of Geospatial Information Technology and Application, Guangdong Key Laboratory of Remote Sensing & GIS Application & Engineer Technology Research Center for Remote Sensing Big Data Application of Guangdong Province, Guangzhou Institute of Geography, Guangzhou 510070, China
css@gdas.ac.cn

Abstract. Leaf area index (LAI) is an important indicator of crop growth status. In this paper, the relationships between canopy reflectance at 400–2500 nm and leaf area index (LAI) in pepper crop were studied. 102 pair of canopy reflectance and LAI of pepper were collected in 2014–2015. Reflectance of canopy were measured in the field over a spectral range of 400–2500 nm. Simultaneously, the LAI were collected by the LAI-2000. Estimation models of LAI were developed based on the whole spectrum range by partial least squares regression (PLSR) and support vector regression (SVR), respectively. Then the field canopy spectra were resampled according to the band response functions of seven satellite sensors. They were the Vegetation and environment monitoring on a new micro-satellite (VENμS), Worldview-2 (WV-2), RapidEye-1 (RE-1), HJ1/CCD1, Sentinel-2, Landsat 8/OLI and GaoFen (GF) 1/WFV1. The values of common used spectral indices were calculated based on the simulated sensor bands, respectively. Prediction models were also developed based on the spectral indices and simulated bands. The results showed that the PLSR model by whole spectrum had the good accuracy of LAI estimation with the $R^2c = 0.726$, RMSEc = 0.462, $R^2cv = 0.635$, RMSEcv = 0.538. For the simulated satellite datasets, the better LAI estimation were obtained by Sentinel-2 and Venμs bands with the R^2cv greater than 0.600 and RMSEcv less than 0.557. The Estimation model by simulated WV-2 bands, and RE-1 bands had the lowest performance with the R^2cv between 0.50 and 0.55, and RMSEcv between 0.600 and 0.623. The inversion results demonstrated the potential of the multispectral remote sensing data to calibrate the LAI estimation model of winter pepper for the precision agriculture application.

Keywords: Canopy reflectance · Winter pepper · LAI · Spectral index · Satellite sensors

---

Work was supported by the National Natural Science Foundation of China (No. 41301401), the Guangdong Natural Science Foundation (No. 2015A030313805) and the Guangdong Science & Technology Plan Foundation (Nos. 2015A030303013 and 2013B020501006).

© Springer Nature Singapore Pte Ltd. 2017
H. Yuan et al. (Eds.): GRMSE 2016, Part I, CCIS 698, pp. 515–526, 2017.
DOI: 10.1007/978-981-10-3966-9_57

1 Introduction

Winter pepper is one widely cultivated crop in winter in Guangdong Province. Guangdong province is an important agricultural bass, where is most abundant in water and heat resources. It is suitable for the pepper planting in winter. The winter pepper harvest in March to May of the coming year. It's just the period of lack of vegetables in Northern China [1].

LAI reflects the biochemical and physiological processes of vegetation. Understanding crop LAI and its dynamics is very important for a wide range of agricultural studies, such as crop growth monitoring and crop yield estimation [2]. Remote sensing techniques are known to be effective but inexpensive tool for crop LAI estimation [3]. The estimation of crop LAI using optical remote sensing data has been based on empirical or semi-empirical statistical relationships formulated between in-situ LAI measurements and the image pixel values, from at-surface spectral or in the form of spectral vegetation indices [4–10]. The indices are designed to enhance the information of vegetation while decreasing the effect of soil background or atmospheric absorption [11, 12].

Previous studies have indicated the successful implementation of moderate/high multispectral optical remote sensing data for crop LAI mapping from various sensors, such as Landsat 8/OLI [13], Landsat 7 ETM+ [14, 15], IKONOS [16–18], SPOT [19], RapidEye [20, 21], Sentinel-2 [22], VENµS [12]. The Chinese HJ1/CCD [23, 24], GF1/WFV [8, 25, 26], and ZY3 mux data [8] were also used in the vegetation LAI estimation. Many of these studies were focused on one type remote sensing data in their research. The increasing number of sensor types for terrestrial remote sensing has necessitated supplementary efforts to evaluate and standardize data from the different available sensors [4]. Some studies adopted two or more remote sensing imageries in LAI estimation, which had different spatial/temporal/spectral resolution [7, 15, 27]. Various studies have documented the differences and consistencies among sensors in LAI inversion. Soudani et al. [4] assessed the potential use of IKONOS, ETM+, and SPOT HRVIR sensors for LAI estimation in forest stands. Masemola et al. [13] compared Landsat 8 OLI and Landsat 7 ETM+ for estimating grassland LAI in South Africa and indicated that Landsat 8 OLI provided an improvement for the estimation of LAI over Landsat 7 ETM+. Li et al. [8] assessed the consistency and applicability of three remote sensing sensors (GF1/WFV, HJ1/CCD, and Landsat 8/OLI) in winter wheat LAI estimation and argued that the effects of spatial resolution are more significant than those of spectral response function and radiation calibration for the different sensors.

In summary, few studies have compared the consistencies of data from Chinese sensor data (HJ1/CCD and GF1/WFV) to determine the LAI for winter pepper. To evaluate the capabilities of these Chinese sensor data, Landsat 8 OLI and other four commercial satellite sensors for monitoring LAI in winter pepper and analyze the main drivers of the differences between different satellites will enhance the reliability of the multi-source remote sensing data for continuous monitoring of the crop LAI.

Thus, the objectives of this paper were to (1) investigate the potential of the reflectance and vegetation indices produced by the GF1/WFV1, HJ1/CCD1, Landsat 8/OLI, VENμS, WV-2, RE-1, and Sentinel-2 for winter pepper LAI monitoring; (2) compare the accuracy of the estimation between the different sensors; and (3) analyze the factors causing differences between the different simulated satellite data.

2   Materials and Methods

2.1   Study Area

The study was conducted in Leizhou Peninsula (located in Zhanjiang City) and Huizhou city, Guangdong Province (Fig. 1). The pepper seedlings were planted in the fields after the second rice harvest. The pepper were harvested from January to April. The total of 102 measured LAI values and reflectance data of winter pepper were collected. The collecting dates and locations were shown in Table 1 and Fig. 1. The range of measured LAI value and the sampling numbers at each winter pepper fields were also presented in Table 1.

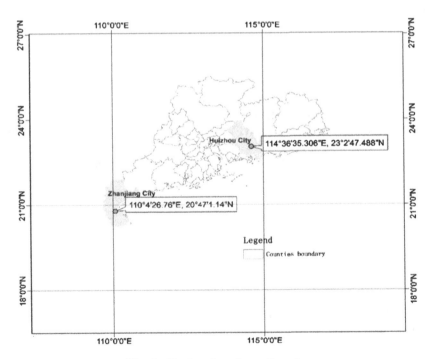

Fig. 1. The location of sampling sites

Table 1. The related information of study area.

Sampling date	Location	Range of measured LAI (m^2/m^2)	Numbers
2015.01.17–2015.01.18	Zhanjiang city	0.72–4.54	41
2015.11.16	Huizhou city	1.06–1.86	16
2015.11.23	Huizhou city	1.11–3.09	17
2015.12.19	Huizhou city	0.89–4.69	12
2015.12.29	Huizhou city	2.05–4.01	16

2.2 Field Measurements

Ground-based observations of effective winter pepper LAI (m^2/m^2) were non-destructive measured by the LAI-2000 plant canopy analyzer (LI-COR Inc., Lincoln, Nebraska, USA).

Canopy reflectance spectra were obtained under clear-sky conditions around mid-day (10:00–15:00 LST) using portable spectra-radiometers (Fieldspec-FR 3, ASD). The spectral range of sensor was 350–2500 nm, with a field of view of 25°. Reflectance measurements were taken at a nadir-looking angle from 2 m above the canopy. More than 5 measurements were made in each observation by moving over each canopy, to derive the representative reflectance spectra for each canopy. Three is 3 observations for each. Spectral reflectance was derived as the ratio of reflected radiance to incident radiance estimated by a calibrated white reflectance (Spectralon, Labsphere).

2.3 Data Preprocessing

The reflectance data at 350–400 nm, 1300–1450 nm, 1800–1900 nm, 2450–2500 nm were removed because of the absorption of water and instrument noise. The spectral response function of seven satellite sensors (VENμS, WV-2, RE-1, HJ1/CCD1, Sentinel-2, Landsat 8/OLI and GF1/WFV1) were downloaded from the corresponding official websites. Then the measured canopy hyperspectral reflectance was used to calculate the broadband reflectance within the visible and near-infrared wavelengths corresponding to seven multispectral sensors. At last some of the following Vis (Table 2) were calculated based on the broadband reflectance values for each satellite sensor.

2.4 Methods

Partial least square regression (PLSR) and Support vector machine regression (SVR) were used in winter pepper LAI calibration and validation. Many researches indicated that PLSR and SVR performed good performance in chemical modeling [43]. Modelling was done in Unscrambler 10.4 (CAMO, Oslo, Norway). Standard goodness-of-fit statistics, such as the coefficient of determination (R^2), and root mean squared error (RMSE) in calibration and cross validation, were used to evaluate the relationships between the measured and estimated LAI values and the estimation performance of the methods and the sensors [44].

Table 2. The spectral indices referenced in this study

Spectral indices	Name	Equations	References
NDVI	Normalized difference vegetation index	$(R_{790} - R_{670})/(R_{790} + R_{670})$	[28]
SR	Simple ratio index	NIR/R	[29]
GNDVI	Green normalized difference vegetation index	$(R_{790} - R_{550})/(R_{790} + R_{550})$	[30]
EVI	Enhanced vegetation index	$2.5((R_{755} - R_{645})/(R_{755} + 6R_{645} + 7.5R_{445} + 1))$	[31]
OSAVI	Optimization of soil adjusted vegetation indices	$(NIR - R)/(NIR + R + 0.16)$	[32]
SIPI	Structural independent pigment index (SIPI)	$(R500 - R445)/(R500 + R650)$	[33]
MTCI	MERIS terrestrial chlorophyll index	$(R750 - R710)/(R710 - R680)$	[34]
AVRI	Atmospherically resistant vegetation index	$ARVI = (NIR - RB)/(NIR + RB)$	[35]
TGI	Triangular greenness index	$-0.5[(675 - 485)(R675 - R555) - (675 - 555)(R675 - R485)]$	[36]
$CI_{red\ edge}$	Chlorophyll index-red edge	$R_{790}/R_{720} - 1$	[37]
CI_{green}	Green chlorophyll index	$R_{790}/R_{550} - 1$	[37]
REP	Red point inflection point (REP)	$708.75 + 45[(R665 + R778.75)/2) - R708.75)/(R753.75 - R708.75)]$	[38]
NDRE	Normalized difference red edge index	$(R_{790} - R_{720})/(R_{790} + R_{720})$	[39]
CCCI	Canopy chlorophyll content index	$(NDRE - NDRE_{min})/(NDRE_{max} - NDRE_{min})$	[40]
NPDI	Nitrogen planar domain index	$(CI_{red\ edge} - CI_{rededgemin})/(CI_{red\ edge\ max} - CI_{red\ edge\ min})$	[41]
ReNDVI	Revised normalized difference vegetation index	$(R_{755} - R_{705})/(R_{755} + R_{705})$	[42]

Finally, calibration models were evaluated based on coefficient of determination (R^2), and root mean square error (RMSE) as discussed later in the results part of this paper. And, in order to enhance the useful information of the raw spectra prior to model calibration, we applied a 10 band averaging to preprocess the remaining data.

3   Results

3.1   Winter Pepper LAI Estimation Using Field Canopy Reflectance

The performance of PLSR model and SVR model for winter pepper LAI estimation by the whole canopy reflectance spectra were listed in Table 3. The R^2 values in

Table 3. The winter pepper LAI estimation result by the whole canopy reflectance spectra

Calibration methods	Data source			
	Whole spectrum			
	R^2c	RMSEc	R^2cv	RMSEcv
PLSR	**0.726**	**0.462**	**0.635**	**0.538**
SVR	0.588	0.578	0.422	0.378

calibration and cross validation were 0.726 and 0.637 for PLSR model, with the RMSE in calibration of 0.462 and RMSE in cross validation of 0.538. The R^2 values in calibration and cross validation were 0.588 and 0.422 for SVR model with the RMSE in calibration of 0.578 and RMSE in cross validation of 0.378. From the calibration results and cross validation results developed by the whole canopy reflectance spectra, we found that the PLSR model performed better than SVR in winter pepper LAI estimation. And the winter pepper LAI estimation model developed by PLSR can be used to quantity the variation of winter pepper LAI.

3.2   Winter Pepper LAI Estimation Using Simulated Satellite Data

We calculated the NDVI, GNDVI, SR, EVI, OSAVI, SIPI, MTCI, and AVRI using the simulated HJ1/CCD1 data. Based on the simulated WV-2 data, the NDVI, SR, EVI, OSAVI, MTCI, SIPI, AVRI, and TGI were obtained. For the simulated RE-1, we calculated 10 spectral indices (NDVI, RNDVI, EVI, OSAVI, MTCI, SIPI, TGI, CI_{green}, $CI_{red\ edge}$, and NPDI). Based on the simulated GF1/WFV1 data and Landsat 8/OLI data, we calculated 8 spectral indices (NDVI, GNDVI, SR, EVI, OSAVI, SIPI, MTCI, and AVRI), respectively.

Based on the simulated VENus data, the SR, NDVI, RNDVI, NVI, EVI, OSAVI, TVI, MTCI, SIPI, REP, TGI, CI_{green}, $CI_{red\ edge}$, NDRE, CCCI, and NDPI were calculated. Based on the simulated Sentinel-2 data, we calculated NDVI, SR, RNDVI, EVI, OSAVI, MTCI, SIPI, REP, TGI, CI_{green}, $CI_{red\ edge}$, and NDRE.

The performance of the PLSR and SVR developed using the simulated sensor bands and the calculated spectral indices were shown in Table 4. For simulated WV-2 bands, the PLSR model performed better than SVR model with the higher R^2 and lower RMSE in calibration and cross validation. Both models can be used to quantify the variation of winter pepper LAI $(0.50 < R^2_{cv} < 0.80)$.

For the simulated Sentinel-2 bands, PLSR model performed better than SVR model in winter pepper LAI estimation. The PLSR model can be used to explain the change of winter pepper LAI with the R^2 of 0.713 and RMSE of 0.472 in calibration and R^2 of 0.634 and RMSE of 0.544 in cross validation. The results showed a good relationship between the predicted and actual values.

According to the simulated Venµs sensor data, the SVR model presented the better performance than that of PLSR with the R^2 of 0.729 and RMSE of 0.473 in calibration and R^2 of 0.610 and RMSE of 0.557 in cross validation.

For simulated HJ 1A/CCD1 data, the correlation between the measured and estimated LAI values in PLSR model is higher than those of SVR, with an R^2 value of

Table 4. The performance of winter pepper LAI estimation model by the simulated satellite wavebands

Calibration methods	Data source			
	R^2c	RMSEc	R^2cv	RMSEcv
Simulated WV-2 bands				
PLSR	**0.645**	**0.526**	**0.546**	**0.600**
SVR	0.644	0.528	0.535	0.603
Simulated Sentinel-2 bands				
PLSR	**0.713**	**0.472**	**0.634**	**0.544**
SVR	0.667	0.511	0.526	0.608
Simulated Venµs bands				
PLSR	0.679	0.501	0.585	0.585
SVR	**0.729**	**0.473**	**0.610**	**0.557**
Simulated HJ1A/CCD1 bands				
PLSR	**0.631**	**0.536**	**0.592**	**0.583**
SVR	0.645	0.530	0.531	0.606
Simulated GF1/WFV1 bands				
PLSR	**0.602**	**0.556**	**0.567**	**0.592**
SVR	0.628	0.543	0.524	0.611
Simulated Landsat8/OLI bands				
PLSR	**0.631**	**0.535**	**0.571**	**0.596**
SVR	0.632	0.538	0.485	0.638
Simulated RE-1 bands				
PLSR	**0.580**	**0.571**	**0.512**	**0.623**
SVR	0.549	0.604	0.435	0.672

0.631 and RMSE of 0.536 in calibration and an R^2 of 0.592 and RMSE of 0.583 in cross validation. There are no significant difference in PLSR model and SVR model for the retrieved LAI based on the simulated GF1/WFV1 data $(0.50 < R^2_{cv} < 0.65)$. For the simulated Landsat 8/OLI data and simulated RE-1 data, the performance of PLSR models are significant better than those of SVR models, because the former model had the R^2 larger than 0.50 in calibration and cross validation.

From the results of Table 3, the PLSR model performed better in most data set except the simulated Venµs bands. And the best winter pepper LAI retrieval model is PLSR model developed by the simulated sentinel-2 data. This model presented the similar performance to that developed by the whole spectrum and PLSR. Generally speaking, the correlation between measured LAI and the simulated sentinel-2 data and HJ 1A/CCD1 data are relatively higher than those by other simulated data.

4   Discussion

We found that the spectral indices calculated by the simulated HJ1/CCD1, GF1/WFV1, RE-1, and Landsat 8/OLI data are mainly used the information in blue, green, red, and near infrared bands. The estimation model developed by the information of these bands can explain the 40% variation of winter pepper LAI in canopy. For the simulated WV-2 dataset, the information of one red edge band was used to calculate the corresponding $CI_{red\ edge}$ index. Comparing the models by the simulated HJ1/CCD1, GF1/WFV1, RE-1, and Landsat 8/OLI data, we found that by the adding of one red-edge band and the corresponding spectral index, the performance of PLSR and SVR model are improved.

For Sentinel-2 and Venμs, both two sensor includes four bands along the red-edge. The adding of four bands along the red-edge and the corresponding spectral indices improved the accuracy of PLSR and SVR LAI estimation models. Because the two sensors have more information of red-edge than those of the other five sensors. For example, in the WV-2 system, a single red-edge band is available but it is not suitable for retrieving the REP. The former satellites are more suitable for precision agriculture tasks. The latter satellites are characterized by a small number of broad spectral bands. While, for other four sensors, there were no bands along the red edge.

The sun-view [8], temporal variation of measured dataset [4], model methods [45], spectral response function, and spatial resolution are the main factors which have the influence on the accuracy of LAI estimation. The averaged NDVI values of different satellite sensors are shown in Fig. 2. We can see that the averaged NDVI values calculated by simulated GF1/WFV1 data was lowest among the seven datasets. The Venμs has the highest averaged NDVI values, which is also similar to those of Sentinel-2, RE-1, WV-2 and Landsat8/OLI and is significantly higher than that calculated by the simulated GF1/WFV1 and HJ-1/CCD1 data. It indicates that there are difference among the seven spectral response functions. In Table 5, the simulated NDVI values of the seven sensors are highly correlated, with R^2 values all greater than 0.97 (at the 0.01 significant level). Consequently, the spectral response function does influence the band reflectance and Vis, but the overall effect is very small and can be ignored during practical applications. This conclusion is also consistent with the studies of Wang et al. [25] and Li et al. [8].

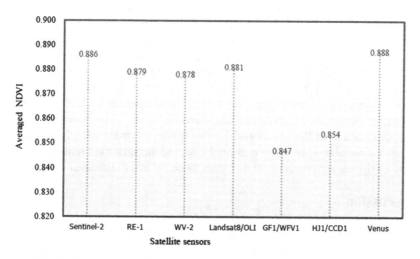

Fig. 2. The variation of averaged NDVI values for seven satellite sensors

Table 5. The correlation between the NDVI values of each simulated dataset and LAI

Dataset	Dataset							
	LAI	Sentinel-2	RE-1	WV-2	Landsat8/OLI	GF1/WFV1	HJ1/CCD1	Venµs
LAI	1							
Sentinel-2	0.451**	1						
RE-1	0.454**	0.997**	1					
WV-2	0.456**	0.995**	1.000**	1				
Landsat8/OLI	0.452**	0.995**	0.999**	1.000**	1			
GF1/WFV1	0.454**	0.978**	0.991**	0.994**	0.994**	1		
HJ1/CCD1	0.454**	0.985**	0.995**	0.997**	0.997**	0.999**	1	
Venµs	0.447**	0.998**	0.992**	0.990**	0.990**	0.970**	0.978**	1

(**Significant at $p < 0.01$; two-tailed)

In this study, we calculated the possible vegetation indices based on the each simulated dataset. Many studies indicated that SVR performed better than PLSR, because SVR are able to handle large input spaces and to deal with noisy patterns and the optimization. While in our study, for most models, PLSR performed better than SVR. Vohland et al. [46] indicated that the performance of multivariate calibration methods was certainly affected by the existence of noise. The variable selection methods may provide more robust models. Thus, the processing of calculation of spectral indices may improve the robustness of the PLSR model, which may can explain the better performance of PLSR than that of SVR in most models. This finding coincides with the results of some other studies. Li et al. [45] also found that the PLSR method had a relatively higher correlation between predicted and actual values than SVR method. Axelsson et al. [43] argued that the PLSR method yielded the highest accuracy when using all bands and was more efficient at suppressing noise in the mangrove foliar nitrogen map.

Meanwhile, growth stage and cultivated varieties are also the assignable causes which have influence on estimation accuracy [24].

5 Conclusions

This study focused on comparing the potential of seven satellite sensors in LAI estimation of winter pepper by simulated data. We further analyzed the possible reasons which influenced the estimation accuracy. We arrived at the following conclusions.

(1) The LAI estimation of winter pepper was developed by PLSR using the simulated Sentinel-2 bands, producing the best performance with the R^2 of 0.634 and RMSE of 0.544. The inversion results demonstrated the potential of the multispectral remote sensing data to calibrate the LAI estimation model of winter pepper for the precision agriculture application.

(2) After comparing the averaged NDVI values calculated by the simulated data and the correlation between LAI and NDVI values, we found that there was strong correlation between the averaged NDVI valued obtained of the simulated dataset

(correlation coefficient >0.97 at the 0.01 significant level). There are no significant difference of NDVI values among seven simulated sensors.

(3) From the performance of different simulated data sets, we found that more red edge bands information can improve the accuracy of LAI estimation model to some extent. Sentinel-2 and Venμs data have more details in red edge region compared with the other satellite data. In our study, the calibration model developed by the simulated data showed their better results than the other simulated datasets.

(4) The calibration methods have the effect on the accuracy of LAI estimation model. The properly processing of spectral data is necessary to improve the robustness of calibration methods.

References

1. Wang, H., Hu, F., Huang, J.: Climatic suitability and regionalization of pepper planted in winter in Guangdong based on GIS. Meteorol. Environ. Sci. **37**(3), 76–80 (2014)
2. Fang, H., Liang, S., Hoogenboom, G.: Integration of MODIS LAI and vegetation index products with the CSM-CERES-Maize model for corn yield estimation. Int. J. Remote Sens. **32**(4), 1039–1065 (2011)
3. Liu, K., Zhou, Q., Wu, W., Xia, T., Tang, H.: Estimating the crop leaf area index using hyperspectral remote sensing. J. Integr. Agric. **15**(2), 475–491 (2016)
4. Soudani, K., François, C., Maire, G.L., Dantec, V.L., Dufrêne, E.: Comparative analysis of IKONOS, SPOT, and ETM+ data for leaf area index estimation in temperate coniferous and deciduous forest stands. Remote Sens. Environ. **102**(1–2), 161–175 (2006)
5. Kobayashi, H., Suzuki, R., Kobayashi, S.: Reflectance seasonality and its relation to the canopy leaf area index in an eastern Siberian larch forest: multi-satellite data and radiative transfer analyses. Remote Sens. Environ. **106**(2), 238–252 (2007)
6. Liu, Q., Liang, S., Xiao, Z., Fang, H.: Retrieval of leaf area index using temporal, spectral, and angular information from multiple satellite data. Remote Sens. Environ. **145**(8), 25–37 (2014)
7. He, Y., Bo, Y., Chai, L., Liu, X., Li, A.: Linking in situ LAI and fine resolution remote sensing data to map reference LAI over cropland and grassland using geostatistical regression method. Int. J. Appl. Earth Obs. Geoinf. **50**, 26–38 (2016)
8. Li, H., Chen, Z., Jiang, Z., Wu, W., Ren, J., Liu, B., Hasi, T.: Comparative analysis of GF-1, HJ-1, and Landsat-8 data for estimating the leaf area index of winter wheat. J. Integr. Agric. **16**, 266–285 (2016). doi:10.1016/S2095-3119(15)61293-X
9. Wu, M., Wu, C., Huang, W., Niu, Z., Wang, C.: High-resolution Leaf Area Index estimation from synthetic Landsat data generated by a spatial and temporal data fusion model. Comput. Electron. Agric. **115**, 1–11 (2015)
10. Viña, A., Gitelson, A.A., Nguy-Robertson, A.L., Peng, Y.: Comparison of different vegetation indices for the remote assessment of green leaf area index of crops. Remote Sens. Environ. **115**(12), 3468–3478 (2011)
11. Nguy-Robertson, A.L., Peng, Y., Gitelson, A.A., Arkebauer, T.J., Pimstein, A., Herrmann, I., Karnieli, A., Rundquist, D.C., Bonfil, D.J.: Estimating green LAI in four crops: potential of determining optimal spectral bands for a universal algorithm. Agric. Forest Meteorol. **192–193**, 140–148 (2014)

12. Herrmann, I., Pimstein, A., Karnieli, A., Cohen, Y., Alchanatis, V., Bonfil, D.J.: LAI assessment of wheat and potato crops by VENμS and Sentinel-2 bands. Remote Sens. Environ. **115**(8), 2141–2151 (2011)

13. Masemola, C., Cho, M.A., Ramoelo, A.: Comparison of Landsat 8 OLI and Landsat 7 ETM + for estimating grassland LAI using model inversion and spectral indices: case study of Mpumalanga, South Africa. Int. J. Remote Sens. **37**(18), 4401–4419 (2016)

14. Ganguly, S., Nemani, R.R., Zhang, G., Hashimoto, H., Milesi, C., Michaelis, A., Wang, W., Votava, P., Samanta, A., Melton, F.: Generating global Leaf Area Index from Landsat: algorithm formulation and demonstration. Remote Sens. Environ. **122**(1), 185–202 (2012)

15. Szporak-Wasilewska, S., Krettek, O., Berezowski, T., Ejdys, B., Łukasz, S., Borowski, M., Będkowski, K., Chormański, J.: Leaf area index of forests using ALS, Landsat and ground measurements in Magura National Park (Se Poland). EARSele Proc. **13**(2), 103–111

16. Kovacs, J.M., Wang, J., Flores-Verdugo, F.: Mapping mangrove leaf area index at the species level using IKONOS and LAI-2000 sensors for the Agua Brava Lagoon, Mexican Pacific. Estuar. Coast. Shelf Sci. **62**(1–2), 377–384 (2015)

17. Colombo, R., Bellingeri, D., Fasolini, D., Marino, C.M.: Retrieval of leaf area index in different vegetation types using high resolution satellite data. Remote Sens. Environ. **86**(1), 120–131 (2003)

18. Gu, Z., Ju, W., Liu, Y., Li, D., Fan, W.: Forest leaf area index estimated from tonal and spatial indicators based on IKONOS_2 imagery. Int. J. Remote Sens. Appl. **3**(4), 175–184 (2013)

19. Aboelghar, M., Arafat, S., Saleh, A., Naeem, S., Shirbeny, M., Belal, A.: Retrieving leaf area index from SPOT4 satellite data. Egypt. J/ Remote Sens. Sci. **13**(2), 121–127 (2010)

20. Kross, A., McNairn, H., Lapen, D., Sunohara, M., Champagne, C.: Assessment of RapidEye vegetation indices for estimation of leaf area index and biomass in corn and soybean crops. Int. J. Appl. Earth Obs. Geoinf. **34**(1), 235–248 (2015)

21. Beckschaefer, P., Fehrmann, L., Harrison, R.D., Xu, J., Kleinn, C.: Mapping leaf area index in subtropical upland ecosystems using RapidEye imagery and the RandomForest algorithm. iForest – Biogeosci. Forest. **7**(1), 1–11 (2013)

22. Richter, K., Hank, T.B., Vuolo, F., Mauser, W., D'Urso, G.: Optimal exploitation of the sentinel-2 spectral capabilities for crop leaf area index mapping. Remote Sens. **4**(3), 561–582 (2012)

23. Chen, B., Wu, Z., Wang, J., Dong, J., Guan, L., Chen, J., Yang, K., Xie, G.: Spatio-temporal prediction of leaf area index of rubber plantation using HJ-1A/1B CCD images and recurrent neural network. ISPRS J. Photogramm. Remote Sens. **102**(11), 148–160 (2015)

24. Li, X., Zhang, Y., Luo, J., Jin, X., Xu, Y., Yang, W.: Quantification winter wheat LAI with HJ-1CCD image features over multiple growing seasons. Int. J. Appl. Earth Obs. Geoinf. **44**, 104–112 (2016)

25. Wang, L., Yang, R., Tian, Q., Yang, Y., Zhou, Y., Sun, Y., Mi, X.: Comparative analysis of GF-1 WFV, ZY-3 MUX, and HJ-1 CCD sensor data for grassland monitoring applications. Remote Sens. **7**(2), 2089–2108 (2015)

26. Gao, L., Li, C., Wang, B., Yang, G., Wang, L., Fu, K.: Comparison of precision in retrieving soybean leaf area index based on multi-source remote sensing data. Chin. J. Appl. Ecol. **27**(1), 191–200 (2016)

27. Claverie, M., Demarez, V., Duchemin, B., Hagolle, O., Ducrot, D., Marais-Sicre, C., Dejoux, J.F., Huc, M., Keravec, P., Béziat, P.: Maize and sunflower biomass estimation in southwest France using high spatial and temporal resolution remote sensing data. Remote Sens. Environ. **124**(6), 844–857 (2012)

28. Rouse, J.W., Haas, R.W., Schell, J.A., Deering, D.W., Harlan, J.C.: Monitoring the vernal advancement and retrogradation (Greenwave effect) of natural vegetation. NASA/GSFCT Type III Final report. Nasa (1974)

29. Gonsamo, A.: Normalized sensitivity measures for leaf area index estimation using three-band spectral vegetation indices. Int. J. Remote Sens. **32**(7), 2069–2080 (2011)

30. Gitelson, A.A., Merzlyak, M.N.: Signature analysis of leaf reflectance spectra: algorithm development for remote sensing of chlorophyll. J. Plant Physiol. **148**(3–4), 494–500 (1996)

31. Jiang, Z., Huete, A.R., Didan, K., Miura, T.: Development of a two-band enhanced vegetation index without a blue band. Remote Sens. Environ. **112**(10), 3833–3845 (2008)

32. Rondeaux, G., Steven, M., Baret, F.: Optimization of soil-adjusted vegetation indices. Remote Sens. Environ. **55**(2), 95–107 (1996)

33. Peñuelas, J., Filella, I., Elvira, S., Inclán, R.: Reflectance assessment of summer ozone fumigated Mediterranean white pine seedlings. Environ. Exp. Bot. **35**(95), 299–307 (1995)

34. Dash, J., Curran, P.J.: Evaluation of the MERIS terrestrial chlorophyll index. Adv. Space Res. **39**, 100–104 (2004)

35. Kaufman, Y.J., Tanré, D.: Atmospherically resistant vegetation index (ARVI) for EOS-MODIS. IEEE Trans. Geosci. Remote Sens. **30**(2), 261–270 (1992)

36. Hunt, E.R., Doraiswamy, P.C., McMurtrey, J.E., Daughtry, C.S.T., Perry, E.M., Akhmedov, B.: A visible band index for remote sensing leaf chlorophyll content at the canopy scale. Int. J. Appl. Earth Obs. Geoinf. **21**(1), 103–112 (2013)

37. Gitelson, A.A., Andrés, V., Verónica, C., Rundquist, D.C., Arkebauer, T.J.: Remote estimation of canopy chlorophyll content in crops. Geophys. Res. Lett. **32**(8), 93–114 (2005)

38. Clevers, J.G.P.W., Jong, S.M.D., Epema, G.F., Meer, F.V.D., Bakker, W.H., Skidmore, A.K., Addink, E.A.: MERIS and the red-edge position. Int. J. Appl. Earth Obs. Geoinf. **3**(4), 313–320 (2001)

39. Fitzgerald, G.J., Rodriguez, D., Christensen, L.K., Belford, R., Sadras, V.O., Clarke, T.R.: Spectral and thermal sensing for nitrogen and water status in rainfed and irrigated wheat environments. Precision Agric. **7**(4), 233–248 (2006)

40. Fitzgerald, G., Rodriguez, D., O'Leary, G.: Measuring and predicting canopy nitrogen nutrition in wheat using a spectral index—the canopy chlorophyll content index (CCCI). Field Crops Res. **116**(3), 318–324 (2010)

41. Li, F., Mistele, B., Hu, Y., Yue, X., Yue, S., Miao, Y., Chen, X., Cui, Z., Meng, Q., Schmidhalter, U.: Remotely estimating aerial N status of phenologically differing winter wheat cultivars grown in contrasting climatic and geographic zones in China and Germany. Field Crops Res. **138**(3), 21–32 (2012)

42. Sims, D.A., Gamon, J.A.: Relationships between leaf pigment content and spectral reflectance across a wide range of species, leaf structures and developmental stages. Remote Sens. Environ. **81**(2–3), 337–354 (2002)

43. Axelsson, C., Skidmore, A., Schlerf, M., Fauzi, A., Verhoef, W.: Hyperspectral analysis of mangrove foliar chemistry using PLSR and support vector regression. Int. J. Remote Sens. **34**(5), 1724–1743 (2013)

44. Rossel, R.A.V., McGlynn, R.N., McBratney, A.B.: Determining the composition of mineral-organic mixes using UV–vis–NIR diffuse reflectance spectroscopy. Geoderma **137** (1–2), 70–82 (2006)

45. Zh, L., Ch, N., Ch, W., Xu, X., Song, X., Wang, J.: Comparison of four chemometric techniques for estimating leaf nitrogen concentrations in winter wheat (TriticumAestivum) based on hyperspectral features. J. Appl. Spectrosc. **83**(2), 1–8 (2016)

46. Vohland, M., Besold, J., Hill, J., Fründ, H.C.: Comparing different multivariate calibration methods for the determination of soil organic carbon pools with visible to near infrared spectroscopy. Geoderma **166**(1), 198–205 (2011)

Geoinformatics in Mapping of Fog-Affected Areas over Northern India and Development of Ion Based Fog Dispersion Technique

Arun K. Saraf[1]([⊠]), Palash Choudhury[1], Josodhir Das[2],
Gaurav Singh[1], Susanta Borgohain[1], Suman Saurav Baral[1],
and Kanika Sharma[1]

[1] Department of Earth Sciences, Indian Institute of Technology Roorkee,
Roorkee, India
arun.k.saraf@gmail.com
[2] Department of Earthquake Engineering, Indian Institute of Technology
Roorkee, Roorkee, India

Abstract. Fog is a phenomenon that affects the Indo-Gangetic Plains every year during winter season (December–January). This fog is sometimes in the form of radiation fog and other also occurs as a mixture with other gases, known as smog (smoke + fog). There are various factors contributing to the formation of fog, that may be either meteorological, topographical or resulting from pollution. Fog has been mapped for the winter seasons of the years 2002–2016. In these winter seasons, fog affected areas were found to be changing significantly. The net cover of fog during a season varies in space, time intensity and frequency of occurrence. Presently, it is now possible to map and to predict fog formation to some extent. However, so far it has not been possible to disperse fog, though theoretically it has been discussed in literature. In the current work, experiments were conducted to find out the possibility and effectiveness of a negative air ionizer for fog dispersion. The experiments were carried out with fog, *dhoop* smoke and a mixture of both to generate smog. Two different glass chambers of different sizes were used in a closed room and the impact of air ionizer on dispersion was studied by testing the time taken for dispersion with or without the ionizer. The results show a significant performance with air ionizer indicating the effectiveness of the ion generator, which reduced the time taken for dispersion (in comparison to without ionizer) by about half.

1 Introduction

Recently, Geoinformatics technique has become one of the foremost techniques in the study of processes and phenomenon that require continuous monitoring over a long period of time. As such, this technique has become one of the core techniques used in analyzing the atmospheric and surface processes. One such peculiar phenomenon that has been plaguing the residents of Northern India, is fog. Fog at times, can be gentle (intensity wise), reducing visibility minutely, while at times can be so dense that even people standing close by are not visible. Sometimes, natural fog can also be replaced by smog (combination of smoke with fog). The problems in visibility can affect a plethora

© Springer Nature Singapore Pte Ltd. 2017
H. Yuan et al. (Eds.): GRMSE 2016, Part I, CCIS 698, pp. 527–534, 2017.
DOI: 10.1007/978-981-10-3966-9_58

of transportation related activities, including road traffic, railways, aviation, naval activities, agriculture, health and culture.

In Northern India, the majority of the region lies in the Indo-Gangetic Plain, which is the worst hit region by the smog in the winter seasons (which are at its worst in Dec.– Jan.). The Indo-Gangetic Plain is a vast stretch of land in South Asia, spread over parts of Pakistan, India and Bangladesh and houses a large portion of the population of South Asia. This forms a trough region, wherein cold air drainage flow from higher plateaus gets collected leading to the enhancement of relative humidity. The two most common types of fog are radiation fog, where air near the ground is cooled at night when there are no clouds, and little wind, and advection fog, when warmer air is cooled by snow or water. The type of fog in the Northern India is radiative fog. The atmosphere is cloudy under a low pressure zone. Thus, a series of foggy and non-foggy periods or episodes are observed (Dutta et al. 2015). Meteorologically, fog is simply a low stratus cloud at or slightly above the surface of the earth (Anthis and Cracknell 1999). Colder air being denser will stay close to the surface of earth while the warmer, less dense air will lie above and hence, overrun the colder air. The air needs to be cooled to sufficiently lower temperatures (preferably below dew point).

Fog has been a major hazard for the civilization since time immemorial. The reduction in visibility hinders progress throughout the world. Almost all types of

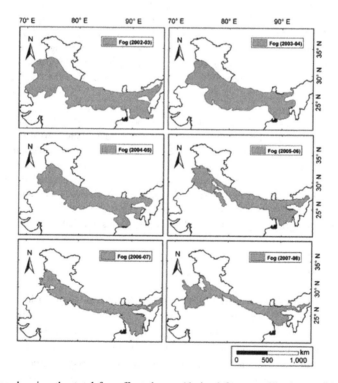

Fig. 1. Maps showing the total fog-affected area (derived from satellite images) in the winter seasons 2002–03 to 2007–08 and also depicting decreasing trend in the fog-affected area during study period (2002–08) (Saraf et al. 2011).

transportation e.g. road, rail and air transportation, agriculture, naval activities etc. get affected. In India, the major brunt is borne by the residents of Northern India, who generally face smog. In recent years, Choudhury et al. (2007) and Saraf et al. (2011) has successfully mapped (2002–2010) (Fig. 1) over Indo-Gangetic Plain. Recently, Chakarvorty et al. (2014) has also done physical characterisation of winter fog of Northern India.

There have been few experiments by NASA (Christensen and Frost 1980; Frost et al. 1981) to find out ways for dispersing fogs in region of United States. Some of the prominent techniques of fog dispersal include use of heated air, electrical discharges, liquids or liquefied gas of greater density, supersonic waves, drying by refrigerating and outdoor air conditioning apparatus and spraying calcium chloride in air (Colbeck 2011). However, the types of fog vary from place to place and as such the effectiveness of any particular method of fog dispersal varies depending on the characteristics of the fog in the region.

2 Data and Methodology

2.1 Mapping

The current work utilizes the data obtained from Moderate-Resolution Imaging Spectro-radiometer (MODIS 2016) which is a key scientific instrument currently onboard Terra and Aqua (these satellites launched by NASA in 1999 and 2002). Among the different bands in the MODIS data, bands 1, 3 and 4, of spectral resolution 500 m, lie in the visible region and correspond to the red, blue and green regions, respectively and are suitable for fog detection. These bands essentially record the surface reflectivity of a region for the respective color and whites in a region almost always correspond to clouds. The clouds at any location can be differentiated from fog by looking at the structure and color intensity of white at that location (Bendix et al. 2015).

The areas depicting the fog have been carefully mapped using polygons and shapefiles have been created for each of them. The current work aims at mapping fog for winter seasons in 2010–11, 2011–12, 2012–13, 2013–14 and 2014–15 (Fig. 2). Earlier work done for the winter fog seasons of 2002–10 (Saraf et al. 2011) using NOAA-AVHRR datasets have been inculcated with the current work. Based on the fog maps created, it will be possible to look at the intensity and frequency of fogs in a region, helping in improving predictive model for the fog forecasting.

The data for the current work has been obtained from NASA 2016 EOSDIS website (https://earthdata.nasa.gov/). As a reference for political boundaries, reference map downloaded from GeoCommunity website (http://data.geocomm.com/) has been used.

Using ArcGIS, regions showing presence of fog were marked. Using channels belonging to infrared, the regions showing higher temperatures than cloud were demarcated and were confirmed using the RGB (red, green and blue) channels. Shapefiles were generated for the fog affected zones, which were used in further analysis.

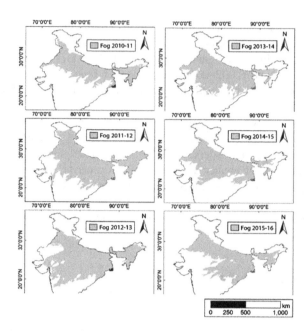

Fig. 2. Maps showing the total fog-affected area (derived from MODIS images) in the winter seasons 2010–11 to 2015–16.

2.2 Dispersion

Atmospheric fog is essentially submicron sized inactivated particles associated with activated droplets of size range up to 10 μm. (Sawant et al. 2012). Given particle size, collision efficiency is proportional to the relative velocity between droplet and particle, while inversely proportional to droplet diameter. Collection efficiency increases if relative humidity decreases for smaller particles, for larger particles – no dependence. Droplet lifetime – effective contact time between droplet and particle depends on temperature and relative humidity. Best collection efficiency is observed when radius of droplets is small enough for adequate spray rate and contact time, but large enough to not evaporate (Mathai 1983). Particle removing mechanism by negative air ions (NAI) is due to particle charging by emitted ions and electro-migration which increases migration velocity of particles (Lee et al. 2004; Mayya et al. 2004; Wu et al. 2005). Too high NAI concentrations will create an electrostatic shield on wall surfaces, prohibiting deposition of charged particles, especially when surface materials have low-level conductivity.

However, in practice, only a part of the ions produced from the ionizer are likely to be available for uniform mixing; the remaining part is likely to get conducted away into the earth through short paths between the ionizer and the nearest point on the ground. Two cases may be considered namely; (a) without a continuous aerosol particle source (source-free removal); and (b) with a steady particle source. Former case is more relevant to chamber experiments wherein one examines the temporal reduction of particles, starting with the injection of an initial concentration. The latter case is applicable to indoor environments wherein a steady particle source of strength (particles cm^{-3} s^{-1}) exists, injecting particles continuously into the room (Mayya et al. 2004). Ejected negative ions migrate in airspace and attach to the aerosol particles, which as a result, experience an enhanced drift velocity towards the walls due to electric fields generated by the space charge. Entire room volume and enclosing surface will participate in the removal of particles.

Harrison (1997) described the aerosol removal process in three steps:

1. Modification of Brownian aerosol coagulation for charged aerosol leading to aerosol growth and subsequent sedimentation
2. Attraction of charged aerosol to charged (or conducting) surfaces, leading to enhanced deposition
3. Electrical repulsion of negatively charged particles by excess of negative space charge, leading to dispersal of aerosol to distant surfaces.
4. In closed chamber conditions, excess negative ions attached to aerosols, settle on surface of chamber.

3   Discussion and Conclusions

The study of maps generated for various fog seasons indicates an erratic pattern in the net area of fog observed (Figs. 2 and 3). There has been no clear trend of either increase or decrease in the total region covered by fog. This indicates that the phenomenon of global warming (during study period 2002–16) doesn't have a significant impact on

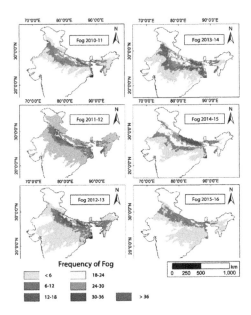

Fig. 3. Maps showing the frequency of the total fog-affected area (derived from MODIS images) in the winter seasons 2010–11 to 2015–16.

formation trends of fog, because otherwise, a drastic increase in temperature due to global warming would result in a drastic change in the fog area. Formation of fog simply depends on weather conditions like rainfall, temperature, humidity, wind speed, etc. Among the winter seasons for which fog was mapped, 2012–13 was affected by fog the most in terms of net area, while, 2013–14 was affected the most in terms of frequency of fog in a particular region, while 2015–16 was least affected (Fig. 3).

Adjudicating by the frequency of occurrence of fog, it can be said that even though the net area covered by fog is vast, yet not whole of it observes fog for prolonged periods. The frequency maps further confirm the fact that the area between longitudes

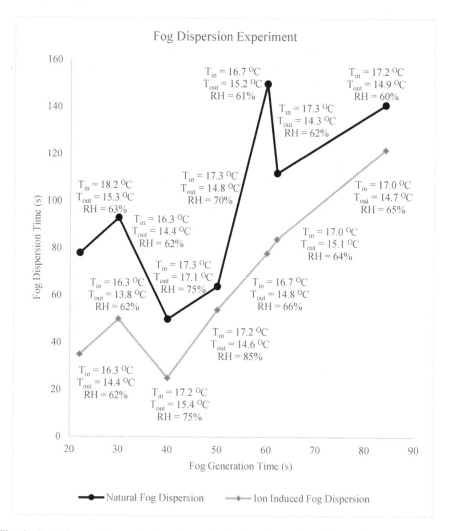

Fig. 4. Fog Dispersal Times in natural as well as ionizer induced conditions. Tin = Temperature recorded inside glass chamber, Tout = Temperature recorded outside glass chamber, RH = Relative Humidity inside glass chamber.

25°N and 30°N and latitudes 80°E and 90°E are most prone to fog. This represents the zone adjacent to foothills of Himalaya and is relatively closer to the Bay of Bengal. The tendency of occurrence of fog increases as one moves towards the north-east and reduces towards south-west. Possible reason for this could be concentration of rain due to blocking by Himalayas providing vast amounts of humidity or moisture.

The ion generator used in the current experiment, generates ions with ion density of over 10^6 ions per cc. When the ion generator is turned off, the visibility in the glass chamber increased gradually, indicating a very slow dispersion of fog and smog. Without the ion generator, the dispersion occurs mostly due to gravitational settling. However, when the negative air ionizer was turned on, the visibility in both glass chambers increased considerably quickly (Fig. 4). This clearly indicates the effectiveness of the ion generator, which reduced the time taken for dispersion by about half.

The experiments were carried on till the visibility increased to about 95% of the initial value. However, the rate of increase in visibility and hence the dispersion of fog was not uniform throughout the experiments. The increase in visibility was more rapid towards the beginning of the experiment than towards the end. This variation in rate of dispersion was most likely due to excess of particulate matter in the beginning compared to towards the end and hence, due to increased repulsion between like charged ions.

Experiments with smog were also carried out using *dhoop* smoke as the source of smoke. This also revealed a similarly improved performance when the ion generator was turned on, than when it was not and the conditions of dispersion were natural. The experiment with smoke was no different and it showed an even better performance with ionizer (Fig. 4). Overall, it was observed that dispersion performance of ionizer with smoke > with smog > with fog. This could have immense benefits, especially in the fields of transportation, as the inputs for the process are economically feasible, have a lower impact on environment, are easy to implement and reduce the intensity of fog in a very short time.

References

Anthis, A.I., Cracknell, A.P.: Use of satellite images for fog detection (AVHRR) and forecast of fog dissipation (METEOSAT) over lowland Thessalia, Hellas. Int. J. Remote Sens. **20**(6), 1107–1124 (1999)

Bendix, J., Thies, B., Cermak, J.: Fog detection with Terra-MODIS and MSG-SEVIRI (2015). CiteSeerX website: http://citeseerx.ist.psu.edu/viewdoc/download?doi=10.1.1.5.7277&rep=rep1&type=pdf. Accessed 13 May 2016

Chakarvorty, M., Pati, J.K., Patil, S.K., Shukla, S., Niyogi, A., Saraf, A.K.: Physical characterization, magnetic measurements, REE geochemistry and biomonitoring of dust load accumulated during a protracted winter fog period and their implications. Environ. Monit. Assess. **186**, 2965–2978 (2014). doi:10.1007/s10661-013-3594-4

Choudhury, S., Rajpal, H., Saraf, A.K.: Mapping and forecasting of North Indian winter fog: an application of spatial technologies. Int. J. Remote Sens. **28**(16), 3649–3663 (2007)

Christensen, L.S., Frost, W.: Fog dispersion, NASA Contractor Report 3225, March 1980

Colbeck, I.: The development of FIDO. In: Aerosol Science and Technology: History and Reviews (2011). Chap. 13. doi:10.3768/rtipress.2011.bk.0003.1109

Dutta, H.N., Singh, B., Kaushik, A.: Characterizing Atmospheric Fog over Northern India (2015). http://www.ursi.org/proceedings/procga05/pdf/AP.4(0804).pdf. Accessed 13 May 2016

Frost, W., Collins, G., Koepf, D.: Charged particle concepts for fog dispersion, NASA Contractor Report 3440, June 1981

Harrison, R.G.: Ionisers and electrical aerosol removal. J. Aerosol Sci. **28**, 333–334 (1997)

Lee, B.U., Yermakov, M., Grinshpun, S.A.: Removal of fine and ultrafine particles from indoor air environments by the unipolar ion emission. Atmos. Environ. **38**, 4815–4823 (2004)

Mathai, C.V.: Charged fog technology part I: theoretical background and instrumentation development. J. Air Pollut. Control Assoc. **33**(7), 664–669 (1983)

Mayya, Y.S., Sapra, B.K., Khan, A., Sunny, F.: Aerosol removal by unipolar ionization in indoor environments. J. Aerosol Sci. **35**, 923–941 (2004)

MODIS support website by NASA Goddard Space Flight Center. http://modis.gsfc.nasa.gov/about/design.php. Accessed 13 May 2016

NASA Earth Observing System Data and Information System (EOSDIS). https://earthdata.nasa.gov/. Accessed 13 May 2016

Saraf, A.K., Bora, A.K., Das, J., Rawat, V., Sharma, K., Jain, S.K.: Winter fog over the Indo-Gangetic Plains: mapping and modelling using remote sensing and GIS. Nat. Hazards **58**, 199–220 (2011)

Sawant, V.S., Meena, G.S., Jadhav, D.B.: Effect of negative air ions on fog and smoke. Aerosol Air Qual. Res. **12**, 1007–1015 (2012)

Wu, C.C., Lee, G.W.M., Cheng, P., Yang, S., Yu, K.P.: Effect of wall surface materials on deposition of particles with the aid of negative air ions. J. Aerosol Sci. **37**, 616–630 (2005)

Ground Subsidence Monitoring in Cheng Du Plain Using DInSAR SBAS Algorithm

Xiaoya Lu[1(⊠)] and Xiaopeng Sun[2]

[1] College of Computer Science and Technology,
Southwest University for Nationalities, Chengdu, Sichuan, China
35519309@qq.com
[2] The Third Engineering Institution of Surveying and Mapping of Sichuan,
Chengdu, Sichuan, China
603400221@qq.com

Abstract. Based on 14 ENVISAT ASAR data sets, the ground deformation information of Chengdu Plain during 2008 to 2010 was acquired using SBAS-InSAR method. The result showed that the average surface deformation was between −8 to 14 mm in major cities in Chengdu Plain during this monitoring period. The subsidence area in the north of Chengdu and south to Deyang City is maximum to −22 mm with an expanding subsidence area as time goes by. Chengdu Plain has no regional tectonic setting of subsidence and has abundant groundwater resources, thus there is no large-scale subsidence funnel. The monitoring results are validated by measured data and the accuracy is 2.9 mm. This outcome can provide a reference for bettering the future surface deformation monitoring for major cities in Chengdu Plain.

Keywords: SBAS-InSAR · Chengdu Plain · Ground deformation monitoring

1 Introduction

Land subsidence is a kind of environmental geological phenomenon by the result of the earth's crust surface layer soil compression results in the decrease of regional ground elevation under the influence of natural and human factors [1]. Land subsidence is a kind of progressive graded geological disasters. Its development process is irreversible, it is difficult to restore once they are formed [2].

A wide range, noncontact and planar millimeter surface deformation monitoring could be carried out by using differential interferometry synthetic aperture radar technology. In recent years, domestic scholars carried out many urban settlements monitoring related work by using satellite radar interferometry. Such as Zheng carried out subsidence monitoring for Beijing during 2003 to 2009 and analyzed control factors by using PS-InSAR technique and ASAR data [3], Luo carried out land subsidence monitoring for Shanghai lujiazui area during 1992 to 2002 by using PS-InSAR technique [4], Yin carried out the mining deformation monitoring by using SBAS technique [5]. The SBAS method is to SAR data consisting of several subsets, uses

Project sources: The central college basic scientific research business expenses special fund project (Project Number: 2014NZYQN28).

© Springer Nature Singapore Pte Ltd. 2017
H. Yuan et al. (Eds.): GRMSE 2016, Part I, CCIS 698, pp. 535–545, 2017.
DOI: 10.1007/978-981-10-3966-9_59

multiple main image, needs less image than PS-InSAR method. The SBAS method can monitor the large-scale deformation. Considering the available data resources, this paper uses the SBAS methods to deformation monitoring.

In addition to natural factors, such as the surface loose strata or half of loose strata in pressure, geological structure, Karst collapse in karst development area and etc., human factors such as excessive exploitation of underground resources (groundwater, oil, gas), solid mineral resources and Geothermal resources, large-scale engineering construction, rail transit, urban underground space development, tall building on the foundation of dynamic loadare are one of the important factors which lead to land subsidence. "The national land subsidence prevention and control planning from 2011 to 2020" pointed out that more than 50 cities of the country which Located in Beijing, Tianjin, Hebei, Shanxi, Inner Mongolia and other 20 provinces suffer from land subsidence disasters at present. The area of cumulative land subsidence rate of more than 200 mm is 7.9 km^2 and has expanding the trend with the time. At the same time, the planning puts forward the overall goal of the future ten years. That is to find out disaster situation, development trend, the reasons and distribution of the land subsidence.

There are rich in groundwater resources and there was no a wide range of land subsidence event caused by natural factors all the year round in Chengdu plain. But land subsidence and surface collapse events caused by man-made factors still occurs, such as a collapse accident occurs caused by heavy rains in a parking lot of Chengdu palm south street on July 9, 2014. Four cars along with a few big trees fell in several meters deep foundation pit. Therefore, it is necessary to overall grasp the surface subsidence of the Chengdu plain, especially it is necessary to carry out land subsidence monitoring work for major cities of the Chengdu plain in the future.

2   Data Processing and Analysis

2.1   The Study Area and Data

Monitoring scope are parts of the Chengdu plain, image coverage is shown in Fig. 1.

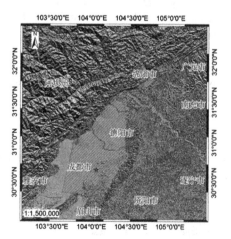

Fig. 1. Spatial region of images.

The monitoring data are ENVISAR ASAR images derived from the European space agency (ESA), Track 18, date: 2008.05.28, 2008.08.06, 2008.09.10, 2008.10.15, 2009.01.28, 2009.03.04, 2009.04.08, 2009.05.13, 2009.09.30, 2010.01.13, 2010.02.17, 2010.03.24, 2010.04.28. The diagram of image sequence and space baseline are shown in Fig. 2.

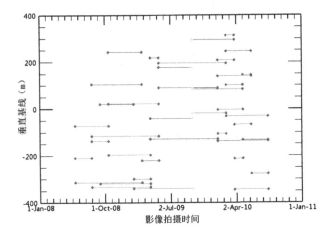

Fig. 2. Space and time baseline.

2.2 The Technology Principle of SBAS

SBAS-InSAR is a kind of time series analysis method proposed by Berardino et al. [6]. The algorithm is based on an appropriate combination of differential interferograms produced by data pairs characterized by a small orbital separation (baseline) which could limit the spatial decorrelation phenomena. Due to SBAS method using the singular value decomposition (SVD) method, therefore, when solving the deformation rate, we can connect isolated SAR data sets separated by a bigger space baseline and improve the data sampling rate. The main steps are as follows:

1. We Access to the N + 1 phase of SAR image for the same area in (t0, t1, ..., tn) period of time, then select one of the images as the main image and take the other SAR image registration on this image. We proposed the assumption that each issue SAR image has at least one image with the interference, N + 1 issues images will be generated M differential interference images. M satisfy the following conditions:

$$\frac{N+1}{2} \leq M \leq N(\frac{N+1}{2}) \tag{1}$$

2. The first j interferometric phase differential interference image generated by SAR images which were get in the subordinate image t_A moment and the main image $t_B(t_B > t_A$ moment (regardless of these factors, such as the atmospheric delay phase, residual terrain phase, phase noise) can be expressed as:

$$\delta\phi_j = \phi_B(x,r) - \phi_A(x,r)$$
$$\approx \frac{4\pi}{\lambda}[d(t_B,x,r) - d(t_A,x,r)] \tag{2}$$

In the formula, $j \in (1,\ldots,M)$, λ is center wavelength of the signal, $d(t_A,x,r)]$ and $d(t_B,x,r)$ are the radar line of sight direction accumulation form variables in the t_A moment and t_B moment when $d(t_0,x,r) = 0$.

3. The phase in the (2) type is expressed as the product of average phase velocity between the two access time and time:

$$v_j = \frac{\phi_j - \phi_{j-1}}{t_j - t_{j-1}} \tag{3}$$

The phase value of first j interference image can be represented as:

$$\sum_{t_{A,j}+1}^{t_{B,j}} (t_k - t_{k-1})v_k = \delta\phi_j \tag{4}$$

That is, the phase value is the velocity's integral in each period on the time interval of the main image and subordinate image. Matrix form is expressed as:

$$Bv = \delta\phi \tag{5}$$

Because differential interference figure of the small baseline set adopted more main image strategy, so matrix B may produce rank-defect in the process of solving the coefficient matrix B. By using SVD method, we can get the generalized inverse matrix of matrix B and the minimum norm solution of the velocity vector. Finally we can get the amount of deformation in each period by the velocity's integral in each period.

2.3   Data Processing Flow

SBAS data processing consists of the following main procedures:

1. SAR pair combination and connection network generation
   Multiple differential interferograms are generated by SAR pairs selected basing on all images according to specific temporal or spatial baseline, and then interferogram flattening, filtering, unwrap processing is performed.
2. Coherent Target Selection
   Using spatial coherence result as a reference, minimum thresholds was set for coherent target analysis and selection.
3. Interferogram unwrap
   Ground control points were selected in the region with high coherence and sparse interference fringes as deformation reference for unwrap using Delaunay 3D option.
4. Time series deformation result generating

Displacement velocity in each time period is computed by unwrapped phase using SVD method, and finally deformation result is achieved by integrating in each period.

The specific process is shown in Fig. 3.

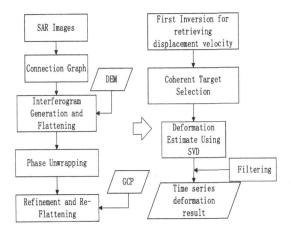

Fig. 3. Flow chat data processing of SBAS.

2.4   Monitoring Results

14 cumulative ground deformation graphs relative to the surface deformation on May 28, 2008 were acquired by using SBAS-InSAR method. Monitoring scope includes: the urban area of chengdu, xindu district, pixian district, wenjiang district, qingbaijiang district, longquanyi district, jintang district, xinjin district, shuangliu district and pengzhou district; jingyang district of deyang city, mianzhu city, shifang city, guanghan city and luojiang county; Fucheng district, youxian district, jiangyou district of mianyang city; Jianyang city. The results are shown in Fig. 4.

It can be seen from the Fig. 4, the subsidence areas within the monitoring scope are mainly distributed in the north of Chengdu and south to Deyang city and the largest settlement is maximum to −22 mm. The subsidence area has expanding trend as time goes by.

Average accumulated deformation which was counted by the county administrative unit during May 28, 2008 to August 11, 2010 are as shown in Fig. 5. It can be seen from Fig. 5, qingbaijiang district, wenjiang district, jinniu district, xindu district and shuangliu country are subsidence area, jingyang district, mianzhu city, luojiang county, fucheng district, youxian district, jiangyou district are uplift area. The deformation of xinjin county, qingyang district, jinjiang district, wuhou district, jintang county, pixian county shifang city, guanghan city and jianyang city is close to 0. Although there are different degrees of subsidence and uplift in study area, but the average accumulated

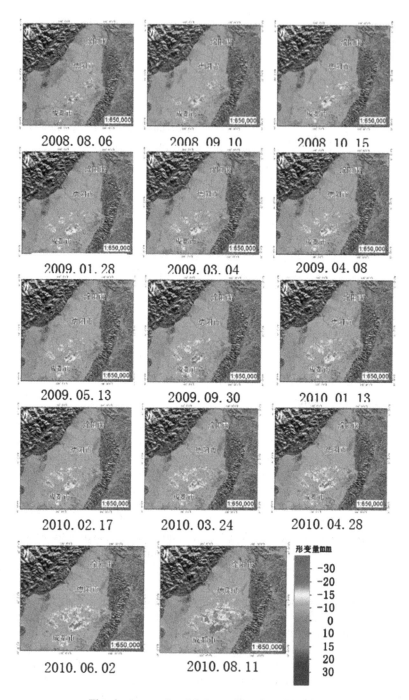

Fig. 4. Accumulated deformation of study area.

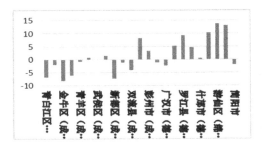

Fig. 5. Average accumulated deformation of countries in study area.

deformation was between −8 to 14 mm in two years. In general, the deformation is not big. Average deformation rat was between −10.3 mm per year to 11.1 mm per year in study area and the diagram is shown in Fig. 6.

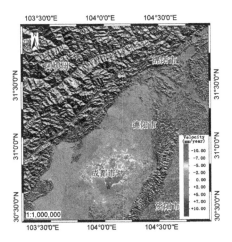

Fig. 6. Average deformation rate.

As shown in Figs. 7, 8 and 9, in the jurisdiction of Chengdu, qingbaijiang district, wenjiang district, jinniu district, chenghua district, xindu district and shuangliu country are subsidence area. Xinjin country is uplift area and the other administrative region's accumulated deformation fluctuate near the value 0 during this monitoring period. In the jurisdiction of mianyang, gaunghan city is subsidence area, its deformation is between 0 to 6 mm and its settlement is small. Shifang city presents the subsidence trend in 2009 and other administrative region are the uplift area. Each administrative region of deyang city is uplift area during the monitoring period and cumulative uplift began to fall in March 2010.

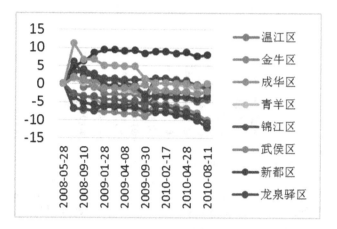

Fig. 7. Average accumulated deformation of countries in Chengdu City.

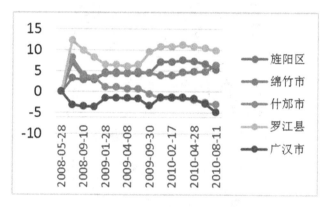

Fig. 8. Average accumulated deformation of countries in Deyang.

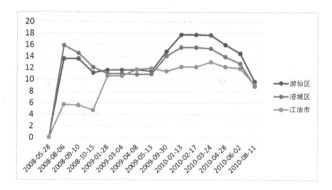

Fig. 9. Average accumulated deformation of countries in Mianyang City.

3 Precision Verification

The results of deformation monitoring were verified by selecting Chengdu CROS (continuous running positioning services integrated system) data and the spatial distribution of verification points is shown in Fig. 9. Four CROS site data of Chengdu area has been collected and the name of the four CROS site is respectively JITA, CDKC, LOQU, PUJI. In the 4 sites, because of PUJI site located outside the monitoring scope, JITA, LOQU two sites located in the mountains, its coherence is poorer, so there is no effective deformation monitoring results. Therefore, Deformation monitoring results were verified by CDKC site data. In addition, the reconstruction after Wenchuan earthquake projects-Special project of surveying and mapping: 1:2000 mapping has been collected and QY10 benchmark was chosen as a verification point in the project. Validation data point is shown in Fig. 10 and comparison results are shown in Table 1. It can be seen from Table 1 that the deformation nearby verification point is small and the measured results and the SBAS monitoring results have consistency on the trend.

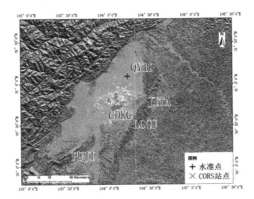

Fig. 10. Distribution of verification points.

Table 1. Differences between SBAS-InSAR result and ground measurement.

The name of the point	The observation time		The measured deformation value (cm)	InSAR results (cm)	Error (cm)
	On-site	InSAR			
DKC	The first time	2008.05.31	2008.05.28	−0.1205	−0.2964	−0.1759
	The second time	2010.08.16	2010.08.11			
QY10	The first time	2010.06.10	2010.06.02	−0.8000	−0.5029	0.2971
	The second time	2010.08.15	2010.08.11			

4 Conclusion and Discussion

The ground deformation information of Chengdu Plain during 2008 to 2010 was acquired by using SBAS-InSAR method. The result showed that the deformation had changed little and the average surface deformation was between −8 to 14 mm in Chengdu Plain after Wenchuan earthquake. Affected by the Wenchuan earthquake and its aftershocks, the western area of Chengdu plain under the influence of thrust role is gradually uplift. The subsidence area are mainly distributed in the north of Chengdu and south to Deyang City. The settlement is small and there is no subsidence funnel in these areas. But the subsidence range has a tendency to gradually expand.

The research of formation and evolution of the Chengdu plain [7–9] shows that Sichuan basin (Chengdu plain) is in the extensional environment under compressive tectonic background since the Himalayan movement. Because rigid block is surrounded by peripheral structural belt and the Chengdu plain is not in foreland depression area, so there is no tectonic conditions of regional subsidence in Chengdu plain. The Chengdu plain has no regional subsidence incentive because Chengdu plain basement fault is in a state of non-development or Inactive. The Chengdu plain of Mesozoic and Cenozoic has experienced a long evolution process of basin, lake basin, continental basin. Its terrigenous clastic is the thick sedimentary and the sedimentary environment is stable. There is no regional subsidence history since Pleistocene. Therefore, the Chengdu plain has no natural cause of subsidence during the monitoring period.

Considering the rapid development of the city construction of Chengdu and the construction activities of subway and the elevated driveway of second ring road are frequently in recent years. Therefore, urban construction activity is likely to become a human cause of land subsidence. It is necessary to pay more attention to these activities in the future.

References

1. Zheng, X., Wu, Q., Hou, Y., Ying, Y.: Some frontier problems on land subsidence research. Acta Geosci. Sin. **23**(3), 279–282 (2002)
2. Yin, Y., Zhang, Z., Zhang, K.: Land subsidence and countermeasures for its prevention in China. Chin. J. Geol. Hazard Control **16**(2), 1–8 (2005)
3. Zheng, J., Gong, H., Li, Q., Cui, Y., Chen, B.: The control factors on subsidence of Beijing plain area in 2003–2009 based on PS-InSAR technology. Bull. Surv. Mapp. **12**, 40–43 (2014)
4. Luo, X., Huang, D., Liu, G.: On urban ground subsidence detection based on PS-DInSAR – a case study for Shanghai. Bull. Surv. Mapp. **4**, 4–8 (2009)
5. Yin, H., Zhu, J., Li, Z., Ding, X., Wang, C.: Ground subsidence monitoring in mining area using DInSAR SBAS algorithm. Acta Geod. Cartogr. Sin. **40**(1), 52–58 (2011)
6. Berardino, P., Fornaro, G., Lanari, R., Sansosti, E.: A new algorithm for surface deformation monitoring based on small baseline differential SAR interferometry. IEEE Aerosp. Electron. **11** (2002)

7. Qian, H., Tang, R.: On the formation and evolution of the Chengdu plain. Sichuan Earthq. **3**, 1–7 (1997)
8. He, Y.: On the formation of the Chengdu basin. Reg. Geol. China **2**, 169–175 (1987)
9. Li, Y., Guo, B.: Cenozonic tectonics of Chengdu plain. J. Chengdu Univ. Technol. (Sci. Technol. Ed.) **35**(4), 371–376 (2008)
10. Zheng, Y.: A preliminary analysis of the distribution of groundwater resources of the Chengdu plain. Sichuan Water Conserv. **1**(4), 28–31 (2005)

GIS in Seismic Hazard Assessment of Shillong Region, India

J.D. Das[1(✉)], A.K. Saraf[2], and V. Srivastava[1]

[1] Department of Earthquake Engineering, Indian Institute of Technology Roorkee, Roorkee, Uttarakhand, India
josoeqiitr@gmail.com
[2] Department of Earth Sciences, Indian Institute of Technology Roorkee, Roorkee, Uttarakhand, India
arun.k.saraf@gmail.com

Abstract. The Shillong region falling in north-eastern part of India is developing fast and has witnessed one of the greatest earthquakes of India in 1897 besides several other earthquakes. Seismic hazard assessment of the Shillong and adjoining regions has become pertinent and been carried out with the help of GIS in view of sharp increase in human activities and the proposed developmental works. For this purpose an area has been selected spanning from 24.2° to 26.8°N latitudes and 89.2° to 92.8°E longitudes and is divided into a 0.2° × 0.2° grid-like framework for analysis. Seismic hazard assessment has been performed by employing deterministic seismic hazard assessment (DSHA) approach. Further, 1897 earthquake rupture zone based ground motion estimations were also done in order to examine and compare the intensity converted ground motion and predicted ground motion in the epicentral and nearby areas. Maximum predicted ground motion is found out to be 0.606 g on the rock sites and 0.634 g on the soil sites. The average ground motion on the rock and the soil sites is 0.433 g and 0.26 g respectively.

Keywords: Shillong Plateau · DSHA · Ground motion · Interpolation

1 Introduction

North-east India and its adjoining region fall in the most seismically active zones of the world. India has experienced four great earthquakes having magnitude more than 8), and out of these great earthquakes experienced by India, two have occurred in the north-eastern part of India. An investigation reveals that Shillong Plateau region is dominated by several seismogenic features. This region also falls in Zone-V (very high risk zone) according to the seismic zoning map of India (Fig. 1). Zone factor of 0.36 is assigned to the area by IS code. In view of above it becomes important to assess the seismic hazard of the area. The 1897 Shillong earthquake and 1950 Assam earthquake had inflicted catastrophic damage to life and property (Nandy 2001). Shillong Plateau earthquake of 12th June, 1897 was one of the most powerful and probably the largest known earthquake in the Indian sub-continent. It caused widespread destruction and devastation of a very large region in north-east India. The high level of ground motion can be judged by the observations that several large boulders were sent off from the

© Springer Nature Singapore Pte Ltd. 2017
H. Yuan et al. (Eds.): GRMSE 2016, Part I, CCIS 698, pp. 546–552, 2017.
DOI: 10.1007/978-981-10-3966-9_60

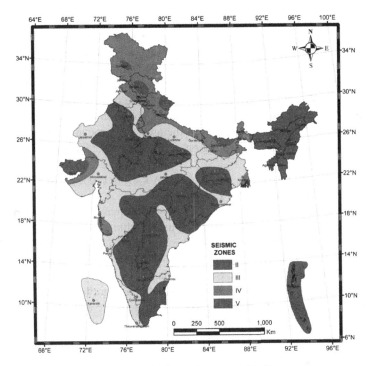

Fig. 1. Seismic zoning map of India

ground. It suggests that at certain locations the ground motion could have even topped the gravitational force (Oldham 1899).

Seismic hazard analysis involves the quantitative estimation of ground shaking hazards at a particular site. Seismic hazards may be analyzed deterministically, as when a particular earthquake scenario is assumed, or probabilistically, in which cumulative effect of earthquake size, location, and time of occurrence are explicitly considered. In the present study, deterministic seismic hazard analysis has been performed for Shillong Plateau region using geospatial approach. The values reported are also correlated with the intensities generated by past earthquakes. Computation of strong ground motions were done at several locations with respect to various potential seismic sources. Though deterministic approach does not provide any information regarding the occurrence frequency of the estimated ground motion but it is widely used in estimation of ground motion hazard level. For example, the design ground motions for structures and facilities which can cause monumental damage to life and property in case of their failure are determined by deterministic method (Panza et al. 2010).

2 Methodology

Study area lies between 24.2° to 26.8°N latitudes and 89.2° to 92.8°E longitudes. The area has been divided into a 0.2° × 0.2° grid-like framework for computation of ground motion parameters. For each grid point the closest distance to the surface projection of each fault in kilometre has been calculated using tools of ArcGIS. Sites are classified as rock and soil sites using visual inspection of the map of study area. A preliminary investigation into the satellite image of study area easily leads to identification of rock and soil sites. Geological and tectonic information is taken from Seismotectonic Atlas of India and its Environs and average shear wave velocity values is taken as 1620 m/s and 290 m/s for rock and soil site respectively (Borcherdt 1994). Various historical sources have also been used to reach a reasonable conclusion regarding the seismic hazard assessment of the area. For ground motion estimation at every grid points in Shillong Plateau region, following predictive equations have been employed:

i. Akkar-Bommer (2010), ii. Boore and Atkinson (2008), iii. Campbell and Bozorgnia (2008), iv. Sharma et al. (2009)

The re-evaluated intensities of 1897 Shillong earthquake (Ambraseys and Bilham 2003) have been converted to PGAs with the using Trifunac and Brady's relationship.

The Shillong plateau is comprised of the Shillong Massif (SM) and Mikir Hills Massif (MHM). The MHM is separated from the SM by an alluvial tract and NW-SE trending Dhansiri Kopli Fault. The Shillong Plateau is mainly represented by oldest Archean landmass with Precambrian deposits. Central and northern parts of the plateau are occupied by Proterozoic granite gneiss and Proterozoic-early Palaeozoic granite

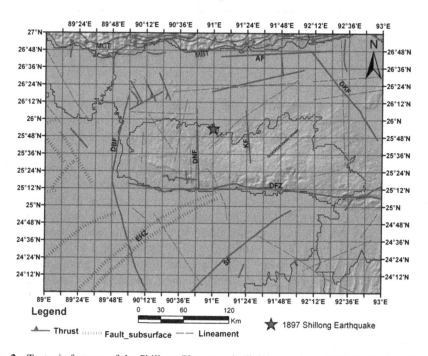

Fig. 2. Tectonic features of the Shillong Plateau and adjoining regions. DFZ-Dauki Fault Zone, SF-Sylhet Fault, EHZ-Eocene Hinge Zone, DBF-Dhubri Fault, DNF-Dudhnoi Fault, KF-Kulsi Fault, DKF-Dhansiri Kopili Fault, AF-Atherkheit Fault, MBT-Main Boundary Fault, MCT-Main Central Thrust.

intrusive. Least disturbed Cretaceous to Recent sediments mark the southern margins of the Plateau (GSI 1974; Nandy 2001). In the vicinity of the study area (Shillong Plateau region) the important tectonic features are the Dauki Fault Zone, Sylhet Fault, Mat Fault, Dhansiri Kopili Fault, Dudhnoi Fault, Kulsi Fault, Naga Disang Thrust, Main Central Thrust, and Main Boundary Thrust. Figure 2 illustrates these features on the map. The Southern margin of Shillong Plateau is marked by the remarkably linear E-W trending Dauki Fault Zone.

3   Isoseismals

Taking into account the revised intensity estimates, isoseismals have been plotted in ArcGIS environment. Geospatial approach has been employed to develop several input parameters which are used as arguments in various ground motion estimation functions. Some of the parameters used in these functions such as dip, strike, maximum magnitude of seismic features to be used in GMPEs. Isoseismals with regard to the Shillong 1897 earthquake is show in Fig. 3.

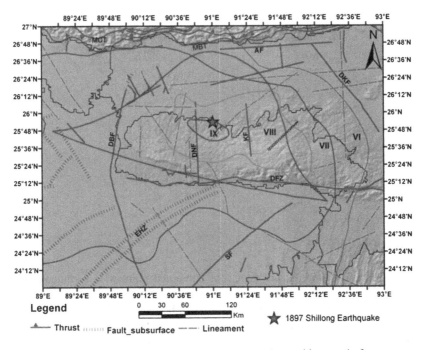

Fig. 3. Isoseismals of 1897 Shillong earthquake along with tectonic features.

The maximum magnitude (Mmax) is calculated by adding magnitude unit of 0.5 to maximum observed magnitude. For the first method as mentioned the maximum magnitude for each zone has been estimated based on past earthquakes. A buffer zone was created around each seismogenic zone and the maximum magnitude was found for each zone by plotting the historic earthquakes. The recurrence intervals for the large earthquakes are generally much longer than the periods for which the historical data are available. Thus, in a given seismic source, there is a fair possibility that the maximum earthquake has not been recorded. In view of the fact that an available catalogue of the recent earthquakes are for a seismic source may not contain the actually possible largest event, magnitude units of 0.5 has been added to the largest earthquake to assume maximum magnitude for the considered seismic source zone. There are various other approaches to assign maximum magnitude to a seismic feature. Empirical relationships developed by Wells and Coppersmith (1994) as shown below can also be used for this purpose.

For faults $M = 5.08 + 1.16\log$ (SRL); For thrust $M = 4.07 + 0.98\log$ (A). Where M is maximum magnitude, SRL is surface rupture length of fault and A is the rupture area of thrust. If calculations are done using this method, then one should take half-length of MCT and MBT in the calculations. This is due to the logic that such a long distance of thrust cannot rupture at the same time. However, maximum values in Table 1 correspond to the former method discussed.

Table 1. Seismic sources and their observed/probable maximum magnitude.

Sr. no.	Seismic sources	Observed Mmax	Probable Mmax
1	MBT	7.0	7.5
2	MCT	7.5	8.0
3	AF	5.5	6.0
4	DKF	6.5	7.0
5	KF	5.5	6.0
6	DFZ	7.5	8.0
7	DNF	5.5	6.0
8	SF	7.5	8.0
9	EHZ	6.0	6.5
10	DBF	7.0	7.5

4 Interpolation

In order to obtain seismic hazard zones for computed point wise earthquake ground motion natural neighbourhood interpolation technique has been employed using Arc-GIS to generate continuous surface. Natural neighbour interpolation finds the closest subset of input samples to a query point and applies weights to them based on proportionate areas in order to interpolate a value. The natural neighbours of any point are those associated with neighbouring Voronoi (i.e. Thiessen) polygons. The proportion of overlap between this new polygon and the initial polygons are then used as the weights (Fig. 4).

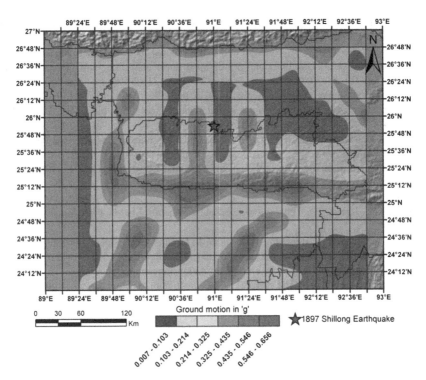

Fig. 4. Peak ground motion zones of the Shillong Plateau and adjoining regions.

5   Conclusions

The seismic hazard in the study area is of considerable importance from seismological and structural engineering perspective.

In the study area, average of PGA values obtained from Sharma et al. (2009) and Akkar-Bommer (2010) (AB) gives the best general results for soil as well as rock sites. For soil (firm) sites, Akkar-Bommer (2010) gives the highest values with Sharma et al. (2009) at the second place. The range of PGAs in the study area is 0.072–0.484 g for soil sites and 0.088–0.497 g for rock sites.

References

Akkar, S., Bommer, J.J.: Empirical equations for the prediction of PGA, PGV and spectral accelerations in Europe, the Mediterranean region and the Middle East. Seismol. Res. Lett. **81**, 195–206 (2010)

Ambraseys, N., Bilham, R.: Reevaluated intensities for the Great Assam earthquake of 12 June 1897, Shillong, India. Bull. Seismol. Soc. Am. **93**, 655–673 (2003)

Boore, D.M., Atkinson, G.M.: Ground motion prediction equations for the average horizontal component of PGA, PGV, and 5%-damped PSA at spectral periods between 0.01 s and 10.0 s. Earthq. Spectra **24**, 99–138 (2008)

Borcherdt, R.D.: Estimates of site dependent response spectra for design (methodology and justification. Earthq. Spectra **10**, 617–653 (1994)

Campbell, K.W., Bozorgnia, Y.: NGA ground motion model for the geometric mean horizontal component of PGA, PGV, PGD and 5%-damped linear elastic response spectra at periods ranging from 0.1 s to 10.0 s. Earthq. Spectra **24**, 139–171 (2008)

GSI: Geology and mineral resources of the states of India, Part IV, Arunachal Pradesh, Assam, Manipur, Meghalaya, Mizoram, Nagaland and Tripura. Geol. Surv. India Misc. Publ. **30**, 124 p. (1974)

Nandy, D.R.: Geodynamics of North-Eastern India and the Adjoining Region. ACB Publications, Kolakata (2001). 209 p.

Oldham, R.D.: Report on the great earthquake of 12th June 1897. Mem. Geol. Surv. India **29**, 1–379 (1899)

Panza, G., et al.: Earthquake scenarios for seismic isolation design and the protection of cultural heritage   (2010).   https://www.researchgate.net/profile/Maurizio_Indirli/publication/259852903

Sharma, M.L., Douglas, J., Bungum, H., Kotadia, J.: Ground motion prediction equations based on data from the Himalayan and Zagros regions. J. Earthq. Eng. **13**, 1191–1210 (2009)

Wells, D.L., Coppersmith, K.J.: New empirical relationships among magnitude, rupture length, rupture width, rupture area, and surface displacement. Bull. Seismol. Soc. Am. **84**, 974–1002 (1994)

Spatial-Temporal Analysis of Soil Erosion in Ninghua County Based on the RUSLE

Ming Yu[1,2(✉)], Yao Huang[1], Chaofeng Sun[1], and Yong Wu[1,2(✉)]

[1] School of Geographical Science,
Fujian Normal University, Fuzhou 350007, China
yumingfz@vip.sina.com, 364361899@qq.com,
449003613@qq.com, Wuyong3216@163.com
[2] Institute of Geography, Fujian Normal University, Fuzhou 350007, China

Abstract. Soil erosion is currently one of the main research topics of environment change, and it has affected the human survival and sustainable development. In this paper, Revised Universal Soil Loss Equation (RUSLE) based on RS/GIS technology was used to estimate soil erosion in years of 2001, 2007 and 2013 in the Ninghua County Fujian Province incorporating spatial analysis method based on GIS, temporal and spatial dynamic changes. For analyzing the temporal variation of soil erosion intensity in different periods in the study area, we investigated the variation tendency of soil erosion intensity each of the land use classes, topographic factors (elevation and slope) and vegetation coverage factors. The results indicated that RUSLE model based on RS/GIS technology could be extended to estimate regional soil erosion, and provided effective technological means for quantitative analysis of regional soil erosion.

Keywords: RUSLE model · Estimation of soil erosion · Temporal and spatial distribution variation · Ninghua County

1 Introduction

The process of urbanization of human beings, the history of soil erosion has lasted nearly a century, among the many complex factors; unreasonable land utilization and vegetation destruction are the main causes. Profound soil erosion not only affects the ecological environment, but also restricts the development of social and economic aspects of the local area for a long time. The current soil erosion forecasting model is at the forefront of the field of soil erosion science. The purpose is to accurately evaluate the degree of soil and its nutrient losses, study and understand the processes and intensity of soil erosion, and reveal the dynamic process of various forms of soil erosion (sheet erosion and gully erosion, etc.). It not only provides the basis for mastering the rational use of land resources, but also the important scientific basis for government departments to formulate the treatment strategy of soil erosion [1]. On the basis of the Universal Soil Loss Equation (Universal Soil Loss Equation, USLE), the amendment of Universal Soil Loss Equation (Revised Universal Soil Loss Equation, RUSLE) [2] was brought forth by Renard et al. These have largely the same structure, but RUSLE made necessary corrections about the significance and algorithms of each

© Springer Nature Singapore Pte Ltd. 2017
H. Yuan et al. (Eds.): GRMSE 2016, Part I, CCIS 698, pp. 553–562, 2017.
DOI: 10.1007/978-981-10-3966-9_61

factor and at the same time it introduced the concept of soil erosion [3]. Kitahara et al. applied RUSLE Model to evaluate the soil erosion of forest area in Japan and the found that the RUSLE model is suitable for a variety of agricultural and forest terrain [4]. With the progress and development of science and geo-information, the combination of RS and GIS technologies has broad application prospects in the field of soil and water conservation. Kouli et al. also used the combination of RUSLE and RS/GIS to predict soil erosion. Despite uncertainties, the RUSLR model still effectively runs the prediction of catchment scale soil erosion [5]. In this study, we take Ninghua for an example, according to the relevant parameters that RUSLE model requires. By means of remote sensing (RS) and geographic information systems (GIS), we create soil erosion estimation model of Ninghua. We also make research about certain topography, land-use pattern soil and water conservation capacity under certain vegetation coverage to provide technical support for scientific management of soil erosion and soil conservation engineering design in the west of Fujian province.

2   Study Areas and Data

Ninghua County, located in the west of Fujian Province and belonging to the Sanming city, which is located in Wuyi Mountain, is one of the border counties to Fujian and Jiangxi provinces (Fig. 1). The land area comprises 2381 km², a mountainous area of 2.7 million microseconds, with forest coverage of 72.99%, and cultivated land area of 427000 μs. The Ninghua is the key forestry and the commodity grain base counties in Fujian province. The main data used in this study include: DEM data, 2001 and 2007 and 2013 remote sensing the original image, Ninghua County 1:25000 soil type distribution map, Ninghua County and its surrounding weather stations monthly rainfall data, the annual statistical yearbook, county annals, soil keep bulletins and other information.

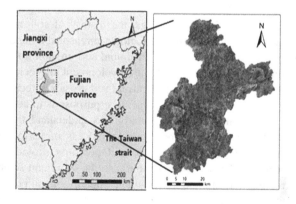

Fig. 1. The geographical location of Ninghua county and its pseudo image of Landsat-8 OLI

3 Methods

3.1 Selecting RUSLE Model

The scope of the study area is a larger county. RUSLE model is suitable for the regional scale of soil erosion estimation. The required parameters of this model could be obtained more conveniently, not limited by the existing water and sediment data of river basin. It is also possible to deal with the existing data through the RS and GIS technology, and it is relatively easier to obtain the vegetation coverage factor and terrain factor parameters that the RUSLE model needed. Moreover, the applicability of this model in China strong, it has a high precision and is easy to deduce. We use the RUSLE model as the method of estimating average annual soil loss. To do so, we need to consider the rainfall (monthly rainfall), the types of soil, topography (slope length and slope), vegetation cover and soil conservation measures. These five major factors define the following equation [6]:

$$A = R * K * LS * C * P$$

wherein A represents the average annual amount of soil erosion, R represents the rainfall erosivity factor, K is the soil erodibility factor, LS to Slope length factor, C is the crop management factor, P is factor for the Conservation Measures (wherein, LS, C, P are dimensionless factors). The calculation methods of each factor are as follows.

The calculation of rainfall erosivity factor R value: due to the scope of the study area being large, it is difficult to obtain the standards of erosive rainfall, and therefore, we use a simple algorithm proposed by Zhoujian; it is also suitable for Fujian to solve R value [7].

The calculation of soil erosion force factor K value: in this study, we refer to the main characteristics of soil K value table [8, 9] of Fujian and use EPIC model to calculate the value of K.

The calculation of topographic factor (LS): in this study, we use the Hickey NCSL slope length calculation algorithm; write the program of automatically extracting the factor of the length and angle of slope [10] in C++ based on the histogram matching principle, to carry out the length and gradient of slope scaling.

The calculation of vegetation coverage and management factor C value: this article intends to adopt the formula [11] proposed by Tsai Chung-ho, using ArcGIS software to statistically classify vegetation management factor C values under each land use/land cover.

The calculation of conservation measures factor P value: according to the current situation of land use/land cover in Ninghua, referring to the USDA Handbook No. 703 and related literature [9, 12, 13] to determine the P value, then connect the value of the property P table to the respective land use/land cover types, using ArcGIS grid computing tools to generate the final factor P layer.

3.2 Results

The RUSLE model was used to obtain the soil erosion modulus average chart of Ninghua in 2001, 2007 and 2013. Seeing Fig. 2(a), (b), (c) and then, according to

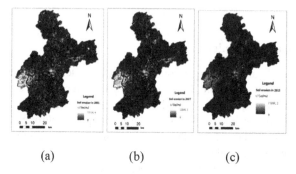

(a)   (b)   (c)

Fig. 2. Result of soil erosion estimation of Ninghua County during 2001 to 2007 to 2013

SL 190-2007 "Soil Erosion standards of classification" issued by People's Republic of China Ministry of Water Resources to divide it, we divide the soil erosion intensity classification into six (Table 1), and we obtain the soil erosion intensity classification diagram during these three times. Seeing Fig. 3(a), (b), (c).

Table 1. Classification standard of national soil erosions 190-2007)

Classification	Average erosion modulus t/(km^2 * a)	Average erosion thickness (mm/a)
1. Micro erosion	<500	<0.37
2. Mild erosion	500–2500	0.37–1.9
3. Moderate erosion	2500–5000	1.9–3.7
4. Strength erosion	5000–8000	3.7–5.9
5. Extreme intensity erosion	8000–15000	5.9–11.1
	>15000	>11.1

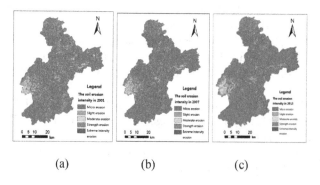

(a)   (b)   (c)

Fig. 3. Classification map of intensity soil erosion for three years (2001, 2007 and 2013)

In order to test the reliability of soil RUSLE based on the supports of GIS and RS Erosion quantitative estimation accuracy verification, in this paper, we use RUSLE model to calculate the soil erosion of three periods, and we also make statistics about the soil erosion area more than mild during these three times and compare with that researched by National Remote sensing of soil erosion quick survey. We then make accurate simulation of the determination results obtained by RUSLE model (Table 2). We find the error of these two methods is less than 5% and therefore combine GIS and RS RUSLE model to estimate soil erosion; it can basically meet the regional scale soil erosion survey.

Table 2. Comparison of two methods for calculating soil erosion (unit: km^2)

Years	Land area	Remote sensing survey	RUSLE model estimation	Area diffidence	Error (%)
		Erosion area	Erosion area		
2001	2381	396.93	408.85	11.92	3.0%
2007		337.79	343.23	5.44	1.6%
2013		331.60	317.16	−14.44	4.6%

3.3 Analysis

Dynamic Space Changes of Soil Erosion Intensity. According to preliminary data processing and the building of the model, we get a summary of different intensity distribution of soil erosion area in these three years (Table 3). From Table 3, we find that the total area of mild erosion decreased significantly from 2001 to 2013, falling from 408.85 km^2 to 317.89 km^2, and the extreme erosion area decreased from 85.02 km^2, 2001 to 27.76 km^2, 2013. In the three periods, pole extreme intensity erosion area is very small, and moderate erosion area mildly increases in soil erosion. Thus, we can see that the overall situation of the Ninghua is improving and putting a high value on science.

Table 3. Area of different intensity soil erosion composition for each year in Ninghua County

Erosion grade	2001 (km^2)	Proportion (%)	2007 (km^2)	Proportion (%)	2013 (km^2)	Proportion (%)
Micro erosion	1970.44	82.76	2041.89	85.81	2063.11	86.65
Mild erosion	227.73	9.56	166.75	7.01	183.01	7.69
Moderate erosion	95.27	4.00	111.09	4.67	106.70	4.48
Strength erosion	85.02	3.57	58.95	2.48	27.76	1.17
Extreme intensity erosion	0.83	0.03	1.00	0.04	0.42	0.02

The Influence of Different Land Use to Water and Soil Erosion Intensity. Further analyzing the relationship of different land use types and the distribution of soil erosion intensity in Ninghua County, this paper adds up the soil erosion area in eight land use types - building, forest, bare land, paddy field, corner, shrub land, bamboo and water - after overlaying land use classification layer and soil erosion intensity layer in same region of three periods. The result is shown in Table 4. In addition, for better analyzing the soil erosion changes of different land use types, this paper calculates soil erosion rate respectively which is shown in Fig. 4. Study show that soil erosion rate of forest and paddy field is relatively low, the average soil erosion rate in three years is 6.46% and 18.21%, the soil erosion intensity of forest is mainly mild and the left is moderate and Strength erosion which validates that forest is beneficial to soil and water conservation. The average soil erosion rate of shrub land is 29.53% where moderate and strength erosion area is the largest of all land use types. As is shown in Fig. 4, the soil erosion rate of bare land, corner and shrub land is relatively high while water and bamboo are relatively low in Ninghua County; the soil erosion rate of bamboo in three years are all less than 10%, the soil erosion situation is better. The building construction land area shows an upward trend from 2001 to 2007. One reason is the increasing city construction land area with the development of urbanization, which leads to the increasing area of soil erosion. This can be seen from the distribution of soil erosion region, which locates in the edge of building construction land. With the city expanding, the government ignores the importance of protecting surrounding vegetation and fails to take relevant protective measures to soil erosion, leaving the risk of soil erosion. The soil erosion area of building construction land shows a decreasing trend from 2007 to 2013, which is relevant to taking positive surface reinforcement measures and adding green land area in Ninghua County.

Table 4. Status of soil erosion distribute in each LUCC

Years	Erosion grade	Building land	Wood land	Bare land	Paddy field	Garden plot	Shrub grassland	Bamboo forest	Water
2001	Micro erosion	18.33	1338.96	17.43	309.96	50.20	256.33	81.98	5.47
	Mild erosion	0.56	104.64	5.33	47.45	9.87	52.10	7.48	0.2
	Moderate erosion	0.82	19.68	5.05	22.65	7.04	38.92	0.91	0.19
	Strength erosion	0.81	11.79	3.52	25.05	4.95	28.22	0.26	0.43
	Extreme intensity erosion	0.02	0.20	0.03	0.45	0.09	0.40	0	0
2007	Micro erosion	35.14	1169.18	12.72	331.45	35.17	366.34	65.30	26.60
	Mild erosion	1.78	64.05	0.74	40.07	5.60	69.62	2.68	1.84
	Moderate erosion	2.74	5.62	2.54	31.08	7.88	57.65	0.36	3.00
	Strength erosion	3.11	3.56	1.71	24.66	4.15	22.86	0.04	1.80
	Extreme intensity erosion	0.11	0	0.03	0.62	0.21	0.03	0	0
2013	Micro erosion	37.21	1386.76	14.13	337.35	45.22	312.32	61.25	18.23
	Mild erosion	2.26	44.75	2.58	30.86	7.75	59.32	2.48	1.02
	Moderate erosion	1.01	4.76	3.87	20.44	6.72	40.34	0.20	0.21
	Strength erosion	1.27	2.11	1.47	20.12	3.90	18.95	0.12	0.34
	Extreme intensity erosion	0.03	0	0.01	0.30	0.04	0.18	0	0

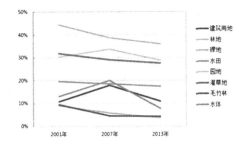

Fig. 4. Curve of erosion ratio for each land use class (Color figure online)

Temporal and Spatial Differentiation of Soil Erosion Under Different Vegetation Coverage. According to the actual situation of Fujian Province, Ninghua's vegetation coverage is divided into six. We evaluate the statistics of the erosion intensity distribution of the soil in three different periods and we calculate the rating of the proportion of erosion area more than mild intensity, namely the erosion rate. The results are shown in Table 5.

Table 5. Statistic of erosion area and erosion ratio in different vegetation covers degree

Vegetation coverage	2001		2007		2013	
	Erosion area (km^2)	Erosion rate (%)	Erosion area (km^2)	Erosion rate (%)	Erosion area (km^2)	Erosion rate (%)
<10%	16.23	29.34	12.64	27.22	10.23	24.52
10%–30%	47.70	38.20	38.70	34.24	35.21	30.24
30%–50%	90.54	36.84	85.57	34.17	78.42	29.17
50%–70%	125.32	24.63	116.18	22.85	101.42	11.67
70%–90%	82.34	10.2	77.29	7.92	72.42	6.23
>90%	5.27	2.23	3.71	1.19	1.43	0.35

From Table 5, the rate change of the area and the erosion of the soil in 2001, 2007 and 2013 showed a certain regularity, the soil erosion rate decreased gradually from vegetation coverage at 10% to 30%, when the vegetation coverage is above 70% or more when, the soil erosion rate reduces to 10%, when the vegetation coverage is above than 90%, the area of soil erosion and erosion rate of arrives the minimum and the soil erosion rates of three years are 2.2% or less. Overall, the soil erosion and vegetation coverage rate of Ninghua show a significant negative correlation; the greater the vegetation coverage is, the less prone the soil erosion is. The largest regional erosion area of vegetation coverage is at 50% to 70%, mainly due to the forest coverage rates up to 75.3% in Ninghua, and the area under the vegetation coverage of 50% to 70% this range is relatively large, the degree of erosion is basically mild intensity. The maximum rate of soil erosion in the Ninghua occurs in 10% to 30% of the vegetation cover, followed by the range of 30% to 50%. This trend can be explained by several factors: this area is the main distribution of sparse shrub grassland in the Ninghua; there is also a small amount of thinning young trees; finally, the soil erosion is mainly moderate and strength erosion. Thus, in the process of carrying out soil and water conservation, in the region of prone to soil erosion, by blockading administration, tree planting and other forest species manner for certain period of time, thereby increasing vegetation coverage, but also effectively alleviating the serious soil erosion.

From Table 5, the rate change of the area and the erosion of the soil in 2001, 2007 and 2013 showed certain regularity. The soil erosion rate decreased gradually from vegetation coverage at 10% to 30%, when the vegetation coverage is above 70% or more when, the soil erosion rate reduces to 10%, when the vegetation coverage is above than 90%, the area of soil erosion and erosion rate of arrives the minimum and the soil erosion rates of three years are 2.2% or less. Overall, soil erosion and vegetation coverage rate of Ninghua show a significant negative correlation, the greater the vegetation coverage is, and the less prone the soil erosion is. The largest regional erosion area of vegetation coverage is at 50% to 70%, mainly due to the forest coverage rates up to 75.3% in the Ninghua, and the area under the vegetation coverage of 50% to 70% this range is relatively large, the degree of erosion is basically mild intensity. Wherein the maximum rate of soil erosion in the Ninghua occurs in 10% to 30% of the vegetation cover, followed by the range of 30% to 50%, the reasons are as follows: this area is the main distribution of sparse shrub grassland in Ninghua, there is also a small amount of thinning young trees, the soil erosion is mainly moderate and strength erosion. Thus, in the process of carrying out soil and water conservation, in the region of prone to soil erosion, by blockading administration, tree planting and other forest species manner for certain period of time, thereby increasing vegetation coverage, but also effectively alleviating the serious soil erosion.

The Relationship Between Soil Erosion and Slope. The slope is directly related to the speed of priorities runoff, according to the National article "Soil erosion intensity criteria for the classification (SL 190-2007)" as the reference slope classification, and related literature, the types of surface slope is divided into six categories (Fig. 5) [14], it can be seen that the terrain in Ninghua is mainly gentle and flat slopes, the steep, acute slope and dangerous slope account for a small proportion. To further explore the relationship between these six Ninghua slope and soil erosion area changes between soil erosion rates, we respectively calculate the soil erosion rate under area with the six slope gradients in three periods (seeing Table 6).

Table 6. Ratio of soil erosion under different slope degree

Slope	2001		2007		2013	
	Erosion area (km^2)	Erosion rate (%)	Erosion area (km^2)	Erosion rate (%)	Erosion area (km^2)	Erosion rate (%)
<5°	93.82	23.53	88.87	22.29	84.56	21.13
5°–8°	79.15	21.40	71.88	19.43	65.35	17.54
8°–15°	136.45	16.35	117.35	15.01	110.63	14.67
15°–25°	67.62	12.17	48.93	8.81	41.63	6.63
25°–35°	18.93	10.08	13.77	7.33	10.34	5.64
>35°	3.14	10.05	2.30	7.35	2.01	5.45

It can be seen from Table 6, the change of soil erosion rate patterns on different slope conditions of these three periods are not big. The basic rule of them is as the slope increased, the soil erosion rate decreased gradually. At the same time, the larger soil erosion rate areas are mainly located in the regions that have the slope under 15°, which their soil erosion rate are greater than 16%, and the highest erosion rate areas are located in the flat areas with the slope less than 5°. The slope range with the largest erosion area of the Ninghua is at the interval of 8°–15°, and from the slope of 15°, with the slope increases, the erosion area is gradually reduced. To explore the reasons, it is development and utilization of human that the slope Under 15°, human disturbance and irrational development of frequent occurrence, causing regional the soil erosion especially serious. Three of each period of slope erosion rate showed a gradual decline law, because in recent years, countries in Ninghua County continue into the soil erosion of fund management support, returning forest, and take the ladder instead of slope for planting crops, the soil erosion area and erosion condition on each slope zone are in good condition.

4 Conclusion

After using the geographic information technology, calculating the soil erosion modulus of the three periods of Ninghua based on the RUSLE model and analyzing the spatial and temporal dynamic changes of soil erosion intensity, we can draw the following conclusions. In the change of soil erosion area, the total area of the soil erosion decreased significantly from 408.85 km^2 down to 317.16 km^2 from 2001 to 2013. (1) During comparative analysis the relationship between soil erosion intensity and utilization pattern of soil, vegetation coverage, topography through three periods, it can be seen that the harnessing of soil erosion in the early period have achieved an obvious success, and the soil erosion has been contain better. The way human use of land are specific performance in utilization pattern of soil, vegetation coverage, terrain slope factor, which can be changed by human activities in the short term, someone who want to effectively control soil and water loss can think about methods from these three aspects. (2) Using the RUSLE model, combined with RS, GIS technology, we can make quantitative evaluation and analysis on the soil erosion condition of Ninghua effectively. The key point of the study are the algorithms selection for model factors and the "localization" for model parameters, the calculation of the parameters of RUSLE model were discussed and analysis in detail in this paper, then combined with the actual situation of the study area, and we finally obtained the calculation methods of parameters of RUSLE model in soil erosion estimation under regional scale, in the future, we hope through the secondary development technology based on the ArcGIS 10.0 Engine, the calculation methods of the model parameters can be integrated into the regional soil erosion evaluation system, thus achieve the data processing, conversion and unification of the study area, and call the calculation functions of model parameters of each modules, then finally achieve the extraction of spatial evolution characteristics of soil erosion and the accuracy analysis of the model.

Acknowledgements. The study was supported by the National Science Foundation of China (No. 41171232) and the National Science Foundation of Fujian Province (No. 2014J01149).

References

1. Yu, X., Qing, F.: Watershed Erosion Dynamics. Science Press, Beijing (2007)
2. Renard, K.G., Foster, G.R., Weesies, G.A., Yoder, D.C., et al.: Predicting soil erosion by water: a guide to conservation planning with the revised universal soil loss equation (RUSLE). In: Agriculture Handbook No. 703 (1997)
3. Zhu, L., Feng, W., Zhu, W.: Progress of "3S" technique application in soil erosion study. Prog. Geogr. **27**(6), 57–62 (2008)
4. Kitahara, H., Okura, Y., Sammori, T., et al.: Application of universal soil loss equation (USLE) to mountainous forests in Japan. J. For. Res. **5**(4), 231–236 (2000)
5. Kouli, M., Soupios, P., Vallianatos, F.: Soil erosion prediction using the revised universal soil loss equation (RUSLE) in a GIS framework. Environ. Geol. **57**, 483–497 (2009)
6. Fan, J., Chai, Z., Liu, S., Tao, H., Gao, P., et al.: Dynamical changes of soil erosion in Lizixi catchment of Sichuan province by remote sensing and geographical information system. J. Soil Water Conserv. **6**(1), 63–70 (1992)
7. Lu, X., Shen, R.: A preliminary study on values K of soil erosibility factor. J. Soil Water Conserv. **6**(1), 63–70 (1992)
8. Fang, G., Ruan, F., Wu, X., et al.: Study on the characteristics of soil erosion in Fujian province. Fujian Soil Water Conserv. **2**, 19–23 (1997)
9. Chen, Y., Pan, W., Cai, Y.: Quantitative study of soil erosion in watershed based on RS/GIS and RUSLE—a case study of the Jixi watershed in Fujian province. J. Geol. Hazards Environ. Preserv. **18**(3), 5–10 (2007)
10. Wischmeier, W.H., Smith, D.D.: Predicting rainfall erosion losses: a guide to conservation planning. In: Agriculture Handbook No. 703. USDA, Washington, D.C. (1997)
11. Cai, C., Ding, S., Shi, Z., Huang, L., et al.: Study of applying USLE and geographical information system IDRISI to predict soil erosion in small watershed. J. Soil Water Conserv. **14**(2), 19–24 (2000)
12. Gao, F., Hu, C., Lu, Y., et al.: Assessment of soil erosion in Qinjiang watershed based on GIS and USLE. J. Soil Water Conserv. **21**(1), 18–22+28 (2014)
13. Huang, J., Hong, H., Zhang, L., et al.: Study on predicting soil erosion in Jiulong River watershed based on GIS and USLE. J. Soil Water Conserv. **18**(5), 75–79 (2004)
14. Ling, H.: The geography of the typical regional soil erosion is analyzed. Fujian Normal University (2009)

Characteristics and Environmental Significance and Physical and Chemical Properties of Karst Cave Water in Shuanghe Cave, Guizhou Province (in China)

Jie Zhang[1,2], Zhongfa Zhou[1,2(✉)], Mingda Cao[1,2], and Yanxi Pan[1,2]

[1] School of Karst Science, Guizhou Normal University, Guiyang, China
975479386@qq.com, 793149352@qq.com, 963860439@qq.com,
fa6897@163.com
[2] State Engineering Technology Center of Karst Rock Desertification
Rehabilitation, Guiyang, China

Abstract. In order to reveal the dolomite strata developed cave system in the cave water gas CO_2 partial pressure of cave water hydro chemical process and its dynamic characteristics and control factors. The dynamic monitoring for a period of year from the water chemistry index of Suiyang Shuanghe cave water from June 2015 to Mar 2016. Results show that: 1. Hydro chemical types of Shuanghedong cave water is mainly $HCO_3^- - Ca^{2+} - Mg^{2+}$ and $SO_4^{2-} - HCO_3^-$ $- Ca^{2+} - Mg^{2+}$ type water and got overlying gypsum strata and SO_4^{2-} become dominant ions and can greatly improve the solution Ca^{2+} ion content. 2. Cave drip water - gas phase CO_2 partial pressure difference (ΔP_{CO2}) between SIc and has good negative correlation at the same time, two important mineral saturation index of SIc, between SId exist a good positive correlation. 3. The analysis of the measured data proves the objective existence of the "actual degassing and deposition" and the difference between the theory and practice of the classical degassing and deposition theory. 4. The liquid phase CO_2 ($P_{CO2(water)}$) in the cave water showed a significant negative correlation with the pH value of the solution, and the CO_2 partial pressure of the cave air in the undeveloped natural closed cavity had little influence on the water of the cave.

Keywords: Karst · Hydrochemistry · Cave water · Partial pressure of carbon dioxide · Dolomite · Shuanghe cave

1 Introduction

Karst cave is one of the important forms of karst landforms, and it is a relatively closed environmental system. As one of the most important components in the karst dynamic system, CO_2 plays an important role in the process of the water in the cave. The dissolution and escape of gas phase P_{CO2} is an important geochemical process in groundwater system. The P_{CO2} difference between the liquid and gas (ΔP) has a decisive effect on the occurrence of water erosion and sedimentation in the cave water, which is influenced by the liquid phase P_{CO2} in the cave water. By the end of the

© Springer Nature Singapore Pte Ltd. 2017
H. Yuan et al. (Eds.): GRMSE 2016, Part I, CCIS 698, pp. 563–571, 2017.
DOI: 10.1007/978-981-10-3966-9_62

twentieth Century, the domestic scholars have studied the karst dynamic system and its operation rules, and gradually formed the basic theory of the karst dynamics. Since 2006, the karst hydrology, water physical and chemical properties and soil air CO_2 were monitored continuously by using the high resolution automatic monitoring instrument [1]. It not only proves the rapid change of the karst action, but also reveals the law of water rock gas coupling. In recent years, the research on the cave water is mainly focused on the water flow rate and the residence time of the rain water in carbonate rocks and the exchange of the water and carbonate rocks [2, 3]. A large number of studies have indicated that the chemical characteristics of the water in the cave, the way of movement and the way of existence directly affect the type and shape characteristics of the sediment [4]. In addition, the CO_2 of the soil covered by the cave has an important influence on the drip and micro climate environment, and shows the close correlation with the air temperature, precipitation, soil depth and biological activities. The influence of the CO_2 concentration of the cave air on the cave environment has been studied by foreign researchers. The effect of CO_2 partial pressure on the physical and chemical properties of the water in the cave was studied by Fairchild [5]. There was a negative correlation between $P_{CO2(air)}$ and mineral saturation index (SI), that is, the higher the air $P_{CO2(air)}$, the stronger inhibition effect on the deposition process of the water drop. In the SI research, there are also some scholars use the measured data to draw that SIc and SId has a good positive correlation. The dissolution of calcite and dolomite has a certain extent of synchronization in the low temperature Karst environmental system [6].

The dolomite areas are often associated with gypsum rock, SO_4^{2-} involvement makes easier dissolution of calcite and dolomite. Whitaker et al. study on the spatial and temporal variations of CO_2 concentration in the air of the cave, and discuss the effect of the solution on the dissolution or sedimentation of the other side. According to the research of single entrance cave, the distance between the CO_2 concentration and the distance from the entrance of the cave is more and more, showing an increasing trend. However, after reaching a certain depth, the increase tends to be gentle until it reaches a steady state that is no longer continue to rise [7]. And put forward the scientific hypothesis that the cave air CO_2 concentration and air temperature and water drop rate of the other side of the calcite deposition or dissolution of the impact of a comprehensive. In summary, most of the domestic and foreign researchers focused on the impact of air CO_2 concentration on the cave environment system, while the CO_2 partial pressure of the liquid phase is rare. This paper tries to from the water - gas CO_2 sub pressure difference angle to explore the effect of CO_2 on cave water hydro geo-chemical processes and their controlling factors, which can provide a scientific basis for the rational development of Shuanghe cave.

2 Study Location

Shuanghe Cave System department is located in the northwest of northern Guizhou Suiyang County town of Wenquan ($28°08'00''$–$26°20'00''$N and $107°02'30''$–$107°25'00''$E). On the geological structure, the cave system in northern Guizhou wide flat box shaped anticline, as affected by different direction of regional tectonic stress, the formation of NE, NW and SN to fault fold belt. The hole area is surrounded by a relatively rising triangular block, and the formation of the rock strata is slightly inclined to the East, and the dip angle is 5 to 7°. Strata exposed to white dolomite of Cambrian Loushanguan group (ϵ2-3ls) and lower Ordovician Tao Tong Zi Group (O_1t) and calcite dolomite and clip of chert and shale dolomite mainly. Mountains belong to the whole of the daloushan mountains of East Branch, the peak in Jinzhongshan elevation 1714.3 m. And the main drainage channel pool Shuanghe water (altitude 670 m) relative height difference of more than 1000 m. On both sides of the two sides of the lower Ordovician shale and clastic rock layers are higher.

The climate in the study area is generally in the sub tropical monsoon climate, and has the characteristics of subtropical mountain climate. Average temperature of the coldest month and the hottest temperature were 1.6 °C and 22.5 °C, the average diurnal temperature less than 10 °C. Precipitation is mainly concentrated in 4–10 months of each year. The western part of the annual precipitation is 1350 mm, and the eastern part is 1160 mm. In the area of 43.7 km^2 basin, there are 7 size of the streams, river density reached 1700 m/km^2. The water leakage is the main source of underground river water [8]. The coverage rate of the forest and shrub vegetation in the area of the area is over 60%.

2.1 Site Description

In Shuanghe cave area, we select 6 water samples include precipitation sampling points, pond water and dripwater. Drip water is divided into fissure water, soda straw water and stalactites dripping water (Table 1). The main in Shuanghe Cave System corresponding to the Dafeng Cave and Pixiao Cave. At the same time, we do chemical monitoring of underground water include the cave system main outlet of the underground river named Dayu Spring and underground river of Xiangshui Cave. Select a dripping point at a distance of about 500 m from entrance of Dafeng Cave (DS2#). Select a dripping point at a distance of about 1000 m from entrance of Pixiao Cave (DS1). And select a dripping point (DS3#) and two pond water (CS1#) (CS2#) inside a large hole in the Pixiao Cave - gypsum crystal flower cave. From the general situation of geological strata, selected sampling points of the environment, including high dry cave drip water, low altitude filled tunnel section of drip water and pool water points and nearly erosion datum underground river system.

Table 1. The characteristics of the studied water samples in Shuanghe cave

Site	Samle type	Detail	Sediment type
DS1#	Drip water	Fast perennial drop	Stalagmite, stalactite
DS2#	Drip water	Slow perennial drop	Stalagmite, stalactite
DS3#	Drip water	Slow perennial drop	Helictite, soda straw
CS1#	Pond water	Stable pond water	Helictite, soda straw
CS2#	Pond water	Stable pond water	Soda straw, dripstone
QS	Spring		
DXH	Underground river		Pothole

In the area of CO_2 monitoring, the main use of the United States telaire-7001 portable infrared CO_2 instrument real-time monitoring of cave air CO_2. Instrument resolution is 1 ppm, range is 0–10000 ppm, measurement precision is ± 50 ppm. Before the experiment with standard gas (0 ppm) calibration, the operation will be placed in the instrument from the operator 2 m to avoid human impact. Real time monitoring of wind speed, temperature, relative humidity, air pressure and altitude in the air of the cave and the cave using the kestrel-4500 Portable Weather Station in the United States. Instrument resolution is 0.1 m/s, 0.1, 0.1%, 0.1 mbar, 1 m, measurement accuracy is $\pm 3\%$, ± 1.0 °C, $\pm 3\%$, ± 1.5 mBar(25 °C), ± 15 m.

3 Methods

The data needed for this experiment were collected 12 times, and the total length of time was one hydrologic year. All monitoring equipment is installed in March 2015 before the completion of the data collected from April 4th onwards. The first sampling time was April 4, 2015, followed by sampling in the 4 day of each month until March 2016. Different collection methods are used for different kinds of cave water samples. In a drop of water, the two 60 ml polyethylene bottles are placed below the dripping point to collect the dripping water. When the amount of water collected enough for on-site titration and sampling can be. As for pond water and spring, we collected them with polyethylene bottles directly. At the same time, the scene with a stopwatch to measure the drop rate, measuring the temperature of water temperature and temperature measuring hole thermometer, and the concentration of Ca^{2+} and HCO_3^- in water sample was measured by using acid and alkali test kit. The electrical conductivity, water temperature and pH value of water samples were measured by the German multi WTW 3430 portable multi parameter water quality analyzer (2fd470). The collected water samples were taken back to the laboratory and sent to the State Key Laboratory of environmental geochemistry, Institute of Geochemistry, CAS. Determination of anions by IS90 type ion chromatograph and determination of cation by MPX VBTA type introductively coupled plasma optical emission spectrometer, both production in the United States. SIc, SId, SIg calculated by the phreeqc3.7 program [9], hardness through the Ca^{2+}, Mg^{2+} ion concentration into a unified CaO mass concentration.

4 Results and Discussion

4.1 Chemical Type Distribution of Cave Water

According to Shuanghe Cave 87 samples of water chemistry data analysis can draw the hydro chemical types mainly $HCO_3^- - Ca^{2+} - Mg^{2+}$ type and $SO_4^{2-} - Ca^{2+} - Mg^{2+}$ type (Fig. 1) The water samples were $HCO_3^- - Ca^{2+} - Mg^{2+}$ type in the absence of overlying Gypsum Beds. Two pond water point and one drip water point in Gypsum crystal flower Cave were affected by overlying gypsum rock, water samples contain a large number of SO_4^{2-} ions, the maximum can even reach 900 mg/L, this lead to HCO_3^- in water samples compared to other water samples point is seriously low, SO_4^{2-} ions

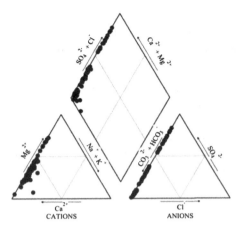

Fig. 1. Piper diagram of 84 water samples in Shuanghe cave

were attacked took advantage of the status of the anion [10]. This fully reflects the dolomite karst area in the strata in a large number of associated gypsum strata on the development of the cave water chemical characteristics of control effect.

4.2   Analysis of the Relationship Between SI and ΔP

Interpretation of Classical Degassing and Deposition Theory. In the karst dynamic system and the relatively closed cave environment, CO_2 plays a very important role in the solvent, and it affects hydrological and geochemical processes of cave water all the time. So CO_2 in air and aqueous solution of CO_2 also exist in the conversion. ΔP and mineral saturation index (SI) control the processes of deposition or erosion of drip water. According to the classical theory,

Setting: $\Delta PCO_2 = PCO_{2(water)} - PCO_{2(air)}$
    There are three kinds of States:

①   when $\Delta PCO_2 > 0$, SI > 0: At this point, CO_2 escapes from the water body, which makes the $CaCO_3$ dissolved in the water of the cave is too saturated, thus causing the $CaCO_3$ deposition, forming the chemical sedimentary landscape of the cave. The resulting reactions are as follows:

②   when $\Delta PCO_2 < 0$, SI < 0: At this time of CO_2 from the air dissolved in the water. The water has the ability to further dissolution of $CaCO_3$, is not conducive to the precipitation of cave landscape, even the formation of dissolution caves of the landscape.

③   when $\Delta PCO_2 = 0$, SI = 0: When water in CO_2 and in the cave air CO_2 balance, water of $CaCO_3$ is saturated, but the state is very ideal, in practice it is difficult to achieve this balance and balance the easily by cave space environment temperature and humidity effects [10].

In summary, there is a close relationship between ΔP and SI, in the dolomite karst area were selected in two *important* mineral saturation indices, namely SI_c, SI_d were saturation index of calcite, dolomite saturation index. And delta P and Si are a value of 0 as judged by dripping whether degassing and sedimentation.

Analysis on the Relationship Between the Actual Δp and SI Changes. Through the calculation of 84 water samples using phreeqc3.7 software P and SI, it can be preliminarily concluded that there is a significant negative correlation between the two. There is a good positive correlation between SI_c and SI_d, R^2 reached 0.98 (Fig. 2). This is from one side of the cold environment in the Karst area of dolomite, dolomite and calcite dissolution simultaneously.

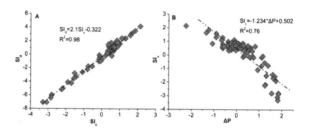

Fig. 2. The linear relationship between the ΔP, SI in cave water

However, in the analysis of the change characteristics of ΔP and SI, it is found that under the premise of good linear relationship, the two parameters are not followed by the 0 values as the standard for judging the actual degassing deposition, but more than 0, about 0.43. This is not entirely consistent with the classical theory mentioned above, but from the curve can be seen, when $\Delta PCO_2 > 0$, the SI curve is significantly decreased. When $\Delta PCO_2 < 0$, SI curve is significantly increased, which is consistent with the classical theory of the core mechanism. The difference between theory and practice is an objective existence. Many scholars believe that the use of SI value to determine the dissolution of minerals is more reliable, and used to determine the precipitation of minerals is often not very reliable. This is because in the calculation of the saturation index does not take into account the kinetics of the reaction and the sampling process due to reduced pressure. CO_2 volatilization and error in laboratory analysis also cause this appearance. And the facts prove that the other minerals, such as calcite, dolomite, gypsum and so on, are dissolving instead of settling. So in a lot of cases, even if the delta $PCO_2 > 0$, it is not necessarily on the occurrence of degassing deposition, Sun Zhanxue et al. (2007) put forward the concept of "apparent saturation index", but there is no further research on the karst area. Under the actual natural conditions, the saturation index of the cave water will have an upper and lower range of 0 values, and the maximum and minimum two thresholds are existed. So in the actual situation, only if the actual saturation index is larger than or smaller than the two threshold value, the real degassing deposition or absorption can occur [11]. And the research in this area is seldom published at domestic and abroad, as previously mentioned in the shuanghedong cave system in the water - degassing value is about 0.43

indicating deposition. As shown in Fig. 3, the black line represents the difference between the "actual degassing deposition indicator line" and the "theory of degassing-sedimentary indicator" represented by the blue line.

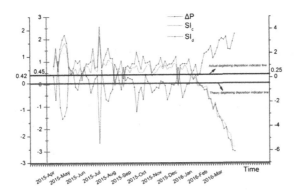

Fig. 3. The relationship between the cave water liquid and gas state changes of PCO_2, SI_c and SI_d (Color figure online)

Difference Between Classical Theory and Practice. Where the blue indicator line is ΔP and SI_c are 0 when the line, and the black indicator line is based on the approximate position of the symmetric axis of the three groups of reverse correlation curves indicate the value of the approximate location. The value is $SI_c = 0.45$, $\Delta P = 0.42$, $SI_d = 0.25$. Because the dolomite is more difficult to dissolve than calcite, its saturation index is slightly lower. If make the ΔP and SIc to judge degas deposition or not, it can be found that the article is worth the "actual degassing - deposition indicator line" of the value of about 0.435, it is far from the classical theory of the blue line which the ΔP and SI_c are 0. But whether the line is stable or not remains to be studied for a long time in the future.

4.3   pH Effect on the PCO_2

According to the theory of karst dynamic system, the pH value of aqueous solution will be changed when the HCO_3^-, SO_4^{2-}, organic acid and other acidic ions add in aqueous solution. The pH value represents the strength of the acidity and alkalinity of the aqueous solution, and the strong acidity indicates that the corrosion ability of the aqueous solution is relatively strong, aqueous solution tends to be not saturated. If the acidity is weak, its dissolution ability is weak, aqueous solution tends to be saturated or over saturated. According to the two pictures (A and B) in Fig. 4, we can see that showed a significant negative correlation between pH and PCO_2, pH and $PCO_{2(water)}$ also showed the same negative relationship. Combined with the analysis of the previous text can be found: without considering the influence of air CO_2, the R^2 between pH and $PCO_{2(water)}$ is 0.89. And in the B diagram, with the air CO_2 effect on the water solution, its R^2 was 0.87, only a slight impact on ΔP. This is due to the high degree of

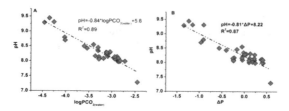

Fig. 4. The relationship between the cave water liquid and gas state changes of P_{CO2} and pH

closure of the cave environment system relative to the outside atmosphere environment, the degree of its natural environment change is relatively low, and its change process is slow, Especially the undeveloped natural caves. For a branch hole in the deep part of the main tunnel, which completely closed at the end Such as Gypsum crystal flower Cave, the concentration change of CO_2 in the air is even more weak, the fluctuation range far less than the range of liquid CO_2 concentration in cave water which through complex migration to the interior of the cave. So it has less influence on it. It also reflects the dominant position of cave water in the cave environment.

With the rise of $PCO_{2(water)}$, the pH of the aqueous solution decreased, and it's acidity was obvious. With high $PCO_{2(water)}$ water solution through the overlying rock or soil and suddenly entered the cave, ΔP was significantly increased. Because the $PCO_{2(air)}$ in the cave air environment is far less than the $PCO_{2(soil)}$ value of the overlying soil layer. In general, in the process of the cave water migration to the cave, mainly occurs degassing and sedimentation. Therefore, if the overlying soil and vegetation conditions are good, it is of great significance to the formation of the sedimentary landscape in the cave.

5 Conclusion

Through the study on physicochemical properties of water to Shuanghe Cave a hydrological year analysis, the preliminary conclusions are as follows:

(a) The dolomite area of Karst water cave cave in overall development belongs to the type of $HCO_3^- - Ca^{2+} - Mg^{2+}$. But in the influence of the overlying gypsum formation, the concentration of SO_4^{2-} ion in water was increased dramatically and the dominant ion position of the original HCO_3^- ion was taken to make the water chemical type changed to $SO_4^{2-} - Ca^{2+} - Mg^{2+}$ type. In addition, gypsum has a high solubility, Ca^{2+} ion concentration in water samples is also significantly increased.

(b) There is a good negative correlation between the CO_2 partial pressure difference (ΔPCO_2) and SI in the cave drip water. At the same time, a good positive correlation between the two kinds of important mineral saturation index also, R^2 reached 0.98. The analysis of the measured data shows that the "actual degassing and deposition" is an objective existence, in which the line is about 0.43, but the threshold value of the fluctuation of SI = 0 is still need to be further studied.)

(c) The liquid phase CO_2 ($PCO_{2(water)}$) in the cave water showed a significant negative correlation with the pH value of the solution, and the $PCO_{2(air)}$ was found to have little effect on the correlation of ΔPCO_2, this result shows that the fluctuation range of air CO_2 concentration is very small, and it is not enough to control the concentration of CO_2 in the liquid drop in the cave. Liquid CO_2 is leading degassing sedimentation or absorption of erosion controlling factors, and cave environment system relative to the overlying soil or rock stratum layer is for cave water provides a liquid gas exchange between the place and its mutation conditions.

Acknowledgments. In this paper, the research was sponsored by National Natural Science Foundation of China (Grant No. 41361081); Guizhou Province Science and Technology Plan (Guizhou S&T Contract G 2014-4004-2); National Natural Science Foundation of China Youth Science Foundation Project (Grant No. 41301504); Major application based research project of Guizhou (Guizhou S&T Contract JZ 2014-0201).

References

1. Qian, H., Liu, G.D.: The calculation of pH values and equilibrium distribution of species in natural water under the conditions of different partial pressure of carbon dioxide. Carsologica Sin. **13**, 133–140 (1994)
2. Yuan, D.X.: Karst Dynamic System in China, pp. 62–108. Geological Publishing House, Beijing (2002)
3. Jiang, Z.C., Pei, J.G., Xia, R.Y., Zhang, M.L., Lei, M.T.: Progresses and important activities of karst research during the 11th five-year plan in China. Carpological Sin. **29**, 349–354 (2010)
4. Zhang, M.L., Zhu, X.Y., Wu, X., Yang, H.P., Pan, M.C.: Characteristics of deposits and the depositional environment in the Shuijinggong cave in Bama, Guangxi. Carpological Sin. **32**, 345–357 (2013)
5. Zhang, M.L., Zhu, X.Y., Wu, X., Zhang, B.Y., Pan, M.C.: Characteristics of cave drip water and modern carbonate (CaCO3) deposits caused by underground river artificial recharge and landscape restoration. Carpological Sin. **34**, 17–26 (2015)
6. Li, T., Cao, J.H., Zhang, M.L., Huang, Y.M., Chen, J.R., Yan, Y.P., et al.: The seasonal variation of soil CO2 concentration in epikarst in the Panlong Cave, Guilin. Carpological Sin. **30**, 348–353 (2011)
7. Fairchild, I.J., Tooth, A.F.: Cave air control on drip water geochemistry, Obir Cave (Austria): implications for speleothem deposition in dynamically ventilated caves. Geochim. Cosmochim. Acta **10**, 2451–2468 (2005)
8. Whitaker, T., Jones, D., Baldini, J.U.L., Baker, A.J.: A high-resolution spatial survey of cave air carbon dioxide concentrations in Scoska Cave (North Yorkshire, UK): implications for calcite deposition and re-dissolution. Cave Karst Sci. **36**, 85–92 (2009)
9. He, W., Li, P., Qian, Z., Bottazzi, J.: Study on Shuanghe Cave Geopark, pp. 21–45. Guizhou People's Publishing House, Guiyang (2008)
10. Chen, J.G., Zhang, Y.J.: Formation and development of Shuanghe Cave system, Suiyang, Guizhou. Carpological Sin. **9**, 247–255 (1994)
11. Hess, J.W., White, W.B.: Groundwater geochemistry of the carbonate karst aquifer, southcentral Kentucky, USA. Appl. Geochem. **8**, 189–204 (1993)

Regional Pollutant Dispersion Characteristics of Weather Systems

Tingshuai Wang[✉], Qi Wang, Yunfeng Ma, Ping Wang, Wei Huang, and Dexin Guan

College of Energy and Environment, Shenyang Aerospace University, Neusoft, Shenyang City, China
tingshuai919@163.com, rcdxph@126.com

Abstract. In recent years, China has been plagued by the problem of air pollution. Considering the great harm of air pollution to human and the global environment, Atmospheric environmental problems have become the most difficult problem to be solved. This article is based on the weather system in July 2, 2014, using the big data cloud technology to show the weather system. The diffusion of pollutants which in the same time was simulated by the CALPUFF model. And then, we can analyze the relationship between the weather system and regional pollutant dispersion.

Keywords: Weather system · Big data cloud technology · CALPUFF model

1 Introduction

The Red River Nuclear Power Plant located in Liaoning Province, located in the east side of Liaodong Bay, site surrounded by the sea, located on the sea two terraces, undulating terrain height is relatively small, most of the area is flat and open. As the weather system has high reproducibility and seasonal, accurate understanding of its relationship to regional pollution has important significance for urban air quality management.

For showing the effect of diffusion of pollutants, use CALPUFF can be simulated image. Xudong Zou used CALPUFF to simulate the diffusion of sulfur dioxide in the atmosphere of Shenyang area [1]. Danli Zhang with CALPUFF model of Chongqing city visibility levels were simulated, these studies are similar, the results show, CALPUFF intuitively, true to the distribution of the concentration of pollutants in the atmosphere to simulate, and concluded that the contribution rate the biggest factor is the meteorological and geographical factors [2]. Not only for pollutant dispersion modeling visualization for visual analysis of weather systems can also be visually manifested by IDV (Integrated Data Viewer). It can be divided on the time scale, spatial scale and choice of meteorological elements, 3-D display provides information on the earth, but also allows users to interactively slice and dice the data probe, and the establishment of a cross-sectional profile, valuable animation and cube read, to meet the different needs of users [3].

© Springer Nature Singapore Pte Ltd. 2017
H. Yuan et al. (Eds.): GRMSE 2016, Part I, CCIS 698, pp. 572–578, 2017.
DOI: 10.1007/978-981-10-3966-9_63

In this paper, the big data cloud computing technology for different time periods in different types of weather systems meteorological elements, analysis, integration and visualization of two-dimensional and three-dimensional space, and big data cloud technology to today's environmental concern protection issues.

2   Method

2.1   WRF Model

With the continuous improvement of science and technology, mesoscale meteorological model in the application it has also made new breakthroughs. Used widely in the three mesoscale meteorological model is: fourth generation mesoscale model (MM4), the fifth generation mesoscale model (MM5) and the Weather Research and Forecasting Model (WRF), which are three ways in our country, the WRF the most widely used. Mainly used in research and simulation of physical parameterization schemes of individual cases. WRF data established in this paper uses a triple-nested mode, for 24 h at 0:00 on July 2, 2014 to at 0:00 on July 3, 2014 during the operation carried out. Set top pressure of 5000 Pa, in the vertical direction, using terrain-following height coordinate, setting the number of layers in accordance with the default pattern layer 28 is calculated. The calculated WRF meteorological diagnosis as a result of the field CALPUFF. As shown in Fig. 1:

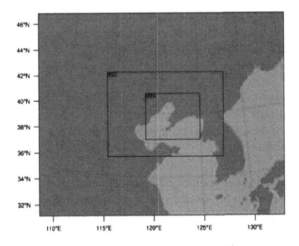

Fig. 1. WRF triple nested schematic

2.2   CALPUFF Model

CALPUFF is a multi-layer, multi-species non-steady-state puff dispersion model which can simulate the effects of time-and space-varying meteorological conditions on pollutant transport, transformation, and removal.

CALPUFF contains algorithms for near-source effects such as building downwash, transitional plume rise, partial plume penetration, subgrid scale terrain interactions as well as longer range effects such as pollutant removal (wet scavenging and dry deposition), chemical transformation, vertical wind shear, overwater transport and coastal interaction effects. It can accommodate arbitrarily-varying point source and gridded area source emissions.

$$Z = z - h_f \tag{1}$$

Z: terrain-following vertical coordinate (m);
z: vertical Cartesian coordinates (m);
hf: terrain height (m);

Terrain following coordinate system, the definition of the vertical velocity W is:

$$W = w - u\frac{\partial h_t}{\partial x} - v\frac{\partial h_t}{\partial y} \tag{2}$$

w: Cartesian coordinate vertical velocity (m/s);
u, v: horizontal wind speed (m/s);

The diagnosis of the core formula is:

$$w = (V \cdot \nabla h_f)\exp(-kz) \tag{3}$$

V: regional average (m/s);
hf: terrain height (m);
k: the stability of the attenuation coefficient;
z: vertical coordinate (m);

CALPUFF is used to simulate unsteady multi-Gaussian puff diffusion model many types of contamination can be simulated in time, spatial variation of meteorological conditions, contaminant migration, conversion and removals, including input and some part of the name of the output file, analog time, pollution, complex terrain and prediction points. CALPUFF use CALMET field related data generated by wind and temperature fields associated data file, transport of pollutants puff of air pollution emissions, so then to puff advection transport model to simulate the diffusion and transformation processes [4].

The basic principle CALPUFF a Gaussian puff model, which uses the method of integrating the sampling period to calculate the center of the grid and each designated point pollution concentrations. When the smoke lifted CALPUFF model to study the use of Briggs uplift formula (momentum and buoyancy uplift), taking into account the stable stratification in the part of the smoke penetration, transition clouds lifting and other factors. Wherein, CALPUFF calculation module, puff receiving point concentration equation is:

$$C = \frac{Q}{2\pi\sigma_x\sigma_y} g \; \exp[\frac{-d_a^2}{2\sigma_x^2}] \exp[\frac{-d_c^2}{2\sigma_y^2}] \tag{4}$$

$$g = \frac{2}{(2\pi)^{1/2}\sigma_z} \sum_{n \to -\infty}^{\infty} \exp[-(H_e + 2nh)^2/(2\sigma_z^2)] \tag{5}$$

CALPOST mainly used by the CALPUFF predicted visibility hour concentrations hour settlement data processing and output, since CALMET output file in binary form, you can use this post-processing module, which then needs to be data See, analysis and verification.

3 Weather Systems and Meteorological Elements CALPUFF Comprehensive Analysis of Simulation Results

3.1 The Relationship Between Temperature and Pollutant Dispersion

Figure 2 shows at 0:00 on July 2nd, 2014, 6:00, 12:00, 18:00, after the temperature change chart at different heights after the highly stratified, respectively 24 points purple, blue, green, red, black line. The figure shows that the temperature increases with height declining at a lower temperature at a height no obvious change in temperature at a relatively low altitude 0:00 and 24 o'clock, 6 o'clock and 12 o'clock high temperature.

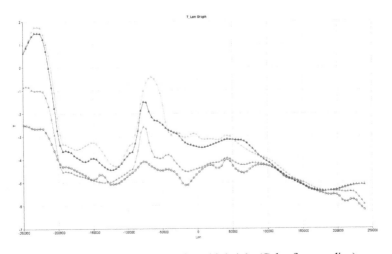

Fig. 2. Figure temperature varies with height (Color figure online)

Below is July 2, 2014 0:00, 6:00, 12:00, 18:24, 00:00 five at 850 hpa temperature profile distribution time Red River area and SO2 concentration diffusion FIG. The figure shows, at 0:00, the temperature of the northeast area shown higher and include Red River nuclear power plant, including the central and southeastern region in the state of lower temperature. Overall, more uniform temperature distribution in the region. And at 6:00, the temperature of the region have shown a slight overall increase, Northeast and Southwest large temperature rise, the temperature rising trend despite the overall presentation, Liaodong Peninsula region where the temperature is not much change, only a small margin promotion. At 12:00 and 18:00, the southwest region shown generally higher temperature, relatively low temperatures in the Northeast, while the Red River region in terms of temperature than the time before, without much change in temperature. At 24 o'clock, the temperature distribution and 0:00 is basically the same, relatively high bias northeast and southern temperature, low temperature of the eastern region, the Red River region presents an intermediate temperature trend.

In such a temperature change, spread of pollutants is first diffused more slowly, concentration change is not particularly large, the morning due to strong variations in temperature, diffusion changes in pollutants than the morning obvious, but in the afternoon to night, the temperature changes small, spread of pollutants is not particularly significant changes.

3.2   Relationship Between Wind Speed and Direction and Diffusion of Pollutants

Figure 3 is a July 2 Red River area in time series record wind field vector. Representative symbols tail wind speed, a wind-driven tail wind every level 2, the rod pointing in the opposite direction to the direction of representation, namely wind. The figure shows that the overall direction of the region from the south, wind speed to level 2–3.

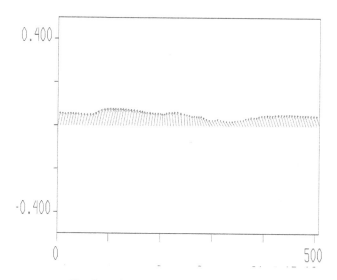

Fig. 3. Wind vector (Color figure online)

Figure 4 is at 0:00 on July 2nd, 2014, 6:00, 12:00, 18:00, 24 points after the wind speed changes after the height of the layered graph under different heights, respectively, with purple, blue, green, red, black wire FIG. The figure shows that at 6:00 when the change of wind at different heights of the most intense, 24 each at the height of the wind basically stable in the southwest wind, in the southeast and northeast monsoon cyclone, the Yellow River cyclones affected areas along the river red wind slightly confusing.

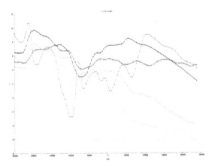

Fig. 4. Wind change map (Color figure online)

Figure 5 is July 2, 2014 0:00, 6:00, 12:00, 18:24:00: wind field characteristics map 00 five time Red River area and SO2 concentration diffusion FIG. We can see by the image, July 2nd malpractice MORALS southerly wind. At 0:00, east of the Liaodong Peninsula and the nearby area mainly southerly trajectory wind is relatively dense, concentrated effects of pollutants by the wind direction to the north west of diffusion, a larger concentration of pollutants, creates a low pressure area northwest of the study area cyclone. At 6:00, a few part of the region southwest wind into wind, wind speed slightly, this time spread range of pollutants that affect migration to the north wind slightly, but still subject to southeast winds spread range is not particularly big change, this pollutant concentrations decreased compared with 0:00. 12 points to

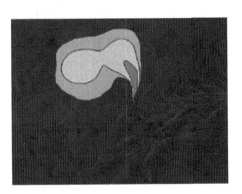

Fig. 5. 2014-07-02:00

the northwest and southeast regional research to produce multiple cyclones, wind slightly confused than before, but still the main southerly wind, wind and cyclones affected by pollutants diffusion range increases, the smoke plume tends to North East the direction in which the concentration of the transition section tends to spread range has undergone very significant changes. Diffusion of pollutants was stable period of time on the 18th, the cyclone is gradually reduced until it disappears at this time, the wind moderated consistent wind speed is relatively stable, diffusion of pollutants are mainly concentrated to the north, when the concentration of pollutants compared with the previous point in time concentration has increased. 24 zone appears when the southwest wind, but the wind is small, then there are slightly to the trend southeast of pollutants diffusion, but the overall diffusion range without much change.

4 Conclusions

Comparison by the temperature distribution map and pollutant dispersion diagram shows that temperature is the reason for the spread of pollutants, but not the main factors. The wind direction and speed distribution and diffusion of pollutants comparison chart shows that the effect of pollutant dispersion and direction of the wind are closely linked to the spread of pollutants and wind direction remained the same, and with the wind speed increases, the rate of diffusion of pollutants will faster. In addition, meteorological factors discussed herein weather system limited, and many other possible impact on the spread of meteorological factors not analyzed, which may have to carry out specific studies in the future.

References

1. Zou, X., Yang, H., Liu, Y.: CALPUFF air pollution simulation in Shenyang area of research. J. Meteorol. Environ. **06**, 24–28 (2008)
2. Zhang, D., Zhou, Z., Zhai, C.: Use CALPUFF model to simulate the main city of Chongqing level of visibility. Three Gorges Environ. Ecol. **01**, 8–11 (2013)
3. Lei, S., He, X., Xue, Q., Qing, X., Zhang, X., Qing, M.: Integrated data viewer and OpenFlashChart in public weather services system. Inf. Syst. Eng. **07**, 14–17 (2011)
4. Bo, X., Ding, F., Xu, H., Shi, S.: Atmospheric diffusion environmental monitoring and management technology. Technol. Rev. CALPUFF Mod. (2009)

Study on the Selection and Moving Model of the Poverty Alleviation and Resettlement in the Typical Karst Mountain Area

—A Case Study of Pan County in Guizhou Province

Yanxi Pan[1,2], Zhongfa Zhou[1,2(✉)], Qian Feng[1,2], and Mingda Cao[1,2]

[1] School of Karst Science, Guizhou Normal University, Guiyang, China
963860439@qq.com, 939268275@qq.com,
793149352@qq.com, fa6897@163.com
[2] State Engineering Technology Center of Karst Rock Desertification Rehabilitation, Guiyang, China

Abstract. Poverty alleviation and relocation is one of the three major measures for poverty alleviation and development, which help the relocation of the population to gradually get rid of poverty and get rich. The application of GIS technology of the poverty alleviation and relocation in the typical Karst mountain area is easy to find the appropriate relocation of residence, for the local government to provide decision-making reference. Resettlement area selection mainly consider the following factors: traffic facilities, the land is rich in resources, rich in water resources, gentle slope, lower elevation, avoid risks of geological disasters, natural protection area and national planning and construction land, ecological fragile area and the environmental carrying capacity overload region. Through the GIS technology, the above factor layer to find a suitable place to place. According to the placement of the situation to choose a suitable and easy to poverty alleviation and relocation model. The results showed that Pan County has a larger area of the resettlement area, There are four villages and towns belongs to the center village settlements, should use the center of the village resettlement mode; and nine villages and towns belongs to small urban settlements, urban commercial relocation mode should be adopted.

Keywords: Karst area · The poverty alleviation and relocation · The model of poverty alleviation and relocation · GIS technology

1 Introduction

The proportion of population, resources and environment in Karst mountainous areas is seriously imbalance, and the ecological problems and the survival problems have become a difficult problem to resolve. Poverty alleviation relocation is an important measure of China's poverty alleviation and development, is also one of the main measures of the construction of new rural areas [1]. It is through the improvement of the living conditions of the resettlement area, adjust the economic structure and expand the channels to help the relocation of the population gradually get rid of poverty and to

© Springer Nature Singapore Pte Ltd. 2017
H. Yuan et al. (Eds.): GRMSE 2016, Part I, CCIS 698, pp. 579–588, 2017.
DOI: 10.1007/978-981-10-3966-9_64

get rich. And it also improve the local ecological environment [2]. The success or failure of resettlement often depends on the analysis and rational utilization of the spatial and geographical environment, and the decision making is closely related to the geographical environment [3]. The problem of how to place and where to place gradually emerged, becoming the focus of research [4]. Such as how to obtain a higher compensation for the resettlement of the poor resettlement [5, 6]. GIS has integrated all kinds of elements and powerful spatial analysis function, therefore, suitable for use in the choice of placement. This paper discusses the application of spatial analysis techniques of GIS spatial analysis, buffer analysis, statistical analysis and other spatial analysis techniques to the selection of Relocation In County in Guizhou Province as an example, through the topography, geomorphology, transport, the environment, characteristics of regional selection and economic and social development and other factors were analyzed, so as to provide a scientific basis for the choice of relocation [7].

2 Study Area

Pan County, belong to Liupanshui City, the cool city in China, is a typical Karst mountainous area. The combination of Yunnan Province, Guizhou Province, Guangxi Province. The specific location between 104°17′46″–104°57′46″E and 25°19′36″–26°17′36″N, is the west gate of Guizhou. East Puan County, South Xingyi County, west of Xuanwei and Fuyuan County, Yunnan County, north of Shuicheng county. Land area 4056 km², accounting for 2.3% of the province's total land area, mountain area accounted for 82.4% of the total area, hilly area accounts for 9.2% and dam occupy 2.4%. Forestry mu land is 2503.3 km², the total land area of 61.3%, is an area of 970 km², the total land area of 23.9%. Wood land has 2503.3 km², accounting for 61.3% of the total land area, arable land area of 970 km², accounting for 23.9% of the total land area. Northwest high terrain, the eastern and southern lower, down from the south. Region is a sub tropical climate, winter without cold, summer without heat, the annual average temperature of 15.2 °C. The county's original 39 townships, followed by a reduction of 27 townships.

In 2012–2015, there are a total of 1785 household of immigrants in Pan County, the number of resettlement of 5989 people, and 7 settlements for relocation of poverty alleviation.

3 Basic Data Sources and the Main Consideration of the Choice of Placement

3.1 Basic Data Sources

Basic research data of this paper include of Pan County: (1) The 1:50000 DEM data and hillshade map; (2) The 1:10 000 land use map; (3) The 1:100 000 administrative zoning map; (4) The 1:18 000 natural village settlements distribution figure; (5) River, traffic data by 1:250000 China basic geographic information layer; (6) The 1:10 000 Nature Reserve distribution map; (7) TIN data.

3.2 The Main Consideration of the Choice of Placement

The choice of the factors that are considered in the selection of poverty alleviation and relocation is the key to the success or failure of the poor, so the choice of the place is very important. Not only to consider whether it is conducive to the lives of immigrants, production recovery, but also to consider the religious beliefs of immigrants, habits and other social factors [8]. The following factors (Table 1) are mainly considered:

Table 1. The mainly factors index considered in the selection of poverty alleviation and resettlement in typical Karst mountain area.

Main consideration factor	Data processing
Convenient traffic	Traffic buffer (1–3 km)
Good condition of water resources	River buffer (0.5–2.5 km)
Slope <25°	Classification of slope
Altitude <2.6 km	Classification of Contour
Outside the existing residence 0.5 km	Eliminating the natural village 0.5 km buffer
Abundant land resources	Extraction of suitable land resources
Avoid nature reserves and the state key construction planning land	Eliminate the protected area
Avoid geological hazards area	Eliminate disaster area
Avoid the ecological vulnerability area and the ecological environment carrying capacity overload area	Evaluation of SRP and evaluation of bearing capacity

After considering all the above indicators in the selection of relocation program for poverty alleviation in typical Karst area, it is concluded that the flow chart (Fig. 1):

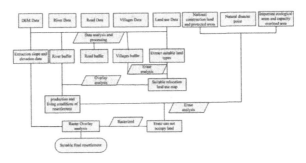

Fig. 1. The flow chart of the selection of relocation program for poverty alleviation in typical Karst area

The slope, slope direction, elevation and so on can be extracted in DEM of Pan County. Ecological vulnerability assessment is a comprehensive evaluation, Its evaluation model is mainly represented by SRP model [9, 10]. At present, the conceptual model of ecological vulnerability assessment is mainly based on the method of SRP (Sensitivity-Recovery-Pressure) evaluation model [11], SRP model is a comprehensive evaluation of the ecological vulnerability in a particular area [12]. SRP model has been widely used in the evaluation of ecological vulnerability in some areas, and has achieved some results [13]. The evaluation of ecological vulnerability in Pan County as follows:

$$E_{EVI} = \sum_{i=1}^{n} F_i \times W_i \qquad (1)$$

In this formula, EEVI for ecological vulnerability index; Fi is the index of the research index; Wi as the index of the corresponding weight value of the research. The main indicators are: Vegetation coverage, soil erosion intensity, GDP density, population density, average annual rainfall, soil type, soil erosion intensity, soil and water loss degree.

4 Analysis of Main Influencing Factors of Relocation and Resettlement in Karst Mountain Area

GIS powerful spatial analysis technology applied to the choice of place, and the space technology analysis, scientific management and assistant decision support can apply in the resettlement work. Application of ArcGIS software analysis functions, such as Buffer, Calculator Surface, Analysis Raster, Erase, and so on, get the above 10 factors, 10 factors are satisfied for the region to be suitable for the relocation of Pan County. The detailed analysis procedure is as follows:

4.1 Analysis of Traffic

Convenient transportation is not only convenient for residents living, production, facilitate the construction of all kinds of facilities, but also the residents and the outside world of information exchange and commodity exchange channels. Therefore, it must be close to the traffic roads, to facilitate the residents living [13]. However, taking into account the impact of debris flow, landslide and other geological disasters, the more close to the highway, the higher sensitivity [14]. The residents are too close to the road and their lives are affected by the noise. The study selects 1–3 km as the buffer zone.

4.2 Analysis of Water Resources

Water resource is one of the factors in the selection of resettlement areas, water resources are the basic guarantee for the survival and development of human beings. Therefore, it is essential to ensure that the residents of the resettlement area have adequate water resources [13]. But to avoid the impact of the main river caused by debris flow and other geological disasters [14, 15]. In the study, the distribution map of the river and the main river in Pan County were taken as the buffer zone of 0.5 km and 2.5 km respectively. By superposition, we can gain a River buffer which 0.5 km and 2.5 km from the river resources.

4.3   Analysis of Slope

After the relocation of migrants, the need for relatively flat land to live and farming. China has more than 25° of arable land to implement the policy of returning farmland to forest, the resettlement area to choose a slope of less than 25°. In ArcGIS9.2, the Analyst Tools Spatial module in ArcToolbox tool was used for slope analysis.

4.4   Analysis of Altitude

With the increase in altitude, the air temperature and the oxygen content in the air are reduced. High altitude and low temperature is not conducive to human habitation. At present, the population distribution of Pan County is mainly in the following 2.6 km, so select 2.6 km as the highest elevation of the suitable living [16]. The resettlement area is selected within the region of the altitude less than 2.6 km.

4.5   Analysis of Natural Village

The resettlement of immigrants can not affect the interests of other non immigrant [17, 18], If the place too close to the existing place of residence, it means that the land, public facilities, employment opportunities and other scarce resources for the competition. Resettlement areas should be away from the non immigrant natural village 0.5 km. Using the ArcGIS buffer command to establish the buffer zone of Pan County natural village 0.5 km, remove the buffer range (0.5 km) from the Pan County with the Erase command, and get the area out of the resident point 0.5 km.

4.6   Analysis of Land Use

Land is the most basic means of production and psychological security of farmers, especially in minority areas, due to historical reasons, the dependence on land is stronger [19]. There are relatively abundant land resources for the development of the relocation of immigrants to help immigrants to resume production and to achieve as soon as possible; So the placement selection eliminates the use of land that is not available on the land type map, such as bare rock, bare land, beaches and other types, Choose a woodland, woodland, shrub land, wasteland and bare land as relocated.

4.7   Analysis of Nature Reserves and National Key Construction Planning

Resettlement areas can not be selected in the nature reserve, the prohibition of the development zone and the state key construction planning land. There are Badashan nature reserve, Laoheishan provincial Forest Park and seven provincial Forest Park peak in Pan County. The placement cannot be selected in this area, when we analysis we must to remove these areas on our Map.

4.8   Analysis of Geological Hazards

Part of Pan County frequent natural disasters, especially the hail disaster. In order to ensure the survival of migrants, the resettlement areas should avoid the hidden dangers of geological hazards, so in the analysis we must to eliminate the scope of natural disaster prone areas.

4.9   Analysis of the Ecological Importance of the Region and the Ecological Environment Carrying Capacity Overload Area

Comprehensive analysis of the degree of ecological vulnerability of the township and classification, eliminate the fragile ecological areas. The ecological environment carrying capacity of each township was evaluated, and the villages and towns which were overloaded with ecological carrying capacity were analyzed to avoid the area which overload of the ecological environment.

5   Resettlement Analysis Results and the Application of the Relocation Model

5.1   Analysis Results of the Placement

After the above layer overlay, get the suitable area of poverty alleviation and relocation in Pan County. But not all places can be used for the placement of the population, it must to reach a certain size that can be used for the resettlement of the population, therefore, the area of a small range will be removed. In addition, it also avoid the geological disaster point range, nature reserves, national key construction sites, ecological vulnerability and resource carrying capacity overload area, through finishing, get the final settlement area distribution map and area.

As we can be seen from Fig. 2, there are a few areas of suitable resettlement areas are located in fourteen towns of Pan County, the rest were scattered in other towns. This is also in line with the actual situation in Pan County. From the figure can be seen there are fourteen villages and towns have larger appropriate placement. The land production value is low in Pan County. It is generally believed that about 3000 m² of land resources to Support a person. The Center Village settlements according to each people need 3500 m² of land resources, that can be placed about 7454 people in these four villages and towns (Table 2). The suburban settlement point according to each people need 2800 m² of land resources, that can be placed about 18175 people in these ten villages and towns (Table 2). There are about 25629 people can be placed in those Suitable settlement area.

Fig. 2. The suitable resettlement of poverty alleviation and relocation in Pan County.

Table 2. The results of the suitable placement of poverty alleviation and relocation in Pan County

Name	Suitable area (m^2)	Suitable for resettlement (person)
WuMeng town	3542893.96	1012
Pingdi town	8673004.41	3098
BaiGuo town	5640287.18	2014
JiChangPing town	6008748.15	2146
BaoJi town	5328512.51	1903
YangChang town	11819094.26	3377
JiuYing town	3018545.65	1078
YingWu town	3301977.17	1179
DanXia town	3911354.93	1397
DaShan town	4563247.42	1304
XinMin town	6164635.49	1761
XiangShui town	7411734.16	2647
Shengjing street	1133726.06	405
HongGuo street	6462238.58	2308

5.2   Corresponding Relocation Model

According to the placement of the analysis of the results, get the appropriate placement of the fourteen towns in the area and the number of resettlement, according to the location of the location and the local socio-economic situation, choice the appropriate relocation model for it.

(a) Center village settlement type

Wumeng Town, Yang Chang Town, Da Shan Town, Xinmin Town belongs to the center village settlements, should use the center of the village resettlement mode. Let the immigrants moved to the nearest villages and towns which with good living conditions, perfect infrastructure, convenient transportation and have good prospects for the farther development. Or rely on the newly reclaimed or adjust the use of arable land in the surrounding counties, towns or administrative villages planning and construction of immigrant village, centralized resettlement of immigrants.

(b) Urban commercial type

PingDi Town, Bai Guo Town, JiChangPing Town, Bao Ji Town, Jiu Ying Town, Ying Wu Town, Danxia Town, Shengjing street, HongGuo street belongs to small urban settlements, urban commercial relocation mode should be adopted. The small towns which with good economic development and better infrastructure should be set as an immigrant settlement point, to help the immigrant to get rich, at the same time also accelerate the development of urbanization.

(c) Park Service Type, Rural tourism type

Xiangshui Town is a resettlement of rural tourism and it belong to the town of park service, we should be rational explore the tourism resources of Ba Ma River. Through construct resettlement nearby the tourist attractions. After immigration, through the creation of farmhouse hotels and tourist services stalls, to carry out ethnic customs tourism, development of tourism products, to join the construction of tourism scenic spots way to solve immigration problems of survival and development, immigrants increase income and living standards.

6   Conclusions

The poverty alleviation and relocation played an important role in solve those who do not have the production, living conditions, as well as the poverty population who live in ecological fragile areas [20]. There are many methods to choice location at present, and there are many advantages based on GIS to choice location [21].

(a) The relocation resettlement selection in the typical Karst mountain area mainly consider the following factors: traffic facilities, the land is rich in resources, rich in water resources, gentle slope, lower elevation, avoid risks of geological disasters, natural protection area and national planning and construction land, ecological fragile area and the environmental carrying capacity overload region.

(b) In Karst mountain area, it is not easy to use the traditional personnel field survey to find the place of the settlement, GIS powerful spatial analysis function is particularly suitable for the choice of settlement in Karst area, and choice the corresponding relocation model for it. In this paper the relocation resettlement select analysis using the vector analysis and raster analysis technology and data conversion, to extract the relevant factor required feature information, finally using raster overlay analysis, find appropriate relocation resettlement in Pan county each villages and towns. Using spatial analysis to making decision will have a more and more important role in the future.

(c) Pan County has a larger area of the resettlement area, There are four villages and towns belongs to the center village settlements, should use the center of the village resettlement mode; and nine villages and towns belongs to small urban settlements, urban commercial relocation mode should be adopted.

(d) This thesis deals with the placement of the selection method can be used not only to poverty alleviation placement choice for immigrants but also to hydropower development and other major national construction projects resettlement choice and the too scattered village relocation work.

Acknowledgments. In this paper, the research was sponsored by the major application foundation research project of Guizhou Province (Guizhou S&T Contract JZ 2014-200201); Guizhou Province Science and Technology Plan (Guizhou S&T Contract G 2014-4004-2); The soft science research project of reservoir and ecological resettlement Bureau of Liupanshui City, Guizhou Province; The soft science research project of Development and Reform Commission of Pan County, Guizhou Province.

References

1. Li, C.N., Chao, M.C.: The characteristics of poor population in Nujiang river and the choice of the way to help the poor. Create **04**, 30–31 (2000)
2. Wang, S.Q.: The role of ease of poverty alleviation and relocation in the development of poverty alleviation. Yunnan Agric. **10**, 34–35 (2006)
3. Yao, H.L.: On the application of GIS in resettlement decision-making process. J. Henan Educ. Inst. (Nat. Sci.) **01**, 63–65 (2000)
4. Kanctuma, N.: Putting houses in place I rebuilding communities in post-tsunami Sri Lanka. Disasters **33**, 436–456 (2009)
5. Yang, H., Huang, J., Li, S.Q., Liu, Y.: Development of Hundred-li tourism corridor of the Dong national culture in Tongdao county Hunan. Trop. Geogr. **05**, 451–454 (2007)
6. Rogecs, S., Wang, M.: Environmental resettlement and social dis/re-articulation in inner Mongolia, China. Popul. Environ. **28**, 41–68 (2006)
7. Li, Y.M., Kong, J.: The selection of resettlement sites based on GIS—a case of Liuku power station in Nujiang river. Geo-Inf. Sci. **02**, 55–59 (2007)
8. Shi, J.H., Gai, Z.Y.: Study on the selection of ecological resettlement in inner Mongolia. Inn. Mong. Agric. Sci. Technol. **01**, 1–4 (2006)
9. Pei, H., Fang, S.F., Qin, Z.H., Hou, C.L.: Method and application of ecological environment vulnerability evaluation in arid oasis-a case study of Turpan oasis, Xinjiang. Geomat. Inf. Sci. Wuhan Univ. **05**, 528–532 (2013)
10. Lu, Y.L., Yan, L., Xu, X.G.: Ecological vulnerability assessment and spatial auto-correlation analysis over the Bohai Rim Region. Resour. Sci. **02**, 303–308 (2010)
11. Qiao, Q., Gao, J.X., Wang, W., Tian, M.R., Lu, S.H.: Method and application of ecological frangibility assessment. Res. Environ. Sci. **05**, 117–123 (2008)
12. Zhou, R.H., Xia, L., Gary, C.H.: Assessment of soil erosion sensitivity and analysis of factors in the Tongbai-Dabie mountainous area of China. Res. Gate **101**, 92–98 (2013)

13. Tan, C., Xu, J.J., Li, L.J., Guo, Q.Y.: Study on social integration of resettlement for reservoir resettlement in the county – a case study of Xiaolangdi Dam reservoir in Yuanqu county Shanxi Province. J. Hohai Univ. (Philos. Soc. Sci.) **02**, 54–57 (2002)

14. Tang, C.: Susceptibility spatial analysis of debris flows in the Nujiang river basin of Yunnan. Geogr. Res. **02**, 178–185 (2005)

15. Zeng, F.W., Xu, G., Li, Q.: Critical gradient and debris flow on mountain slope. Sci. Geogr. Sin. **02**, 244–247 (2005)

16. Hao, X.Z., Li, Y.M.: Reasons analyses and solutions discussions of poverty in Nujiang prefecture. J. Yunnan Univ. **01**, 166–172 (2007)

17. Li, Y.M., Hao, X.Z., Li, Y.M.: An analysis of poverty reasons and aid-the-poor strategy for Nujiang prefecture, Yunan. Trop. Geogr. **01**, 63–67 (2008)

18. Chen, X.Y.: Resettled backward and resettled outward. Water Econ. **03**, 53–57 (2002)

19. Huang, S.W.: Thoughts rural anti poverty in Western ethnic areas. Acad. Forum **04**, 95–98 (2004)

20. Wang, Y.P., Yuan, J.Y., Zeng, F.Q., Chen, N.: Discussion on the resettlement mode of the poverty peasants in the ecologically fragile areas in Guizhou province. Ecol. Environ. (Acad. Ed.) **01**, 400–401 (2008)

21. Chen, Y.: Based on GIS location space decision support system. China Water Transp. **04**, 120–121 (2007)

Assessment of Flood Hazard Based on Underlying Surface Change by Using GIS and Analytic Hierarchy Process

Lin Lin, Caihong Hu[(⊠)], and Zening Wu

School of Water Conservancy and Environment,
Zhengzhou University, Zhengzhou, China
Linlin_1577@163.com, hucaihong@zzu.edu.cn

Abstract. Flood hazard is one of the most common natural hazard in plain urban area. The location chosen for the study is Zhengzhou city, provincial capital, China. The model incorporates four factors: rainstorm with increased frequency and intensity, land subsidence, density of population, impermeable land surface, land use change. In this study, the coupling of geographical information system (GIS) and the analytical hierarchy process (AHP) were used, GIS analysis urban flood hazard vulnerability based on the underlying change and AHP was used in order to calculate the weighting values of each factor. A hazard map, based on the historical data in the past ten years, was obtained by spatial analysis technology. The hazard map was compared with the actual flood area, and good coincidence was found between them. The flood hazard assessment method presented here is meaningful for the local government to improve flood risk management and protecting environment.

Keywords: Flood hazard · GIS · Analytic hierarchy process · Hazard analysis

1 Introduction

Flood is one of the most frequently immeasurable disasters in China, affecting more than 1.3 billion people in the last century (Liu et al. 2012). The dramatic and ongoing urbanization process in plains city is leading to an increase of flood hazard and an increase of population an infrastructure in flood-prone areas (Muller et al. 2011). The integrated risk of flood disaster in urban areas has presented spatial disparities evidently at present. With the more complicated surface environment and, inland urban is at higher risk compared with other areas. The consequences are not only environmental but economic as well, while they may cause damages to urban areas and agricultural lands and may even result in loss of lives (Stefanidis et al. 2013). Thus, it is very necessary to understand the historical evolution of flood disasters in the past several decades to control flood risk effectively.

Frequent extreme weather events and increased flood risk cities have become the social focus in the worldwide and academia. Assessment of flood risk has been widely studied. Generally, the primary methods for flood hazard assessment are divided into 3 types including historical flood data-based assessment method, index system-based

© Springer Nature Singapore Pte Ltd. 2017
H. Yuan et al. (Eds.): GRMSE 2016, Part I, CCIS 698, pp. 589–599, 2017.
DOI: 10.1007/978-981-10-3966-9_65

assessment method, hydrologic-hydraulic model and scenario simulation-based assessment method. Benito (2004) build the assessment method of urban flood risk, according to the historical disasters data during the past 1000 years and the 50 years of specification. This method is simple and convenient to understand the variation of flood risk, with numerous urban waterlogging disaster observation data, but it's hard to get the historical damage data, and it has certain imitation in application. For index system-based assessment method, several indicators should be selected from numerous impact factors of flood risk, a series of mathematical methods are used to deal with the indicators and the regional disaster risk. This method focuses on the selection of the indicators and the optimization of weights, wide range of spatial scales is involved, Hotspots and American plans are suitable for the global indicator plan of disaster risk assessment (Dongdong et al. 2014). The method of flood hazard prediction by applying hydrologic and hydraulic models and scenarios simulation can derive the flood range of urban waterlogging, submerged depth and submerged duration, caused by different scenarios storms, by using watershed runoff model and numerical simulation of flood routing model (Booij 2005). However, these models need large-scale data, which are often unavailable.

Alternatively, flood hazard assessment methods can use multi criteria analysis. Analytic hierarchy process is the most widely used multi criteria analysis method and has been widely used to a wide range of scientific fields to calculate the relative weight of each factor. Furthermore, as a decision support system, GIS often integrates the spatially referenced data in different problem-solving environment (Cowen 1988). The method based on GIS-AHP has not only widely used in the practical multi criteria decision analysis, but also been widely applied to suitability analysis and natural hazard analysis (Chang et al. 2008). This study is designed to develop a GIS-AHP flood hazard model for an urban area, which is meaningful for the flood management and environmental protection in the area under the similar condition as this study.

2   Study Area and Materials

2.1   Study Area

Zhengzhou is a transportation hub city in China and sits in the middle of Henan province, Zhengzhou covers an area of 1700 km^2, where is bordered by Yellow river to the north (Fig. 1), located in upstream of Jia Lu river, the area is characterized as lowland with an average altitude of 50 m (maximum 340 m), from the southwest to northeast stepped down. The study area is in a warm temperate zone continental monsoon climate, the mean annual precipitation is 625.9 mm, Zhengzhou has suffered heavy rainstorm more than 15 times per year since 2006, each time flood disaster caused more than 200 million economic losses. From June to September, the area gets a great level of rainfall, i.e., 70% of annual precipitation occurs in this period. The changes of flood risk are closely related to the combined effect of the expansion of impermeable land surface, climate change, urbanization, land subsidence and sea-level rise, resulting in heavy losses of property and seriously affecting sustainable development of society and economy. Zhengzhou has experienced a rapid urbanization

Fig. 1. The study area. (Color figure online)

during the past 20 years. Moreover, its urbanization levels will continue to increase in the future several 20 years. Coupled with rapid population growth and economic aggregation, as the extreme weather increased, Zhengzhou is facing more and more serious flood disaster.

2.2 Data Sources

The associated data source utilized in this research includes foundation geographical data and the statistical data of flood hazard. Data supported by GIS technology are about vector and raster format, such as the infrastructures, roads, residents and administrative map of Zhengzhou in vector format, high resolution terrain data (8 m resolution digital elevation model) in raster format. The historical flood data between 2009 and 2015 are provided by hydrographic office of Zhengzhou city, recorded the numbers and the losses of each flood disaster event, from start and end dates, affected areas by districts. In this paper, district was taken as basic unit to analyze the spatial-temporal and characteristics and established the flood disaster database.

3 Methodology

3.1 Flood Hazard Assessment Function

The existing method to evaluate flood hazard risk, most of which could partly due to the knowledge of experts (Furdada et al. 2008) and based on numerical modeling to investigate how controlling factors contribute to flood. Hydrologic and hydraulic models are two main types of numerical models for evaluating the risk. These methods are useful for regional studies (Dewan et al. 2007). In this study, a semi quantitative method was selected to investigate the flood hazard assessment in the study area Fig. 2

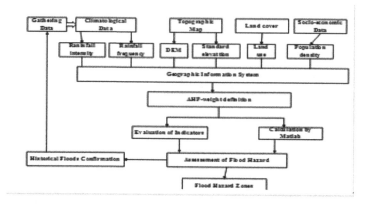

Fig. 2. Flow chart of the GIS-AHP flood hazard assessment method.

shows the steps used in this study including data gathering, GIS manipulation, and AHP. The assessment of flood hazard incorporates risk analysis of disaster factors, sensibility of hazard-formative environment, and vulnerability of disaster victims, along with selection and definition of the parameters' weights, the flood hazard zones are the outcome of the assessment. The risk assessment can be calculated using Eq. (1)

$$R = f(F, S, V) \tag{1}$$

3.2   Selection of Relevant Variables

The identification of the flood relevant indicators is the most important step ate the assessment of flood hazard. In the present study, an index model has been developed in a GIS environment aiming to define flood hazard areas. The review of relevant literature and the field surveys and interviews carried out in the scope of this research showed that the variables compiled in Fig. 1 are the most relevant for flood vulnerability analysis in this study area. Main data are climatological data, topographic map, land cover and socio-economic data. Thus the index model comprises six criteria-parameters: rainfall intensity and frequency, DEM and topographic standard deviation, land use and population density. These indicators fits the condition and the flood characteristics well. All these data for this assessment are available.

3.3   Relative Weights of the Criteria

The template is used to format your paper and style the text. The GIS-AHP flood hazard assessment method considers the above hydrogeological, morphological and socio-economic parameters and the weights of each factor determines its role in the final result. The weights are defined following the AHP (Saaty 2008). These weights of the criteria are defined after they are ranked by according to their relative importance.

Thus the criteria are sorted in a hierarchical manner. The relative significance between the criteria is evaluated from 1 to 9 indicating less important to much more important criteria. A comparison of the criteria significance resulted to the principal eigenvalues of Tables 1 and 2. Table 3 includes the weight of each index assignment by MATLAB computational procedure.

Table 1. Judgement marix table of flood hazard

Criterion name	Disaster factors	Hazard-formative environment	Disaster victims
Disaster factors	1	5/7	5/9
Hazard-formative environment	7/5	1	7/9
Disaster victims	9/5	9/7	1

Table 2. Judgement marix table of index layers

Criterion Layer	Parameters		
Disaster factors		Rainf.intensity	Rainf.frequency
	Rainf.intensity	1	8/9
	Rainf.frequency	9/8	1
Hazard-formative environment		DEM	Stand. elevation
	DEM	1	5/7
	Stand.elevation	7/5	1
Disaster victims		Land use	Pop. density
	Land use	1	3/4
	Pop.density	4/3	1

Table 3. Index weights of relevant variables

Weights	Disaster factors	Hazard-formative environment	Disaster victims	
	0.4406	0.3147	0.2448	
Rainf.intensity	0.5294			0.2333
Rainf.frequency	0.4706			0.2073
DEM		0.4167		0.1311
Stand.elevation		0.5833		0.1836
Land use			0.4286	0.1049
Pop.density			0.5714	0.1988

Following the calculating of the weights of indicators by AHP, its consistency needs to be evaluated.

3.4   Consistency Check

Following the creation of the eigenvector matrix of the AHP, its consistency needs to be evaluated. The required level of consistency is evaluated using the following index:

$$CR = \frac{CI}{RI} \tag{2}$$

Where:

CR   the consistency ratio
CI   the consistency index
RI   the random index

AHP's theory suggests that the consistency ratio (CR) must be <0.1. CI is calculated using Eq. (3), with λ_{max} being the maximum eigenvalue of the comparison matrix and n the number of criteria. RI values are given in specific tables.

$$CI = \frac{\lambda_{max} - n}{n - 1} \tag{3}$$

For the values of Table 2, CI was calculated for: λ_{max} = 6, RI = 0.52. Eventually, the consistency ratio has been calculated CR = 0. Since CR's value is lower than the threshold (0.1) the weights' consistency is affirmed.

4   Analysis and Results

4.1   GIS-AHP Flood Hazard Assessment

Based on factors' relevance with respect to the flood susceptibility in the study area and on the data quality, six factors were selected. These are rainfall intensity and frequency, DEM and standard elevation, land use and population density. The resolution of this hazard assessment is 8 m * 8 m, following normalizing the value of the six indexes, the analysis of the risk of disaster factors, fragility of the hazard-formative environment and the vulnerability of disaster victims could be adopted by spatial analysis technology of GIS.

4.2   Analysis of Parameters

Risk Analysis of Disaster Factors. Zhengzhou city affected by continental monsoon climate, more rain in summer and rain the same period. Frequent short heavy rainfall is the main cause of urban rainstorm floods. Following analyzing rain intensity and rain frequency details of Zhengzhou city from 2009 to 2015, rainfall intensity can be expressed in maximum daily precipitation, rainstorm frequency can be represented with the number of storm rainfall. The two factors are calculated by the hierarchical analysis and the weighted synthesis method. Specific steps are as follows:

(a) Statistical calculating maximum daily precipitation and number of storm rainfall.
(b) Normalizing factors based on the 100 m * 100 m mesh grid data.
(c) Calculating the relative weights of the disaster factors by AHP.
(d) Conclude spatial distribution characteristics of two kinds of factor influence with spatial analysis kriging interpolation technique.
(e) Draw risk zoning of disaster factors using natural breakpoint classification method, in which risk can be divided into 5 grades (very low, low, moderate, high, very high), as shown in Fig. 3.

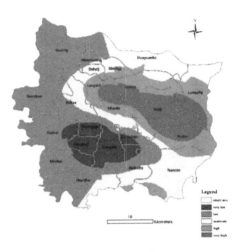

Fig. 3. Map of risk analysis of disaster factors.

The results show that the precipitation is strong in the eastern part of Zhengzhou City, and weak in the western part, and the heavy rainfall in Zhengzhou is concentrated in the middle and eastern part of the city. It is found that, besides the influence of climate, it may be influenced by the topography factors.

Sensibility of Hazard-Formative Environment. Terrain condition is the most important element of underlying surface, water flows from higher to lower elevation DEM influences the amount of surface runoff and infiltration. Lower altitude in the urban area and the area of gently topography is easy to form water area caused by urban waterlogging. Therefore vulnerability analysis of subsequently environment could select absolute height and relative standard deviation as evaluation indexes.

(a) Calculating standard deviation based on 100 m * 100 m of DEM grid in the window size of the 3 * 3 lattice with the surrounding eight level lattice.
(b) Integrated terrain factors effect as is shown in Table 4.

(c) Elevating and standardization of standard deviation and DEM, adopt the method of fuzzy membership.
(d) Classify the integrated terrain effect according to natural breakpoint classification method of GIS. Zhengzhou city rainstorm floods subsequently environment vulnerability is divided into five grades, as shown in Fig. 4.

Table 4. Integrated terrain factors effect

Elevation (m)	Standard deviation(m)				
	High risk (≤ 3.5)	Higher risk (≤ 5.35)	Moderate risk (≤ 8.2)	Lower risk (≤ 12.9)	Low risk (12.9)
High risk (≤ 95.9)	0.9	0.8	0.7	0.6	0.5
Higher risk (≤ 121.2)	0.8	0.7	0.6	0.5	0.4
Moderate risk (≤ 154.9)	0.7	0.6	0.5	0.4	0.3
Lower risk (≤ 197.1)	0.6	0.5	0.4	0.3	0.2
Low risk (197.1–289)	0.5	0.4	0.3	0.2	0.1

Fig. 4. Map of Sensibility of hazard-formative environment.

The results show that the smaller the terrain relief, the smaller the impact, the greater the risk of flooding. The terrain of Zhengzhou City decline from southeast to northeast, and the east and north part of the city is more sensitive than other areas.

Vulnerability of Disaster Victims. In addition to topographic relief, some other underlying surface conditions such as the population density and land use condition also have greater impact on flood risk. As an inland city, Zhengzhou located in the central plains, when suffered heavy rains, the greater the population density and building in the greater the ratio of the land use types, the more economic losses. Thus the population density and land use type were chose as evaluation indexes. Land use types in Zhengzhou mainly include building lands, roads, drainage, forest land and grassland, etc. Therefore the expert scoring method is adopted for each land use type of empowerment, and rasterize processing. Expert scoring method for weight assignment of fragility of land use type as shown in Table 5. In Zhengzhou city storm flood vulnerability analysis of hazard-affected victims, using natural discontinuous point method and the comprehensive weighted technique of GIS will standard after the normalization of population density and grading index of land use types in zoning. Vulnerability zoning of flood victims in Zhengzhou as is shown in Fig. 5.

Table 5. Landuse sensitive assignment

Risk body	Building	Water	Roads	Grass	Woodland
Flood	0.43	0.27	0.2	0.07	0.03

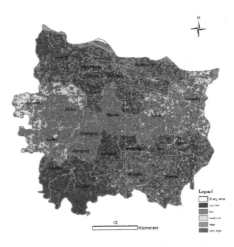

Fig. 5. Map of vulnerability of disaster victims.

4.3 Comprehensive Analysis of Flood Risk and Zoning

The proposed methodology linearly combines the selected parameters, taking into account the relevant weights. According to the risk evaluation index system and evaluation model of storm and flood disaster in Zhengzhou City, through the analysis of risk factors, sensitivity of disaster environment and vulnerability of the disaster victims. The flood hazard map is created (Fig. 6), the result of risk evaluation is classified by using the method of natural breakpoint, defining 5 classes of flood risk (very low, low, moderate, high, and very high).

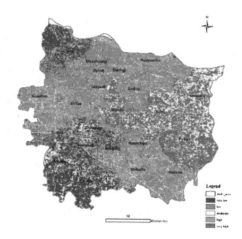

Fig. 6. Map of comprehensive analysis of flood risk and zoning.

4.4   Validation of the Results

According to results, the risk of rainstorm and flood disaster descended from west to east in Zhengzhou City. The disaster is partly bent on central and eastern regions in Zhengzhou city. The high risk areas were located in the middle and eastern areas of Zhengzhou City, including Miaoli Town, Jicheng Road Block and Longzihu of Jinshui District and Putian and Nancao of Guancheng District. The results of evaluation model have highly fitness with the historical recorded flood events. According to the historical disaster statistics of Zhengzhou City from 2009 to 2015, the historical disasters occurred in the central and eastern areas of Zhengzhou, and the high incidence occurred in Jinshui and Zhengdong New Area, which was basically consistent with the results of the model evaluation.

5   Discussion and Summary

According to the results of flood disaster risk zoning of Zhengzhou, high risk area mainly concentrated in the lower altitudes of most east-central of Jinshui district in town with road blocks and CBD having inland lake. The risk increases from west to east.

The study is the first attempt to assess the inland storm flood risk by GIS-AHP method and the first production of rainstorm flood risk zoning map. Studies have found that high risk area of Zhengzhou city in the historical disaster data have good fitness with evaluating results based on assessment model. The results can support for Zhengzhou city storm flood prevention work and urban rain flood model refinement provide strong scientific support for the establishment of rules, to some degree.

Acknowledgment. This work is supported by the key research project in the thirteenth Five-Year Plan (2016YFC0402402), Key project of water conservancy science and technology by Henan province (GG201518). The National Natural Science Foundation of China (No. 51079131, and

No. 51509223). The authors are grateful to colleagues and friends who shared their meteorological and hydrological data with us. We also thank reviewer for insightful comments that improved an earlier version of this manuscript.

References

Liu, M., Quan, R., Xu, S.: Risk Assessment of Urban Rainstorm Waterlogging Disaster: Theory, Method and Practice. Science Press, Beijing (2012). (in Chinese)

Chen, H., Ito, Y., Sawamukai, M., Tokunaga, T.: Flood hazard assessment in the Kujukuri Plain of Chiba Prefecture, Japan, based on GIS and multicriteria decision analysis. Nat. Hazards **78**, 105–120 (2015)

Muller, A., Reiter, J., Weiland, U.: Assessment of urban vulnerability towards floods using an indicator-based approach – a case study for Santiago de Chile. Nat. Hazards Earth Syst. **11**, 2107–2123 (2011)

Stefanidis, S., Stathis, D.: Assessment of flood hazard based on natural and anthropogenic factors using analytic hierarchy process (AHP). Nat. Hazards **68**, 569–585 (2013)

Benito, G., Lang, M., Barriendos, M., et al.: Use of systematic, palaeoflood and historical data for the improvement of flood risk estimation, review of scientific methods. Nat. Hazard **31**(3), 623–643 (2004)

Dongdong, Z., Denghua, Y., Yicheng, W.: Research progress on risk assessment and integrated strategies for urban pluvial flooding. J. Catastrophology **29**(1), 144–149 (2014)

Booij, M.: Impact of climate change on river flooding assessed with different spatial model resolutions. J. Hydrol. **303**, 176–198 (2005)

Cowen, D.: GIS versus CAD versus DBMS: what are the differences. Photogramm. Eng. Remote Sens. **54**, 1551–1555 (1988)

Chang, N., Parvathinathan, G., Breeden, J.: Combining GIS with fuzzy multi criteria decision-making for landfill siting in a fast-growing urban region. J. Environ. Manag. **87**(1), 139–153 (2008)

Furdada, G., Calderon, L.E., Marques, M.A.: Flood hazard map of La Trinidad (NW Nicaragua). Method and results. Nat. Hazards **45**, 183–195 (2008)

Dewan, A.M., Islam, M.M., Kumamoto, T., Nishigaki, M.: Evaluating flood hazard for land-use planning in Greater Dhaka of Bangladesh using remote sensing and GIS techniques. Water Resour. Manag. **21**, 1601–1612 (2007)

Saaty, T.L.: Decision making with the analytic hierarchy process. Int. J. Serv. Sci. **1**(1), 83–98 (2008)

Author Index

Printed in the United States
By Bookmasters